METHODS FOR LINEAR AND QUADRATIC PROGRAMMING

STUDIES
IN MATHEMATICAL
AND MANAGERIAL
ECONOMICS

Editor

HENRI THEIL

VOLUME 17

NORTH-HOLLAND PUBLISHING COMPANY–
AMSTERDAM · OXFORD

AMERICAN ELSEVIER PUBLISHING COMPANY, INC.–
NEW YORK

METHODS FOR LINEAR AND QUADRATIC PROGRAMMING

C. VAN DE PANNE

University of Calgary

1975

NORTH-HOLLAND PUBLISHING COMPANY–
AMSTERDAM · OXFORD

AMERICAN ELSEVIER PUBLISHING COMPANY, INC.–
NEW YORK

Library of Congress Catalog Card Number: 73 87371
North-Holland ISBN for this Series: 0 7204 3300 2
North-Holland ISBN for this Volume: 0 7204 3316 9
American Elsevier ISBN: 0 444 10589 1

PUBLISHERS:
NORTH-HOLLAND PUBLISHING COMPANY – AMSTERDAM
NORTH-HOLLAND PUBLISHING COMPANY, LTD. – OXFORD

SOLE DISTRIBUTORS FOR THE U.S.A. AND CANADA:
AMERICAN ELSEVIER PUBLISHING COMPANY, INC.
52 VANDERBILT AVENUE, NEW YORK, N.Y. 10017

PRINTED IN THE NETHERLANDS

INTRODUCTION TO THE SERIES

This is a series of books concerned with the quantitative approach to problems in the social and administrative sciences. The studies are in particular in the overlapping areas of mathematical economics, econometrics, operational research, and management science. Also, the mathematical and statistical techniques which belong to the apparatus of modern social and administrative sciences have their place in this series. A well-balanced mixture of pure theory and practical applications is envisaged, which ought to be useful for universities and for research workers in business and government.

The Editor hopes that the volumes of this series, all of which relate to such a young and vigorous field of research activity, will contribute to the exchange of scientific information at a truly international level.

THE EDITOR

To Anja, Job, and Michiel

PREFACE

This book is the outcome of a number of years of work in the area of linear and quadratic programming. Most of it is based on articles and papers written together by Andrew Whinston or by this author only.

While writing this book, I had two objectives in mind. The first one was to provide a detailed exposition of the most important methods of linear and quadratic programming which will introduce these methods to a wide variety of readers. For this purpose, numerical examples are given throughout the book. The second objective was to relate the large number of methods which exist in both linear and quadratic programming to each other. In linear programming the concepts of dual and parametric equivalence were useful and in quadratic programming that of symmetric and asymmetric variants.

Throughout the book, the treatment is in terms of methods and tableaux resulting from these methods rather than mathematical theorems and proofs, or, in other terms, the treatment is constructive rather than analytical. The main advantage of such a constructive approach is thought to be in the accessibility of the methods. Whereas an analytical approach introduces a method through a maze of theorems after which the methods appear as an afterthought, a constructive approach first states the main principles of the method, after which obstacles and implications are dealt with one by one.

A disadvantage of detailed exposition and of numerical examples is lack of conciseness. This has resulted in the limitation of the number of topics treated in this book. Hence such topics as decomposition methods for linear and quadratic programming, quadratic transportation problem and integer linear and quadratic programming are missing in this book, though most concepts on which methods for these problems are based follow rather naturally from the methods which are treated. However, inclusion of a number of these subjects would have increased the size of this book unduly.

The linear complementarity problem is treated in some detail, not only because it is an immediate generalization of linear and quadratic programming but also because it is amenable to an interesting generalization of the parametric methods which form the core of this book.

The book can be used for graduate or senior undergraduate courses in mathematical programming. In case of a one-year course, it could serve as a basis for the first half; the second half would then deal with general non-linear programming and integer programming.

vii

The mathematical prerequisite of the book is that the reader should have a working knowledge of elementary matrix algebra.

Thanks are due to Henri Theil, who has introduced me to the subject of quadratic programming by asking me to participate in the development of a method for quadratic programming (which can be found in chapter 12) and to Andrew Whinston, with whom I had a number of years of fruitful cooperation and who was a coauthor of a number of the articles on which this book is based.

Thanks are also due to the National Research Council of Canada for research support and to *Econometrica*, *The International Economic Review*, and *Operations Research* for allowing me to use material published in these journals for parts of chapters 6, 9 and 12.

C. van de Panne

CONTENTS

CHAPTER 1

THE SOLUTION OF LINEAR SYSTEMS

1.1. General linear systems and canonical forms

Let us consider an equation system of the following form:

$$a_{11}x_1 + a_{12}x_2 + ... + a_{1n}x_n = b_1,$$
$$a_{21}x_1 + a_{22}x_2 + ... + a_{2n}x_n = b_2,$$
$$...$$
$$a_{m1}x_1 + a_{m2}x_2 + ... + a_{mn}x_n = b_m. \tag{1}$$

This is a system of m linear equations in n variables $x_1, ..., x_n$. The system (1) may have no solution, one solution or a number of solutions; this depends on the coefficients of the system $a_{11}, ..., a_{mn}$ and $b_1, ..., b_m$.

The solutions, if any, of a general system like (1) cannot immediately be found from the system. Systems having some special form may provide more information about their solutions. Consider a system of the following form:

$$x_1 \qquad + a_{1,m+1}x_{m+1} + a_{1,m+2}x_{m+2} + ... + a_{1n}x_n = b_1,$$
$$\qquad x_2 \quad + a_{2,m+1}x_{m+1} + a_{2,m+2}x_{m+2} + ... + a_{2n}x_n = b_2,$$
$$...$$
$$\qquad x_m + a_{m,m+1}x_{m+1} + a_{m,m+2}x_{m+2} + ... + a_{mn}x_n = b_m. \tag{2}$$

One solution of this system is, for example,

$$x_1 = b_1, x_2 = b_2, ..., x_m = b_m, x_{m+1} = x_{m+2} = ... = x_n = 0. \tag{3}$$

Other solutions of this system may be found by choosing any values for the variables $x_{m+1}, x_{m+2}, ..., x_n$ and computing the corresponding values for the first m x-variables. If the particular values chosen for the last $n-m$ x-variables are indicated by bars, we have for instance for the corresponding value of x_1, indicated by $\bar{x}_1$:

$$\bar{x}_1 = b_1 - a_{1,m+1}\bar{x}_{m+1} - a_{1,m+2}\bar{x}_{m+2} - ... - a_{1n}\bar{x}_n. \tag{4}$$

In fact, any solution satisfying (2) may be obtained in this way. Hence it may be said that systems having forms like (2) give immediate access to any solution of the system.

1

The system (2) is said to be in *ordered canonical form*. A system is in ordered canonical form if the first variable only appears in the first equation, having a unity coefficient, the second variable only appears in the second equation, having a unity coefficient, and so on. The variables which appear in one equation only with a unity coefficient in that equation are called the *basic variables* of the system; each basic variable is connected with one equation. Hence in a system in ordered canonical form, the first variable is a basic variable and is connected with the first equation; the second variable is a basic variable and is connected with the second equation, and so on. In a system of m equations in ordered canonical form, the first m variables are the basic variables; the remaining $n-m$ variables are called the *nonbasic variables*.

Consider the following system:

$$2x_1 + 5x_2 \qquad + x_4 = 10,$$
$$3x_1 - 4x_2 + x_3 \qquad = 12. \tag{5}$$

In this system x_4 is a basic variable in the first equation and x_3 in the second equation, but the system is not ordered. Such a system is said to be in *canonical form*. A system in canonical form can easily be brought into *ordered* canonical form by renaming and rearranging variables. For example, in the above system, x_4 may be renamed as x_1^*, x_3 as x_2^*, x_1 as x_3^*, and x_2 as x_4^*. Rearranging, we find

$$x_1^* \qquad + 2x_3^* + 5x_4^* = 10,$$
$$x_2^* + 3x_3^* - 4x_4^* = 12, \tag{6}$$

which is a system in ordered canonical form. It may therefore be concluded that the difference between an ordered and a general canonical form is a rather trivial one. In the following we shall in most cases deal with general canonical forms, but in some cases it will be convenient to write such a system in ordered canonical form.

1.2. Reduction to canonical form

Two linear equation systems which have the same solutions are called *equivalent*. Because it is much easier to find solutions of systems in canonical form than solutions of general equation systems, it is desirable to find a system in canonical form which is equivalent to a given general system. First it will be shown how an equivalent system can be derived from a given system, or, which is the same, how a given system may be transformed into an equivalent one. After that, it is shown how the transformations can be used to obtain from any given system an equivalent canonical form, if this is at all possible.

The solutions of a system of equations remain the same if an equation of that system is replaced by c times that equation, where c is a nonzero constant. This can be shown as follows. Let the equation concerned be the first one of (1). This equation may be replaced by

$$c(a_{11}x_1 + a_{12}x_2 + ... + a_{1n}x_n) = cb. \tag{7}$$

Any solution satisfying (1) will also satisfy the modified system (1), in which the first equation has been replaced by (7). Furthermore, any solution satisfying the modified system must also satisfy the original system, since the first equation of (1) and equation (7) have the same solutions. Hence the original and the modified system must have the same solutions.

The solutions to an equation system also remain the same if an equation is replaced by the sum of that equation and a multiple of another equation of the system. Let us, for example, add to the second equation of (1) c times the first equation. We then have

$$a_{21}x_1 + a_{22}x_2 + ... + a_{2n}x_n + c(a_{11}x_1 + a_{12}x_2 + ... + a_{1n}x_n) = b_2 + cb_1. \tag{8}$$

Any solution which satisfied (1) will satisfy (1) with the second equation replaced by (8); on the other hand, solutions satisfying (1) with (8) instead of the second equation will also satisfy (1) because, if the first equation of (1) is satisfied, the corresponding parts of (8) cancel. Hence also this operation leaves the solutions of the system unchanged.

We may therefore multiply (or divide) any equation by a nonzero constant or add to an equation a multiple of another equation without altering the solutions of the system. These two operations will be used to transform a general equation system into one in canonical form, if that is possible.

These two standard transformations are used to transform a general system into one in canonical form. This is done as follows. Suppose we want to bring the system (1) in canonical form. The system is first transformed in such a way that x_1 appears in the first equation only, and with a unit coefficient; after that, the resulting system is transformed in such a way that x_2 appears only in the second equation, with a unit coefficient, and so on, until finally a system in canonical form is obtained.

The first transformation is started by dividing the first equation by a_{11}, in order to give x_1 a unit coefficient in this equation. If a_{11} is zero, the first equation and another equation with a nonzero x_1-coefficient are interchanged. The resulting equation is

$$x_1 + c_i \; a_{12}x_2 + a_{11}^{-1}a_{13}x_3 + \ldots + a_{11}^{-1}a_{1n}x_n = a_{11}^{-1}b_1. \tag{9}$$

This may be written as

$$x_1 + b_{12}x_2 + b_{13}x_3 + \ldots + b_{1n}x_n = b_{10}. \tag{10}$$

with

$$b_{12} = a_{11}^{-1}a_{12}, \; b_{13} = a_{11}^{-1}a_{13}, \ldots, \; b_{1n} = a_{11}^{-1}a_{1n}, \; b_{10} = a_{11}^{-1}b_1.$$

The coefficients of x_1 in the other equations are made zero by adding to these equations appropriate multiples of (9). Hence we add to the second equation $-a_{21}$ times (9). The second equation then becomes

$$(a_{22} - a_{21}a_{11}^{-1}a_{12})x_2 + (a_{23} - a_{21}a_{11}^{-1}a_{13})x_3 + \ldots + (a_{2n} - a_{21}a_{11}^{-1}a_{1n})x_n$$
$$= b_2 - a_{21}a_{11}^{-1}b_1; \tag{11}$$

this can be written as

$$b_{22}x_2 + b_{23}x_3 + \ldots + b_{2n}x_n = b_{20}, \tag{12}$$

with

$$b_{22} = a_{22} - a_{21}a_{11}^{-1}a_{12} = a_{22} - a_{21}b_{12},$$
$$b_{23} = a_{23} - a_{21}a_{11}^{-1}a_{13} = a_{23} - a_{21}b_{13},$$
$$\ldots$$
$$b_{2n} = a_{2n} - a_{21}a_{11}^{-1}a_{1n} = a_{2n} - a_{21}b_{1n},$$
$$b_{20} = b_2 - a_{21}a_{11}^{-1}b_1 = b_2 - a_{21}b_{10}.$$

The other equations are transformed in the same way. After the first transformation the system has the following form:

$$x_1 + b_{12}x_2 + b_{13}x_3 + \ldots + b_{1n}x_n = b_{10},$$
$$b_{22}x_2 + b_{23}x_3 + \ldots + b_{2n}x_n = b_{20},$$
$$\ldots$$
$$b_{m2}x_2 + b_{m3}x_3 + \ldots + b_{mn}x_n = b_{m0}. \tag{13}$$

The transformation formulas for the coefficients, if we denote b_1 by a_{10}, b_2 by a_{20}, and so on, are

$$b_{ij} = a_{11}^{-1}a_{ij}, \qquad\qquad i = 1,$$
$$= a_{ij} - a_{i1}a_{11}^{-1}a_{1j} = a_{ij} - a_{i1}b_{1j}, \; i \neq 1. \tag{14}$$

The second expression for $i \neq 1$ is computationally more efficient than the first one.

We have now obtained a system which is equivalent to the original system and in which x_1 is a basic variable. Now the second transformation starts which has as its objective to transform the system (13) into a system in which x_2 is also a basic variable with a unit coefficient in the second equation and zeros in the other equations. Hence we divide the second equation of (13) by b_{22}; if b_{22} is zero, the second equation and one of the remaining equations having a nonzero x_2-coefficient are interchanged. After this, suitable multiples of the resulting equation are added to the other equations of (13), the first equation included. The resulting system then has the following form

$$
\begin{aligned}
x_1 \quad + \; c_{13}x_3 + \ldots + c_{1n}x_n &= c_{10}, \\
x_2 + \; c_{23}x_3 + \ldots + c_{2n}x_n &= c_{20}, \\
c_{33}x_3 + \ldots + c_{3n}x_n &= c_{30}, \\
\ldots \\
c_{m3}x_3 + \ldots + c_{mn}x_n &= c_{m0};
\end{aligned}
\tag{15}
$$

the transformation formulas are

$$
\begin{aligned}
c_{ij} &= b_{22}^{-1}b_{ij}, && i = 2, \\
&= b_{ij} - b_{i2}b_{22}^{-1}b_{2j} = b_{ij} - b_{i2}c_{2j}, && i \neq 2.
\end{aligned}
\tag{16}
$$

Note that the second transformation preserves the effect of the first transformation in the sense that the coefficients of x_1 in equations other than the first stay zero. This can be seen as follows. We have

$$
\begin{aligned}
c_{i1} &= b_{22}^{-1}b_{i1}, && i = 2, \\
c_{i1} &= b_{i1} - b_{i2}b_{22}^{-1}b_{21}, && i \neq 2,
\end{aligned}
\tag{17}
$$

From (13) it is obvious that $b_{i1} = 0$ for $i \neq 1$, so that c_{i1} for $i \neq 1$ must be zero. Furthermore, $c_{11} = b_{11} - b_{12}b_{22}^{-1}b_{21} = 1 - 0 = 1$.

The other transformations are performed in a similar way. If no complications occur, a system in canonical form as given in (2) may be obtained after m transformations. Before dealing with possible complications, a numerical example of the transformation of a general system into one in canonical form will be given.

1.3. Example of a reduction to canonical form

Let us consider the following general linear equation system:

$$2x_1 + 3x_2 - x_3 \qquad + 5x_5 = 2,$$
$$-2x_1 + x_2 - 3x_3 - x_4 \qquad = 6,$$
$$x_1 + 2x_2 \qquad - x_4 + 4x_5 = 3. \tag{18}$$

In the first transformation x_1 is made a basic variable. The first equation is therefore divided by 2; the result is

$$x_1 + 1\tfrac{1}{2}x_2 - \tfrac{1}{2}x_3 + 2\tfrac{1}{2}x_5 = 1. \tag{19}$$

To the second equation we add 2 times (19), so that we obtain

$$4x_2 - 4x_3 - x_4 + 5x_5 = 8. \tag{20}$$

To the third equation we add -1 times (19), obtaining

$$\tfrac{1}{2}x_2 + \tfrac{1}{2}x_3 - x_4 + 1\tfrac{1}{2}x_5 = 2. \tag{21}$$

Hence after the first transformation we have obtained the equivalent system (19)–(21).

In the second transformation x_2 is made a basic variable. We start with dividing (20) by 4. The result is

$$x_2 - x_3 - \tfrac{1}{4}x_4 + 1\tfrac{1}{4}x_5 = 2. \tag{22}$$

To (19) we add $-1\tfrac{1}{2}$ times (22) and to (21) we add $-\tfrac{1}{2}$ times (22). Hence (19) and (21) are replaced by

$$x_1 + x_3 + \tfrac{3}{8}x_4 + \tfrac{5}{8}x_5 = -2, \tag{23}$$

$$x_3 - \tfrac{7}{8}x_4 + \tfrac{7}{8}x_5 = 1. \tag{24}$$

In the third transformation x_3 is made a basic variable. Since in (24) x_3 has a unit coefficient, we can leave this equation as it is. To (22) we add 1 times (24) and to (23) -1 times (24). The following system in ordered canonical form is then obtained:

$$x_1 + 1\tfrac{1}{4}x_4 - \tfrac{1}{4}x_5 = -3,$$
$$x_2 - 1\tfrac{1}{8}x_4 + 2\tfrac{1}{8}x_5 = 3,$$
$$x_3 - \tfrac{7}{8}x_4 + \tfrac{7}{8}x_5 = 1. \tag{25}$$

It is somewhat more convenient to write down all equations in detached coefficient form, that is listing for each system the coefficients of the various variables in separate columns. This leads to a representation of the equations and their transformations in tableau-form. Table 1 gives the transformations for the equation system (18) in tableau-form.

In tableau 0 the original system (18) is given with the constant term of the equations on the left side instead of on the right-hand side. In the first transformation the coefficients of the first equation are divided by the coefficient of x_1, which is 2. This coefficient 2 is called the *pivot* of the first transformation; it is underlined in order to indicate what happens during the transformation of which this element is the pivot. The row in which the pivot occurs is called the *pivot-row* and the column in which it occurs, the *pivot-column*.

The transformation formulas (14) and (16) may then be interpreted as follows. To transform a tableau into an equivalent one with a given pivot,

(1) Divide the elements of pivot-row by the pivot;

(2) Subtract from the elements of other rows the product of the corresponding element in the pivot-column and the corresponding elements of the transformed pivot row.

Hence, in the first transformation the first row of tableau 0 is divided by 2; the result is given in the first row of tableau 1. The other elements of the tableau are only slightly more difficult to transform. In order to obtain

Table 1

Reduction to a canonical system in tableau form

Tableau	Equation	c-term	x_1	x_2	x_3	x_4	x_5
0	1	2	$\underline{2}$	3	-1	0	5
	2	6	-2	1	-3	-1	0
	3	3	1	2	0	-1	4
1	1	1	1	$1\frac{1}{2}$	$-\frac{1}{2}$	0	$2\frac{1}{2}$
	2	8	0	$\underline{4}$	-4	-1	5
	3	2	0	$-\frac{1}{2}$	$\frac{1}{2}$	-1	$1\frac{1}{2}$
2	1	-2	1	0	1	$\frac{3}{8}$	$\frac{5}{8}$
	2	2	0	1	-1	$-\frac{1}{4}$	$1\frac{1}{4}$
	3	1	0	0	$\underline{1}$	$-\frac{7}{8}$	$\frac{7}{8}$
3	1	-3	1	0	0	$1\frac{1}{4}$	$-\frac{1}{4}$
	2	3	0	1	0	$-1\frac{1}{8}$	$2\frac{1}{8}$
	3	1	0	0	1	$-\frac{7}{8}$	$\frac{7}{8}$

the transformed constant term of the second equation, we have to subtract from the original tableau element, which is 6, the product of the
corresponding element in the pivot-column, which is -2, and the corresponding element of the first row of tableau 1, which is 1. We have therefore

$$6 - (-2)(1) = 8.$$

All other elements of tableau 1 can be obtained in this manner. In the
second transformation the underlined element 4 of tableau 1 is the pivot; in
the third transformation the underlined element 1 of tableau 2 is the pivot.
In tableau 3 we have obtained the desired canonical form.

1.4. Dependent equations and cases of no solution

A system in canonical form has at least one solution, namely the solution
given in (3), where all the basic variables are equal to the constant terms
in (2) and all the nonbasic variables zero. If there are no nonbasic variables, this is the unique solution of the system. It may be that a general
system has no solution. Hence, if a general system has no solution, it
must not be possible to transform this system into canonical form; the
transformations must therefore at a certain point break down. This may
occur as follows.

   While the transformations transforming a general system into a canonical system are carried out, it may happen that a certain coefficient which
we want to use as a pivot is zero. If the coefficients of the variable we
want to make basic in the following equations are all zero, this variable
cannot be made basic in this situation. In this case some other variable
may be made basic, interchanging the columns of the variables concerned, if at least this variable has some nonzero coefficient in the equations which have not yet a basic variable. Only if all coefficients of the
variables which are not yet basic are zero in all equations which have not
yet a basic variable, it is not possible to transform the system any further.

   Let us, for example, consider the following partly transformed system:

$$x_1 + b_{12}x_2 + b_{13}x_3 + b_{14}x_4 = b_{10},$$
$$0x_2 + \quad 0x_3 + \quad 0x_4 = b_{20},$$
$$0x_2 + \quad 0x_3 + b_{34}x_4 = b_{30}. \tag{26}$$

If $b_{34} \neq 0$, the second and third equation and x_2 and x_4 may be interchanged after which b_{34} can be used as a pivot for the next transformation. However, if $b_{34} = 0$, no further transformation can follow.

There are then two possibilities. If b_{20} and b_{30} are both zero, the last two equations of the system have zero coefficients only; such equations are called *vacuous equations*. Vacuous equations do not restrict the solution of the system in any way and may hence be deleted. The occurrence of a vacuous equation indicates that in the original system an equation may be represented as a linear combination of the other equations; hence the original equations are *dependent*. If b_{20} and b_{30} are not both zero, the system has no solution, since for $b_{34} = 0$ the left-hand side of the second and third equation of (26) is zero for any solution, whereas the right-hand side is not. The original system must therefore have been *inconsistent*.

In case the original system has more equations than variables, that is, if $m > n$, the system must either be inconsistent or must have dependent equations. This can be seen as follows. Suppose we have made all n variables basic, so that n out of the m equations must have a basic variable. The remaining $m-n$ equations must contain only zero coefficients on the left-hand side. If the constant terms of these equations are all zero, all these equations are vacuous and may be deleted. If one or more of these constant terms are nonzero, the original system must have contained at least one inconsistent equation, and has therefore no solution. The case in which less than n variables can be made basic is similar.

In the equation systems which were presented above it was implied that n, the number of variables, exceeds the number of equations. This situation is characteristic for programming problems. If, in the case $n = m$, it is possible to make all n variables basic, we have a canonical form in which each of the basic variables is equal to the corresponding constant term in the canonical form, as in (3). If it is not possible to make all n variables basic, the equations which have no basic variables must all have zero coefficients for the nonbasic variables in the remaining equations. If the corresponding constant terms are all zero, these equations are vacuous and can be deleted. If one or more of these constant terms are nonzero, the system has no solution.

In the previous sections a method was indicated for finding a canonical form with given basic variables for cases in which a solution exists in these variables. We shall now prove that for a given set of basic variables, the ordered canonical form is *unique*. This means that, no matter how the canonical form is generated, the result is the same. For instance, a canonical form of the system (1) in which the first m variables are basic may be found in various ways. We may proceed as indicated before by first making x_1 a basic variable in the first equation, then x_2 a basic variable in the second equation, and so on. Alternatively, we may make first x_m a basic variable in the first equation, then x_{m-1} a basic variable in the second equation, and so on and then rearrange the equations. Provided the transformations preserve equivalence of solutions, the final result should be the same.

The proof runs as follows. Suppose two canonical forms of the same original system were to be found. Let the first one be

$$x_1 + d_{1,m+1}x_{m+1} + \dots + d_{1n}x_n = d_1,$$

$$\dots$$

$$x_m + d_{m,m+1}x_{m+1} + \dots + d_{mn}x_n = d_m, \tag{27}$$

and the second

$$x_1 + d^*_{1,m+1}x_{m+1} + \dots + d^*_{1n}x_n = d^*_1,$$

$$\dots$$

$$x_m + d^*_{m,m+1}x_{m+1} + \dots + d^*_{mn}x_n = d^*_m. \tag{28}$$

Suppose the constant terms of both forms are different. The form (27) has the basic solution

$$x_1 = d_1, \dots, x_m = d_m.$$

If this solution is substituted into (28), we find

$$d_1 = d^*_1, \dots, d_m = d^*_m.$$

If $d_i \neq d^*_i$ for some i, then a solution of (27) is not a solution of (28), so that the systems are not equivalent; hence both these systems cannot have been generated from the same general system by equivalence-preserving transformations.

Consider now the case in which some of the coefficients of the variables are different, say those of x_{m+1}. For (27) the following solution is found if $x_{m+1} = 1$:

$$x_1 = d_1 - d_{1,m+1}, \dots, x_m = d_m - d_{m,m+1},$$

$$x_{m+1} = 1;$$

$$x_{m+2} = \dots = x_n = 0.$$

If this solution is substituted into (28), we find

$$d_1 - d_{1,m+1} + d^*_{1,m+1} = d^*_1,$$

$$\ldots$$

$$d_m - d_{1,m+1} + d^*_{1,m+1} = d^*_m.$$

Hence, if $d_{i,m+1} \neq d^*_{i,m+1}$, the forms (27) and (28) are not equivalent.

1.5. Relations with other solution methods

It was observed earlier that transforming a general system in canonical form amounts to solving the system if the system has a unique solution. The method described for transforming a system into canonical form, interpreted as a solution method, is known as the *Gauss–Jordan* or *complete elimination method*. Almost all procedures for mathematical programming are based on this elimination method. However, it is only one of the many methods available for solving linear systems. The most basic of the other methods is the *Gaussian elimination method*. Here we shall restrict ourselves to giving a slightly different interpretation of the complete elimination method, to an exposition of the Gaussian elimination method and to some remarks on the relationship between the two methods.

Let us consider the example of section 1.3 again, where the system (18) had to be solved. The first equation of (18) can be used to express x_1 in terms of the other variables:

$$x_1 = 1 - 1\tfrac{1}{2}x_2 + \tfrac{1}{2}x_3 - 2\tfrac{1}{2}x_5. \tag{29}$$

Substituting x_1 in the other equations, we find

$$4x_2 - 4x_3 - x_4 + 5x_5 = 8, \tag{30}$$

$$\tfrac{1}{2}x_2 + \tfrac{1}{2}x_3 - x_4 + 1\tfrac{1}{2}x_5 = 2. \tag{31}$$

Note that these equations are the same as those obtained in section 1.3, (19)–(21). x_1 has been eliminated from eqs. (30) and (31). Eq. (30) is used to obtain an expression for x_2 in terms of the remaining variables:

$$x_2 = 2 + x_3 + \tfrac{1}{4}x_4 + 1\tfrac{1}{4}x_5 \tag{32}$$

Substituting this into the other equations, (29) and (31), we find

$$x_1 = -2 - x_3 - \tfrac{3}{8}x_4 - \tfrac{5}{8}x_5, \tag{33}$$

and

$$x_3 - \tfrac{7}{8}x_4 + \tfrac{7}{8}x_5 = 1. \tag{34}$$

Note that the result, eqs. (32)–(34), is again the same as that found earlier, see (22)–(24). On somewhat closer consideration this is not surprising since what is happening in both cases is essentially the same. Eq. (34) may now be used to obtain an expression of x_3:

$$x_3 = 1 + \tfrac{7}{8}x_4 - \tfrac{7}{8}x_5. \tag{35}$$

On substituting this into (32) and (34), we find

$$
\begin{aligned}
x_1 &= -3 - 1\tfrac{1}{4}x_4 + \tfrac{1}{4}x_5, \\
x_2 &= 3 + 1\tfrac{1}{8}x_4 - 2\tfrac{1}{8}x_5, \\
x_3 &= 1 + \tfrac{7}{8}x_4 - \tfrac{7}{8}x_5.
\end{aligned}
\tag{36}
$$

This is the same as the system (25). In the case x_4 and x_5 are zero, or, what amounts to the same, not present in the equation system, we have found a unique solution to the system.

This was an application of the Gauss–Jordan or complete elimination method to the system. The difference between this method and the reduction to canonical form is restricted to writing some terms at a different side of equality-sign.

The Gaussian elimination method is very similar. Just as in the complete elimination method, the first equation of (18) is used to obtain an expression for x_1 in terms of the other variables, which are then substituted in the other equations. The result would again be (29)–(31). Eqs. (30) and (31) are now considered a new system; in case x_4 and x_5 are zero or do not occur in the equation system, it is a system with two equations in the two unknowns x_2 and x_3. This equation system is solved first. The system (30) and (31) is solved by using (30) to obtain an expression of x_2 in terms of the remaining variables; the result is (32). This expression is substituted in (31). We then obtain (34). This is an equation which gives x_3 in terms of a constant and x_4 and x_5; if x_4 and x_5 are zero, it gives the solution $x_3 = 1$.

The method consisted up to now of deriving smaller equation systems in fewer unknowns by using each time an equation to eliminate a certain variable until an equation is obtained from which the remaining variable, in our example x_3, is solved. This phase of the method is called the *forward solution*. In the next phase, which is called the *back solution*, the other variables are solved successively from the equations which were used to eliminate them from the system. Hence we find, substituting for x_3 according to (35) in (32):

$$x_2 = 3 + 1\tfrac{1}{8}x_4 + 2\tfrac{1}{8}x_5. \tag{37}$$

Substituting for x_2 and x_3 according to (35) and (37) in (29), we find

$$x_1 = -3 - 1\tfrac{1}{4}x_4 + \tfrac{1}{4}x_5. \tag{38}$$

The difference between the complete elimination method and the Gaussian elimination method is that the latter method does not substitute into the equations which have already been used for the elimination process; the complete elimination method substitutes into all equations. This has as a result that the complete elimination method, after the last equation has been treated, finds the complete solution, whereas the Gaussian elimination method finds, after the treatment of the last equation, only the solution for one variable (in the example x_3); the solution for the other variables must be found in the back solution.

1.6. The value of a linear function

Linear programming is concerned with the value of a linear function of variables which, among other things, must be solutions of a linear equation system. It is then required to compute for each solution of the system the value of the linear function. This can be done as follows.

Let the equation system be again the general system (1) and let the linear function be

$$f(x_1, x_2, ..., x_n) = p_0 + p_1x_1 + p_2x_2 + ... + p_nx_n, \tag{39}$$

where $p_0, p_1, p_2, ..., p_n$ are given coefficients. (39) may be written as follows

$$f - p_1x_1 - p_2x_2 - ... - p_nx_n = p_0. \tag{40}$$

Now eq. (40) may be considered an ordinary equation and added to the equations of (1). f is then considered a variable on a par with the x-variables. The only difference of (40) with other equations is that in the system (1) augmented by (40), f is a basic variable associated with (40), since it occurs in this equation only. The augmented system can then be reduced to canonical form as indicated before, where f plays the role of the first basic variable. The constant term in the transformed eq. (40) gives in the canonical form the value of the function f for the basic solution of the canonical form. The value of f for any other solution of the system can be easily found using the transformed equation of the function in the canonical form.

As an example the value of the function

$$f(x_1, ..., x_5) = x_1 + 2x_2 - x_3 - 4x_4 + x_5 \tag{41}$$

will be computed for the basic solution of the canonical form of the system (18). These computations can, of course, be performed in tableau-form as in table 1.

Eq. (41) is written as

$$f - x_1 - 2x_2 + x_3 + 4x_4 - x_5 = 0, \tag{42}$$

and is considered the first equation of the system; this equation has already f as its basic variable. In the first transformation, x_1 is made a basic variable, associated with the second equation of the system. We obtain then, in addition to eqs. (19)–(21), the equation

$$f - \tfrac{1}{2}x_2 + \tfrac{1}{2}x_3 + 4x_4 + 1\tfrac{1}{2}x_5 = 1. \tag{43}$$

In the second transformation, when x_2 becomes a basic variable connected with the third equation, the system consists of eqs. (22)–(24) plus the equation

$$f + 3\tfrac{7}{8}x_4 + 2\tfrac{1}{8}x_5 = 2. \tag{44}$$

This equation is not altered during the third transformation in which x_3 becomes a basic variable associated with the third equation, since (44) does not contain a term in x_3.

Hence the augmented system in canonical form consists of (44) plus the equations of (25). The constant term 2 in (44) gives the value of f for the basic solution of (25). The value of f for any other solution of the system (18)–(20) can be found immediately from (44) by substitution of the appropriate values of x_4 and x_5.

For instance, if we take $x_4 = 1$ and $x_5 = 1$, we must have, according to (25), $x_1 = -4$, $x_2 = 2$ and $x_3 = 1$. For this solution the value of f is, according to (44), -4, which can easily be checked from (41).

1.7. A matrix formulation of transformations

Linear equation systems and their transformations can both be represented in terms of vectors and matrices. Such a representation is of great theoretical importance and it can also be of considerable use in computations.

The system (1) can be represented in matrix form as follows:

$$Ax = b, \tag{45}$$

where A is the $m \times n$ matrix of the coefficients of the variables in the system,

$$A = \begin{bmatrix} a_{11} & a_{12} & \cdot & a_{1n} \\ a_{21} & a_{22} & \cdot & a_{2n} \\ \cdot & \cdot & \cdot & \cdot \\ a_{m1} & a_{m2} & \cdot & a_{mn} \end{bmatrix},$$

and b is a column of m elements containing the constant term coefficients,

$$b = \begin{bmatrix} b_1 \\ b_2 \\ \cdot \\ b_m \end{bmatrix},$$

while x is the column vector of the n x-variables,

$$x = \begin{bmatrix} x_1 \\ x_2 \\ \cdot \\ x_n \end{bmatrix}.$$

The matrix A may be partitioned as follows:

$$[A_1 \quad A_2], \tag{46}$$

where A_1 contains the first m columns of A and A_2 the remaining $n-m$ ones. This system may then be transformed into a canonical system in which the first m x-variables are basic, if this is possible; we assume that this is the case. The resulting canonical system must have the following form,

$$[I \quad D]x = d, \tag{47}$$

where I is the $m \times m$ unit matrix, D an $m \times (n - m)$ matrix and d a column vector of m elements.

The transformations which were used to bring a general system in canonical form can be represented as the premultiplication of both sides of (45) by a number of matrices. For example, the first transformation of the general system (45) which makes x_1 a basic variable, is, if a_{11} is nonzero, equivalent to a premultiplication by the matrix

$$
E_1 = \begin{bmatrix}
a_{11}^{-1} & 0 & 0 & \cdot & 0 \\
-a_{21}a_{11}^{-1} & 1 & 0 & \cdot & 0 \\
-a_{31}a_{11}^{-1} & 0 & 1 & \cdot & 0 \\
\cdot & & \cdot & \cdot & \cdot \\
-a_{m1}a_{11}^{-1} & 0 & 0 & \cdot & 1
\end{bmatrix}. \tag{48}
$$

Note that the operation of the first row of E_1 on the system (45) amounts to a division of the first equation of (45) by a_{11}; the operation of the second row of E_1 amounts to adding $-a_{21}a_{11}^{-1}$ times the first equation to the second one; the operation of the third row amounts to adding $-a_{31}a_{11}^{-1}$ times the first equation to the third one, and so on.

The resulting system is then transformed by the second transformation, which can be represented by a premultiplication of the system by

$$
E_2 = \begin{bmatrix}
1 & -b_{12}b_{22}^{-1} & 0 & \cdot & 0 \\
0 & b_{22}^{-1} & 0 & \cdot & 0 \\
0 & -b_{32}b_{22}^{-1} & 1 & \cdot & 0 \\
\cdot & \cdot & & \cdot & \cdot \\
0 & -b_{m2}b_{22}^{-1} & 0 & \cdot & 1
\end{bmatrix}. \tag{49}
$$

The other transformations can be represented in a similar manner. Since there are m transformations, we have a series of corresponding matrices $E_1, E_2, ..., E_m$. All transformations together amount to the premultiplication of the original system (45) by a matrix, denoted by X, which is the product of the matrices of the successive transformations:

$$
X = E_m E_{m-1} ... E_2 E_1. \tag{50}
$$

Hence the original system, premultiplied by X, must give the equivalent system in canonical form (47). Premultiplying the system in partitioned form (46) by X, we find

$$
[XA_1 \quad XA_2]x = Xb. \tag{51}
$$

The corresponding matrices and vectors in (51) and (47) must be identical. We have therefore

$$
XA_1 = I, \tag{52}
$$

and

$$
XA_2 = D, \quad Xb = d. \tag{53}
$$

From (52) we find

$$X = A_1^{-1};$$ (54)

we find then from (53)

$$D = A_1^{-1}A_2, \quad d = A_1^{-1}b.$$ (55)

According to (54), the transformation of a general system into one in canonical form amounts to a premultiplication of the general system by the inverse of a matrix consisting of the columns of the basic variables. This suggests that the desired canonical form may also be obtained by the following procedure. Compute, in some way, A_1^{-1}; after that compute $D = A_1^{-1}A_2$ and $d = A_1^{-1}b$. This procedure has the advantage that, if only some particular elements of D and d are required, these may be computed without computing the elements which are not required. For instance, if the element in the mth row and the first column of D is required, this element may be found by multiplying the mth row of A_1^{-1} into the first column of A_2. This approach proves to be useful in practical computational methods. The computation of A_1^{-1} will be considered in more detail below.

1.8. Numerical example

The ideas put forward will be applied to the numerical example of section 1.3. The matrix A and the vectors b and x are in this case

$$A = \begin{bmatrix} 2 & 3 & -1 & 0 & 5 \\ -2 & 1 & -3 & -1 & 0 \\ 1 & 2 & 0 & -1 & 4 \end{bmatrix}, \quad b = \begin{bmatrix} 2 \\ 6 \\ 3 \end{bmatrix}, \quad x = \begin{bmatrix} x_1 \\ x_2 \\ x_3 \\ x_4 \\ x_5 \end{bmatrix}.$$

Table 2 gives the same reduction to canonical form as table 1, but it also contains additional columns denoted by z_1, z_2, z_3, which in tableau 0 together form the unit matrix.

In the first transformation, the first equation is divided by 2; we add to the second equation 2 times the transformed first equation and subtract from the third equation 1 times the transformed first equation. This is equivalent to a multiplication of the first row of the tableau by $\frac{1}{2}$, which in its turn can be represented as a premultiplication of the tableau by the row vector:

$$[\tfrac{1}{2} \quad 0 \quad 0];$$

Table 2

Reduction in tableau-form with transformation matrices

Tableau	Basic var.	c-term	x_1	x_2	x_3	x_4	x_5	z_1	z_2	z_3
						Variables				
0	z_1	2	2	3	-1	0	5	1	0	0
	z_2	6	-2	1	-3	-1	0	0	1	0
	z_3	3	1	2	0	-1	4	0	0	1
1	x_1	1	1	$1\frac{1}{2}$	$-\frac{1}{2}$	0	$2\frac{1}{2}$	$\frac{1}{2}$	0	0
	z_2	8	0	4	-4	-1	5	1	1	0
	z_3	2	0	$\frac{1}{2}$	$\frac{1}{2}$	-1	$1\frac{1}{2}$	$-\frac{1}{2}$	0	1
2	x_1	-2	1	0	1	$\frac{3}{8}$	$\frac{5}{8}$	$\frac{1}{8}$	$-\frac{3}{8}$	0
	x_2	2	0	1	-1	$-\frac{1}{4}$	$1\frac{1}{4}$	$\frac{1}{4}$	$\frac{1}{4}$	0
	z_3	1	0	0	1	$-\frac{7}{8}$	$\frac{7}{8}$	$-\frac{5}{8}$	$-\frac{1}{8}$	1
3	x_1	-3	1	0	0	$1\frac{1}{4}$	$-\frac{1}{4}$	$\frac{3}{4}$	$-\frac{1}{4}$	-1
	x_2	3	0	1	0	$-1\frac{1}{8}$	$2\frac{1}{8}$	$-\frac{3}{8}$	$\frac{1}{8}$	1
	x_3	1	0	0	1	$-\frac{7}{8}$	$\frac{7}{8}$	$-\frac{5}{8}$	$-\frac{1}{8}$	1

the transformation of the second row is equivalent to adding to the second row 1 times the first row, which can be represented as a pre-multiplication of the tableau by

$$[1 \quad 1 \quad 0];$$

finally, the transformation of the third row is equivalent to adding to the third row $-\frac{1}{2}$ times the first row which is equivalent to pre-multiplication of the tableau by

$$[-\tfrac{1}{2} \quad 0 \quad 1].$$

Hence the first transformation is equivalent to a premultiplication of the tableau by the matrix

$$E_1 = \begin{bmatrix} \frac{1}{2} & 0 & 0 \\ 1 & 1 & 0 \\ -\frac{1}{2} & 0 & 1 \end{bmatrix}. \tag{56}$$

Tableau 1 of table 2 can therefore be obtained by a premultiplication of tableau 0 by E_1. Because the z-columns form together the unit matrix in tableau 0, these columns must give E_1 in table 1, since $E_1 I = E_1$.

In the second transformation x_2 is made a basic variable by dividing the second row by 4 and adding to the first row $-1\frac{1}{2}$ times the resulting second row and to the third row $-\frac{1}{2}$ times the resulting second row. This turns out to be equivalent to a premultiplication of tableau 1 by the matrix

$$E_2 = \begin{bmatrix} 1 & -\frac{3}{8} & 0 \\ 0 & \frac{1}{4} & 0 \\ 0 & -\frac{1}{8} & 1 \end{bmatrix}. \tag{57}$$

The result is tableau 2. The product $E_2 E_1$ appears in tableau 2 in the z-columns, since we multiplied the unit-matrix in tableau 0 first by E_1 and then by E_2. $E_2 E_1 I = E_2 E_1$. Hence we must find in the z-columns

$$E_2 E_1 = \begin{bmatrix} 1 & -\frac{3}{8} & 0 \\ 0 & \frac{1}{4} & 0 \\ 0 & -\frac{1}{8} & 1 \end{bmatrix} \begin{bmatrix} \frac{1}{2} & 0 & 0 \\ 1 & 1 & 0 \\ -\frac{1}{2} & 0 & 1 \end{bmatrix} = \begin{bmatrix} \frac{1}{8} & -\frac{3}{8} & 0 \\ \frac{1}{4} & \frac{1}{4} & 0 \\ -\frac{5}{8} & -\frac{1}{8} & 1 \end{bmatrix}. \tag{58}$$

In the same way it is found that the transformation of tableau 2 into tableau 3 is equivalent to a premultiplication of tableau 2 by

$$E_3 = \begin{bmatrix} 1 & 0 & -1 \\ 0 & 1 & 1 \\ 0 & 0 & 1 \end{bmatrix}. \tag{59}$$

Hence we find in the z-columns of tableau 3 the matrix

$$E_3 E_2 E_1 = \begin{bmatrix} \frac{3}{4} & -\frac{1}{4} & -1 \\ -\frac{3}{8} & \frac{1}{8} & 1 \\ -\frac{5}{8} & -\frac{1}{8} & 1 \end{bmatrix}. \tag{60}$$

This matrix is, according to (54), the inverse of the matrix consisting of the columns of x_1, x_2 and x_3 in tableau 0. Hence we must have

$$A_1^{-1} = \begin{bmatrix} 2 & 3 & -1 \\ -2 & 1 & -3 \\ 1 & 2 & 0 \end{bmatrix}^{-1} = \begin{bmatrix} \frac{3}{4} & -\frac{1}{4} & -1 \\ -\frac{3}{8} & \frac{1}{8} & 1 \\ -\frac{5}{8} & -\frac{1}{8} & 1 \end{bmatrix}, \tag{61}$$

which can be checked immediately.

If the matrix A_1^{-1} is known, any element of the canonical form of tableau 3 may be generated immediately from tableau 0 by a premultiplication of the appropriate column of tableau 0 by the appropriate row of A_1^{-1}. For instance, the element in the last row in the column of x_5 may be found from

$$\begin{bmatrix} -\frac{5}{8} & -\frac{1}{8} & 1 \end{bmatrix} \begin{bmatrix} 5 \\ 0 \\ 4 \end{bmatrix} = \frac{7}{8}. \tag{62}$$

For the constant terms in the canonical form we have

$$
\begin{bmatrix} \frac{3}{4} & -\frac{1}{4} & -1 \\ -\frac{3}{8} & \frac{1}{8} & 1 \\ -\frac{5}{8} & -\frac{1}{8} & 1 \end{bmatrix} \begin{bmatrix} 2 \\ 6 \\ 3 \end{bmatrix} = \begin{bmatrix} -3 \\ 3 \\ 1 \end{bmatrix}. \tag{63}
$$

1.9. Artificial variables

In the numerical example the z-columns were interpreted as the columns of a unit matrix; in transformed systems these columns represent the matrix by which the original system has been premultiplied. It is also possible to interpret the z's as being variables. Instead of eqs. (18), we then have the system

$$
\begin{aligned}
2x_1 + 3x_2 - x_3 \quad &+ 5x_5 + z_1 \qquad\qquad = 2, \\
-2x_1 + x_2 - 3x_3 - x_4 \quad &\qquad\quad + z_2 \qquad = 6, \\
x_1 + 2x_2 \quad - x_4 + 4x_5 \quad &\qquad\qquad\quad + z_3 = 3.
\end{aligned} \tag{64}
$$

The system (64) is a canonical system with the z-variables as basic variables. The transformations of the system performed before can be interpreted as a transformation of one canonical system with z_1, z_2, and z_3 as basic variables into another canonical system with x_1, x_2, and x_3 as basic variables.

Still another, more realistic interpretation is possible. The system (64) is not equivalent to the system (18), but it will be equivalent if the z-variables are zero, because in that case the z-variables play no role in the system. Now the z-variables may be interpreted as being variables which do not belong to the original system, but which have been introduced into it in order to facilitate its solution. Such variables are called *artificial variables*; they play an important role in programming methods. If this interpretation is used, the transformations of the numerical example and of the Gauss–Jordan elimination method can be viewed as follows. To each of the equations of the original system an artificial variable is added; the result is a canonical system with the artificial variables as basic variables. This system then is transformed in such a way that in each transformation one variable of the original system replaces one artificial variable; after m transformations (in the numerical example 3) a canonical system is obtained with ordinary basic variables only; all artificial variables are then nonbasic and can be given a zero value.

Hence any transformation of a system may be interpreted as one which transforms one canonical system into another canonical system;

the basic variables of these systems may be artificial variables or ordinary variables. For solutions of a system we want, of course, canonical forms without any basic artificial variable; in fact, all transformations can be viewed as having the aim of making the artificial variables nonbasic.

If it is not possible to obtain a solution in which all artificial variables are nonbasic, the original system must contain dependent equations or must have no solution. The first case corresponds with the remaining artificial variables having a zero value in the basic solution, the second with a nonzero value in a basic solution.

1.10. Relations between equivalent canonical forms

The relations between equivalent canonical forms will now be analyzed in general. First the case is considered in which all basic variables are different.

Let in the first canonical system the last m variables be basic and in the second canonical system the first m variables be basic. As indicated before, the first system can then be represented by

$$[A_1 \quad A_2 \quad I] \begin{bmatrix} x^1 \\ x^2 \\ x^3 \end{bmatrix} = b, \tag{65}$$

where the vector x is partitioned in three subvectors, the first of which contains the first m x-variables, the third the last m x-variables, and the second the remaining x-variables.

The second system may be represented by

$$[I \quad D_1 \quad D_2] \begin{bmatrix} x^1 \\ x^2 \\ x^3 \end{bmatrix} = d. \tag{66}$$

If both canonical forms exist, the second one can be obtained by transformations of the first one. The transformations are equivalent to a premultiplication by the inverse of the matrix consisting of the columns of the variables which are to become basic; hence the system (65) must have been premultiplied by A_1^{-1}. The second system can therefore be represented by

$$[I \quad A_1^{-1}A_2 \quad A_1^{-1}] \begin{bmatrix} x^1 \\ x^2 \\ x^3 \end{bmatrix} = A_1^{-1}b. \tag{67}$$

The system (66) can be transformed into (65) by a premultiplication of the inverse of the matrix consisting of the columns in (66) of the variables which are basic in (65), hence by

$$D_2^{-1} = (A_1^{-1})^{-1} = A_1. \tag{68}$$

Hence, in table 2 tableau 3 is obtained from tableau 0 by a premultiplication of

$$\begin{bmatrix} 2 & 3 & -1 \\ -2 & 1 & -3 \\ 1 & 2 & 0 \end{bmatrix}^{-1} = \begin{bmatrix} \frac{3}{4} & -\frac{1}{4} & -1 \\ -\frac{3}{8} & \frac{1}{8} & 1 \\ -\frac{5}{8} & -\frac{1}{8} & 1 \end{bmatrix}, \tag{69}$$

while tableau 0 can be obtained from tableau 3 by a premultiplication by

$$\begin{bmatrix} \frac{3}{4} & -\frac{1}{4} & -1 \\ -\frac{3}{8} & \frac{1}{8} & 1 \\ -\frac{5}{8} & -\frac{1}{8} & 1 \end{bmatrix}^{-1} = \begin{bmatrix} 2 & 3 & -1 \\ -2 & 1 & -3 \\ 1 & 2 & 0 \end{bmatrix}. \tag{70}$$

The case in which two canonical forms have basic variables in common can be treated in the same manner. Suppose that the first system has the last m variables in the basis, as in (65), but that in a second equivalent system the first q and the last $m-q$ variables are basic, so that both systems have the last $m-q$ variables in the basis. As an example the systems of tableau 0 and of tableau 2 in table 2 may be taken, where $q = 2$.

The first system can be represented by

$$[A \quad I]x = b, \tag{71}$$

but it may be partitioned as follows

$$\begin{bmatrix} A_{11} & A_{12} & I & 0 \\ A_{21} & A_{22} & 0 & I \end{bmatrix} \begin{bmatrix} x^1 \\ x^2 \\ x^3 \\ x^4 \end{bmatrix} = \begin{bmatrix} b^1 \\ b^2 \end{bmatrix}, \tag{72}$$

x is partitioned in 4 subvectors, the first of which contains the first q x-variables; x^3 and x^4 contain the last m x-variables, which are basic in the first canonical form, the $m-q$ x-variables in x^4 contain the variables which are also basic in the second canonical system, whereas the q variables in x^3 are not basic in the second system. A is partitioned in q and $m-q$ rows and q and $n-q$ columns, so that A_{11} is a $q \times q$ matrix, and so on. b is partitioned in b_1 and b_2, which contain q and $m-q$ elements.

For tableau 0 we have

$$x^1 = \begin{bmatrix} x_1 \\ x_2 \end{bmatrix}, \qquad x^2 = x_3, \qquad x^3 = \begin{bmatrix} z_1 \\ z_2 \end{bmatrix}, \qquad x^4 = z_3,$$

$$A_{11} = \begin{bmatrix} 2 & 3 \\ -2 & 1 \end{bmatrix}, \qquad A_{12} = \begin{bmatrix} -1 \\ -3 \end{bmatrix}, \qquad A_{21} = [1 \ \ 2], \qquad A_{22} = 0,$$

$$b^1 = \begin{bmatrix} 2 \\ 6 \end{bmatrix}, \qquad b^2 = 3. \tag{73}$$

The second system can be represented as

$$\begin{bmatrix} I & D_{12} & D_{13} & 0 \\ 0 & D_{22} & D_{23} & I \end{bmatrix} \begin{bmatrix} x^1 \\ x^2 \\ x^3 \\ x^4 \end{bmatrix} = \begin{bmatrix} d^1 \\ d^2 \end{bmatrix}, \tag{74}$$

where the dimensions of the vectors and matrices correspond to those in (72). This system can be obtained from the system (72) by a premultiplication by the inverse of the matrix consisting of the columns in (72) of the variables which are basic in (74). Since the variables in x^1 and x^4 are basic in (74), this inverse matrix is

$$\begin{bmatrix} A_{11} & 0 \\ A_{21} & 1 \end{bmatrix}^{-1} = \begin{bmatrix} A_{11}^{-1} & 0 \\ -A_{21}A_{11}^{-1} & I \end{bmatrix}. \tag{75}$$

The second system may therefore be written as

$$\begin{bmatrix} I & A_{11}^{-1}A_{12} & A_{11}^{-1} & 0 \\ 0 & A_{22} - A_{21}A_{11}^{-1}A_{12} & -A_{21}A_{11}^{-1} & I \end{bmatrix} \begin{bmatrix} x^1 \\ x^2 \\ x^3 \\ x^4 \end{bmatrix} = \begin{bmatrix} A_{11}^{-1}b^1 \\ b^2 - A_{21}A_{11}^{-1}b^1 \end{bmatrix}. \tag{76}$$

Note that though this case seems more complicated than that in which the systems have no basic variables in common, it is in fact a particular case of it, in which the special form of the inverse matrix (75) permits breaking down the inverse into submatrices. The system (72) may, of course, be obtained from the system (74) by a premultiplication of

$$\begin{bmatrix} D_{13} & 0 \\ D_{23} & I \end{bmatrix}^{-1} = \begin{bmatrix} A_{11}^{-1} & 0 \\ -A_{21}A_{11}^{-1} & I \end{bmatrix}^{-1} = \begin{bmatrix} A_{11} & 0 \\ A_{21} & I \end{bmatrix}. \tag{77}$$

In the numerical example, tableau 2 may be obtained from tableau 0 by premultiplication by

$$\begin{bmatrix} A_{11} & 0 \\ A_{21} & I \end{bmatrix}^{-1} = \begin{bmatrix} \begin{bmatrix} 2 & 3 \\ -2 & 1 \end{bmatrix} & \begin{bmatrix} 0 \\ 0 \end{bmatrix} \\ [\ 1 \quad 2] & 1 \end{bmatrix}^{-1}$$

$$= \begin{bmatrix} \begin{bmatrix} 2 & 3 \\ -2 & 1 \end{bmatrix}^{-1} & \begin{bmatrix} 0 \\ 0 \end{bmatrix} \\ -[1 \quad 2]\begin{bmatrix} 2 & 3 \\ -2 & 1 \end{bmatrix}^{-1} & 1 \end{bmatrix} = \begin{bmatrix} \frac{1}{8} & -\frac{3}{8} & 0 \\ \frac{1}{4} & \frac{1}{4} & 0 \\ -\frac{5}{8} & -\frac{1}{8} & 1 \end{bmatrix}. \tag{78}$$

As a special case two canonical systems may be considered which have only one basic variable different. In the numerical example any two successive tableaux have this property. In this case the matrix A_{11} is a scalar and A_{21} is a column vector. The inverse matrix may then be written as

$$\begin{bmatrix} a_{11} & 0 & 0 & \cdot & 0 \\ a_{21} & 1 & 0 & \cdot & 0 \\ a_{31} & 0 & 1 & \cdot & 0 \\ \cdot & \cdot & \cdot & & \cdot \\ a_{m1} & 0 & 0 & \cdot & 1 \end{bmatrix}^{-1} = \begin{bmatrix} a_{11}^{-1} & 0 & 0 & \cdot & 0 \\ -a_{21}a_{11}^{-1} & 1 & 0 & \cdot & 0 \\ -a_{31}a_{11}^{-1} & 0 & 1 & \cdot & 0 \\ \cdot & & \cdot & \cdot & \cdot \\ -a_{m1}a_{11}^{-1} & 0 & 0 & \cdot & 1 \end{bmatrix}. \tag{79}$$

1.11. The value of a linear function in matrix formulation

Let us consider again, but now in terms of vectors and matrices, the value of a linear function of variables which must be solutions of a general linear equation system. Let the linear function be

$$f = p_0 + p'x \tag{80}$$

and the general linear system

$$Ax = b. \tag{81}$$

Then (80) and (81) may be considered one system of equations

$$\begin{bmatrix} 1 & -p' \\ 0 & A \end{bmatrix}\begin{bmatrix} f \\ x \end{bmatrix} = \begin{bmatrix} p_0 \\ b \end{bmatrix}. \tag{82}$$

Suppose now that the first m x-variables are made basic, so that the system (82) is transformed into a canonical system. Partitioning A, p and

x and reordering, we have

$$\begin{bmatrix} 1 & -p^{1\prime} & -p^{2\prime} \\ 0 & A_1 & A_2 \end{bmatrix}\begin{bmatrix} f \\ x^1 \\ x^2 \end{bmatrix} = \begin{bmatrix} p_0 \\ b \end{bmatrix}. \tag{83}$$

The desired canonical form is found by premultiplication of (83) by the matrix

$$\begin{bmatrix} 1 & -p^{1\prime} \\ 0 & A_1 \end{bmatrix}^{-1} = \begin{bmatrix} 1 & p^{1\prime}A_1^{-1} \\ 0 & A_1^{-1} \end{bmatrix}; \tag{84}$$

the result is

$$\begin{bmatrix} 1 & 0 & p^{1\prime}A_1^{-1}A_2 - p^{2\prime} \\ 0 & I & A_1^{-1}A_2 \end{bmatrix}\begin{bmatrix} f \\ x^1 \\ x^2 \end{bmatrix} = \begin{bmatrix} p_0^+p^{1\prime}A_1^{-1}b \\ A_1^{-1}b \end{bmatrix}. \tag{85}$$

In equational form this may be written as

$$f = p_0^+p^{1\prime}A_1^{-1}b - (p^{1\prime}A_1^{-1}A_2 - p^{2\prime})x^2, \tag{86}$$

$$x^1 = A_1^{-1}b - A_1^{-1}A_2x^2. \tag{87}$$

The value of f for this basic solution can be found in various ways. Firstly, the value of x^1 may be substituted in (80). Secondly, the f-equation may be transformed along with the rest of the system and the value of f is then given by the constant term in its row. Thirdly, (85) is used by first computing $p^{1\prime}A_1^{-1}$ which is then multiplied into b; the result is added to p_0.

1.12. Canonical systems and determinants

Determinants play a very small part in mathematical programming, but since they arise rather naturally in the course of the transformations of linear systems and since positive definite forms which arise in quadratic programming are frequently defined in terms of determinants they are treated here.

Consider again the reduction to canonical form of the system

$$Ax = b. \tag{88}$$

It is assumed that the first m x-variables can be made basic without difficulty. It was established before that the first transformation which makes x_1 a basic variable is equivalent to a premultiplication of the system by the matrix

$$E_1 = \begin{bmatrix} a_{11}^{-1} & 0 & 0 & \cdot & 0 \\ -a_{21}a_{11}^{-1} & 1 & 0 & \cdot & 0 \\ -a_{31}a_{11}^{-1} & 0 & 1 & \cdot & 0 \\ \cdot & & \cdot & \cdot & \cdot \\ -a_{m1}a_{11}^{-1} & 0 & 0 & \cdot & 1 \end{bmatrix}. \tag{89}$$

In the resulting system, let b_{i2} be the coefficient of x_2 in the ith equation. It is assumed that b_{22} is nonzero, so that this coefficient can be used as pivot in the second transformation. This second transformation amounts to a premultiplication of the present system by

$$E_2 = \begin{bmatrix} 1 & -b_{12}b_{22}^{-1} & 0 & \cdot & 0 \\ 0 & b_{22}^{-1} & 0 & \cdot & 0 \\ 0 & -b_{32}b_{22}^{-1} & 1 & \cdot & 0 \\ \cdot & \cdot & & \cdot & \cdot \\ 0 & -b_{m2}b_{22}^{-1} & 0 & \cdot & 1 \end{bmatrix}. \tag{90}$$

As explained before, we have now

$$E_2E_1 = \begin{bmatrix} a_{11} & a_{12} & 0 & \cdot & 0 \\ a_{21} & a_{22} & 0 & \cdot & 0 \\ a_{31} & a_{32} & 1 & \cdot & 0 \\ \cdot & \cdot & & \cdot & \cdot \\ a_{m1} & a_{m2} & 0 & \cdot & 1 \end{bmatrix}^{-1} = \begin{bmatrix} A_{11}^{-1} & 0 \\ -A_{21}A_{11}^{-1} & I \end{bmatrix}, \tag{91}$$

where

$$A_{11} = \begin{bmatrix} a_{11} & a_{12} \\ a_{21} & a_{22} \end{bmatrix}, \quad A_{21} = \begin{bmatrix} a_{31} & a_{32} \\ \cdot & \cdot \\ a_{m1} & a_{m2} \end{bmatrix}.$$

Let us consider the determinant of E_2E_1. Using the properties of determinants of matrix products, we find

$$|A_{11}^{-1}| = |E_2E_1| = |E_2|\,|E_1| = b_{22}^{-1}a_{11}^{-1}; \tag{92}$$

the first equality is derived from (91), the third from (90) and (89). From (92) follows

$$b_{22} = |A_{11}|/a_{11}. \tag{93}$$

Hence, it is found that the pivot of the second transformation can be written as the ratio of the second and the first principal determinant of A.

A similar property can be proved for the pivot of the kth transformation. After the $(k-1)$st transformation the matrix by which the system is premultiplied is

$$E_{k-1} \ldots E_2 E_1 = \begin{bmatrix} A_{11}^{-1} & 0 \\ -A_{21}A_{11}^{-1} & I \end{bmatrix}, \tag{94}$$

where A_{11} is now a $(k-1) \times (k-1)$ submatrix of A and A_{21} an $(m-k+1) \times (k-1)$ submatrix. After k transformations the matrix by which the system is premultiplied is

$$E_k E_{k-1} \ldots E_2 E_1 = \begin{bmatrix} A_{11}^{*-1} & 0 \\ -A_{21}^{*}A_{11}^{*-1} & I \end{bmatrix}, \tag{95}$$

where A_{11}^{*} is a $k \times k$ submatrix of A and A_{21}^{*} a $(m-k) \times k$ submatrix. Let us denote the ith element of the kth column after the $(k-1)$st transformation by d_{ik}. We have then

$$E_k = \begin{bmatrix} 1 & 0 & \cdot & d_{1k} & \cdot & 0 \\ 0 & 1 & \cdot & d_{2k} & \cdot & 0 \\ \cdot & \cdot & & \cdot & & \cdot \\ 0 & 0 & \cdot & d_{kk} & \cdot & 0 \\ \cdot & \cdot & & \cdot & & \cdot \\ 0 & 0 & \cdot & d_{mk} & \cdot & 1 \end{bmatrix}^{-1} = \begin{bmatrix} 1 & 0 & -d_{1k}d_{kk}^{-1} & \cdot & 0 \\ 0 & 1 & -d_{2k}d_{kk}^{-1} & \cdot & 0 \\ \cdot & \cdot & \cdot & & \cdot \\ 0 & 0 & d_{kk}^{-1} & \cdot & 0 \\ \cdot & \cdot & \cdot & & \cdot \\ 0 & 0 & -d_{mk}d_{kk}^{-1} & \cdot & 1 \end{bmatrix}. \tag{96}$$

Substituting (94) and (96) into (95) and taking the determinants of the result, we find

$$d_{kk}^{-1}|A_{11}^{-1}| = |A_{11}^{*-1}|, \tag{97}$$

which can be written as

$$d_{kk} = |A_{11}^{*}|/|A_{11}|. \tag{98}$$

This formula gives the pivot of the kth transformation as the ratio of the kth and $(k-1)$st principal determinants of A.

This result provides an easy way for computing determinants, which is more efficient than the usual evaluation according to cofactors. Suppose we want to compute the determinant of a square matrix A of order m. This matrix is transformed as usual, using as pivots the diagonal elements of the matrix which is gradually transformed into a unit matrix. Let the ith principal determinant of A be D_i and let us write p_i for the pivot in the ith transformation. Eq. (98) is then equivalent to

$$p_i = D_i/D_{i-1}. \tag{99}$$

which can be written as

$$D_i = p_i D_{i-1}. \tag{100}$$

Using this formula recursively, we find

$$D_i = p_i p_{i-1} \dots p_2 p_1. \tag{101}$$

Hence we have for the determinant of the matrix itself

$$|A| = p_1 p_2 \dots p_{m-1} p_m. \tag{102}$$

If the determinant is nonzero, it is always possible to find nonzero pivots by interchanging rows and columns as described before.

Using this formula, we can easily find the determinant of the matrix of the first three columns of the matrix of coefficients in the numerical example. We have

$$\begin{bmatrix} 2 & 3 & -1 \\ -2 & 1 & -3 \\ 1 & 2 & 0 \end{bmatrix} = 2 \times 4 \times 1 = 8. \tag{103}$$

It is possible to interpret any coefficient of any canonical form of a system as a ratio of two determinants of coefficients of the original system or any other canonical form of that system. Suppose that the system $Ax = b$ is transformed in such a way that the first m x-variables are basic. Any other canonical form of the system can be rearranged in such a way that this is true. The inverse of the columns of basic variables is then

$$A_1^{-1} = \begin{bmatrix} a_{11} & a_{12} & \cdot & a_{1m} \\ a_{21} & a_{22} & \cdot & a_{2m} \\ \cdot & \cdot & \cdot & \cdot \\ a_{m1} & a_{m2} & \cdot & a_{mm} \end{bmatrix}^{-1}. \tag{104}$$

Now let us consider the coefficient of the jth nonbasic x-variable of the ith equation in this canonical form and let us denote it by d_{ij}; the elements in the same column are denoted in a similar way. If d_{ij} is non-zero, it can be used as a pivot for a following transformation. This transformation amounts to a premultiplication of the canonical system by

$$E_j = \begin{bmatrix} 1 & 0 & \cdot & -d_{1j}d_{ij}^{-1} & \cdot & 0 \\ 0 & 1 & \cdot & -d_{2j}d_{ij}^{-1} & \cdot & 0 \\ \cdot & \cdot & \cdot & \cdot & & \cdot \\ 0 & 0 & \cdot & d_{ij}^{-1} & \cdot & 0 \\ \cdot & \cdot & \cdot & \cdot & & \cdot \\ 0 & 0 & \cdot & -d_{mj}d_{ij}^{-1} & \cdot & 1 \end{bmatrix}. \tag{105}$$

After the transformation, the inverse of the basis is

$$E_j A_1^{-1} = A_1^{*-1}, \tag{106}$$

where A_1^* is equal to A_1 except that it has the jth column of A instead of the ith column. Taking the determinants of both sides of (106), we find

$$d_{ij}^{-1} |A_1^{-1}| = |A_1^{*-1}|, \tag{107}$$

which can be written as

$$d_{ij} = |A_1^*|/|A_1|. \tag{108}$$

Note that this result is also valid for $d_{ij} = 0$.

Returning to our numerical example, we find for instance that the coefficient of x_5 in the second equation of tableau 3 in table 2 may be expressed as

$$\begin{vmatrix} 2 & 5 & -1 \\ -2 & 0 & -3 \\ 1 & 4 & 0 \end{vmatrix} \Big/ \begin{vmatrix} 2 & 3 & -1 \\ -2 & 1 & -3 \\ 1 & 2 & 0 \end{vmatrix} = 2\tfrac{1}{8}. \tag{109}$$

The result (108) is also valid for the constant term of the systems. In this case pivoting is considered on coefficients of the constant term of the system, something which does not occur in the usual transformations; however, there is nothing which prevents us from doing this.

Let us now consider the special case in which A is square. If A is non-singular, the unique solution of the system is $x = A^{-1}b$. This result may be obtained by transforming the system until the following canonical form is found

$$\begin{bmatrix} 1 & 0 & \cdot & 0 \\ 0 & 1 & \cdot & 0 \\ \cdot & \cdot & \cdot & \cdot \\ 0 & 0 & \cdot & 1 \end{bmatrix} \begin{bmatrix} x_1 \\ x_2 \\ \cdot \\ x_m \end{bmatrix} = \begin{bmatrix} d_1 \\ d_2 \\ \cdot \\ d_m \end{bmatrix}, \tag{110}$$

where $d = A^{-1}b$. Consider now pivoting on d_i, the solution for x_i. We find then, analogous to (108)

$$d_i = x_i = |A_i^*|/|A|, \tag{111}$$

where A_i^* is equal to A with its ith column replaced by b. This result is known as *Cramer's rule*.

As an example of application of (111), let us consider the solution of the numerical example with x_4 and x_5 equal to zero. We have then for x_2:

$$x_2 = \begin{vmatrix} 2 & 2 & -1 \\ -2 & 6 & -3 \\ 1 & 3 & 0 \end{vmatrix} \Big/ \begin{vmatrix} 2 & 3 & -1 \\ -2 & 1 & -3 \\ 1 & 2 & 0 \end{vmatrix} = 3 \qquad (112)$$

CHAPTER 2

THE SIMPLEX METHOD FOR LINEAR PROGRAMMING

2.1. The linear programming problem: an example of production planning

Formally, the linear programming problem can be thought of as a purely mathematical problem. It then can be stated as the problem of maximizing or minimizing a linear function subject to a number of linear equality and inequality constraints. Characteristic for programming problems are the inequality constraints; without these, a problem can be analyzed and in principle solved using the ordinary calculus methods.

Linear programming has found a very wide field of applications. Dominant among these are economic applications in which the optimal allocation of scarce resources or the least cost of production are sought. For this reason and also because a more concrete interpretation of abstract mathematical concepts greatly improves the understanding of the problem and the solution methods, it is useful to deal with linear programming at least partly in economic terms. Hence we shall first describe a typical economic problem, which will be formulated as a linear programming problem.

The problem is concerned with planning production in a firm over a given period. A number of products can be produced, each of which can be sold for a given price per unit. However, there are limitations to the amounts of these products which can be produced. These limitations may be of different kinds. The machines and installations which are used for production usually have a given capacity which cannot be exceeded, so that the number of machine-hours for the various machines is limited during the period. The same is true for manpower, if only normal working hours are considered. In the following we shall, for shortness' sake, use the word machine for any fixed limitation of production. It is assumed that the capacity of a machine in use for a certain product is always proportional to the quantity of that product. The problem then is to find a production program which maximizes profits (or, in terms of the business economist, contribution to fixed costs and profits), and which does not exceed the various fixed capacities.

The following numerical example will be used. There are three machines with limited fixed capacities; these capacity limits are 100, 40, and 60 hours per period for machine 1, 2, and 3. Five products can be produced. Product 1 requires per unit of product 4 hours on machine 1, 1 hour on machine 2, and 2 hours on machine 3. The other products use different quantities of the capacity of the three machines, as given in table 1. Selling price minus variable cost which, for shortness' sake, is called profit per unit, is 3 monetary units for product 1, and similar amounts for other products, as given in table 1.

Inequalities may be constructed by formulating the requirements that the sum of the capacities of a machine used by the various products should be less than or equal to the available machine capacity. Let us denote the number of units produced of product 1 by x_1, that of product 2 by x_2, and so on. The time on machine 1 used for producing 1 is $4x_1$, that for producing product 2 is $3x_2$, and so on. The inequality of required and available machine-time for machine 1 becomes

$$4x_1 + 3x_2 + x_3 + 2x_4 \leq 100. \tag{1}$$

For the machines 2 and 3 we have in the same way

$$x_1 + \qquad x_4 + 2x_5 \leq 40, \tag{2}$$

$$2x_1 + 4x_2 + x_4 + 2x_5 \leq 60. \tag{3}$$

Furthermore, it is required that the quantities in production are non-negative, since negative production is not possible. Hence we have the requirements

$$x_1, x_2, x_3, x_4, x_5, \geq 0. \tag{4}$$

Table 1

Data for production planning example

	Machine capacity per period	Capacity utilization				
		Product 1	Product 2	Product 3	Product 4	Product 5
Machine 1	100	4	3	1	2	0
Machine 2	40	1	0	0	1	2
Machine 3	60	2	4	0	1	2
Profit per unit of product		3	1	$\frac{1}{2}$	2	2

Indicating the profit from a production program $x_1, ..., x_5$ as $f(x_1, ..., x_5)$, or just f, we have

$$f = 3x_1 + x_2 + \tfrac{1}{2}x_3 + 2x_4 + 2x_5. \tag{5}$$

If profits are to be maximized, the problem may be formulated as that of maximizing the function f given in (5) subject to the inequalities (1)–(4). The function to be maximized is called the *objective function*; the requirements (1)–(4) are the *constraints* of the problem.

2.2. General formulation of linear programming problems

The problem considered above can be stated in a more general way. We then arrive at the following general formulation of a linear programming problem. Maximize the function[1]

$$f = p_1x_1 + p_2x_2 + ... + p_nx_n \tag{6}$$

subject to the constraints

$$a_{11}x_1 + a_{12}x_2 + ... + a_{1n}x_n \leq b_1,$$
$$a_{21}x_1 + a_{22}x_2 + ... + a_{2n}x_n \leq b_2,$$
$$...$$
$$a_{m1}x_1 + a_{m2}x_1 + ... + a_{mn}x_n \leq b_m, \tag{7}$$

and

$$x_1, x_2, ..., x_n \geq 0. \tag{8}$$

The last constraints are called *nonnegativity constraints*.

In terms of vectors and matrices the problem may be stated as follows. Maximize

$$f = p'x \tag{9}$$

subject to

$$Ax \leq b, \tag{10}$$

$$x \geq 0. \tag{11}$$

This is but one of the possible general formulations of a linear programming problem. Other formulations include constraints which have instead of the $\leq$-sign in (7), an $=$-sign or a $\geq$-sign. Furthermore, there may

---

[1] A possible constant term p_0 is deleted.

be an objective function, which is to be minimized instead of maximized. All these different formulations can be shown to be equivalent to one another in the sense that a problem formulated in one way may be given in one of the other formulations.

Any inequality constraint may be replaced by an equation plus a nonnegativity constraint by the introduction of a new variable. For instance, instead of the first inequality of (7), we may write

$$a_{11}x_1 + a_{12}x_2 + \ldots + a_{1n}x_n + y_1 = b_1, \tag{12}$$

$$y_1 \geq 0. \tag{13}$$

The newly introduced variable y_1 is called a *slack variable*. All constraints of (7) may be dealt with in this way, so that the inequalities of (10) may be replaced by

$$Ax + y = b, \tag{14}$$

$$y \geq 0, \tag{15}$$

where y is a vector of m slack variables. In the same way the inequalities

$$Ax \geq b \tag{16}$$

can be reformulated as

$$Ax - y = b, \tag{17}$$

$$y \geq 0. \tag{18}$$

The problem given in (9)–(11) may therefore be reformulated as follows. Maximize

$$f = p^{*\prime}x^*, \tag{19}$$

subject to

$$A^*x^* = b, \tag{20}$$

$$x^* = 0, \tag{21}$$

with

$$x^* = \begin{bmatrix} x \\ y \end{bmatrix}, \qquad p^* \begin{bmatrix} p \\ 0 \end{bmatrix}, \qquad A^* = [A \quad I].$$

On the other hand, any equation may be stated as an inequality by elimination of a variable. For instance, consider the equation

$$a_{11}x_1 + a_{12}x_2 + \ldots + a_{1n}x_n = b_1. \tag{22}$$

Take any nonzero coefficient of an x-variable, for instance a_{11}. After division of both sides of (22) the result may be written as

$$x_1 = a_{11}^{-1} b_1 - a_{11}^{-1} a_{12} x_2 - \dots - a_{11}^{-1} a_{1n} x_n. \qquad (23)$$

This may be combined with $x_1 \geq 0$, to give

$$a_{11}^{-1} a_{12} x_2 + \dots + a_{11}^{-1} a_{1n} x_n \leq a_{11}^{-1} b_1. \qquad (24)$$

This equation replaces (22) and $x_1 \geq 0$. Furthermore, (23) is used to substitute for x_1 in the objective function and the remaining constraints.

The example given above happened to be a maximization problem. Other linear programming problems may require a minimization of a linear function. However, minimization problems may be easily formulated as maximization problems, by multiplication of the objective function by -1, and vice versa; for instance, the minimization of a function f is equivalent to the maximization of $-f$.

2.3. Introduction to the simplex method

The simplex method for linear programming was the first, and is the most well-known and basic method for solving linear programming problems; most other methods for solving linear programming problems are closely related to this method. The method is due to Dantzig, who developed it in 1947.[2]

For an application of the simplex method, the form in which the linear programming problem is stated must satisfy certain requirements. The first requirement is that the constraints of the linear programming problem, apart from the nonnegativity constraints, must be given in the form of equations. Hence (14) and (15) should be used instead of (10). A second requirement is that the system of equations of the constraints and the objective function must be in canonical form. This is true for the equations of (9) and (14), since they may be written as

$$\begin{bmatrix} -p' & 1 & 0 \\ A & 0 & I \end{bmatrix} \begin{bmatrix} x \\ f \\ y \end{bmatrix} = \begin{bmatrix} 0 \\ b \end{bmatrix}. \qquad (25)$$

A third requirement is that the basic solution of the canonical form for basic variables other than f must satisfy the nonnegativity constraints. In

---

[2] His fundamental paper on the subject was published in KOOPMANS (ed.)[10].

(25) the basic solution is

$$x = 0, \qquad f = 0, \qquad y = b.$$

If $b \geq 0$, that is, if all elements of b are nonnegative, the requirement is satisfied.

Any solution of a linear programming which satisfies the constraints of the problem is called a *feasible solution*; it can therefore be said that the simplex method requires as an initial solution a *basic feasible solution* to the constraints. First we shall assume that the problem has the form (9), (14), and (15), and that $b \geq 0$, so that the basic feasible solution $x = 0$, $f = 0$, $y = b$ is immediately available; later we shall deal with other cases.

The simplex method is based on a system of equations, consisting of the equations of the constraints and the objective function. For the problem as formulated in (9)–(11), this system can be stated as follows:

$$0 = -p'x + f,$$
$$b = Ax + y. \tag{26}$$

For the example presented in section 2.1, this system is

$$
\begin{aligned}
0 &= -3x_1 - x_2 - \tfrac{1}{2}x_3 - 2x_4 - 2x_5 + f, \\
100 &= 4x_1 + 3x_2 + x_3 + 2x_4 + 0x_5 + y_1, \\
40 &= x_1 + 0x_2 + 0x_3 + x_4 + 2x_5 + y_2, \\
60 &= 2x_1 + 4x_2 + 0x_3 + x_4 + 2x_5 + y_3.
\end{aligned}
\tag{27}
$$

Note that the system is in canonical form and that the basic solution of this form is

$$f = 0,$$
$$y_1 = 100, \; y_2 = 40, \; y_3 = 60,$$
$$x_1 = x_2 = x_3 = x_4 = x_5 = 0.$$

This solution has to be nonnegative as far as the variables other than f are concerned if it is to be used as a starting-solution for the simplex method; it evidently satisfies this requirement. The slack variables y_1, y_2, and y_3 can in this case be interpreted as the idle capacity on the three machines. This basic solution describes the situation in which all available machine capacity is left idle and nothing is produced. The profit, given by the value of f, is zero. This obviously is not a situation in which profits are maximized.

Let us now consider what happens if one of the products is produced in a positive quantity, say the first product. For this purpose (27) is re-written as follows

$$3x_1 = -x_2 - \tfrac{1}{2}x_3 - 2x_4 - 2x_5 + f,$$
$$100 - 4x_1 = 3x_2 + x_3 + 2x_4 + 0x_5 \qquad + y_1,$$
$$40 - x_1 = 0x_2 + 0x_3 + x_4 + 2x_5 \qquad\qquad + y_2,$$
$$60 - 2x_1 = 4x_2 + 0x_3 + x_4 + 2x_5 \qquad\qquad\qquad + y_3. \qquad (28)$$

Since only a change in x_1 is considered, and x_2, x_3, x_4, and x_5, have a zero value, the terms in these variables can, for the moment, be deleted. We obtain

$$3x_1 = f,$$
$$100 - 4x_1 = y_1,$$
$$40 - x_1 = y_2,$$
$$60 - 2x_1 = y_3. \qquad (29)$$

In this simple equation system we observe the following. If x_1 is increased from a zero level to, say, 1, then according to the last equation the value of the objective function f increases from 0 to 3. Hence it is a good thing to increase x_1. We immediately generalize that an increase in any nonbasic variable which has a negative coefficient in the row of the objective function in (27), increases the objective function. Furthermore, we observe that the larger, in absolute value, the negative coefficient, the larger the increase in the objective function per unit increase of the nonbasic variable. Hence a reasonable choice for the nonbasic variable to be increased is the one with the largest negative coefficient in the objective function row. This happens in (27) to be x_1.

How far should x_1 be increased? According to the first equation of (29), in order to maximize f, x_1 should be increased as much as possible. But the other eqs. (29) give the values of the basic variables which result from an increase in x_1. According to the first equation, y_1 is positive for $x_1 > 25$, zero for $x_1 = 25$ and negative for $x_1 < 25$. Since, according to the constraints, negative values of the variables are not allowed and since the simplex method tries to find successive feasible solutions with increasing values of the objective function, x_1 cannot be increased beyond 25. From the second equation it can be observed that the highest value of x_1 which does not make y_2 negative is 40, and from the third equation it is obvious that x_1 cannot be increased beyond 30 without making y_3 negative. Hence the highest value to which x_1 can be increased, without making any of the basic variables negative, is 25. For $x_1 = 25$, (29) becomes

$$75 = f,$$
$$0 = y_1,$$
$$15 = y_2,$$
$$10 = y_3. \tag{30}$$

In this solution y_1 is zero, because it was the first variable which became zero when x_1 was increased. Eqs. (30), $x_1 = 25$ and all other variables zero give another solution to the system (27). We see that in this solution the basic variable y_1 in (27) has become zero, while the nonbasic variable x_1 has become nonzero. In fact, this solution can be generated from (27) if the whole system is transformed in such a way that x_1 replaces y_1 as a basic variable. The pivot of the transformation is the coefficient of x_1 in the first equation, which is 4. The result of the transformation is the following system:

$$75 = \quad 1\tfrac{1}{4}x_2 + \tfrac{1}{4}x_3 - \tfrac{1}{2}x_4 - 2x_5 + f + \tfrac{3}{4}y_1,$$
$$25 = x_1 + \tfrac{3}{4}x_2 + \tfrac{1}{4}x_3 + \tfrac{1}{2}x_4 + 0x_5 \quad + \tfrac{1}{4}y_1,$$
$$15 = \quad - \tfrac{3}{4}x_2 - \tfrac{1}{4}x_3 + \tfrac{1}{2}x_4 + 2x_5 \quad - \tfrac{1}{4}y_1 + y_2,$$
$$10 = \quad 2\tfrac{1}{2}x_2 - \tfrac{1}{2}x_3 + 0x_4 + 2x_5 \quad - \tfrac{1}{2}y_1 \quad + y_3. \tag{31}$$

The basic solution of this canonical system is that given in (30). We have now increased the objective function from 0 to 75 by producing 25 units of the first product. Though we would have liked to produce more of this product, because according to the last equation of (28) it yields 3 per unit, this cannot be done because the capacity of machine 1 proves to be a bottleneck.

However, we may now try to produce other products, by increasing the quantities produced from zero. Let us rewrite the first equation of (31) with the terms of the nonbasic variables taken to the left-hand side:

$$75 - 1\tfrac{1}{4}x_2 - \tfrac{1}{4}x_3 + \tfrac{1}{2}x_4 + 2x_5 - \tfrac{3}{4}y_1 = f. \tag{32}$$

We then see that an increase of x_4 by one unit gives an increase in profits of $\tfrac{1}{2}$, while an increase of x_5 by one unit yields an increase of 2. An increase in the other nonbasic variables decreases the objective function. Hence we decide to increase x_5. Writing the x_5-terms at the left-hand side and deleting the other nonbasic variables, since they are kept at a zero value, we have

$$75 + 2x_5 = f,$$
$$25 - 0x_5 = \quad x_1,$$
$$15 - 2x_5 = \quad\quad y_2,$$
$$10 - 2x_5 = \quad\quad\quad y_3. \tag{33}$$

An increase of x_5 does not have any effect on x_1, but x_5 can only be increased to $7\frac{1}{2}$ without making y_2 negative and to 5 without making y_3 negative. Hence we find that the highest value which x_5 can take without making any basic variable negative is 5; for this value y_3 becomes zero.

The same solution can be obtained from (31) by making x_5 a basic variable in the third equation instead of y_3. The corresponding canonical system is generated by transforming (31) using the coefficient of x_5 in the third equation as a pivot. The following system is obtained:

$$85 = \quad 3\tfrac{3}{4}x_2 - \tfrac{1}{4}x_3 - \tfrac{1}{2}x_4 \quad + f + \tfrac{1}{4}y_1 \quad + y_3,$$
$$25 = x_1 + \tfrac{3}{4}x_2 + \tfrac{1}{4}x_3 + \tfrac{1}{2}x_4 \quad + \tfrac{1}{4}y_1,$$
$$5 = \quad -3\tfrac{1}{4}x_2 + \tfrac{1}{4}x_3 + \tfrac{1}{2}x_4 \quad + \tfrac{1}{4}y_1 + y_2 - y_3,$$
$$5 = \quad 1\tfrac{1}{4}x_2 - \tfrac{1}{4}x_3 + 0x_4 + x_5 \quad - \tfrac{1}{4}y_1 \quad + \tfrac{1}{2}y_3. \tag{34}$$

The objective function has now increased to 85 by producing products 1 and 5; no spare capacity on the machines 1 and 3 is left, but there are still 5 units of spare capacity on machine 2. According to the first equation there is still scope for increasing profit, since increasing the production of x_3 yields $\tfrac{1}{4}$ per unit and an increase of x_4 yields $\tfrac{1}{2}$ per unit. Hence it is decided to increase x_4 from zero. Putting the terms in x_4 at the left-hand side and deleting the terms in the other nonbasic variables, we have

$$85 + \tfrac{1}{2}x_4 = f,$$
$$25 - \tfrac{1}{2}x_4 = x_1,$$
$$5 - \tfrac{1}{2}x_4 = y_2$$
$$5 - 0x_4 = x_5. \tag{35}$$

From this we see that increasing x_4 leaves x_5 unchanged, but that it decreases x_1 and y_2. x_1 becomes zero for $x_4 = 50$, and y_2 becomes zero for $x_4 = 10$. Hence we should not increase x_4 any further than 10, at which value y_2 is zero. Hence in the system (34) x_4 should replace y_2 as a basic variable; (34) is therefore transformed with the coefficient of x_4 in the second equation, $\frac{1}{2}$, as a pivot. The result is

$$
\begin{aligned}
90 &= \quad\quad \tfrac{1}{2}x_2 + 0x_3 \quad\quad\quad\quad + f + \tfrac{1}{2}y_1 + \quad y_2 + 0y_3, \\
20 &= x_1 + \ 4x_2 + 0x_3 \quad\quad\quad\quad + 0y_1 - \quad y_2 + \quad y_3, \\
10 &= \quad -6\tfrac{1}{2}x_2 + \tfrac{1}{2}x_3 + x_4 \quad\quad + \tfrac{1}{2}y_1 + 2y_2 - 2y_3, \\
5 &= \quad\quad 1\tfrac{1}{4}x_2 - \tfrac{1}{4}x_3 \quad + x_5 \quad - \tfrac{1}{4}y_1 + 0y_2 + \tfrac{1}{2}y_3.
\end{aligned}
\tag{36}
$$

For the basic solution of this system, f has the value 90. In the first equation there are no nonbasic variables with negative coefficients, so that no increase of the objective function is possible. In fact, it will be proved that the present basic solution is an optimal solution of the problem.

2.4. Simplex tableaux and rules of the simplex method

In section 2.3 the equation systems used in the method were written down explicitly. It is much more convenient to present the successive equation systems in tableau-form, as was shown earlier. Table 2 gives the systems (27), (31), (34) and (36) in detached coefficient form as tableaux 0, 1, 2, and 3.

It is even possible to present these tableaux in a more compact form by deleting the unit columns of the basic variables because they give no extra information, since we know already which basic variable belongs to an equation because this is given in the second column of each tableau. Table 3 gives the simplex tableaux of table 2 in such a compact form. Note that in this case the column of the nonbasic variable which becomes basic is replaced by that of the basic variable which becomes nonbasic; for example, in table 3 the column of x_1 in tableau 0 is replaced by that of y_1 in tableau 1. Because the positions of the nonbasic variables change in the various tableaux, the nonbasic variables have to be indicated separately for each tableau.

The simplex method turned out to be an iterative method, which means that the method operates by steps which are repeated. Each iteration or step of the method consisted in generating from a given canonical system with a nonnegative basic solution, another canonical system with a nonnegative basic solution and a higher value of the objective function.[3]

---

[3] So-called degenerate systems can be exceptions to this; these cases are treated in section 3.5.

Table 2

Simplex tableaux for the production planning example

Tableau	Basic variables	Values of basic variables	Ratio	x_1	x_2	x_3	x_4	x_5	f	y_1	y_2	y_3
0	f	0		-3	-1	$-\frac{1}{2}$	-2	-2	1	0	0	0
	y_1	100	25	4	3	1	2	0	0	1	0	0
	y_2	40	40	1	0	0	1	2	0	0	1	0
	y_3	60	30	2	4	0	1	2	0	0	0	1
1	f	75		0	$1\frac{1}{4}$	$-\frac{1}{4}$	$-\frac{1}{2}$	-2	1	$\frac{3}{4}$	0	0
	x_1	25		1	$\frac{3}{4}$	$\frac{1}{4}$	$\frac{1}{2}$	0	0	$-\frac{1}{4}$	0	0
	y_2	15	$7\frac{1}{2}$	0	$-\frac{3}{4}$	$-\frac{1}{4}$	$\frac{1}{2}$	2	0	$-\frac{1}{4}$	1	0
	y_3	10	5	0	$2\frac{1}{2}$	$-\frac{1}{2}$	0	$2\frac{1}{4}$	0	$-\frac{1}{2}$	0	-1
2	f	85		0	$3\frac{3}{4}$	$-\frac{1}{4}$	$-\frac{1}{2}$	0	1	$-\frac{1}{4}$	0	-1
	x_1	25	50	1	$\frac{3}{4}$	$\frac{1}{4}$	$\frac{1}{2}$	0	0	$-\frac{1}{4}$	0	0
	y_2	5	10	0	$-3\frac{3}{4}$	$-\frac{1}{4}$	$\frac{1}{2}$	0	0	$-\frac{1}{4}$	1	-1
	x_5	5		0	$1\frac{1}{4}$	$-\frac{1}{4}$	0	1	0	$-\frac{1}{4}$	0	$-\frac{1}{2}$
3	f	90		0	$-\frac{1}{2}$	0	0	0	1	$-\frac{1}{2}$	-1	0
	x_1	20		1	4	0	0	0	0	0	-1	-1
	x_4	10		0	$-6\frac{1}{2}$	$-\frac{1}{2}$	1	0	0	$-\frac{1}{2}$	2	-2
	x_5	5		0	$1\frac{1}{4}$	$-\frac{1}{4}$	0	1	0	$-\frac{1}{4}$	0	$-\frac{1}{2}$

Table 3

Compact simplex tableaux for the production planning example

Tableau	Basic var.	Values of bas. var.	Ratio	\multicolumn{5}{c}{Nonbasic variables}
				x_1	x_2	x_3	x_4	x_5
0	f	0		-3	-1	$-\frac{1}{2}$	-2	-2
	y_1	100	25	4	3	1	2	0
	y_2	40	40	$\underline{1}$	0	0	1	2
	y_3	60	30	2	4	0	1	2
				y_1	x_2	x_3	x_4	x_5
1	f	75		$\frac{3}{4}$	$1\frac{1}{4}$	$\frac{1}{4}$	$-\frac{1}{2}$	-2
	x_1	25		$\frac{1}{4}$	$\frac{3}{4}$	$\frac{1}{4}$	$\frac{1}{2}$	0
	y_2	15	$7\frac{1}{2}$	$-\frac{1}{4}$	$-\frac{3}{4}$	$-\frac{1}{4}$	$\frac{1}{2}$	2
	y_3	10	5	$-\frac{1}{2}$	$2\frac{1}{2}$	$-\frac{1}{2}$	0	2
				y_1	x_2	x_3	x_4	y_3
2	f	85		$\frac{1}{4}$	$3\frac{3}{4}$	$-\frac{1}{4}$	$-\frac{1}{2}$	1
	x_1	25	50	$\frac{1}{4}$	$\frac{3}{4}$	$\frac{1}{4}$	$\frac{1}{2}$	0
	y_2	5	10	$\frac{1}{4}$	$-3\frac{1}{4}$	$\frac{1}{4}$	$\frac{1}{2}$	-1
	x_5	5		$-\frac{1}{4}$	$1\frac{1}{4}$	$-\frac{1}{4}$	0	$\frac{1}{2}$
				y_1	x_2	x_3	y_2	y_3
3	f	90		$\frac{1}{2}$	$\frac{1}{2}$	0	1	0
	x_1	20		0	4	0	-1	1
	x_4	10		$\frac{1}{2}$	$-6\frac{1}{2}$	$\frac{1}{2}$	2	-2
	x_5	5		$-\frac{1}{4}$	$1\frac{1}{4}$	$-\frac{1}{4}$	0	$\frac{1}{2}$

Each step can be divided into three parts, namely the selection of a nonbasic variable which is to become basic, the selection of a basic variable which is to become nonbasic and the transformation of the system. We shall now state the rules for the three parts of an iteration of the simplex method in a more formal manner.

We have shown in the previous section that an increase in a nonbasic variable which has a negative coefficient in the objective function equation increases the objective function if the new basic variable is positive. Hence it is possible to choose any nonbasic variable with a negative coefficient in the last row of the tableau as the new basic variable. However, the following rule has some advantages and is the usual one.[4]

---

[4] In chapter 3 some alternative rules are discussed.

Rule 1: *Select as the nonbasic variable to enter the basis the one with the most negative coefficient in the objective function row. If this row does not contain a negative coefficient, an optimal solution has been obtained.* If the jth element in the f-row is indicated by d_{0j}, the new basic variable is determined by

$$\underset{j=1,\,...,\,n}{\text{Min}}\ (d_{0j}) = d_{0k}$$

if $d_{0k} \geq 0$, the current basic solution is optimal.

The rule for the basic variable which becomes nonbasic is based on the consideration that no basic variable may attain negative values, see (29), (33) and (35). The values of the new basic variable for which a basic variable becomes zero were determined by bringing the terms in the new basic variable to the left-hand side, after which the left-hand side is put equal to zero in order to find the value of the new basic variable for which the basic variable of the equation concerned becomes zero, see e.g. (29). Quite the same is achieved by determining the ratios in each row of the values of the basic variables and the corresponding positive coefficients in the column of the new basic variable. For instance, in tableau 1, we find that y_1 becomes zero for $x_1 = 25$ by taking the ratio 100/4, y_2 becomes zero for $x_1 = 40$ by taking the ratio 40/1, and y_3 becomes zero for $x_1 = 30$ by taking the ratio 60/2. These ratios are to be found in the column Ratios in tables 2 and 3.

In the example the terms in the new basic variable in the equations other than that of the objective function happened to be positive. Let us find out what would happen if one or more of these terms were negative. Suppose, for example, that we had in the third equation of (27) $-2x_1$ instead of $2x_1$. The third equation of (29) would then become

$$60 + 2x_1 = y_3. \tag{37}$$

We see that y_3 increases for increasing values of x_1. There is therefore no danger that y_3 becomes negative when x_1 is increased. Hence the ratios in rows in which the coefficients of the new basic are negative may be disregarded when determining which basic variable leaves the basis. The same is true when the coefficient of the new basic variable is zero in a certain row, as is, for example, the case with the coefficient of x_5 in the first equation of (31).

Let us now consider the case in which the new basic variable has only zero or negative coefficients of the new basic variable in the equations of the ordinary basic variables. For instance, we could consider the case in which the coefficients of x_5 in the second and third equation of (31) are -2 rather than 2. In this case we have instead of (33)

$$75 + 2x_5 = f,$$
$$25 + 0x_5 = x_1,$$
$$15 + 2x_5 = y_2,$$
$$10 + 2x_5 = y_3. \tag{38}$$

From this it is obvious that an increase in x_5 leaves x_1 at the same value, but increases y_2 and y_3. The last equation indicates that an increase of x_1 by 1 unit increases the objective function by 2 units. Hence the objective function may be increased without bounds by increasing x_5, since there is no basic variable which becomes negative for any positive value of x_5. In this case the problem is said (somewhat improperly) to have an *infinite optimal solution*. In practical problems this means that the problem has not been well formulated; for instance, it might be that an existing constraint has not been included in the formulation.

For the second part of an iteration of the simplex method, the following rule may therefore be given:

Rule 2: *Select as the variable to leave the basis the one corresponding to the smallest ratio of the value of the basic variable and the corresponding coefficient of the new basic variable in rows in which these coefficients are positive. If the new basic variable has no positive coefficients, the problem has an infinite solution. If there is a tie, select any of the tied variables.*

An alternative description of this rule in terms of a formula can be given as follows. If d_{i0} stands for the value of the basic variable in the ith row and d_{ik} stands for the corresponding coefficient of the new basic variables x_k, the leaving basic variable is the one associated with

$$\underset{i}{\text{Min}} \ (d_{i0}/d_{ik} \,|\, d_{ik} > 0). \tag{39}$$

After the new basic variable and the leaving basic variable have been selected, the system has to be transformed in order to obtain a canonical system with the new basis. This is the part of a simplex iteration which requires the most computational work. How such a transformation is performed was described extensively before. The rule for the third part of a simplex iteration is as follows.

Rule 3: *Transform the tableau with the coefficient of the new basic variable in the row of the leaving basic variable as a pivot.*

2.5. Termination of the simplex method

A basic solution is called *degenerate* if one or more of its basic variables has a zero value. In this section we shall prove that the simplex method finds the optimal solution of a linear programming in a finite number of iterations, provided an initial basic feasible solution to the problem is known, and provided no degenerate basic solutions occur. In chapter 3 it is shown that an initial basic solution can always be found if one exists and also that the non-degeneracy requirement is superfluous if an additional rule is followed in case the choice of the leaving basic variable is not unique.

First it is shown that the value of the objective function f increases in any iteration if there is no degeneracy. Let the canonical system before the transformation be represented by

$$f_0 = p_1 x_1 + \ldots + p_{n-m} x_{n-m} + f,$$
$$b_1 = a_{11} x_1 + \ldots + a_{1,n-m} x_{n-m} + x_{n-m+1},$$

$$\ldots$$

$$b_m = a_{m1} x_1 + \ldots + a_{m,n-m} x_{n-m} + x_n. \tag{40}$$

The variables have been, if necessary, reordered in such a way that the basic variables are $x_{n-m+1}, \ldots, x_n$. The new basic variable in an iteration should have a negative coefficient in the f-equation; let the new basic variable be x_k with $p_k < 0$. The leaving basic variable is selected according to the criterion

$$\text{Min}_i (b_i / a_{ik} | a_{ik} > 0);$$

let this be x_r. The value of f for the new solution is then

$$f_0 - p_k a_{rk}^{-1} b_r. \tag{41}$$

Because of the nondegeneracy assumption, $b_r > 0$, so that $b_r a_{rk}^{-1} > 0$; since $p_k < 0$, the value of f must have increased. Hence the value of the objective function increases in each iteration in nondegenerate cases.

An upper bound for the number of successive tableaux of the simplex method is $\binom{n}{m}$, since this is the number of different ways in which m basic variables can be picked from n variables. Each tableau stands for a canonical form with given basic variables which corresponds with a unique solution of the problem. Since the value of f increases, no canonical form or its corresponding tableau can reoccur in the simplex method.

The maximum number of tableaux being $\binom{n}{m}$, which is a finite number, the simplex method must find the optimal solution in a finite number of iterations.

Rule 1 of the simplex method implies that if no basic variables can be found because all coefficients of nonbasic variables in the objective function equation are nonnegative, the present solution is an optimal solution. This may be easily proved as follows. Let the equation of the objective function be

$$f_0 = p_1 x_1 + p_2 x_2 + \dots p_{n-m} x_{n-m} + f, \tag{42}$$

with

$$p_1, \dots, p_{n-m} \geq 0,$$

where again the variables are, if necessary, reordered in such a way that the basic variables are $x_{n-m+1}, \dots, x_n$. This equation may be written as

$$f = f_0 - p_1 x_1 - p_2 x_2 - \dots - p_{n-m} x_{n-m}. \tag{43}$$

The value of f for any feasible solution is found by substitution in (43) of the (nonnegative) values of the nonbasic variables $x_1, \dots, x_{n-m}$. It is obvious that for any positive p_j, $j = 1, \dots, n - m$, any positive value of x_j will decrease f; for any p_i which is zero, a positive value of x_i has no effect on f. Hence another feasible solution can never give a higher value of f than f_0, which is found by using the basic solution leading to (43). It must be realized that this is valid because the present canonical system is equivalent with the original one. The basic solution of (43) is the unique optimal solution of $p_1, p_2, \dots, p_{n-m} > 0$. If one or more of the coefficients of nonbasic variables are zero, there usually are multiple optimal solutions, all leading to the same value of the objective function.

2.6. Alternative optimal solutions

Let us consider tableau 3 of table 3, also reproduced in table 4, in which the elements of the f-row are all nonnegative. The element of this row in the column of x_3 is zero, which means that x_3 can be increased without changing the value of f. The elements of the other rows in the column of x_3 are not zero, which means that the values of basic variables are changing as a consequence of changing values of x_3. We find that the highest value that x_3 can take is 20, because x_4 becomes zero for that value of x_3. Hence there is also another basic feasible solution which is optimal, namely the one which is obtained by replacing x_4 as a basic

Table 4

Alternative optimal solutions for the production planning example

Tableau	Basic var.	Values of bas. var.	Ratio	y_1	x_2	x_3	y_2	y_3
3	f	90		$\frac{1}{2}$	$\frac{1}{2}$	0	1	0
	x_1	20		0	4	0	-1	1
	x_4	10	20	$\frac{1}{2}$	$-6\frac{1}{2}$	$\underline{\frac{1}{2}}$	2	-2
	x_5	5		$-\frac{1}{4}$	$1\frac{1}{4}$	$-\frac{1}{4}$	0	$\frac{1}{2}$

Tableau	Basic var.	Values of bas. var.	Ratio	y_1	x_2	x_4	y_2	y_3
4	f	90		$\frac{1}{2}$	$\frac{1}{2}$	0	1	0
	x_1	20	20	0	4	0	-1	1
	x_3	20		1	-13	2	4	-4
	x_5	10		0	-2	$\frac{1}{2}$	1	$-\frac{1}{2}$

Tableau	Basic var.	Values of bas. var.	Ratio	y_1	x_2	x_4	y_2	x_1
5	f	90		$\frac{1}{2}$	$\frac{1}{2}$	0	1	0
	y_3	20		0	4	0	-1	1
	x_3	100	50	1	3	2	0	$\frac{1}{4}$
	x_5	20	40	0	0	$\frac{1}{2}$	$\frac{1}{2}$	$\frac{1}{2}$

Tableau	Basic var.	Values of bas. var.	Ratio	y_1	x_2	x_5	y_2	x_1
6	f	90		$\frac{1}{2}$	$\frac{1}{2}$	0	1	0
	y_3	20	20	0	4	0	-1	1
	x_3	20	10	1	3	-4	-2	2
	x_4	40	40	0	0	$\underline{2}$	1	1

Tableau	Basic var.	Values of bas. var.	Ratio	y_1	x_2	x_5	y_2	x_3
7	f	90		$\frac{1}{2}$	$\frac{1}{2}$	0	1	0
	y_3	10		$-\frac{1}{2}$	$2\frac{1}{2}$	2	0	$-\frac{1}{2}$
	x_1	10		$\frac{1}{2}$	$1\frac{1}{2}$	-2	-1	$\frac{1}{2}$
	x_4	30		$-\frac{1}{2}$	$-1\frac{1}{2}$	4	2	$-\frac{1}{2}$

variable by x_3. Tableau 3 is then transformed into tableau 4, by pivoting on the underlined element $\frac{1}{2}$.

In tableau 3, also the element in the f-row in the column of y_3 is zero. We may therefore also increase y_3 without changing f; the basic variable which first becomes zero is then x_5, so that x_5 would be replaced by y_2 as a basic variable. Pivoting on the corresponding element $\frac{1}{2}$, we would obtain tableau 7, apart from a different arrangement of rows and columns.

In tableau 4, we may repeat the same procedure by selecting nonzero elements in the f-row. The first zero-element is found in the column of x_4, but when introducing x_4 into the basis, we find that x_3 leaves it, so that the solution of tableau 3 is found again. Another nonbasic variable which has a zero-element in the f-row of tableau 4 is y_3. It turns out that y_3 replaces x_1 as a basic variable. Pivoting on the underlined element 1, we find tableau 5. In this tableau, both x_1 and x_4 have zero's in the f-row. If x_1 is introduced into the basis, tableau 4 is found again. Introducing x_4 into the basis, we find that it replaces x_5; after pivoting on the underlined element $\frac{1}{2}$ tableau 6 is found. In tableau 6, x_1 and x_5 have zero's in the f-row. If x_5 is introduced into the basis, tableau 5 is found again. If x_3 is introduced into the basis, a new tableau, tableau 7 is found. If in this tableau x_5 and x_3 are introduced into the basis, tableaux 3 and 6 are found. We have therefore exhausted the number of optimal basic feasible solutions.

But there are also *nonbasic* optimal feasible solutions which are linear combinations of basic optimal solutions. The problem can be formulated in terms of equality constraints as in (19)–(21). Maximize

$$f = p'x \tag{44}$$

subject to

$$Ax = b, \tag{45}$$

$$x \geq 0. \tag{46}$$

Now if the vectors $x^1, ..., x^s$ are optimal basic solutions of the above problem, any nonnegative linear homogeneous combination $\bar{x}$ of these solutions

$$\bar{x} = \theta_1 x^1 + \theta_2 x^2 + ... + \theta_s x^s, \tag{47}$$

with

$$\theta_1 + \theta_2 + ... + \theta_s = 1,$$

and

$$\theta_1, \theta_2, ..., \theta_s \geq 0.$$

The proof is easy. Let the value of f for the optimal solutions be $\hat{f}$. Then

$$p'\bar{x} = p'(\theta_1 x^1 + ... + \theta_s x^s) = \theta_1 p'x^1 + ... + \theta_s p'x^s$$
$$= \theta_1 \hat{f} + ... + \theta_s \hat{f} = \hat{f}, \tag{48}$$

so that the solution has the same value of the objective function as the optimal solutions. The solution $\bar{x}$ is also feasible since

$$A\bar{x} = A(\theta_1 x^1 + ... + \theta_s x^s) = \theta_1 Ax^1 + ... + \theta_s Ax^s$$
$$= \theta_1 b + ... + \theta_s b = b. \tag{49}$$

Furthermore, $\bar{x} = \theta_1 x^1 + \ldots + \theta_s x^s \geq 0$ since $\theta_1, \ldots, \theta_s \geq 0$ and $x^1, \ldots, x^s \geq 0$.

Hence the optimal solution of the problem used as an example can be written as follows:

$$
\begin{bmatrix} x_1 \\ x_2 \\ x_3 \\ x_4 \\ x_5 \\ y_1 \\ y_2 \\ y_3 \end{bmatrix} = \theta_1 \begin{bmatrix} 20 \\ 0 \\ 0 \\ 10 \\ 5 \\ 0 \\ 0 \\ 0 \end{bmatrix} + \theta_2 \begin{bmatrix} 20 \\ 0 \\ 20 \\ 0 \\ 10 \\ 0 \\ 0 \\ 0 \end{bmatrix} + \theta_3 \begin{bmatrix} 0 \\ 0 \\ 100 \\ 0 \\ 20 \\ 0 \\ 0 \\ 20 \end{bmatrix} + \theta_4 \begin{bmatrix} 0 \\ 0 \\ 0 \\ 20 \\ 40 \\ 0 \\ 0 \\ 20 \end{bmatrix} + \theta_5 \begin{bmatrix} 10 \\ 0 \\ 0 \\ 30 \\ 0 \\ 0 \\ 0 \\ 10 \end{bmatrix}. \tag{50}
$$

In this small example, only two nonbasic variables had zero coefficients in the f-row in each example. In cases in which more nonbasic variables have nonbasic coefficients it is difficult to keep track of all solutions to be generated, especially if the choice of leaving basic variables is not unique. Special methods for finding all basic optimal solutions of larger problems in a systematic manner will be given later.

2.7. Matrix formulation of the simplex method

Since the simplex method generates successive canonical systems with corresponding sets of basic variables, it can be put into matrix formulation as explained before. This matrix formulation can be further extended with the product form of the inverse. Using the inverse of the columns of basic variables in the initial tableau as it is or in the product form leads to considerable savings for larger problems; furthermore it is of theoretical interest.

Let us again use the following problem formulation. Maximize

$$f = p'x \tag{51}$$

subject to

$$Ax + y = b, \tag{52}$$

$$x, y \geq 0. \tag{53}$$

Let us assume that $y = b \geq 0$ is the initial feasible solution and that after a number of iterations of the simplex method the first m x-variables, denoted by a vector x^b, are in the basis; x^n will indicate the vector of

Table 5

Matrix formulation of simplex tableau

Tableau	Basic var.	Values of bas. var.	x^b	x^n	f	y
I	f	0	$-p^{1'}$	$-p^{2'}$	1	0
	y	b	A_1	A_2	0	I
II	f	$p^{1'}A_1^{-1}$	0	$p^{1'}A_1^{-1}A_2 - p^{2'}$	1	$p^{1'}A_1^{-1}$
	x^b	$A_1^{-1}b$	I	$A_1^{-1}A_2$	0	A_1^{-1}

x-variables which are not in the basis. The matrix A is correspondingly partitioned in A_1 and A_2. The initial tableau is then given by tableau I of table 5 and the tableau after a number of iterations is presented in tableau II. It is clear that tableau II can be obtained from tableau I by premultiplication by the matrix

$$B^{-1} = \begin{bmatrix} 1 & -p^{1'} \\ 0 & A_1 \end{bmatrix}^{-1} = \begin{bmatrix} 1 & p^{1'}A_1^{-1} \\ 0 & A_1^{-1} \end{bmatrix}. \tag{54}$$

Before dealing with the main subject of this section, it is useful to indicate an alternative way of computing the elements in the row of the objective function. The elements of the other rows are found in the manner described before, so that the element of the vectors and matrices $A_1^{-1}b$, $A_1^{-1}A_2$ and A_1^{-1} are known. Let the jth column of $A_1^{-1}A_2$ or A_1^{-1} be indicated by a_{ij}^*. The corresponding element of the objective function row is computed as

$$\sum_i p_i a_{ij}^* - p_j = z_j - p_j, \tag{55}$$

where the p_i's are the coefficients of the objective function for the basic variable; z_j is defined as the summation at the left side. Comparing this way of computing the objective function with treatment as any other row, we note that (55) requires m multiplications and one subtraction for each element, against one multiplication and one addition for the ordinary method, so that (55), which is found in some textbooks, is considerably less efficient.

The matrix B^{-1} is called the *inverse of the basis*; more explicitly, it is the inverse of the matrix consisting of the columns in the set-up tableau, of the variables which are basic in the current tableau.

The fact that a tableau can be generated by premultiplication of the set-up tableau by the inverse of the basis lies at the basis of another computational form of the simplex method; this form is called the *revised simplex method* or *inverse matrix method*.

The simplex iterations do not require all elements in each successive simplex tableau; in fact, we only need for each iteration the elements in the first row, those in the column values of basic variables and in the column of the new basic variable. The other elements of the tableau are not needed in the particular iteration of the method, but they may be necessary in later iterations if the variable of their column becomes basic. However, it is possible to use the matrix formulation to generate each element of the tableau when it is needed by premultiplying the corresponding column in the initial tableau by the corresponding row of the inverse of the columns of basic variables; for tableau II this inverse is the matrix given in (54) which occurs also in the columns of the f and y in this tableau. For instance, the first element in the column of the first variable of x'' is generated by premultiplying the corresponding column in tableau I by the first row of B^{-1}, and the last element by premultiplying the corresponding column by the last row of B^{-1}. It is therefore not necessary to transform in each iteration the entire tableau; it is sufficient to transform only the columns of the basic variables in the initial tableau, because the matrix formed by these columns may be used to generate any element of a tableau when needed.

This can be illustrated by treating the application of the simplex method to the production planning example in this manner. Table 6 gives the computations. The set-up tableau is entirely the same as before; we shall return to it in each iteration. As indicated before, the simplex method requires for each iteration the matrix B^{-1}, the objective function row and the columns of the new basic variable and the values of basic variables. All these are given in table 6 for the three iterations; the row of the objective function is given separately for typographical reasons.

The first matrix B^{-1} consists of the elements of the columns of f, y_1, y_2, and y_3 and a unit matrix. This unit matrix is multiplied into the first row of the set-up tableau and is therefore just the first row of the tableau, which is reproduced as such. The most negative element is -3 (which is indicated by a star), so that x_1 enters the basis. The column of x_1 is then generated by premultiplying its column in the set-up tableau by the inverse of the basis:

$$\begin{bmatrix} 1 & 0 & 0 & 0 \\ 0 & 1 & 0 & 0 \\ 0 & 0 & 1 & 0 \\ 0 & 0 & 0 & 1 \end{bmatrix} \begin{bmatrix} -3 \\ 4 \\ 1 \\ 2 \end{bmatrix} = \begin{bmatrix} -3 \\ 4 \\ 1 \\ 2 \end{bmatrix};$$

Table 6

Computations for simplex method with explicit inverse

	Basic var.	Values bas. var.	x_1	x_2	x_3	x_4	x_5	f	y_1	y_2	y_3
	f	0	-3	-1	$-\frac{1}{2}$	-2	-2	1	0	0	0
Set-up	y_1	100	4	3	1	2	0	0	1	0	0
tableau	y_2	40	1	0	0	1	2	0	0	1	0
	y_3	60	2	4	0	1	2	0	0	0	1
Objective	0	0	-3^*	-1	$-\frac{1}{2}$	-2	-2	1	0	0	0
function	1	75	0	$1\frac{1}{4}$	$\frac{1}{4}$	$-\frac{1}{2}$	-2^*	1	$\frac{3}{4}$	0	0
rows	2	85	0	$3\frac{3}{4}$	$-\frac{1}{4}$	$-\frac{1}{4}^*$	0	1	$\frac{1}{4}$	0	1
	3	90	0	$\frac{1}{2}$	0	0	1	1	$\frac{1}{2}$	1	0

It.	Basic var.		Inverse	of basis		New bas. var.	Values bas. var.	Ratio
	f	1	0	0	0	-3	0	
0	y_1	0	1	0	0	$\underline{4}$	100	25
	y_2	0	0	1	0	1	40	40
	y_3	0	0	0	1	2	60	30
	f	1	$\frac{3}{4}$	0	0	-2	75	
1	x_2	0	$\frac{1}{4}$	0	0	0	25	
	y_2	0	$-\frac{1}{4}$	1	0	2	15	$7\frac{1}{2}$
	y_3	0	$-\frac{1}{2}$	0	1	2	10	5
	f	1	$\frac{1}{4}$	0	1	$-\frac{1}{2}$	85	
2	x_2	0	$\frac{1}{4}$	0	0	$\frac{1}{2}$	25	50
	y_2	0	$\frac{1}{4}$	1	-1	$\frac{1}{2}$	5	10
	x_5	0	$-\frac{1}{4}$	0	$\frac{1}{2}$	0	5	
	f	1	$\frac{1}{2}$	1	0		90	
3	x_2	0	0	-1	1		20	
	x_4	0	$\frac{1}{2}$	2	-2		10	
	x_5	0	$-\frac{1}{4}$	0	$\frac{1}{2}$		5	

this result is written down in the table. Then the values of basic variables are generated in the same manner:

$$\begin{bmatrix} 1 & 0 & 0 & 0 \\ 0 & 1 & 0 & 0 \\ 0 & 0 & 1 & 0 \\ 0 & 0 & 0 & 1 \end{bmatrix} \begin{bmatrix} 0 \\ 100 \\ 40 \\ 60 \end{bmatrix} = \begin{bmatrix} 0 \\ 100 \\ 40 \\ 60 \end{bmatrix};$$

which is also written down. The ratios are found as usual and the pivot is determined by the minimum ratio; in this case it is 4. The computations done so far could just as well have been performed in the set-up tableau.

Then the inverse of the basis is transformed by pivoting on the pivot 4, which results in the matrix given in Iteration 1. This matrix is first used to generate the objective function row. For instance, the element in this row in the column of x_2 has been calculated as

$$[1 \quad \tfrac{3}{4} \quad 0 \quad 0] \begin{bmatrix} -1 \\ 3 \\ 0 \\ 4 \end{bmatrix} = 1\tfrac{1}{4}.$$

All elements of the objective function row are calculated in this manner. The most negative element is found to be -2, which is starred, so that x_5 is the new basic variable. Then the column of the new basic variable and that of the values of basic variables are found from

$$\begin{bmatrix} 1 & \tfrac{3}{4} & 0 & 0 \\ 0 & \tfrac{1}{4} & 0 & 0 \\ 0 & -\tfrac{1}{4} & 1 & 0 \\ 0 & -\tfrac{1}{2} & 0 & 1 \end{bmatrix} \begin{bmatrix} -2 & 0 \\ 0 & 100 \\ 2 & 40 \\ 2 & 60 \end{bmatrix} = \begin{bmatrix} -2 & 75 \\ 0 & 25 \\ 2 & 15 \\ 2 & 10 \end{bmatrix}.$$

The ratios and the minimum ratio are found as usual, after which the pivot is determined, which is the underlined element 2. This pivot is used to transform the inverse. The following iterations are similar. When the objective function row has no negative elements left, an optimal solution has been found; this happens after the third iteration.

The revised simplex method can be used as described, which is then called the revised simplex method with *explicit form of the inverse*, or it may be modified by computing the matrix B^{-1} each time as the product of the matrices $E_1, E_2, \ldots$ (see section 1.7); the method is then called the revised simplex method with *product form of the inverse*. This means that in each iteration only the nontrivial column of the corresponding E-matrix, together with its position is stored. For instance, after three iterations for the example we would have stored the following columns:

$$\begin{matrix} \tfrac{3}{4} & 1 & 1 \\ \tfrac{1}{4}* & 0 & -1 \\ -\tfrac{1}{4} & -1 & 2* \\ -\tfrac{1}{2} & \tfrac{1}{2}* & 0 \end{matrix} ;$$

where the stars indicate the position of the column in the matrices. The inverse matrix is then built up in the following manner:

$$\begin{bmatrix} 1 & 0 & 1 & 0 \\ 0 & 1 & -1 & 0 \\ 0 & 0 & 2 & 0 \\ 0 & 0 & 0 & 1 \end{bmatrix} \begin{bmatrix} 1 & 0 & 0 & 1 \\ 0 & 1 & 0 & 0 \\ 0 & 0 & 1 & -1 \\ 0 & 0 & 0 & \frac{1}{2} \end{bmatrix} \begin{bmatrix} 1 & \frac{3}{4} & 0 & 0 \\ 0 & \frac{1}{4} & 0 & 0 \\ 0 & -\frac{1}{4} & 1 & 0 \\ 0 & -\frac{1}{2} & 0 & 1 \end{bmatrix}.$$

The elements in the f-row for x_2 in the third tableau would for instance, be generated by postmultiplication of the above matrix product by

$$\begin{bmatrix} -1 \\ 3 \\ 0 \\ 4 \end{bmatrix}.$$

First this vector is premultiplied with the last factor of the product, then the resulting vector with the middle factor and then this result with the first factor. It can be checked that each multiplication of a vector and a matrix requires $m + 1$ times a multiplication and an addition.

In the product form of the inverse as indicated above, the number of matrix factors grows with the number of iterations, so that it may easily exceed m. This may require excessive storage capacity and it may also affect the accuracy of computation. It is then advantageous to recompute the inverse from the set-up tableau. This is called *reinversion*, which is done periodically, usually after m iterations.

Though in the revised simplex method with explicit or product form of the inverse fewer elements are computed, the number of computations is not necessarily smaller. In the ordinary or standard simplex method, where all elements of the tableau are transformed, the transformation of an element requires usually a multiplication and an addition. For a problem with m constraints and n x-variables there are $(m + 1) \times (n + 1)$ in such transformations. In the revised simplex method with explicit form of the inverse, the generation of an element of the f-row requires the multiplication of a row with $m + 1$ elements into a column of $m + 1$ elements, which is $m + 1$ times a multiplication and an addition. For $n + 1$ elements in the f-row, this amounts to $(m + 1) \times (n + 1)$ computations involving a multiplication and an addition. This is already the same as in the standard simplex method. The explicit form requires furthermore the transformation of the inverse of the basis and the generation of the columns of the leaving basic variable and the values of basic variable.

The revised simplex method with product form of the inverse involves the same amount of computations as the explicit form. From these considerations it seems that the revised simplex method in both forms is useless.

However, we have not taken into account the effects of zeros. If in the standard simplex method elements of the pivot row or column are zero, no transformation is necessary for the elements in the corresponding columns or rows. Similarly, if in the revised simplex method some elements of the inverse are zero or some elements of the set-up tableau are zero, some computations are not necessary. Hence it depends on the occurrence of zeros which of the three methods is more efficient.

Wolfe and Cutler[30] have performed experiments with 9 realistic linear programming problems. In their problems, which were of various sizes, the ratio n/m range from 1.67 to 3.43, with an average of 2.31. They programmed iterations for all three variants of the simplex method and found the following average numbers of operations (one multiplication + one addition or equivalent computation):

Standard Simplex method	$1.23(m + 1)^2$
Revised Simplex method with Explicit Form	$0.57(m + 1)^2$
Revised Simplex method with Product Form	$0.42(m + 1)^2$.

From this, it is obvious that the product form of the inverse (PFI) has the smallest number of operations. Furthermore, there was evidence that the advantage of PFI was greater for larger than for smaller problems. Most existing computer programs for linear programming use PFI.

Among the combinations of the various choices for the new basic variables, which were discussed earlier in this section, and the various forms of the simplex method, Wolfe and Cutler found that the choice of the most negative f-row element and PFI lead to the smallest number of operations. As a rough guide, it can be said that m^3 operations are needed for the solution of a linear programming problem with m constraints.

An additional useful feature is the following. If the revised simplex method is used in either form, we may not at each iteration wish to generate all elements of the f-row. We could, for instance in the first iteration generate all f-row elements, then from these pick the 7 most negative ones, and then in the following iterations only generate these elements, deleting them if the corresponding variable becomes basic or if the element becomes nonnegative. When of these 7, no negative element is left, all f-row elements are generated, then again the 7 most negative elements or less if there are less than 7, are selected, and only these are used in the following iterations until there is no negative element, and so on. Obviously fewer computations are required per minor iteration since only at most 7 elements of the f-row are generated. On the other hand, the number of iterations may be somewhat larger.

CHAPTER 3

VARIANTS OF THE SIMPLEX METHOD

3.1. A typical minimization problem: the composition of minimum-cost cattlefeed

The general form used for a linear programming problem in section 2.2 was: Maximize

$$f = p'x \tag{1}$$

subject to

$$Ax \le b, \tag{2}$$

$$x \ge 0. \tag{3}$$

If all elements of b are nonnegative, as they were in the production planning example, a basic feasible solution is immediately available by using the slack variables, which can be introduced in (2) as basic variables:

$$f = 0, \qquad y = b. \tag{4}$$

Since this solution is feasible, it is used as an initial solution for the simplex method. If in a problem the constraints have the form (2) with $b \ge 0$, this problem is called a *typical maximization problem*. Usually, but not necessarily always, the vector p has nonnegative elements. At first sight, the word maximization does not seem appropriate, since the properties refer to the constraints, but this will become clear later. The production planning example was a typical maximization problem.

Consider the following problem: Minimize

$$f = c'x \tag{5}$$

subject to

$$Ax \ge b, \tag{6}$$

$$x \ge 0. \tag{7}$$

If the elements of c are nonnegative, the problem is called a *typical minimization problem*. Usually, but not necessarily, the elements of b in

57

(6) are nonnegative. In a typical minimization problem, the slack variables generally cannot be made basic for an initial basic feasible solution of the simplex method since the solution $y = -b$ is not feasible unless all elements of b are nonpositive. Hence, for an application of the simplex method to such a problem, first an initial basic feasible solution should be found. However, it will be shown in chapter 4 that this basic solution can be used as an initial basic solution for another linear programming method, namely the dual method.

An example of a typical minimization problem is that of the least-cost composition of cattlefeed from a number of raw materials; the feed must satisfy a number of nutritional requirements. For the human diet the same problem can be formulated; this problem is known as the diet problem, but, since humans are more fastidious, a proper formulation of this problem is more complicated than the cattlefeed problem.

We consider the case in which the daily feed of an animal must be determined in such a way that costs are minimal and certain nutritional requirements are met. In order to keep the example as simple as possible, only four raw materials are considered, namely barley, oats, sesame flakes and groundnut meal. The number of nutritional requirements is limited to two, the first one being that the feed should contain at least 20 units of protein and the second that it should contain at least 5 units of fat. The protein and fat content per unit of each of the raw materials and the costs per unit are given in table 1.

Let us indicate the amounts of barley, oats, sesame flakes and groundnut meal in the daily feed by x_1, x_2, x_3 and x_4, respectively. The protein and fat requirements then may be stated as follows:

$$12x_1 + 12x_2 + 40x_3 + 60x_4 \geq 20,$$
$$2x_1 + 6x_2 + 12x_3 + 2x_4 \geq 5. \tag{8}$$

After introduction of nonnegative slack variables y_1 and y_2, which stand

Table 1

Protein and fat content and cost of raw materials

	Protein content	Fat content	Cost per unit
Barley	12	2	24
Oats	12	6	30
Sesame flakes	40	12	40
Groundnut meal	60	2	50

for the overfulfillment of the protein and fat requirements, the inequalities can be formulated as equalities:

$$12x_1 + 12x_2 + 40x_3 + 60x_4 - y_1 = 20,$$
$$2x_1 + 6x_2 + 12x_3 + 2x_4 - y_2 = 5. \tag{9}$$

The objective is to minimize the costs of the feed, so that we have for the equation of the objective function:

$$f = 24x_1 + 30x_2 + 40x_3 + 50x_4. \tag{10}$$

Eqs. (9) and (10) may be put into canonical form with f, y_1 and y_2 as basic variables by multiplying the equations of (9) by -1 and by moving in (10) the terms in the x-variables to the other side of the equality sign. The following system is then obtained:

$$0 = -24x_1 - 30x_2 - 40x_3 - 50x_4 + f,$$
$$-20 = -12x_1 - 12x_2 - 40x_3 - 60x_4 + y_1,$$
$$-5 = -2x_1 - 6x_2 - 12x_3 - 2x_4 + y_2. \tag{11}$$

This system may be put into a simplex tableau, see tableau 0 of table 2.

The difficulty is that the solution of tableau 0 is not a feasible one, because the basic variables y_1 and y_2 are negative. Since the simplex

Table 2
Simplex tableaux for minimum-cost cattlefeed problem

Tableau	Basic variables	Values basic variables	Ratio	x_1	x_2	x_3	x_4
0	f	0		-24	-30	-40	-50
	y_1	-20		-12	-12	-40	-60
	y_2	-5		-2	-6	-12	-2
				x_1	x_2	y_1	x_4
1	f	20		-12	-18	-1	10
	x_3	$\frac{1}{2}$	$\frac{1}{3}$	$\frac{3}{10}$	$\frac{3}{10}$	$-\frac{1}{40}$	$1\frac{1}{2}$
	y_2	1	$\frac{1}{16}$	$1\frac{3}{5}$	$-2\frac{2}{5}$	$-\frac{3}{10}$	16
				x_1	x_2	y_1	y_2
2	f	$19\frac{3}{8}$		-13	$-16\frac{1}{2}$	$-\frac{13}{16}$	$-\frac{5}{8}$
	x_3	$\frac{13}{32}$		$\frac{3}{20}$	$\frac{21}{40}$	$\frac{1}{320}$	$-\frac{3}{32}$
	x_4	$\frac{1}{16}$		$\frac{1}{10}$	$-\frac{3}{20}$	$-\frac{3}{160}$	$\frac{1}{16}$

method requires a feasible solution to start with, such a solution must first be found. A general method to find a feasible solution of a problem will be treated in section 3.2. In this section it will be shown that if a basic feasible solution to the problem is known, this solution may be generated using simplex tableaux by introducing the variables of this solution one by one into the basis. Once this solution is generated, the simplex method may be applied as described before.

Finding a basic feasible solution for our example is very simple. Half a unit of sesame flakes will precisely meet the protein requirement; it contains 6 units of fat, which is more than the requirement of 5 units. This solution is generated from tableau 0 by pivoting on the element in the row of y_1 and the column of x_3. The result is tableau 1, which contains the desired feasible solution. The simplex method now may be applied, but it must be kept in mind that we are minimizing the objective function. Hence the variable corresponding with the most *positive* element in the f-row is chosen as the new basic variable; this is immediately clear by writing the equation corresponding with the first row in tableau 1 of table 2 as

$$20 - 10x_4 = -12x_1 - 18x_2 - y_1 + f. \tag{12}$$

x_4 is therefore introduced into the basis and y_2 leaves it. The result is tableau 2, the last row of which does not contain any positive elements. The minimum-cost solution must therefore have been reached, which is $x_3 = \frac{13}{32}$ and $x_4 = \frac{1}{16}$; the costs of this solution are $19\frac{3}{8}$.

3.2. Starting methods using artificial variables

In the cases presented before, an initial basic feasible solution was given by the formulation of the problem, or, as in the last section, the basic variables of such a solution were known. Now the general case in which no basic feasible solution is known will be treated. First it will be indicated how linear constraints of any form may be put into a canonical form with a basic variable in each equation. After that, a method is given for obtaining a basic feasible solution to the problem (if such a solution exists).

The constraints of a linear programming problem have one of the following forms:

$$a_{i1}x_1 + \dots + a_{in}x_n \leq b_i, \tag{13}$$

$$a_{i1}x_1 + \dots + a_{in}x_n = b_i, \tag{14}$$

$$a_{i1}x_1 + \ldots + a_{in}x_n \geq b_i. \tag{15}$$

We assume that the constant terms b_i are nonnegative; if any of them is negative, a multiplication by -1 renders it nonnegative. After the introduction of a nonnegative slack variable y_i in (13), the following equation is obtained:

$$a_{i1}x_1 + \ldots + a_{in}x_n + y_i = b_i. \tag{16}$$

In this equation, y_i is a basic variable with a nonnegative value b_i.

General equations like (14) have no basic variable, but we may insert an *artificial variable* z_i into the equation, so that we obtain

$$a_{i1}x_1 + \ldots + a_{in}x_n + z_i = b_i. \tag{17}$$

If the original eq. (14) is to be satisfied, z_i should take a value zero. Hence any solution of (17) with a nonzero artificial variable is unfeasible. A basic solution with $z_i = b_i$ clearly is unfeasible for $b_i \neq 0$.

Constraints of the type (14) can be formulated as an equation by the introduction of a nonnegative slack variable y_i; this results in

$$a_{i1}x_1 + \ldots + a_{in}x_n - y_i = b_i. \tag{18}$$

In this equation y_i is not a basic variable, but it could be made basic by multiplication by -1; however, this would result in a basic solution with $y_i = -b_i$, which is unfeasible for $b_i > 0$. Hence, if $b_i > 0$, (18) is treated as a general equation like (14), so that an artificial variable z_i is introduced:

$$a_{i1}x_1 + \ldots + a_{in}x_n - y_i + z_i = b_i. \tag{19}$$

In this manner all constraints of a linear programming problem may be formulated as a linear system in canonical form with slack variables and artificial variables having nonnegative values. Though the values of basic variables are all nonnegative, the solution is not feasible if artificial variables appear in it with nonzero values. Our first objective is to obtain a basic feasible solution, that is a basic solution which has nonnegative values of basic variables without any artificial variables. Once such a solution is obtained, the simplex method for linear programming may be applied as described before.

The artificial variables may be made nonbasic or basic with a zero value by minimizing the sum of the artificial variables, or, what amounts to the same, maximizing minus this sum. Hence instead of the original objective function we use

$$f^* = -\sum_i z_i. \tag{20}$$

The simplex method may be used with this objective function. If a basic feasible solution without artificial variables or with basic artificial variables with a value zero exists, it must be found by an application of the simplex method with the objective function (20). After such a solution is found, the simplex method may be used with the original objective function, with the solution obtained as a starting-point. Artificial variables with a zero value will be treated in a manner to be indicated below.

An equivalent formulation is obtained if the following combination of the original objective function and its preliminary substitute is used

$$f^{**} = p_1 x_1 + \ldots + p_n x_n - M \sum_i z_i, \tag{21}$$

where M is a very large number, so that terms with M always dominate terms without M. In this case only one objective function is used, so that formally the methods consist of only one phase, but since M dominates the p's, $\sum z_i$ will first be minimized.

Using matrix notation, a general linear programming problem can be formulated as follows. Maximize

$$f = p'x \tag{22}$$

subject to

$$A_1 x \leq b^1, \tag{23}$$

$$A_2 x = b^2, \tag{24}$$

$$A_3 x \geq b^3, \tag{25}$$

$$x \geq 0. \tag{26}$$

All elements of the b-vectors are assumed to be nonnegative. After introduction of slack variables and artificial variables in the indicated manner, we obtain for (23)–(26):

$$A_1 x + y^1 = b^1, \tag{27}$$

$$A_2 x + z^2 = b^2, \tag{28}$$

$$A_3 x - y^3 + z^3 = b^3, \tag{29}$$

$$x, y^1, y^3, z^2, z^3 \geq 0. \tag{30}$$

The objective function is changed into

$$f^{**} = p'x - M(e^{2\prime} z^2 + e^{3\prime} z^3), \tag{31}$$

where e^2 and e^3 are column vectors of units of appropriate orders.

Equations (27)–(31) can be put into detached coefficient form as in tableau 0 of table 3. In this tableau the row of the objective function is split into two rows, one indicated by f_c, which contains the coefficients without M, and one indicated by f_M, containing the coefficients which should be multiplied by M. It is easily observed that the system apart from the f-rows is in canonical form, but that, if the f-rows are also included, it is not in canonical form, because the f_M-row contains nonzero elements in the z^2- and z^3-columns. This can be changed by adding to it the z_2-rows premultiplied by $-Me^{2\prime}$ and the z_3-rows premultiplied by $-Me^{3\prime}$. This may also be interpreted as block-pivoting on the unit-matrices in the columns of z^2 and z^3. The result is tableau 1 of table 3, which is entirely in canonical form and has a nonnegative basic solution.

Now the simplex method may be applied in two phases; at phase I it is used with the f_M-row as objective function row and in phase II with the f_c-row as objective function row. In phase I the new basic variables are selected by using the f_M-row; if according to this row no new basic variable is eligible, a solution without nonzero z-variables should have been obtained; if not, the problem has no feasible solution since if a feasible solution exists having $\Sigma_i z_i = 0$ it should be found by applying the simplex method to the constraints with the objective function $f_M = \Sigma_i z_i$ to be minimized. After that, the f_M-row can be deleted; in phase II, the new basic variables now are selected according to the f_c-row which contains the coefficients of the original objective function.

Table 3

Set-up and initial tableau for the simplex method using artificial variables

Tableau	Basic variables	Values basic variables	x	y^3	y^1	z^2	z^3
0	f_c	0	$-p'$	0	0	0	0
	f_M	0	0	0	0	$e^{2\prime}$	$e^{3\prime}$
	y^1	b^1	A_1	0	I	0	0
	z^2	b^2	A_2	0	0	I	0
	z^3	b^3	A_3	$-I$	0	0	I
1	f_c	0	$-p'$	0	0	0	0
	f_M	$-e^{2\prime}b^2 - e^{3\prime}b^3$	$-e^{2\prime}A_2 - e^{3\prime}A_3$	$e^{3\prime}$	0	0	0
	y^1	b^1	A_1	0	I	0	0
	z^2	b^2	A_2	0	0	I	0
	z^3	b^3	A_3	$-I$	0	0	I

If after phase I artificial variables occur in the basis at a zero value, we may proceed in two ways. Firstly, the artificial variables may all be removed immediately by pivoting on any nonzero element in the row of the artificial variable and the column of an ordinary variable; since the value of the artificial variable is zero, such iterations do not change the values of the other basic variables. Secondly, we may proceed as in the ordinary simplex method, but if the element in the column of the new basic variable and in the row of one or more of the artificial variables is nonzero, one of these elements is used as a pivot; the reason for this is that if the element is positive, the artificial basic variable has the ratio 0, which is the minimum ratio possible and if the element is negative, the artificial variable would obtain a nonzero value if another pivot were chosen belonging to a nonzero ratio.

Table 4 gives an application of the simplex method using artificial variables to the minimization problem of the previous section. The modified objective function which is to be maximized is in this case

$$f = -24x_1 - 30x_2 - 40x_3 - 50x_4 - M(z_1 + z_2), \tag{32}$$

Table 4

An application of the simplex method using artificial variables to a typical minimization problem

Tableau	Basic variables	Values basic variables	Ratio	x_1	x_2	x_3	x_4	y_1	y_2
0	f_c	0		-24	-30	-40	-50	0	0
	f_M	25		14	18	52	62	-1	-1
	z_1	20	$\frac{1}{3}$	12	12	40	$\underline{60}$	-1	0
	z_2	5	$2\frac{1}{2}$	2	6	12	$\underline{2}$	0	-1
				x_1	x_2	x_3		y_1	y_2
1	f_c	$16\frac{2}{3}$		-14	-20	$-6\frac{2}{3}$		$-\frac{5}{6}$	0
	f_M	$4\frac{1}{3}$		$1\frac{3}{5}$	$5\frac{3}{5}$	$10\frac{2}{3}$		$\frac{1}{30}$	1
	x_4	$\frac{1}{3}$	$\frac{1}{2}$	$\frac{1}{5}$	$\frac{1}{5}$	$\frac{2}{3}$		$-\frac{1}{60}$	0
	z_2	$4\frac{1}{3}$	$\frac{13}{32}$	$1\frac{3}{5}$	$5\frac{3}{5}$	$10\frac{2}{3}$		$\frac{1}{30}$	-1
				x_1	x_2			y_1	y_2
2	f_c	$19\frac{3}{8}$		-13	$-16\frac{1}{2}$			$-\frac{13}{16}$	$-\frac{5}{8}$
	f_M	0		0	0			0	0
	x_4	$\frac{1}{16}$		$\frac{1}{10}$	$-\frac{3}{20}$			$-\frac{3}{160}$	$-\frac{1}{16}$
	x_3	$\frac{13}{32}$		$\frac{3}{20}$	$\frac{21}{40}$			$\frac{1}{320}$	$-\frac{3}{32}$

which may be written as

$$0 = 24x_1 + 30x_2 + 40x_3 + 50x_4 + Mz_1 + Mz_2 + f. \qquad (33)$$

This equation corresponds with the f_c- and f_M-row of tableau 0 of table 3. The elements of the vector $-p$ are then 24, 30, 40, and 50. The elements of the f_M-row of the initial tableau are easily found as

$$-e^{3\prime}A_3 = -[1 \quad 1]\begin{bmatrix} 12 & 12 & 60 & 60 \\ 2 & 6 & 12 & 2 \end{bmatrix}$$
$$= -[14 \quad 18 \quad 52 \quad 62]. \qquad (34)$$

Since in the original problem the objective function is minimized instead of maximized, the interpretation of the tableaux is slightly facilitated if the objective function in the tableau is minimized too. Hence all elements in the f_c- and f_M-row which would follow from tableau 1 of table 3 are multiplied by -1; see the f_c- and f_M-row of tableau 0 of table 4.

The simplex method may now be applied by selecting the largest positive element in the f_M-row, which is 62, so that x_4 enters the basis. The leaving basic variable is determined as usual and turns out to be z_1. Since if an artificial variable is nonbasic, we do not want it to enter the basis again, we may cease to update its column once it is nonbasic.[1] Hence the column of z_1 is deleted. The remainder of the tableau is transformed in the usual manner. In tableau 2 the largest positive element in the f_M-column is $10\frac{2}{3}$, so that x_3 is the new basic variable. z_2 leaves the basis and its column is deleted. All artificial variables have left the basis in tableau 2, so that we may start to use the original objective function, the coefficients of which are given in the f_c-row. Since in this case it happens that all elements in the f_c-row in the columns of nonbasic variables are negative, no further iterations are necessary; the optimal solution has been obtained.

In the method described above, an artificial variable was added to each constraint which did not have a basic variable with a feasible solution. The number of iterations in phase I must then be at least equal to the number of artificial variables, since in each iteration at most one artificial variable is eliminated.

Let us consider more closely the case of a $\geq$-constraint, in which both

---

[1] This is only valid if the standard form of the simplex method is used; if the inverse of columns of basic variables is used to generate other parts of the tableau when needed, the columns of the z-variables are always transformed.

a slack variable and an artificial variable have been inserted:

$$a_{i1}x_1 + \dots + a_{in}x_n - y_i + z_i = b_i. \tag{35}$$

If in an application of phase I the variable z_i leaves the basis, this is because the increase of the new basic variable decreases z_i and threatens to make it negative. However, a negative value of z_i corresponds with a positive value of y_i, so that a negative value of z_i does not really create an unfeasibility.

This motivates the following approach. Artificial variables are added to equality constraints but not to $\geq$-constraints; in the latter constraints, the slack variables are made basic so that we have unfeasibilities of the type $y_i = -b_i$ for rows corresponding to $\geq$-constraints. The objective function for a maximization problem is then formulated as

$$f = p'x - M\left(\sum_{i \in E} z_i - \sum_{i \in N} y_i\right), \tag{36}$$

where E denotes the set of rows having artificial basic variables and N the set of rows having a negative basic variable. Hence in phase I, the sum of unfeasibilities is minimized:

$$\sum_{i \in E} z_i - \sum_{i \in N} y_i. \tag{37}$$

The advantage of this formulation is that for an increase in the new basic variable, one or more of the negative y_i may become positive and hence feasible.

Some modifications are necessary for the formulation of the objective function and the choice of the leaving basic variable. The f_M-part of the objective function changes whenever in an iteration one or more negative y-variables obtain nonnegative values. The f_M-row may be computed in each iteration from the other rows using the expression (55) of chapter 2. This means that in a minimization problem the element in the f_M-row in column j is computed as

$$\sum_{i \in E} a_{ij} - \sum_{i \in N} a_{ij}. \tag{38}$$

For an explanation of the choice of the leaving basic variable, figure 1 may be useful, which gives the values of the basic variables, $y_1, \dots, y_6$ as a function of the row basic variable x_k. The values of y_1, y_2, and y_3 are nonnegative and should remain so, while y_4, y_5, and y_6 have negative val-

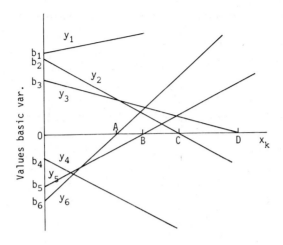

Fig. 1.

ues which are quite welcome to become nonnegative. We assume that x_k has been chosen as the new basic variable by means of (38); the elements in its column and in the rows of the six variables are supposed to have the following signs: $a_{1k} < 0$, $a_{2k} > 0$, $a_{3k} > 0$, $a_{4k} > 0$, $a_{5k} < 0$, $a_{6k} < 0$. Since

$$y_i = b_i - a_{ik}x_k, \qquad (39)$$

the slopes in figure 1 have the opposite sign of a_{ik}.

From the figure it is obvious what happens if x_k is increased. At A, y_6 becomes nonnegative, which is all-right; at B, y_5 becomes nonnegative, which is all-right; but at C, y_2, which was nonnegative, threatens to become negative, and the same happens at D for y_3. Hence x_k should not be increased beyond

$$\operatorname*{Min}_{i \in P} (b_i / a_{ik} | a_{ik} > 0), \qquad (40)$$

where P is the set of rows with nonnegative basic variables.

However, we could have stopped increasing x_k at B, since increasing x_k beyond B only serves to make y_4 more negative. Point B corresponds with

$$\operatorname*{Max}_{i \in N} (b_i / a_{ik} | a_{ik} < 0). \qquad (41)$$

The leaving basic variable is now chosen to be the one corresponding to the minimum of (40) and (41); in figure 1 this would be y_5.

Let us apply this to a variant of the cattlefeed problem, in which it is required that a cattlefeed be mixed containing at least 20 units of protein and 5 units of fat *per unit* of feed. The equations for this problem are the same except that the following constraint is added:

$$x_1 + x_2 + x_3 + x_4 = 1; \qquad (42)$$

the x-variables are now interpreted as the fractions of the feed.

Tableau 0 of table 5 gives the set-up tableau for this problem. Note that the f_M-row is equal to the z-row minus the sum of the rows having negative basis variables. x_4 is the new basic variable.

The leaving basic variable is chosen as the minimum of

$$1/1 = 1$$

and

$$\text{Max}(\tfrac{20}{60}, \tfrac{5}{2}),$$

so that z is the leaving basic variable. All rows are transformed as usual, except the f_M-row. Note that in addition to z, also the negative value of

Table 5

Modified first phase of simplex method

Tableau	Basic variables	Values basic variables	x_1	x_2	x_3	x_4
0	f_c	0	-24	-30	-40	-50
	f_M	26	15	19	53	63
	y_1	-20	-12	-12	-40	-60
	y_2	-5	-2	-6	-12	-2
	z	1	1	1	1	$\underline{1}$
			x_1	x_2	x_3	
1	f_c	50	26	20	10	
	f_M	3	0	4	10	
	y_1	40	48	48	20	
	y_2	-3	0	-4	-10	
	x_4	1	1	1	$\underline{1}$	
			x_1	x_2	y_2	
2	f_c	47	26	16	1	
	y_1	34	48	40	2	
	x_3	$\frac{3}{10}$	0	$\frac{2}{5}$	$-\frac{1}{10}$	
	x_4	$\frac{7}{10}$	1	$\frac{3}{5}$	$\frac{1}{10}$	

y_1 has been eliminated. The f_M-row is again constructed as the sum of rows of artificial basic variables minus the sum of rows of negative basic variables. In this case there is only one row of the latter type, so that the f_M-row is equal to minus the y_2-row. x_3 is chosen as the new basic variable. For the leaving basic variable we compare

$$\text{Min } (40/20, 1/1)$$

and $\frac{3}{10}$, so that y_2 leaves the basis. The next solution is feasible, so that phase II of the simplex method may be applied.

Note that if artificial variables had been used in the $\geq$-constraints, we would have had 3 artificial variables which require at least three iterations to be eliminated; in table 5, only two were needed.

3.3. Alternative methods for finding a feasible solution

The method of finding a feasible solution to the constraints outlined in the previous section is not the only one, but, as will be indicated later, it is the best for most practical problems. This section will deal with some alternative methods.

The first alternative method to be treated here is called the *sequential method*. In this case the unfeasibilities are eliminated one by one. Let us return to the modified cattlefeed problem, in which again all slack variables are made basic and the artificial variables are inserted in equations. The set-up tableau is then again tableau 0 of table 5, in which the f_M-row should be disregarded. First the unfeasibility of y_1 is eliminated, which means that y_1 should be increased. This means in effect that our first objective function is

$$\text{Max } f^* = y_1.$$

The corresponding f^*-row is then identical to the y_1-row, so that it does not have to be written down.

The simplex method is then applied as usual by selection of the most negative element in the row of y_1, which is -60, so that x_4 enters the basis. The leaving basic variable is chosen in the same manner as in the last section, so that again z leaves the basis. Tableau 1 results, of which the f_M-row should be disregarded. The only remaining unfeasibility is that of y_2, so that the y_2-row is used as objective function row. x_3 is then selected as the new basic variable, again the same variable as found when using the f_M-row. The leaving basic variable is found as before and a feasible solution results in tableau 2. The same iterations were performed in the ordinary phase I method and in the sequential method; the explanation of this is that in each iteration the largest element in the f_M-row and the most negative element in the row with an unfeasible basic variable happened to be the same.

A procedure proposed by Gass, and named the *fudge method* by Wolfe and Cutler[30], formulates constraints of the $\geq$-type in such a manner that only one artificial variable is added for all constraints of this type. Let us consider the following linear programming problem. Minimize

$$f = c_1 x_1 + c_2 x_2 \tag{43}$$

subject to

$$a_{11}x_1 + a_{12}x_2 - y_1 = b_1,$$
$$a_{21}x_2 + a_{22}x_2 - y_2 = b_2, \tag{44}$$
$$a_{31}x_1 + a_{32}x_2 - y_3 = b_3,$$
$$x_1, x_2, y_1, y_2, y_3 \geq 0. \tag{45}$$

It is assumed that b_1, b_2 and b_3 are positive, so that taking the y-variables as basic variables does not give a basic feasible solution. Let us now introduce in each of the eqs. (44) the same artificial variable z_1, so that we have

$$a_{11}x_1 + a_{12}x_2 - y_1 + z_1 = b_1,$$
$$a_{21}x_1 + a_{22}x_2 - y_2 + z_1 = b_2,$$
$$a_{31}x_1 + a_{22}x_2 - y_3 + z_1 = b_3. \tag{46}$$

The objective function is modified as usual by adding M times the artificial variable to it:

$$f^* = c_1 x_1 + c_2 x_2 + M z_1. \tag{47}$$

Tableau 0 of table 6 gives eqs. (46) and (47) in tableau-form with the y-variables as basic variables. Now let $b_3 = \mathrm{Max}\,(b_1, b_2, b_3)$. Then, if x_1 replaces y_3 as basic variable, tableau 1 is generated, in which the remaining y-variables all have nonnegative values. If the problem also includes equality constraints, additional artificial variables are required for each such constraint.

Another method worth mentioning is based on the idea that it is not so much the amount of unfeasibility which matters, but the number of them. The choice of the new basic variable is the same as in the modified ordi-

Table 6

Transformation for the fudge method

Tableau	Basic variables	Values basic variables	x_1	x_2	y_1	y_2	y_3	z_1
0	f_c	0	$-c_1$	$-c_2$	0	0	0	0
	f_M	0	0	0	0	0	0	-1
	y_1	$-b_1$	$-a_{11}$	$-a_{12}$	1	0	0	-1
	y_2	$-b_2$	$-a_{21}$	$-a_{22}$	0	1	0	-1
	y_3	$-b_3$	$-a_{31}$	$-a_{32}$	0	0	1	-1
1	f_c	0	$-c_1$	$-c_2$	0	0	0	0
	f_M	b_3	a_{31}	a_{32}	0	0	-1	0
	y_1	b_3-b_1	$a_{31}-a_{11}$	$a_{32}-a_{12}$	1	0	-1	0
	y_2	b_3-b_2	$a_{31}-a_{21}$	$a_{32}-a_{22}$	0	1	-1	0
	z_1	b_3	a_{31}	a_{32}	0	0	-1	1

nary method described in the previous section, but the leaving basic variable is determined in such a manner that after pivoting the number of unfeasibilities is as small as possible. This method has been proposed by Wolfe, who named it the *extended composite method*.

Before giving some computational comparisons of the above methods, something should be said about problem preparation. So far, any slack variable with a positive sign was always used as an initial basic variable for an equation. This can be extended slightly by using as a basic variable any variable which has only one positive element which happens to be in one equation which does not yet have a basic variable (we assume that all constant terms are nonnegative). Equations which do not yet have a basic variable are augmented with artificial variables, and in case of constraints of the $\geq$-type, slack variables with a negative value are used. The resulting basis is called a *singleton basis*.

Instead of inserting artificial variables in constraints which do not have basic variables, existing variables can be made basic in such constraints by pivoting on any nonzero element. The result is then a basic solution which has a number of negative variables in most cases. This is called a *full basis*. Note that for obtaining a full basis more computational effort is required.

Now some results of Wolfe and Cutler[30], who used a number of realistic linear programming problems, will be mentioned. It turned out that if the ordinary phase I simplex method as described in section 3.2 is used the average number of iterations for obtaining a basic feasible solution with a full basis is 0.63 of the average number of iterations with a singleton basis. For the sequential method this figure is 0.84. The full basis therefore leads to substantial savings in terms of iterations, though the work involved in obtaining this basis should be taken into account.

Wolfe and Cutler have also compared the average number of iterations for the various methods. Some of their results are given in table 7. From this it should be concluded that the ordinary phase I method is the best method for obtaining an initial basic feasible solution.

Table 7

Average number of iterations for obtaining a basic feasible
solution (ordinary phase I method = 1.00)

Singleton basis:	
Ordinary phase I method	1.00
Sequential method	1.06
Full basis:	
Ordinary phase I method	1.00
Sequential method	1.39
Fudge method	1.27
Extended composite method	1.03

3.4. Alternatives for selection of the new basic variable

In chapter 2, the nonbasic variable with the most negative element in the f-row was indicated as the choice for the new basic variable, but it was pointed out that any nonbasic variable with a negative f-element could be selected. We shall now indicate some alternatives, which are based on this freedom of choice.

Since the objective function should be maximized, that nonbasic variable could be chosen which would lead to the *greatest* change (increase) in the objective function. The exact formulation is left as an exercise.

It can be argued that the choice of the most negative f-element is rather arbitrary, since by scaling the variables the coefficients can be changed. If a variable is measured in units which are 10 times as large, the elements in its column are multiplied by 10. If the elements in a column are normalized in some way, the selection of the most negative f-element is less arbitrary. Two normalizations for this purpose have been proposed:

$$\text{I:} \quad q_j = \frac{p_j}{\sum_{i \in P_j} a_{ij}}, \, p_j < 0, \qquad \text{II:} \quad q_j^* = \frac{p_j^2}{\sum_{j \in P_j} a_{ij}^2}, \, p_j < 0;$$

P_j is the set of rows for which $a_{ij} > 0$. In I, for a $p_j > 0$, p_j is divided by the sum of positive elements in its column; after this, the most negative q_j is selected, so that its corresponding variable is the new basic variable. In II, the maximum q_j^* is selected. The criteria will be called 'positive normalized methods I and II'.

Since, as stated in the last section, the ordinary f_M-method works best for obtaining a feasible solution, then various criteria can be compared for the solution in both phases. Again the results of the experiments of Wolfe and Cutler are used; these are given in table 8.

Table 8

Comparison of average number iterations for different criteria for the new basic variable (ordinary rule with singleton basis = 1.00)

Singleton basis:	
Ordinary rule	1.00
Greatest change	0.91
Positive normalized method I	0.82
Positive normalized method II	0.83
Full basis:	
Ordinary rule	0.83
Greatest change	0.64
Symmetric	0.75

From this, it may be concluded that the normalization and the greatest change criteria are most effective. It should, of course, be realized that they require more computations per iteration than the ordinary rule. The symmetric method will be explained in chapter 4.

Consider now the total number of iterations in relation to n and m. As was noted before there is an upper bound of $\binom{n+m}{m}$ on the number of iterations in each phase of the simplex method. Wolfe and Cutler found that the number of iterations is related to the number of constraints; it varies for the ordinary simplex method with singleton basis from m to $3m$ with an average of $1.71m$. For the greatest change criteria with full basis, an average of $0.98m$ was found, which is quite low, considering the fact that an equation system with m variables requires m transformations for its solution.

3.5. Degeneracy in the simplex method

A basic solution is said to be *degenerate*, if one or more of its basic variables has a zero value. If there are degenerate solutions, the proof of termination of the simplex method in a finite number of iterations, breaks down, because the increase in the value of the objective function is zero if a basic variable with a zero value leaves the basis, which can be deduced from the expression for the increase

$$- p_k a_{rk}^{-1} b_r, \tag{48}$$

where p_k is the negative element in the objective function row in the column of the new basic variable, a_{rk} the pivot and b_r the value of the leaving basic variable. If the value of the objective function does not change, it is conceivable that the same solution reappears after a number of iterations; the solutions are then said to *cycle*.

An example of cycling constructed by Beale is given in table 9. In tableau 0, x_1 is taken as the new basic variable. The leaving basic variable should be found by comparison of the ratios

$$0/\tfrac{1}{4} \quad \text{and} \quad 0/\tfrac{1}{2}.$$

Table 9

An example of cycling in the simplex method

Tableau	Basic variable	Values basic variables	x_1	x_2	x_3	x_4
0	f	0	$-\tfrac{3}{4}$	20	$-\tfrac{1}{2}$	6
	y_1	0	$\tfrac{1}{4}$	-8	-1	9
	y_2	0	$\tfrac{1}{2}$	-12	$-\tfrac{1}{2}$	3
	y_3	1	0	0	1	0
			y_1	x_2	x_3	x_4
1	f	0	3	-4	$-3\tfrac{1}{2}$	33
	x_1	0	4	-32	-4	36
	y_2	0	-2	4	$1\tfrac{1}{2}$	-15
	y_3	1	0	0	1	0
			y_1	y_2	x_3	x_4
2	f	0	1	1	-2	18
	x_1	0	-12	8	8	-84
	x_2	0	$-\tfrac{1}{2}$	$\tfrac{1}{4}$	$-\tfrac{3}{8}$	$-3\tfrac{3}{4}$
	y_3	1	0	0	1	0

Table 9   (*continued*)

Tableau	Basic variable	Values basic variables	y_1	y_2	x_1	x_4
3	f	0	-2	3	$\frac{1}{4}$	-3
	x_3	0	$-1\frac{1}{2}$	1	$\frac{1}{8}$	$-10\frac{1}{2}$
	x_2	0	$\frac{1}{6}$	$-\frac{1}{8}$	$-\frac{3}{64}$	$\frac{3}{16}$
	y_3	1	$1\frac{1}{2}$	-1	$-\frac{1}{8}$	$10\frac{1}{2}$

Tableau	Basic variable	Values basic variables	y_1	y_2	x_1	x_2
4	f	0	-1	1	$-\frac{1}{2}$	16
	x_3	0	2	-6	$-2\frac{1}{2}$	56
	x_4	0	$\frac{1}{3}$	$-\frac{2}{3}$	$-\frac{1}{4}$	$5\frac{1}{3}$
	y_3	1	-2	6	$2\frac{1}{2}$	-56

Tableau	Basic variable	Values basic variables	x_3	y_2	x_1	x_2
5	f	0	$\frac{1}{2}$	-2	$-1\frac{3}{4}$	44
	y_1	0	$\frac{1}{2}$	-3	$-1\frac{1}{4}$	28
	x_4	0	$-\frac{1}{6}$	$\frac{1}{3}$	$\frac{1}{6}$	-4
	y_3	1	1	0	0	0

Tableau	Basic variable	Values basic variables	y_2	x_2	x_3	x_4
1a	f	0	$1\frac{1}{2}$	2	$-1\frac{1}{4}$	$10\frac{1}{2}$
	y_1	0	$-\frac{1}{2}$	-2	$-\frac{3}{4}$	$7\frac{1}{2}$
	x_1	0	2	-24	-1	6
	y_3	1	0	0	1	0

Tableau	Basic variable	Values basic variables	y_2	x_2	y_3	x_4
2a	f	$1\frac{1}{4}$	$1\frac{1}{2}$	2	$1\frac{1}{4}$	$10\frac{1}{2}$
	y_1	$\frac{3}{4}$	$-\frac{1}{2}$	-2	$\frac{3}{4}$	$7\frac{1}{2}$
	x_1	1	2	-24	1	6
	x_3	1	0	0	1	0

Since both ratios are 0, we can take any of the two; let us take the first one, so that y_1 leaves the basis. In tableau 1 the value of f is the same as in tableau 0, namely 0. In tableau 1, x_2 enters the basis, y_2 leaves it, and tableau 2 results, again with no increase in f. In tableau 2, x_3 enters the basis. Either x_1 or x_2 can leave the basis; we select again the first variable, which is x_1. In tableau 3, the value of f is still 0. Now x_4 enters the basis, x_2 leaves it and tableau 4 is generated with no increase in f. Tableau 5 is generated in the same fashion. In this tableau, y_2 enters the basis, x_4

leaves it and then a tableau is generated which is entirely equal to tableau 0 (apart from the arrangement of columns).

If cycling can occur in the simplex method, this means that the simplex method is not a general method for solving linear programming problems, since it then does not necessarily solve problems with degenerate solutions, which occur frequently in practice. However, no cycling has ever been found for any practical problems; the examples of cycling have all been specially constructed. Furthermore, if some additional rule is given for the selection of the leaving basic variable whenever there is a tie between the ratios, it cannot occur even in theory; this will be shown in the following. The approach was first developed by Charnes[3].

Let us consider the canonical form of the set-up tableau which may be written as

$$
\begin{aligned}
0 &= \quad -p_1 x_1 - \ldots - p_n x_n + f \\
b_1 &= \quad a_{11} x_1 + \ldots + a_{1n} x_n \quad + y_1 \\
&\ldots \\
b_m &= \quad a_{m1} x_1 + \ldots + a_{mn} x_n \quad\quad + y_m.
\end{aligned}
\tag{49}
$$

The constant terms are now changed slightly, or *perturbed* in the following manner:

$$
\begin{aligned}
&b_1 + \epsilon, \\
&b_2 + \epsilon^2, \\
&\ldots \\
&b_m + \epsilon^m.
\end{aligned}
\tag{50}
$$

where ϵ is a very small positive number.[2] For $\epsilon = 0$, the constant terms are unchanged and the solution of the problem with the ϵ's is the same as that of (49). In tableau-form, the perturbed system may be represented as in tableau I of table 10.

Tableau II gives a general notation for any other basic solution generated by the simplex method. The columns of $\epsilon, \ldots, \epsilon^m$ must be the same as the columns of $y_1, \ldots, y_m$, so that these columns do not require any additional computations. Furthermore, the matrix

$$
\begin{bmatrix}
c_{11} & \cdots & c_{1m} \\
\cdot & \cdots & \cdot \\
c_{m1} & \cdots & c_{mm}
\end{bmatrix}
$$

is nonsingular since it is the inverse of the basis.

---

[2] The superindices are exponents.

Table 10

Tableaux for perturbed problem

Tableau	Basic var.	Values bas. var.				x_1	·	x_n	f	y_1	·	y_m
		1	ϵ	·	ϵ^m							
I	f	0	0	·	0	$-p_1$	·	$-p_n$	1	0	·	0
	y_1	b_1	1	·	0	a_{11}	·	a_{1n}	0	1	·	0
	·	·				·		·		·		·
	y_m	b_m	0	·	1	a_{m1}	·	a_{mn}	0	0	·	1
II	f	c_0	c_{01}	·	c_{0m}	d_{01}	·	d_{0n}	1	c_{01}	·	c_{0m}
	c_1	c_{11}	·	c_{1m}	d_{11}	·	d_{1n}	0	c_{11}	·	c_{1m}
	·	·	·	·	·	·		·		·		·
	c_m	c_{m1}	·	c_{mm}	d_{m1}	·	d_{mn}	0	c_{m1}	·	c_{mm}

Let us now consider the value of a basic variable in tableau II, say the first one after f. This value is

$$c_1 + c_{11}\epsilon + c_{12}\epsilon^2 + ... + c_{1m}\epsilon^m. \qquad (51)$$

In degenerate problems, c_1 may be zero. The value of the basic variable is then determined by the ϵ-terms. Since ϵ is a very small positive number, the term $c_{11}\epsilon$ will dominate all following terms if $c_{11} \neq 0$, and will determine the sign of the value of the variable. If $c_{12} = 0$ and $c_{13} \neq 0$, the term $c_{13}\epsilon$ determines the sign, and so on. The sign of the variable will therefore be determined by the first nonzero element of the vector

$$[c_1 \quad c_{11} \quad c_{12} \quad \cdots \quad c_{1m}].$$

Since the value of the basic variable should be nonnegative, the first nonzero element of this vector should be positive; it is not possible that all elements are zero because then the vector

$$[c_{11} \quad c_{12} \quad \cdots \quad c_{1m}]$$

is zero, which cannot happen since it is a row of a nonsingular matrix.

A vector of which the first nonzero element is positive is called *lex-icographically positive*. The values of basic variables of tableau I are polynomials in ϵ with coefficients which form vectors that are lexicographically positive.

Let us now apply the rule of the simplex method to the leaving basic variable of the perturbed problem. Let the new basic variable be x_k. The leaving basic variable is now connected with

$$\text{Min}_i \left(\frac{c_1 + c_{i1}\epsilon + c_{i2}\epsilon^2 + \ldots + c_{im}\epsilon^m}{d_{ik}} \middle| d_{ik} > 0 \right) \tag{52}$$

This amounts to first taking

$$\text{Min}_i \, (c_i/d_{ik} \,|\, d_{ik} > 0);$$

if this does not yield a unique minimum, let the rows involved in the minimum form a set I_1; we determine

$$\text{Min}_{i \in I_1} c_{i1}/d_{ik} \, ;$$

if this does not yield a unique minimum, let the rows of this minimum form a set I_2; we then determine

$$\text{Min}_{i \in I_2} c_{i2}/d_{ik},$$

and so on, until we have found a unique minimum. The procedure we first described can be summarized by saying that we determined the *lexico-graphic minimum* of the vectors

$$1/d_{ik} \, [c_i \quad c_{i1} \quad c_{i2} \quad \cdots \quad c_{im}] \text{ for } d_{ik} > 0.$$

The lexicographic is in this case always unique. If it were not, for instance if rows 1 and 2 were connected with all minima, we would have

$$c_1/d_{1k} = c_2/d_{2k}$$
$$c_{11}/d_{1k} = c_{21}/d_{2k}$$
$$\cdots$$
$$c_{1m}/d_{1k} = c_{2m}/d_{2k}, \tag{53}$$

which implies

$$\begin{bmatrix} c_{11} \\ \cdot \\ c_{1m} \end{bmatrix} = \frac{d_{1k}}{d_{2k}} \begin{bmatrix} c_{21} \\ \cdot \\ c_{2m} \end{bmatrix}. \tag{54}$$

This is impossible since the inverse of the basis is nonsingular.

If a basic solution is lexicographically positive, then if the leaving basic variable is chosen according to the lexicographic minimum, the next basic solution is also lexicographically positive. The proof is a simple generalization of that for a simple basic variable. Let the lexicographic minimum be connected with the rth row. The row-vectors connected with the new basic solution are then for the rth row

$$1/d_{rk} [c_r \quad c_{r1} \quad \cdot \quad c_{rm}]$$

which is lexicographically positive, and for $i = 1, ..., m, i \neq r$:

$$[c_i - d_{ik}d_{rk}^{-1}c_r \quad c_{i1} - d_{ik}d_{rk}^{-1}c_{r1} \quad \cdot \quad c_{im} - d_{ik}d_{1k}^{-1}c_{rm}]$$
$$= d_{ik} \{1/d_{ik} [c_i \quad c_{i1} \quad \cdot \quad c_{im}] - 1/d_{rk} [c_r \quad c_{r1} \quad \cdot \quad c_{rm}]\}. \quad (55)$$

For $d_{ik} > 0$, this is lexicographically positive because row r gave the lexicographic minimum and for $d_{ik} \leq 0$ it is obvious from the left-hand side that the vector is lexicographically positive ($d_{rk} > 0$, $c_i > 0$, or $c_r > 0$, or both, or both are zero; in the latter case, $c_{i1} > 0$, or $c_{r1} > 0$, or both, or both are zero; in the latter case ..., and so on).

Let us now look at the objective function. Before the transformation it is

$$c_0 + c_{01}\epsilon + c_{02}\epsilon^2 + ... + c_{0m}\epsilon^m,$$

which is represented by the vector

$$[c_0 \quad c_{01} \quad c_{02} \quad \cdot \quad c_{0m}].$$

After the transformation, it is represented by the vector

$$[c_0 - d_{0k}d_{rk}^{-1}c_r \quad c_{01} - d_{0k}d_{rk}^{-1}c_{r1} \quad \cdot \quad c_{0m} - d_{0k}d_{rk}^{-1}c_{rm}]$$
$$= [c_0 \quad c_{01} \quad c_{02} \quad \cdot \quad c_{0m}] - d_{0k}/d_{rk} [c_r \quad c_{r1} \quad c_{r2} \quad \cdot \quad c_{rm}]. \quad (56)$$

Since $- d_{0k}/d_{rk} > 0$, and the vector $[c_r \quad c_{r1} \quad \cdot \quad c_{rm}]$ is lexicographically positive, the value of f has increased lexicographically. Coming back to the ϵ-formulation, the objective function has been increased in terms of ϵ; it has been increased by $- d_{0k}/d_{rk}$ times the first nonzero (positive) coefficient in the rth row times the corresponding ϵ-term; later ϵ-terms are dominated. Since there has been a strict increase in the objective function, cycling is not possible if the lexicographic minimum is used for the choice of the leaving basic variable.

Let us apply this to the example of table 9. The values of basic variables should be thought of as rows, the elements of which are the coefficients of the ϵ-terms. These rows are

$$\begin{array}{cccc} 0 & 1 & 0 & 0 \\ 0 & 0 & 1 & 0 \\ 1 & 0 & 0 & 1 \end{array}$$

We compare

$$0/\tfrac{1}{4}, \quad 0/\tfrac{1}{2},$$

which gives no unique minimum. We then compare

$$1/\tfrac{1}{4}, \quad 0/\tfrac{1}{2},$$

which has a minimum in the second row, so that y_2 leaves the basis, with $\tfrac{1}{2}$ being the pivot. This results in tableau 1a. The value of f which was $0 + 0\epsilon + 0\epsilon^2 + 0\epsilon^3$ has now been increased to $0 + 0\epsilon + 1\tfrac{1}{2}\epsilon^2 + 0\epsilon^3$. In tableau 1a, x_3 enters the basis, y_3 leaves it, and tableau 2a gives the optimal solution.

The perturbation or lexicographic choice does not only make the simplex method a theoretically perfect method for linear programming; it also serves a useful part in post-optimization methods, as shown in chapter 7.

3.6. The simplex method for problems with upper-bound constraints[3]

In linear programming problems, constraints of the following type frequently occur:

$$x_j \le d_j, \tag{57}$$

d_j is called the upper bound of x_j. These constraints may be treated as ordinary constraints; however, the corresponding row of the set-up tableau has a very simple structure, since it consists of zero elements apart from the elements in the constant-term column and the columns of x_j and the slack variable. Due to the simplicity of such constraints, it is possible to deal with them without having rows in the tableaux for any of the upper-bound constraints, which reduces the size of the tableau considerably; furthermore, as we shall see, it reduces the number of tableaux.

Suppose we have the following system with a basic feasible solution.

$$
\begin{aligned}
f_0 &= p_1 x_1 + p_2 x_2 & + f, \\
b_1 &= a_{11} x_1 + a_{12} x_2 + x_3, \\
b_2 &= a_{21} x_1 + a_{22} x_2 & + x_4.
\end{aligned} \tag{58}
$$

We assume that the variables x_1, x_2, x_3, and x_4 have upper bounds of d_1, d_2, d_3, and d_4, but that the values of x_3 and x_4 in (58), b_1 and b_2 are not

---

[3] This section is identical with section 5.2 in [24].

exceeding the upper bounds d_3 and d_4, so that the basic solution of (58) is also feasible for the upper-bound constraints.

According to the rules of the ordinary simplex method, the nonbasic variable with the most negative element in the objective function row is introduced into the basis; let this be x_1. Hence x_1 should be increased from zero to a positive value. The system (58) may therefore be written as

$$f_0 - p_1 x_1 = f,$$
$$b_1 - a_{11} x_1 = x_3,$$
$$b_2 - a_{21} x_1 = x_4. \tag{59}$$

The maximum value which x_1 can take is in the ordinary simplex method determined by

$$x = \operatorname*{Min}_i (b_i / a_{i1} | a_{i1} > 0), \tag{60}$$

but now, we should also take into account the upper bound of x_1 which is d_1; for the moment we disregard the other upper bounds. Hence the maximum value of x_1 is determined by

$$x = \operatorname*{Min}_i (b_i / a_{i1} | a_{i1} > 0, \, d_1). \tag{61}$$

If the minimum is not connected with d_1, the system is transformed as usual with either a_{11} or a_{21} as a pivot.

In case the minimum is equal to d_1, x_1 should in the next iteration become equal to d_1. This can be done by keeping x_1 as a nonbasic variable but putting it at a constant value d_1. Let us measure x_1 as a deviation from d_1. This amounts to replacing x_1 by x_1^*, which is defined as

$$x_1^* = x_1 - d_1. \tag{62}$$

Note that $x_1^* = 0$ for $x_1 = d_1$. We may substitute for x_1 in (58) according to (62). The result is

$$f_0 - p_1 d_1 = p_1 x_1^* + p_2 x_2 \qquad\qquad + f,$$
$$b_1 - a_{11} d_1 = a_{11} x_1^* + a_{12} x_2 + x_3,$$
$$b_2 - a_{21} d_1 = a_{21} x_1^* + a_{22} x_2 \qquad + x_4. \tag{63}$$

The only difference from (58), apart from having x_1^* instead of x_1, is that d_1 times the coefficient of x_1 is subtracted from the value of the basic variables. The values of basic variables are still nonnegative since they may be written as

$$a_{i1}(b_i/a_{i1} - d_1), \tag{64}$$

which is nonnegative for $a_{i1} > 0$ since d_1 was the minimum in (61) and which is also nonnegative for $a_{i1} < 0$.

The variable x_1^* differs from other nonbasic variables since it can be *decreased* but not increased; any increase would mean that x_1 would exceed its upper bound. The lower bound for a decrease of x_1^* is of course $-d_1$. The relation between x_1 and x_1^* may conveniently be indicated by the following figure:

$$x_1 \text{-----} \ \underline{\qquad} \ \text{-----}$$
$$\quad 0 \qquad\quad d_1$$

$$x_1^* \text{-----} \ \underline{\qquad} \ \text{-----}$$
$$\quad -d_1 \qquad 0$$

where the full line indicates the feasible range of both variables.

Starred variables like x_1^* are eligible as new basic variables. Let us consider a general system as (58) with x_1^* instead of x_1:

$$
\begin{aligned}
f_0 &= p_1 x_1^* + p_2 x_2 &&+ f, \\
b_1 &= a_{11} x_1^* + a_{12} x_2 + x_3, \\
b_2 &= a_{21} x_1^* + a_{22} x_2 &&+ x_4.
\end{aligned} \tag{65}
$$

If x_1^* is considered a new basic variable we write

$$
\begin{aligned}
f_0 &- p_1 x_1^* = f. \\
b_1 &- a_{11} x_1^* = x_3, \\
b_2 &- a_{12} x_1^* = x_4.
\end{aligned} \tag{66}
$$

Since x_1^* is decreased, the objective function increases for $p_1 > 0$. Hence starred nonbasic variables should have positive elements in the last row to be eligible as the new basic variable. Let us assume that $p_1 > 0$ and that x_1^* is selected as the new basic variable. Since x_1^* is decreased, x_3 increases if $a_{11} > 0$, but it decreases if $a_{11} < 0$. Only in the latter case it becomes zero and may become negative. Hence we should in this case determine the leaving basic variable as follows:

$$\underset{i}{\text{Min}} \ (b_i/-a_{i1}|a_{i1} < 0, d_1). \tag{67}$$

If d_1 is the minimum, x_1^* reaches its lower bound, which means that

$x_1 = 0$, hence x_1^* may be replaced by x_1 as nonbasic variable by substituting in (65), according to (62):

$$f_0 + p_1 d_1 = p_1 x_1 + p_2 x_2 \qquad\qquad + f,$$
$$b_1 + a_{11} d_1 = a_{11} x_1 + a_{12} x_2 + x_3,$$
$$b_2 + a_{21} d_1 = a_{21} x_1 + a_{22} x_2 \qquad + x_4. \qquad (68)$$

This means that apart from deleting the star from x_1, we should add the coefficients of the column of x_1^* multiplied by d_1 to the values of basic variables.

If the minimum in (67) is not d_1, but, say, $b_1/-a_{11}$, the system is transformed with a_{11} as a pivot. The values of basic variables are after this transformation:

$$f_0 - p_1 a_{11}^{-1} b_1 = \qquad (p_2 - p_1 a_{11}^{-1} a_{12}) x_2 - p_1 a_{11}^{-1} x_3,$$
$$a_{11}^{-1} b_1 = x_1^* \qquad + a_{11}^{-1} a_{12} x_2 + \qquad a_{11}^{-1} x_3,$$
$$b_2 - a_{21} a_{11}^{-1} b_1 = \quad (a_{22} - a_{21} a_{11}^{-1} a_{12}) x_2 - a_{21} a_{11}^{-1} x_3 + x_4. \qquad (69)$$

Since $a_{11} < 0$, x_1^* is negative, but it exceeds its lower bound $-d_1$, since $b_1/-a_{11} \leq d_1$. It is now convenient to replace in (69) x_1^* by x_1 which is done by substituting for x_1^* according to (62); the second equation of (69) is then

$$d_1 + a_{11}^{-1} b_1 = x_1 + a_{11}^{-1} a_{12} x_2 + a_{11}^{-1} x_2 \qquad (70)$$

the value of x_1 is obviously positive. The other values of basic variables are unchanged.

We have not considered the possibility that due to an increase of the new basic variable, a basic variable will reach its upper bound; we shall do this now. Returning to (59); we see that the values of x_3 and x_4 increase for negative values of a_{11} and a_{12}; the upper bound for x_3, d_3, can be reached if $b_3 - a_{11} x_1 = x_3 = d_3$ or

$$x_1 = (d_3 - b_1)/a_{11}. \qquad (71)$$

Hence the maximum value which x_1 can take for a system with a number of rows of basic variables with upper bounds d_i is

$$\operatorname*{Min}_i \left(\frac{b_i}{a_{i1}} \middle| a_{11} > 0, \frac{d_i - b_i}{-a_{i1}} \middle| a_{i1} < 0, d_1 \right). \qquad (72)$$

Let us assume that in the system (58) with x_1 as the new basic variable, the minimum is connected with the upper bound of x_3. Hence x_3 should be put at its upper bound and should become a nonbasic variable. Firstly, we measure x_3 from its upper bound d_3:

$$x_3^* = x_3 - d_3. \tag{73}$$

This is substituted in the first equation of (58):

$$b_1 - d_3 = a_{11}x_1 + a_{12}x_2 + x_3^*. \tag{74}$$

Note that $b_1 - d_3 < 0$ and also $a_{11} < 0$. Next x_1 is made a basic variable replacing x_3^* by pivoting on a_{11}. In the resulting solution all basic variables have nonnegative values.

Let us now consider the system (65) and assume that x_1^* is the new basic variable (having a positive element in the row of the objective function) and that x_3 and x_4 have upper bounds d_3 and d_4. The system again may be written as in (66). Since x_1^* decreases, the value of x_3 increases if $a_{11} > 0$; $x_3 = d_3$ for

$$b_1 - a_{11}x_1^* = d_3 \tag{75}$$

or

$$x_1^* = (d_3 - b_1)/-a_{11}. \tag{76}$$

Hence the leaving basic variable should be determined by

$$\operatorname*{Min}_i \left(\frac{b_i}{-a_{i1}} \middle| a_{i1} < 0, \frac{d_i - b_i}{a_{i1}} \middle| a_{i1} > 0, d_1 \right). \tag{77}$$

Assuming that the minimum is connected with the upper bound of x_3, d_3, we should first put x_3 as its upper bound by substituting in (65) for x_3 according to (73) and transform the tableau with a_{11} as pivot. After that x_1^* is replaced by x_1 adding d_1 to its value.

It may be formally proved that the rules for the simplex method for variables with upper bounds implied by the above are equivalent to an application of the ordinary simplex method to the problem with explicit upper-bound constraints. This, however, will not be done here.

The rules for the simplex method for variables with upper bounds may be summarized as follows. Let a_{ij} be the element in a tableau with ith row and the jth column; 0 is the row of the objective function. The values of basic variables are indicated by b_i. The indices of row and columns with variables having an upper bound are said to belong to a set U. If the nonbasic variable in column j has an upper bound, this bound is indicated by d_j; similarly, an upper bound of the basic variable in row i is indicated by d_i. Columns of nonbasic variables put at their upper bound (starred variables) are said to belong to a set S; the other columns are said to belong to a set N.

SELECTION OF THE NEW BASIC VARIABLE: Select the column connected with

$$\text{Min}_j(a_{0j}|a_{0j} < 0, j \in N, -a_{0j}|a_{0j} > 0, j \in S). \tag{78}$$

Let k be the column of the new basic variable. If $a_{0j} \geq 0$ for all $j \in N$ and $a_{0j} \leq 0$ for all $j \in S$, the optimal solution has been obtained.

SELECTION OF THE LEAVING BASIC VARIABLE: If $k \in N$, select the row connected with

$$\text{Min}_i \begin{cases} \text{(i)} & b_i/a_{ik}|a_{ik} > 0, \\[1mm] \text{(ii)} & (d_i - b_i)/-a_{ik}|a_{ik} < 0,\ i \in U, \\[1mm] \text{(iii)} & d_k|k \in U; \end{cases} \tag{79}$$

if $k \in S$, select the row connected with

$$\text{Min}_i \begin{cases} \text{(iv)} & b_i/-a_{ik}|a_{ik} < 0, \\[1mm] \text{(v)} & (d_i - b_i)/a_{ik}|a_{ik} > 0,\ i \in U, \\[1mm] \text{(vi)} & d_k. \end{cases} \tag{80}$$

Let the row connected with the minimum in other cases than (iii) and (vi) be the rth row.

TRANSFORMATION OF THE TABLEAU:

Case (i): Transform the tableau with a_{rk} as a pivot.

Case (ii): Subtract d_r from b_r, transform with a_{rk} as a pivot and include the column of the leaving basic variable in S.

Case (iii): Subtract $a_{ik}d_k$ from b_i for all rows; include k in S.

Case (iv): Transform with a_{rk} as a pivot and add d_k to the value of the new basic variable. Include the row of the new basic variable in N.

Case (v): Subtract d_r from b_r; transform with a_{rk} as a pivot and add d_k to the value of the new basic variable. Include the column of the leaving basic variable in S.

Case (vi): Add $a_{ik}d_k$ to b_i for all rows. Include k in N.

As an example of application of the upper-bounds technique, the following problem is used. Maximize

$$f = 2\tfrac{1}{2}x_1 + \tfrac{1}{2}x_2 + 2x_3 + 2x_4 \tag{81}$$

subject to

$$4x_1 + x_2 + 2x_3 \qquad\ \leq 100,$$
$$x_1 \qquad + x_3 + 2x_4 \leq 40,$$
$$\tfrac{1}{2}x_1 \qquad + x_3 + 2x_4 \leq 60, \tag{82}$$

$$0 \le x_j \le 15, \quad j = 1, 2, 3, 4. \tag{83}$$

The set-up tableau is tableau 0 of table 11, in which the second and third columns of the values of basic variables and the stars attached to x_1 and x_3 should for a moment be disregarded. The solution of this tableau is feasible, so that it can be used as an initial solution. Since initially there are no starred variables, we take for the new basic variable simply the variable with the most negative coefficient in the first row, which is x_1. For the selection of the leaving basic variable, we compare the ratios

$$100/4, \qquad 40/1, \qquad 60/\tfrac{1}{2},$$

with the upper bound of x_1, which is 15; this upper bound is the minimum so that x_1 should be put at its upper bound. This is done by subtracting from the values of basic variables 15 times the x_1-column; the result is

Table 11

An application of the upper-bounds technique

Tableau	Basic variables	Values basic variables			x_1^*	x_2	x_3^*	x_4
0	f	0	$37\frac{1}{2}$	$67\frac{1}{2}$	$-2\frac{1}{2}$	$-\frac{1}{2}$	-2	-2
	y_1	100	40	10	4	1	2	0
	y_2	40	25	10	1	0	1	2
	y_3	60	$52\frac{1}{2}$	$37\frac{1}{2}$	$\frac{1}{2}$	0	1	2
					x_1^*	x_2	x_3^*	y_2
1	f	$77\frac{1}{2}$			$-1\frac{1}{2}$	$-\frac{1}{2}$	-1	1
	y_1	10			4	1	2	0
	x_4	5			$\frac{1}{2}$	$\bar{0}$	$\frac{1}{2}$	$\frac{1}{2}$
	y_3	$27\frac{1}{2}$			$-\frac{1}{2}$	0	0	-1
					x_1^*	y_1	x_3^*	y_2
2	f	$82\frac{1}{2}$			$\frac{1}{2}$	$\frac{1}{2}$	0	1
	x_2	$10(-5)$			4	1	2	0
	x_4	5			$-\frac{1}{2}$	0	$\frac{1}{2}$	$\frac{1}{2}$
	y_3	$27\frac{1}{2}$			$-\frac{1}{2}$	0	0	-1
					x_2^*	y_1	x_3^*	y_2
3	f	$83\frac{1}{8}$			$-\frac{1}{8}$	$\frac{3}{8}$	$-\frac{1}{4}$	1
	x_1	$(-1\frac{1}{4})$	$13\frac{3}{4}$		$\frac{1}{4}$	$\frac{1}{4}$	$\frac{1}{2}$	0
	x_5	$5\frac{5}{8}$			$-\frac{1}{8}$	$-\frac{1}{8}$	$\frac{1}{4}$	$\frac{1}{2}$
	y_3	$26\frac{7}{8}$			$\frac{1}{8}$	$\frac{1}{8}$	$\frac{1}{4}$	-1

the second column of values of basic variables of tableau 0; the nonbasic variable x_1 is given a star to indicate that it has reached its upper bound. The next new basic variable is selected by comparing in the first row negative elements for columns of nonstarred variables and positive elements for starred variables; x_3 is then selected as the new basic variable. Since none of the basic variables has an upper bound we compare

$$40/2, \quad 25/1, \quad 52\tfrac{1}{2}/1,$$

with the upper bound of x_3, 15. Again the upper bound is the minimum and x_3 is put at its upper bound by subtracting from the values of basic variables 15 times the column of x_3; the result is the third column of values of basic variables; to x_3 a star is attached. The next variable selected to come into the basis is x_4. The leaving basic variable is determined by comparing

$$10/2, \quad 37\tfrac{1}{2}/2,$$

with the upper bound of x_4, which is 15. It is found that y_2 leaves the basis. Since case (i) applies, tableau 0 can be transformed into tableau 1 without any adjustment.

In tableau 1, x_2 is the only variable eligible for new basic variable and y_1 is the variable leaving the basis since $10 < 15$. Again case (i) applies; after transformation of the tableau, tableau 2 is generated. The variable to enter the basis is now x_1, a variable put at its upper bound. To find the leaving basic variable, we should compare according to (80)

(iv) $27\tfrac{1}{2}/-(-\tfrac{1}{2})$,

(v) $(15-10)/4, (15-5)/\tfrac{1}{2}$,

(vi) 15.

The minimum ratio $1\tfrac{1}{4}$, occurs in (v) and is connected with x_2. Hence x_2 is the leaving basic variable; this variable is put at its upper bound. Before transforming tableau 2, we subtract 15 from the value of x_2; the result, -5, is indicated within brackets. After that, a simplex transformation follows, which results in tableau 3. The value of x_1 in this tableau is indicated within brackets, since we should adjust it for the fact that x_1 is no longer at its upper bound, which is done by adding 15 to it. Tableau 3 is found to be the optimal tableau.

In some problems variables occur with lower bounds:

$$x_j \geq c_j. \tag{84}$$

These lower bounds usually occur in combination with upper bounds:

$$c_j \le x_j \le d_j, \tag{85}$$

so that x_j is restrained to the range of values between c_j and d_j. Lower bounds can be treated by simply measuring the variable concerned from its lower bound. We define a new variable x_j^*

$$x_j^* = x_j - c_j, \tag{86}$$

which should be nonnegative, (86) should be substituted into the equation system; the upper bound for x_j^* is now $d_j - c_j$. If x_1 in the system (58) would have lower and upper bounds as in (85), it would be reformulated as

$$\begin{cases} b_1 - a_{11}c_1 = a_{11}x_1^* + a_{12}x_1 + x_3, \\ b_2 - a_{21}c_1 = a_{21}x_1^* + a_{22}x_2 \quad\;\; + x_4, \\ f_0 - p_1c_1 = p_1x_1^* + p_2y_2 \qquad\quad + f. \end{cases} \tag{87}$$

For x_1^* we now have

$$0 \le x_1^* \le d_1 - c_1. \tag{88}$$

In this manner problems with lower bounds on some variables may be reformulated as problems without these bounds.

CHAPTER 4

DUALITY IN LINEAR PROGRAMMING AND THE DUAL METHOD

4.1. Opportunity costs

The simplex method was presented mainly as a computational device for maximizing a linear function, subject to linear equations and inequalities. However, it is also possible, and it will turn out to be useful, to give an economic interpretation of the simplex method. By doing so, the important properties of duality in linear programming will be uncovered in a natural fashion. As an illustration we shall use again the example of production planning presented in chapter 2.

The starting-point in the simplex method was the situation in which no products were produced and all machines were idle; the only nonzero variables were the slack variables representing idle capacity: $y_1 = 100$, $y_2 = 40$, $y_3 = 60$. Though all products have a positive profit, the actual profit is in this situation zero, since none of the products are produced.

Next we consider the increase in profit which occurs by taking in production one of the products. The product with the highest profit per unit obviously is product 1 with a profit of 3 per unit. The maximum quantity of product 1 to be produced is determined by the minimum of the ratios:

$$\frac{100}{4}, \quad \frac{40}{1}, \quad \frac{60}{2}. \tag{1}$$

Machine 1 turns out to be the bottleneck for the production of product 1. The new solution is $x_1 = 25$, $y_2 = 15$ and $y_3 = 10$. Total profits are $25 \times 3 = 75$.

Since machine 1 is the bottleneck for the production of product 1, we may ask ourselves what the effect of a change in capacity on this machine will be on total profits. Let us denote the change in capacity on machine 1 by Δ_1.

$$100 + \Delta_1 = 4x_1$$

or

$$x_1 = \tfrac{1}{4}(100 + \Delta_1) = 25 + \tfrac{1}{4}\Delta_1. \tag{2}$$

89

This solution is feasible if x_1 does not exceed the next lowest ratio of (1), which is 30, which means that Δ_1 should not exceed 20. Furthermore, Δ_1 should not be smaller than -100, since otherwise capacity on machine 1 becomes negative. Total profits for the solution (2) are

$$3x_1 = \tfrac{3}{4}(100 + \Delta_1) = 75 + \tfrac{3}{4}\Delta_1. \tag{3}$$

This means that for the present solution any change Δ_1 in the capacity on machine 1 within the range -100 and 20 results in a change of profits $\tfrac{3}{4}\Delta_1$. We may therefore say that capacity on machine 1 is worth 0.75 per unit in this solution.

Total profits may be evaluated in this manner. In eq. (3), the present capacity, 100 and Δ_1 have the same factor. Hence we have for $\Delta_1 = 0$,

$$3 \times 25 = \tfrac{3}{4} \times 100 = 75. \tag{4}$$

This means that total profits may also be computed by multiplying the capacity of the machine by the value of this capacity per unit.

This value may also be obtained by equating profits per unit of product 1 to the required capacity on machine 1 per unit of product times this value which is denoted by u_1:

$$3 = 4u_1, \tag{5}$$

from which we obtain

$$u_1 = \tfrac{3}{4}. \tag{6}$$

This is the same as the figure found in the objective function row in the column of y_1 in tableau 1 of table 2 in chapter 2. This is no coincidence, since this row can be formulated in equational form as follows:

$$75 - \tfrac{3}{4}y_1 = f, \tag{7}$$

from which we note that increasing the idle capacity on machine 1 leads to a decrease in profits of $\tfrac{3}{4}$ per unit of increase; an increase in idle capacity has the same effect on the solution as a decrease of capacity used for production.

Now this value of capacity in machine 1 can be used to evaluate the profitability of other products, since we may subtract from the profit of a product the capacity used per unit times the value of this capacity. For instance, for product 2 we have

$$1 - 3 \times \tfrac{3}{4} = -1\tfrac{1}{4}. \tag{8}$$

Increasing the production of product 2 would decrease profit by 1.25. This is the same result we found before.

It must be noted that the value of capacity on machine 1 we found plays the role of costs; in economics the term *opportunity costs* would be used, at least if the present situation would have been the optimal one, since scarce resources are evaluated in terms of the profits on the product(s) produced, see eq. (5). Other terms are *imputed costs* or *shadow prices*.

All other products not in the production program may be evaluated in this manner; the profitability per unit for each of the products will be found equal to minus the corresponding elements in the objective function row in the simplex tableau. For product 5 we find

$$2 - 0 \times \tfrac{3}{4} = 2. \tag{9}$$

This turns out to be the product with the highest positive profitability and it is therefore decided that it should be produced alongside (or instead of) product 1. In the usual manner it is found that while increasing x_5, y_3 becomes zero first, so that capacities on machines 1 and 3 are both in short supply. For this new solution the value of capacities on machines 1 and 3 may be found by determining the effect of changes in the capacities on these machines on profits. The effect of changes Δ_1 and Δ_2 on the production program are found from

$$100 + \Delta_1 = 4x_1,$$
$$60 + \Delta_3 = 2x_1 + 2x_5, \tag{10}$$

which results in

$$\begin{bmatrix} x_1 \\ x_5 \end{bmatrix} = \begin{bmatrix} 4 & 0 \\ 2 & 2 \end{bmatrix}^{-1} \left\{ \begin{bmatrix} 100 \\ 60 \end{bmatrix} + \begin{bmatrix} \Delta_1 \\ \Delta_3 \end{bmatrix} \right\}. \tag{11}$$

This solution is only feasible for certain ranges of Δ_1 and Δ_3 around zero. The effect on total profit is given by

$$3x_1 + 2x_5 = \begin{bmatrix} 3 & 2 \end{bmatrix} \begin{bmatrix} 4 & 0 \\ 2 & 2 \end{bmatrix}^{-1} \left\{ \begin{bmatrix} 100 \\ 60 \end{bmatrix} + \begin{bmatrix} \Delta_1 \\ \Delta_3 \end{bmatrix} \right\}$$

$$= \begin{bmatrix} \tfrac{1}{4} & 1 \end{bmatrix} \left\{ \begin{bmatrix} 100 \\ 60 \end{bmatrix} + \begin{bmatrix} \Delta_1 \\ \Delta_3 \end{bmatrix} \right\} = 85 + \tfrac{1}{4}\Delta_1 + \Delta_3. \tag{12}$$

Hence, the imputed costs of capacity on machine 1 and machine 3 are $\tfrac{1}{4}$ and 1, which agrees with the elements in the last row in the columns of y_1 and y_3 in tableau 2 of table 2 in chapter 2.

Note that total profits can be evaluated in terms of imputed costs and available capacity; in (12) we have

$$\tfrac{1}{4} \times 100 + 1 \times 60 = 85. \tag{13}$$

The same results could have been obtained by evaluating the profits of products 1 and 5 in terms of the imputed costs of the capacities on machines 1 and 3, u_1 and u_3:

$$3 = 4u_1 + 2u_3,$$

$$2 = \qquad 2u_3, \tag{14}$$

which yields

$$\begin{bmatrix} u_1 \\ u_3 \end{bmatrix} = \begin{bmatrix} 4 & 2 \\ 0 & 2 \end{bmatrix}^{-1} \begin{bmatrix} 3 \\ 2 \end{bmatrix} = \begin{bmatrix} \tfrac{1}{4} \\ 1 \end{bmatrix}. \tag{15}$$

The same reasoning can be used for the next solution in which capacity on all machines is in short supply and in which products 1, 4 and 5 are produced. However, we shall now turn to the general case, using the matrix formulation of the simplex method. A general maximization problem is considered of which the set-up tableau is given in tableau I of table 1. Tableau II gives the system in which the first m x-variables, which are arranged in a vector x^b, are basic, and the remaining ones, arranged in a vector x^n, are nonbasic; all slack variables are nonbasic. In terms of the production planning example this means that the present production program consists of the first m products and that all m machines are fully occupied.

Table 1

A basic solution in matrix formulation

Tableau	Basic variables	Values basic variables	x^b	x^n	f	y
I	f	0	$-p^{1\prime}$	$-p^{2\prime}$	1	0
	y	b	A_1	A_2	0	1
II	f	$p^{1\prime}A_1^{-1}b$	0	$p^{1\prime}A_1^{-1}A_2 - p^{2\prime}$	1	$p^{1\prime}A_1^{-1}$
	x^b	$A_1^{-1}b$	I	$A_1^{-1}A_2$	0	A_1^{-1}

Consider for this solution changes in capacity on all machines given by the vector Δ:

$$\Delta = \begin{bmatrix} \Delta_1 \\ \Delta_2 \\ \vdots \\ \Delta_m \end{bmatrix}.$$

The production program is related to these changes by

$$b + \Delta = A_1 x^b, \tag{16}$$

from which we find

$$x^b = A_1^{-1}(b - \Delta). \tag{17}$$

Total profits for this solution are

$$p'x = p^{1\prime}x^b = p^{1\prime}A_1^{-1}(b + \Delta) = p^{1\prime}A_1^{-1}b + p^{1\prime}A_1^{-1}\Delta. \tag{18}$$

The vector of imputed costs u is therefore

$$u = A_1^{\prime -1}p^1. \tag{19}$$

Total profits can alternatively be viewed as the summation of the revenues of the products or as the summation of the values of the scarce resources, in our example capacity on the machines:

$$p^{1\prime}x^b = p^{1\prime}A_1^{-1}b = b'u. \tag{20}$$

The same results can be obtained by evaluating the profits of the products in the production program, given by the vector p^1, in terms of the capacity on the machines used for these products. We then obtain

$$p^1 = A_1'u, \tag{21}$$

from which we find

$$u = A_1^{\prime -1}p^1. \tag{22}$$

The profitability of products not in the production program is found by subtracting from the profits per unit of these products the capacity required on the machines times the imputed costs of capacity:

$$p^2 - A_2'u = p^2 - A_2'A_1^{\prime -1}p^1. \tag{23}$$

We may define the vector v as minus the profitability for all products. Partitioning v into subvectors v^1 and v^2, according to basic and nonbasic variables, we have

$$v^2 = A_2'A_1^{\prime -1}p^1 - p^2. \tag{24}$$

The products in the production program have a profitability of zero:

$$v^1 = p^1 - A_1'u = p^1 - A_1'A_1'^{-1}p^1 = 0. \tag{25}$$

Comparing (22), (24) and (25) with tableau II of table 1, we find that the vector u of imputed costs appears in the objective function row in the matrix of the inverse and that the elements in this row in the other columns are the v-variables which stand for minus the profitability of the various products. The requirement for an optimal solution to the linear programming problem that all elements in the objective function row be nonnegative is therefore equivalent to the condition that the u- and v- variables should be nonnegative for the optimal solution.

4.2. Dual problems

Closely related to each linear programming problem there is a so-called *dual problem*. Consider the typical maximization problem: Maximize

$$f = p'x \tag{26}$$

subject to

$$Ax \leq b, \tag{27}$$

$$x \geq 0. \tag{28}$$

The dual problem of this problem is: Minimize

$$g = b'u \tag{29}$$

subject to

$$A'u \geq p, \tag{30}$$

$$u \geq 0. \tag{31}$$

The original problem is now called the *primal problem*. Note that the dual problem of a typical maximization problem is obtained from the primal one by the following changes:

(1) the objective function is minimized instead of maximized;

(2) the coefficients of the objective function of the primal problem appear in the constant term of the dual problem and the coefficients of the constant term of the primal problem become the coefficients of the objective function of the dual problem;

(3) instead of $\leq$-signs in the constraints of the primal problem there are $\geq$-signs in the constraints of the dual problem;

(4) the matrix A is transposed.

Note that with each variable of the primal problem there corresponds a constraint in the dual problem and with each variable of the dual problem there corresponds a constraint of the primal problem. Writing the constraints of both problems as equalities by the introduction of vectors of slack variables y and v, we have

$$Ax + y = b,$$
$$x, y \geq 0, \tag{32}$$

and

$$A'u - v = p,$$
$$u, v \geq 0. \tag{33}$$

From this we conclude that for each x-variable in the primal problem there is a corresponding v-variable in the dual problem and for each y-variable there is a corresponding u-variable.

Let us now determine the dual problem of the dual problem. For this purpose, the dual problem is rewritten as a typical maximization problem: Maximize

$$-g = -b'u \tag{34}$$

subject to

$$-A'u \leq -p, \tag{35}$$
$$u \geq 0. \tag{36}$$

The dual problem of this problem is: Minimize

$$h = -p'z \tag{37}$$

subject to

$$-Az \geq -b, \tag{38}$$
$$z \geq 0. \tag{39}$$

Using x-variables instead of z-variables, we find that this problem may be reformulated as: Maximize

$$-h = p'x \tag{40}$$

subject to

$$Ax \leq b, \tag{41}$$
$$x \geq 0, \tag{42}$$

which is exactly the primal problem. Hence the dual of the dual problem is the primal problem. In the same way it can be shown that the dual of a typical minimization problem is a typical maximization problem.

What is the effect of equalities in the constraints (27) of the primal problem on the dual problem? We consider the following problem: Maximize f,

$$f = p'x,$$
$$Ax = b,$$
$$x \geq 0. \tag{43}$$

Let us assume that the equations of the equality constraints are independent; if one or more equations are dependent, they can be deleted. Furthermore, it is assumed that the first m columns of the matrix A are independent; this can always be made true by rearrangement. The problem may then be partitioned as follows: Maximize f,

$$f = p^{1\prime}x^1 + p^{2\prime}x^2,$$
$$A_1 x^1 + A_2 x^2 = b,$$
$$x^1 \geq 0, \ x^2 \geq 0. \tag{44}$$

From the second set of equations we obtain

$$x^1 = A_1^{-1}b - A_1^{-1}A_2 x^2. \tag{45}$$

This is substituted into the objective function and into $x^1 \geq 0$, after which the problem can be written as: Maximize f,

$$f = p^{1\prime}A_1^{-1}b + (p^{2\prime} - p^{1\prime}A_1^{-1}A_2)x^2,$$
$$A_1^{-1}A_2 x^2 \leq A_1^{-1}b,$$
$$x^2 \geq 0. \tag{46}$$

The dual problem of this problem is: Minimize g,

$$g = p^{1\prime}A_1^{-1}b + b^{1\prime}A_1^{-1}u,$$
$$A_2'A_1^{\prime-1}u \geq p^2 - A_2'A_1^{\prime-1}p^1,$$
$$u \geq 0. \tag{47}$$

Let us define

$$u^* = A_1^{\prime-1}u + A_1^{\prime-1}p^1 \quad \text{or} \quad u = A_1'u^* - p^1. \tag{48}$$

The problem (47) may then be written as: Minimize g,

$$g = b'u^*,$$
$$A_2'u^* \geq p^2,$$
$$A_1'u^* \geq p^1, \tag{49}$$

or: Minimize g,

$$g = b'u^*,$$
$$A'u^* \geq p, \tag{50}$$

where u^* is a vector of unconstrained variables. We therefore conclude that if there are equalities in the primal problem, the corresponding constraints in the dual problem are unrestricted. In the same way it may be shown that if there are unrestricted variables in the primal problem, the corresponding constraints in the dual problem are equalities.

4.3. Interpretation of dual problems

Let us reconsider the production planning problem. The primary objective was to find a production program which, given the prices of the various products and given available capacities of the machines, would maximize profits. We have seen that as a by-product the maximum profit and the opportunity costs related to the machines were obtained. The maximal profit can be interpreted as the total value of available capacity on the machines for that period; it has been proved that the maximal profit is equal to the opportunity cost of capacity on each machine times the available capacity on each machine summed over all machines. This is the production and value problem as seen from the *inside* of the enterprise.

It is also possible to take the viewpoint of an *outsider* who wants to take over the enterprise for a period by making an offer for capacity on each of the machines. In our example there were three machines, the first of which had 100 units of capacity, the second 40 units and the third 60 units. If the outsider offers u_1 (e.g., dollars) for each unit of capacity on machine 1, u_2 for each unit of capacity on machine 2 and u_3 for each unit of capacity on machine 3, the total amount which the outsider has to pay for the capacity on all machines is

$$g = 100u_1 + 40u_2 + 60u_3. \tag{51}$$

It is obvious that he wants to *minimize* this amount.

It is assumed that the owners of the enterprise are free to accept or to refuse the deal as a whole or in part; they may, for instance, sell only part of the capacity on one machine. The owners will never yield capacity on any machine if this is going to cost money instead of bringing a revenue.

Hence the prices offered for a unit capacity on each machine should be nonnegative:

$$u_1, u_2, u_3 \geq 0. \tag{52}$$

Furthermore, the prices should be such that the owners have no advantage in keeping the machines for the production of the various products; that is, the prices should be such that there is no profit on any of the products if the required productive-capacity for each product is evaluated at the prices offered. For instance, take product 1 of the example. The costs of the product, evaluated at the prices offered for the capacity on the machines are

$$4u_1 + u_2 + 2u_3. \tag{53}$$

In order to induce the owners to sell the capacity rather than to use it to produce product 1, these costs should be at least equal to the net revenue of product 1, which is 3. We have for this product and for the other products the following constraints:

$$\begin{aligned}
4u_1 + u_2 + 2u_3 &\geq 3, \\
3u_1 \qquad + 4u_3 &\geq 1, \\
u_1 \qquad\qquad &\geq \tfrac{1}{2}, \\
2u_1 + u_2 + u_3 &\geq 2, \\
2u_2 + 2u_3 &\geq 2.
\end{aligned} \tag{54}$$

The problem for the outsiders is then to find prices u_1, u_2 and u_3 which minimize (51), subject to (52) and (54). From the previous section it is obvious that this problem is the dual of the production planning problem.

For the dual problem of a typical minimization problem a similar interpretation can be given. Let us consider the minimum-cost cattlefeed problem, in which the owner of an animal wanted to mix a number of ingredients in such a manner that the animal obtains at least 20 units of protein and 5 units of fat per day. Once again we introduce an outsider who is offering to supply units of protein and fat in some manner at prices to be determined. The manner in which the units of protein and fat are delivered does not concern us at the moment. It may be imagined that the outsider has "pure" units of protein and fat at his disposal. Alternatively, we may think that he has materials in unlimited supply which he can mix in such a manner that the required units of protein and fat become available. The outsider wants to *maximize* the revenue he obtains by supplying the owner of the animal with the required units of protein and fat. Denoting the price for one unit of protein by u_1 and that for one unit of fat by u_2, we find that the total revenue which he wants to maximize is:

$$g = 20u_1 + 5u_2. \tag{55}$$

However, he must be sure that it is not advantageous for the animal-owner to use one of the available feeding materials to supply partly or wholly the required protein and fat units. Evaluating barley in the terms of the prices of protein and fat we have for its worth:

$$12u_1 + 2u_2. \tag{56}$$

This should be less than or at most equal to the market price 24, in order to induce the animal-owner not to use barley; otherwise barley is a cheaper supplier of protein and fat than the outsider. For oats, sesame flakes and groundnut meal similar inequalities are valid, so that we obtain:

$$
\begin{aligned}
12u_1 + \ 2u_2 &\le 24, \\
12u_1 + \ 6u_2 &\le 30, \\
40u_1 + 12u_2 &\le 40, \\
60u_1 + \ 2u_2 &\le 50.
\end{aligned} \tag{57}
$$

Furthermore the protein and fat prices should be nonnegative, because otherwise the animal-owner would react by disposing excess quantities of protein and fat. Hence we have

$$u_1, u_2 \ge 0. \tag{58}$$

The problem of the outsider is then to find prices u_1 and u_2 which maximize his revenue (55) subject to the constraints of (57) and (58). This problem clearly is the dual problem of the minimum-cost cattle feed problem.

4.4. Relations between primal and dual problems

Let us consider the dual problem of the typical maximization problem given by (29) and (33), and let us partition the equality constraints of (33) and the u-variables into the first q and the remaining constraints or variables. The matrix A and the vectors u, v, p and b are then partitioned as follows:

$$A = \begin{bmatrix} A_{11} & A_{12} \\ A_{21} & A_{22} \end{bmatrix}, \qquad u = \begin{bmatrix} u^1 \\ u^2 \end{bmatrix}, \qquad v = \begin{bmatrix} v^1 \\ v^2 \end{bmatrix},$$

$$p = \begin{bmatrix} p^1 \\ p^2 \end{bmatrix}, \qquad b = \begin{bmatrix} b^1 \\ b^2 \end{bmatrix}. \tag{59}$$

A_{11} is a $q \times q$ matrix, A_{12} a $q \times (n-q)$ matrix, and so on. The equations of (29) and (33) are put into a simplex tableau format with the v-variables as basic variables, see tableau IIa of table 2.

Table 2

Simplex tableaux for corresponding solutions of the primal and the dual problem

Tab.	Basic var.	Values basic variables	x_1	x_2	f	y^1	y^2
Ia	f	0	$-p^{1\prime}$	$-p^{2\prime}$	1	0	0
	y^1	b^1	A_{11}	A_{12}	0	I	0
	y^2	b^2	A_{21}	A_{22}	0	0	I
Ib	f	$p^{1\prime}A_{11}^{-1}b^1$	0	$-p^{2\prime}+p^{1\prime}A_{11}^{-1}A_{12}$	1	$p^{1\prime}A_{11}^{-1}$	0
	x^1	$A_{11}^{-1}b^1$	I	$A_{11}^{-1}A_{12}$	0	A_{11}^{-1}	0
	y^2	$b^2-A_{21}A_{11}^{-1}b^1$	0	$A_{22}-A_{21}A_{11}^{-1}A_{12}$	0	$-A_{21}A_{11}^{-1}$	I
			u^1	u^2	g	v^1	v^2
IIa	g	0	$-b^{1\prime}$	$-b^{2\prime}$	1	0	0
	v^1	$-p^1$	$-A_{11}'$	$-A_{21}'$	0	I	0
	v^2	$-p^2$	$-A_{12}'$	$-A_{22}'$	0	0	I
IIb	g	$b^{1\prime}A_{11}'^{-1}p^1$	0	$-b^{2\prime}+b^{1\prime}A_{11}'^{-1}A_{21}$	1	$-b^{1\prime}A_{11}'^{-1}$	0
	u^1	$A_{11}'^{-1}p^1$	I	$A_{11}'^{-1}A_{21}$	0	$-A_{11}'^{-1}$	0
	v^2	$-p^2+A_{12}'A_{11}'^{-1}p^1$	0	$-A_{22}'+A_{12}'A_{11}'^{-1}A_{21}'$	0	$-A_{12}'A_{11}'^{-1}$	I

We now wish to consider a basic solution of this dual problem, in which the first q u-variables and the last $n-q$ v-variables are basic. Any basic solution of the dual problem can always be re-arranged in such a way that this is true. Special cases are $q = 0$, for which the basic solution is that of tableau IIa, and $q = m$, in which case the rows of A are not partitioned. This basic solution is generated from tableau IIa by premultiplying its columns by the inverse of the matrix consisting of the columns of the basic variables:

$$
\begin{bmatrix} 1 & -b^{1\prime} & 0 \\ 0 & -A'_{11} & 0 \\ 0 & -A'_{12} & I \end{bmatrix}^{-1} = \begin{bmatrix} 1 & -b^{1\prime}A'^{-1}_{11} & 0 \\ 0 & -A'^{-1}_{11} & 0 \\ 0 & -A'_{12}A'^{-1} & I \end{bmatrix}.
\tag{60}
$$

The result is tableau IIb.

Let us also consider the original primal problem; its set-up tableau is given in tableau Ia of table 2. We wish to consider the basic solution in which the first q x-variables and the last $m-q$ y-variables are basic. The matrix A and the vectors x, y, p and b may be partitioned accordingly.

$$
\begin{bmatrix} 1 & -p^{1\prime} & 0 \\ 0 & A_{11} & 0 \\ 0 & A_{21} & I \end{bmatrix}^{-1} = \begin{bmatrix} 1 & p^{1\prime}A^{-1}_{11} & 0 \\ 0 & A^{-1}_{11} & 0 \\ 0 & -A_{21}A^{-1}_{11} & I \end{bmatrix}.
\tag{61}
$$

The result is tableau Ib of table 2.

Now the solutions of tableaux Ib and IIb (and also those of tableaux Ia and IIa) correspond to each other in the following manner. If an x-variable is basic in the primal tableau (tableau Ib), its corresponding v-variable is nonbasic in the dual tableau (tableau IIb); if an x-variable is nonbasic in the primal tableau, its corresponding v-variable is basic in the dual tableau. The same is valid for the y-variables in the primal tableau and the u-variables in the dual tableau. Hence if a primal variable is basic (non-basic) in a basic solution of the primal problem, its corresponding dual variable is nonbasic (basic) in the corresponding basic solution of the dual problem. Hence, in corresponding primal and dual solutions either a primal variable or its corresponding dual variable is zero (or both in cases of degeneracy). This property is called *complementary slackness*. Hence we have for the corresponding primal and dual solutions

$$
u'y = 0, \qquad v'x = 0.
\tag{62}
$$

In the first row of the primal tableau the values of the basic variables of the corresponding dual solution appear; in the first row of the dual problem we find the values of basic variables of the primal problem with a minus sign. In the body of the tableaux of both problems we find also the same elements, apart from sign; furthermore, the value of the objective function is the same for corresponding solutions of both problems. It may therefore be concluded that both tableaux each give complete information about the solutions of either problems.

Further it may be concluded from a comparison of both tableaux that if a basic solution of the primal problem is feasible because $x^1 = A_{11}^{-1}b^1 \geq 0$ and $y^2 = b^2 - A_{21}A_{11}^{-1}A_{12} \geq 0$ (but not necessarily optimal, which would imply $-p^{2\prime} + p^{1\prime}A_{11}^{-1}A_{12} \geq 0$ and $p^{1\prime}A_{11}^{-1} \geq 0$), then the corresponding solution of the dual problem is optimal (but not necessarily feasible), since the elements of the last row in tableau IIb of table 2 are all nonpositive, which is the requirement for an optimal solution of a minimization problem. Conversely, if the primal problem has an optimal solution (but not necessarily a feasible one) because the elements in the last row are all nonnegative:

$$p^{1\prime}A_{11}^{\prime-1} \geq 0, \qquad p^{1\prime}A_{11}^{\prime-1}A_{12}' - p^{2\prime} \geq 0,$$

the corresponding solution of the dual problem is a feasible one (which is not necessarily optimal), see tableau IIb of table 2.

Finally, if a basic solution to the primal problem is feasible as well as optimal, the corresponding solution of the dual problem is feasible and optimal. The last remark implies that, when solving a linear programming problem, we may solve the problem itself or its dual problem.

The following theorem which is called the *duality theorem* follows from the above observations and characterizes the relationship between the two problems: *The value of the objective function of the dual problem for a feasible solution of this problem is greater than or equal to the value of the objective function of the primal problem for a feasible solution of that problem. The values of the objective functions of both problems for the optimal solutions of these problems are equal.*

The last statement is proved by relationships between corresponding basic primal and dual solutions as shown in tableaux Ib and IIb of table 2. It can easily be shown to be valid also for multiple optimal solutions. The first statement then follows, since the dual problem is a minimization problem, so that any feasible solution to the problem must have at least as great a value of the objective function as the minimum solution; any feasible solution to the primal problem in which the objective function is maximized can at most give as large a value of the objective function as the optimal solution.

An alternative proof is found by premultiplication of the equations

$$Ax + y = b, \tag{63}$$

$$A'u - v = p, \tag{64}$$

by the vectors u' and x', respectively; the result is

$$u'Ax + u'y = b'u, \tag{65}$$

$$x'A'u - x'v = p'x. \tag{66}$$

Substituting $g = b'u$ and $f = p'x$, we find, after considering that for feasible solutions we have $u'y \geq 0$ and $v'x \geq 0$,

$$f \leq x'A'u \leq g, \tag{67}$$

which proves the first part of the theorem. If $u'y = 0$ and $v'x = 0$ as we have for corresponding primal and dual solutions, we find from combining (65) and (66):

$$f = x'A'u = g, \tag{68}$$

which proves the latter part of the theorem. If there are multiple optimal solutions to the primal and to the dual problem, the relations $u'y = 0$ and $v'x = 0$ are still valid for any pair of multiple primal and dual solutions, so that (68) is also valid in this case.

We shall now consider the case in which one of the problems has an infinite solution. Using the presentation given in tableau Ib of table 2, we assume that the primal problem has an infinite solution because an element of the vector $-p^{2'} + p^{1'}A_{11}^{-1}A_{12}$ is negative (the case in which an element of $p^{1'}A_{11}^{-1}$ is negative is similar), while the elements of the corresponding column in the matrices $A_{11}^{-1}A_{12}$ and $A_{22} - A_{21}A_{11}^{-1}A_{12}$ are all nonpositive. In the corresponding row of the dual solution in tableau IIb, the dual variable is negative and the elements in the columns of nonbasic variable are all nonnegative. This row may be written in equational form as follows:

$$\bar{p}_j = \ldots + \bar{a}_{ij}u_i + \ldots + \bar{a}_{ik}v_k + \ldots + v_j \tag{69}$$

with $\bar{p} < 0$ and the $\bar{a}$'s nonnegative. This equation is written as

$$\bar{p}_j - \ldots - \bar{a}_{ij}u_i - \ldots - \bar{a}_{ik}v_k = v_j ; \tag{70}$$

it is obvious that, because the $\bar{a}$'s are nonnegative, there is no nonnegative solution in the v's and the u's, which satisfies this equation. This means that the dual problem is unfeasible. Hence, if the primal problem has an infinite solution, the dual problem has no feasible solution. In the same way it can be shown that if the dual problem has an infinite solution, the primal problem has no feasible solution. It is also possible that neither problem has a feasible solution.

4.5. The dual method

The typical minimization problem is formulated as follows. Minimize

$$f = c'x \tag{71}$$

subject to

$$Ax \geq b, \tag{72}$$

$$x \geq 0. \tag{73}$$

It is assumed that the elements of c are nonnegative. The introduction of a vector of slack variables y leads to

$$Ax - y = b, \tag{74}$$

$$x, y \geq 0. \tag{75}$$

The basic solution of this problem with the y-variables as basic variables

$$f = 0,$$

$$y = -b, \tag{76}$$

is not feasible if b has some positive elements. On the other hand, eq. (71) indicates that for this solution the f-row coefficients have the right sign; if the solution was feasible, it would be the optimal solution.

We shall call solutions in which the coefficients of nonbasic variables in the f-row of the tableau have the right sign (≥ 0 in the case of maximization, ≤ 0 in the case of minimization) but which are not necessarily feasible, *optimal in the narrow sense*. Solutions which are really optimal (f-row coefficients have the right sign and the solution is feasible) will be called *optimal in the wide sense*.[1]

---

[1] It should be noted that the terms narrow and wide refer to the set of properties of the tableaux; if 1 stands for the property of a feasible basis solution and 2 for the property of having coefficients in the f-row of the right sign, then the optimality in the narrow sense consists of property 2 and optimality in the wide sense consists of properties 1 and 2. We could also consider the set of basic solutions and its subsets. Let A be the set of basic solutions with nonnegative f-row coefficients and B the set of basic solutions with nonnegative b's:

A ∩ B

Then A is the set of optimal solutions in the narrow sense and the intersection of A and B the set of solutions optimal in the wide sense. It is obvious then that the words narrow and wide refer to the set of properties and not to the sets of basic solutions.

Apparently typical minimization problems provide no natural starting point for the simplex method. However these problems do provide a natural starting point for an alternative linear programming method which is called *dual method*, which was discovered by Lemke[21]. The adjective "dual" is related to the fact that an application of the dual method to a problem is equivalent to an application of the simplex method to the dual of this problem. This will be explained later; first the dual (or, as it is sometimes called, the dual simplex) method will be explained. This method requires as an initial solution one which is optimal in the narrow sense, which means that the coefficients of the f-row for nonbasic variables should be nonpositive for a minimization problem and nonnegative for a maximization problem.

As an example for the explanation of the dual method the minimum-cost cattlefeed problem will be used, which may be represented as follows. Minimize

$$f = 24x_1 + 30x_2 + 40x_3 + 50x_4 \qquad (77)$$

subject to

$$12x_1 + 12x_2 + 40x_3 + 60x_4 \geq 20,$$
$$2x_1 + 6x_2 + 12x_3 + 2x_4 \geq 5, \qquad (78)$$

$$x_1, x_2, x_3, x_4 \geq 0. \qquad (79)$$

After introduction of the slack variables y_1 and y_2 the following equation system may be formulated

$$0 = -24x_1 - 30x_2 - 40x_3 - 50x_4 + f,$$
$$-20 = -12x_1 - 12x_2 - 40x_3 - 60x_4 \qquad + y_1,$$
$$-5 = -2x_1 - 6x_2 - 12x_3 - 2x_4 \qquad + y_2. \qquad (80)$$

The basic solution of this system is

$$f = 0,$$
$$y_1 = -20,$$
$$y_2 = -5. \qquad (81)$$

so that this solution is unfeasible. But this solution is optimal in the narrow sense since the coefficients of the x-variables in the first equation of (80) are all nonpositive.

The principle of the dual method is that the unfeasibility of a solution should be decreased by making a basic variable with a negative value, usually the basic variable with the most negative value, nonbasic in such a way that the optimality in the narrow sense is preserved. Hence first the leaving basic variable of an iteration is determined by selecting the basic variable with the most negative value. Then the new basic variable is determined in such a way that the coefficients in the f-row remain non-positive.

Let us consider the solution of (80). It can be interpreted as using none of the raw materials, resulting in a shortage of protein of 20 units and a shortage of fat of 5 units. These unfeasibilities will be eliminated one by one, starting with the largest, which is connected with the protein shortage, showing up in the equation

$$-20 = -12x_1 - 12x_2 - 40x_3 - 60x_4 + y_1. \tag{82}$$

y_1 can be increased from -20 to 0 by increasing a nonbasic variable with a negative coefficient. An increase of x_1 by 1 increases y_1 by 12, and the same increases in x_2, x_3, and x_4 increase it by 12, 40, and 60. However, while eliminating the unfeasibility, we wish to preserve the optimality in the narrow sense.

For the objective function we have the equation

$$0 = -24x_1 - 30x_2 - 40x_3 - 50x_4 + f. \tag{83}$$

Hence an increase in x_1 increases costs by 24, while the same increase in x_2, x_3 and x_4 increases costs by 30, 40, and 50. An obvious choice of the variable to be increased is that one which increases y_1 at least costs per unit of increase in y_1. For x_1 we find $\frac{24}{12} = 2$, and for x_2, x_3, and x_4, $\frac{30}{12} = 2\frac{1}{2}$, $\frac{40}{40} = 1$, and $\frac{50}{60} = \frac{5}{6}$. The minimum is $\frac{5}{6}$, which is connected with x_4. Hence, x_4 is the new basic variable, which replaces y_1, so that the pivot is -60. See tableau 0 of table 3.

More formally the rule for the new basic variable can be indicated as follows. Let the f-equation and that of the leaving basic variable be denoted as

$$c_0 = -c_1x_1 - c_2x_2 - \ldots - c_nx_n + f,$$
$$a_{r0} = a_{r1}x_1 + a_{r2}x_2 + \ldots + a_{rn}x_n + y_r. \tag{84}$$

The new basic variable is then determined by

$$\text{Min} \, (-c_j/a_{rj}|a_{rj} < 0). \tag{85}$$

Let the minimum be found in the kth column; then a_{rk} is used as a pivot. The next solution is then again optimal in the narrow sense, which is

Table 3

An application of the dual method to the minimum-cost cattlefeed problem

Tableau	Basic variables	Val. basic var.	x_1	x_2	x_3	x_4
0	f	0	-24	-30	-40	-50
	y_1	-20	-12	-12	-40	-60
	y_2	-5	-2	-6	-12	-2
	Ratio		2	$2\frac{1}{2}$	1	$\frac{5}{6}$
			x_1	x_2	x_3	y_1
1	f	$16\frac{2}{3}$	-14	-20	$-6\frac{2}{3}$	$-\frac{5}{6}$
	x_4	$\frac{1}{3}$	$\frac{1}{5}$	$\frac{1}{5}$	$\frac{2}{3}$	$-\frac{1}{60}$
	y_2	$-4\frac{1}{3}$	$-1\frac{3}{5}$	$-5\frac{3}{5}$	$-10\frac{2}{3}$	$-\frac{1}{30}$
	Ratio		$8\frac{3}{4}$	$3\frac{4}{7}$	$\frac{5}{8}$	25
			x_1	x_2	y_2	y_1
2	f	$19\frac{3}{8}$	-13	$-16\frac{1}{2}$	$-\frac{5}{8}$	$-\frac{13}{16}$
	x_4	$\frac{1}{16}$	$\frac{1}{10}$	$-\frac{3}{20}$	$\frac{1}{16}$	$-\frac{3}{160}$
	x_3	$\frac{13}{32}$	$\frac{3}{20}$	$\frac{21}{40}$	$-\frac{3}{32}$	$\frac{1}{320}$

shown as follows. The elements in the f-row of the next tableau are

$$c_j a_{rk}^{-1} \qquad j = k,$$
$$c_j - c_k a_{rk}^{-1} a_{rj} \qquad j \neq k. \tag{86}$$

The element in the kth column is nonpositive, since $-c_k \leq 0$ and $a_{rk} < 0$. For the other elements, let us first consider the case $a_{rj} \geq 0$. Since $-c_j \leq 0$, $-c_k \leq 0$, $a_{rk} < 0$ and $a_{rj} > 0$, the element is nonpositive. In the case $a_{rj} < 0$, let us rewrite the element as

$$a_{rj}[(-c_j/a_{rj}) - (-c_k/a_{rk})]. \tag{87}$$

The element within brackets is nonnegative because of (85); since $a_{rj} < 0$, the product must be nonpositive.

In tableau 1 of table 3, the elements in the f-row are indeed nonpositive. In the solution of this tableau, another unfeasibility is left, $y_2 = -4\frac{1}{3}$, so that this is the leaving basic variable for the next iteration. The new basic variable is chosen by comparison of the ratios

$$\text{Min}\ (-14/-1\tfrac{3}{5},\quad -20/-5\tfrac{3}{5},\quad -6\tfrac{2}{3}/-10\tfrac{2}{3},\quad -\tfrac{5}{6}/-\tfrac{1}{30}) = \tfrac{5}{8},$$

so that x_3 is the new basic variable. In tableau 2, the solution is feasible as well as optimal in the narrow sense, so that it is optimal in the wide sense.

It is interesting to compare the successive solutions of the dual method with those of the simplex method for the same problem; these were given in table 2 of the previous chapter. In a minimization problem the simplex method generates successive solutions in which the objective function approaches its optimal value from above; in the dual method the value of the objective function is increased towards its optimal value. Figure 1 gives the values of the objective function for the solutions of table 2 in the previous chapter and for those of table 6 of the present chapter. In a maximization problem the simplex method approaches the optimal value of the objective function from below, while the dual method does this from above. Figure 2 gives a typical pattern of iterations of both methods for a maximization problem. All this is in accordance with the statement proved before that in a maximization problem the value of the objective function for a feasible solution of the dual problem is at least as large as the value for a feasible solution of the primal problem and that these values are the same for optimal solutions to both problems.

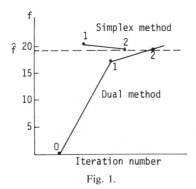

Fig. 1.

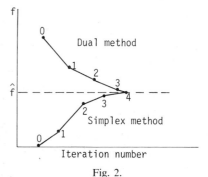

Fig. 2.

Note that the dual method does not preserve existing feasibility, since nonnegative basic variables may become negative in the course of the iterations. However, the dual method will eventually restore its feasibility if it has not come along of its own.

If there are no negative coefficients of nonbasic variables in the row of the leaving basic variable, no feasible solution exists, which is shown as follows. Let us rewrite the second equation of (84):

$$a_{r0} - a_{r1}x_1 - a_{r2}x_2 - \ldots - a_{rn}x_n = y_r. \tag{88}$$

If $a_{r0} < 0$ and $a_{r1}, a_{r2}, \ldots, a_{rn} \geq 0$, obviously no nonnegative values of $x_1, \ldots, x_n$ exist which can make y_r nonnegative.

The rules for the dual method may be stated as follows. We assume that the objective function is to be minimized and that the set-up tableau as well as any subsequent system may be represented as

$$b_0 = -c'x + f,$$
$$b = Ax + y. \tag{89}$$

x stands for the vector of nonbasic variables, y for that of basic variables. Since we have an optimal solution in the narrow sense, the elements of the vector $-c'$ are all nonpositive, but the elements of b may be nonnegative.

SELECTION OF THE LEAVING BASIC VARIABLE: *Select the variable connected with*

$$\operatorname*{Min}_{i}(b_i | b_i < 0). \tag{90}$$

If the elements of b are nonnegative, a feasible as well as an optimal solution has been found.

It is not necessary to take as leaving basic variable the one with the most negative value, but it is an obvious choice. Let the leaving basic variable be that of the rth row.

SELECTION OF THE NEW BASIC VARIABLE: *Select the variable connected with*

$$\operatorname*{Min}_{j}(-c_j/a_{rj} | a_{rj} < 0). \tag{91}$$

If $a_{rj} \geq 0$ for each j, the problem has no feasible solution. Let the new basic variable be that in the kth column.

TRANSFORMATION OF THE TABLEAU: *Transform the tableau with a_{rk} as a pivot.*

Finally, we wish to show that an application of the dual method to the primal problem and an application of the simplex method to the dual problem are equivalent. The system of the dual problem of the problem in (89) is

$$b_0 = b'u + g,$$
$$c = -A'u + v. \tag{92}$$

The basic solution of (92) corresponds with the basic solution of (89), since the basic v-variables correspond with the nonbasic x-variables in (89) and the nonbasic u-variables correspond with the basic y-variables in (89). Note that (92) is a maximization problem.

In the dual method the variable y_r with a value $b_r = \underset{i}{\text{Min}}\,(b_i | b_i < 0)$ leaves the basis and the x-variable x_k with

$$- c_k / a_{rk} = \underset{j}{\text{Min}}\,(- c_j / a_{rk} | a_{rj} < 0) \tag{93}$$

enters the basis. In the simplex method applied to (92) the u-variable with $b_r = \text{Min}\,(b_i | b_i < 0)$ enters the basis and the v-variable corresponding with

$$c_k / - a_{rk} = \underset{i}{\text{Min}}\,(c_i / - a_{ri} | - a_{ri} > 0) \tag{94}$$

leaves it. Hence in the dual method the variable y_r leaves the basis, while the corresponding variable u_r enters the basis in the simplex method applied to (92); in the dual method x_k enters the basis, while in the simplex method applied to (92) the corresponding variable v_k leaves the basis. This means that the systems after the transformations again must be equivalent.

Though the dual method usually is applied to unfeasible solutions in which the unfeasibility takes the form of negative values of basic variables, it can also be applied for solutions containing artificial variables. Suppose that a solution which is optimal in the narrow sense has an artificial basic variable z_i in the sth row with a positive value b_s. The equations of the f-row and that of the artificial variable can be written as

$$b_0 = - c_1 x_1 - c_2 x_2 - \ldots - c_n x_n + f,$$
$$b_s = \quad a_{s1} x_1 + a_{s2} x_2 + \ldots + a_{sn} x_n \quad + z_i; \tag{95}$$

Suppose that $b_s > 0$. z_i can be made zero by increasing some nonbasic variable x_j with $a_{sj} > 0$. Again the choice of the new basic variable should be such as to increase the objective function least, so that it is determined by

$$\underset{j}{\text{Min}}\,(c_j / a_{sj} | a_{sj} > 0). \tag{96}$$

It is easy to prove that this choice leads to a solution which again is optimal in the narrow sense.

In the dual method extended in this manner any nonartificial basic variable with a negative value or any artificial variable irrespective of the sign of its value may be taken as the leaving basic variable. To make a definite choice, we may take among these variables the one with the largest absolute value.

As an example, the minimum-cost cattlefeed is considered in which a feed is required that contains at least 20 units of protein and 5 units of fat *per unit of feed*. This problem can be formulated as in (77)–(79) adding to it the constraint

$$x_1 + x_2 + x_3 + x_4 = 1. \tag{97}$$

After insertion of the artificial variable z this equation becomes

$$x_1 + x_2 + x_3 + x_4 + z = 1. \tag{98}$$

Tableau 0 of table 4 gives the set-up tableau for this problem. The first two iterations take place as in table 3, but in tableau 2 it is found that z should leave the basis. The new basic variable is determined using (96). In tableau 4 the optimal solution is found.

This solution has $f = 28\frac{3}{7}$ against $19\frac{3}{8}$ in table 3, which can be imputed to the addition of the constraint (97). The coefficient of z in the f-row of tableau 4, $20\frac{4}{7}$, is the dual variable of constraint (97). This dual variable corresponds with z, and z can be interpreted as the amount of a kind of filling material, containing no protein and fat, which can be mixed with the feed (see (98)). Such a filling material, if it were available, would be worth $20\frac{4}{7}$ per unit in this situation.

The convergence of the dual method to a basic solution which is optimal in the wide sense, rests on the monotonic increase of the objective function in the minimization case. If f increases in each iteration, no solution can recur. However, if $-c_k = 0$, the objective function remains the same, so that in principle cycling can occur. An anticycling method is in this case based on a perturbation of the coefficients of the objective function. We shall not go into details, but an example is given, both of cycling and of application of an anti-cycling method in table 5. The problem is the dual problem of the one used for demonstration of cycling and an anti-cycling method in the simplex method.

Talacko and Rockafeller (see [30]) have proposed a procedure which is called the *symmetric method*, in which steps of the simplex method and the dual method are mixed. The procedure starts with a basic solution, which does not have to be either primally or dually feasible. In a maximization problem, first that pivot is determined which gives the greatest increase in the objective function while preserving existing primal feasibility. After that, the pivot is determined which gives the greatest

Table 4

An application of the dual method to a problem containing artificial variables

	Basic variables	Values basic variables	x_1	x_2	x_3	x_4
	f	0	-24	-30	-40	-50
	y_1	-20	-12	-12	-40	-60
0	y_2	-5	-2	-6	-12	-2
	z	1	1	1	1	1
	Ratio		24	30	40	50

			x_1	x_2	x_3	y_1
	f	$16\frac{2}{3}$	-14	-20	$-6\frac{2}{3}$	$-\frac{5}{6}$
	x_4	$\frac{1}{3}$	$\frac{1}{5}$	$\frac{1}{5}$	$\frac{2}{3}$	$-\frac{1}{60}$
1	y_2	$-4\frac{1}{3}$	$-1\frac{3}{5}$	$-5\frac{3}{5}$	$-10\frac{2}{3}$	$-\frac{1}{30}$
	z	$\frac{2}{3}$	$\frac{4}{5}$	$\frac{4}{5}$	$\frac{1}{3}$	$\frac{1}{60}$
	Ratio		$8\frac{3}{4}$	$3\frac{4}{7}$	$\frac{5}{8}$	25

			x_1	x_2	y_2	y_1
	f	$19\frac{3}{8}$	-13	$-16\frac{1}{2}$	$-\frac{5}{8}$	$-\frac{13}{16}$
	x_4	$\frac{1}{16}$	$\frac{1}{10}$	$-\frac{3}{20}$	$\frac{1}{16}$	$-\frac{3}{160}$
2	x_3	$\frac{13}{32}$	$\frac{3}{20}$	$\frac{21}{40}$	$-\frac{3}{32}$	$\frac{1}{320}$
	z	$\frac{17}{32}$	$\frac{3}{4}$	$\frac{5}{8}$	$\frac{1}{32}$	$\frac{1}{64}$
	Ratio		$17\frac{1}{3}$	$26\frac{2}{5}$	20	40

			z	x_2	y_2	y_1
	f	$28\frac{7}{12}$	$17\frac{1}{3}$	$-5\frac{2}{3}$	$-\frac{1}{12}$	$-\frac{13}{24}$
	x_4	$-\frac{1}{120}$	$-\frac{2}{15}$	$-\frac{7}{30}$	$\frac{7}{120}$	$-\frac{1}{48}$
3	x_3	$\frac{3}{10}$	$-\frac{1}{5}$	$\frac{2}{5}$	$-\frac{1}{10}$	0
	x_1	$\frac{17}{24}$	$1\frac{1}{3}$	$\frac{5}{6}$	$\frac{1}{24}$	$\frac{1}{48}$
	Ratio			$24\frac{2}{7}$		26

			z	x_4	y_2	y_1
	f	$28\frac{3}{7}$	$20\frac{4}{7}$	$-24\frac{2}{7}$	$-1\frac{1}{2}$	$-\frac{1}{28}$
	x_2	$\frac{1}{28}$	$\frac{4}{7}$	$-4\frac{2}{7}$	$-\frac{1}{4}$	$\frac{5}{56}$
4	x_3	$\frac{2}{7}$	$-\frac{3}{7}$	$1\frac{5}{7}$	0	$-\frac{1}{28}$
	x_1	$\frac{19}{28}$	$\frac{6}{7}$	$3\frac{4}{7}$	$\frac{1}{4}$	$-\frac{3}{56}$

Table 5

Example of cycling in the dual method

Tableau	Basic variables	Values basic variables	x_1	x_2	x_3
0	f	0	0	0	-1
	y_1	$-\frac{3}{4}$	$-\frac{1}{4}$	$-\frac{1}{2}$	0
	y_2	20	8	12	0
	y_3	$-\frac{1}{2}$	1	$\frac{1}{2}$	-1
	y_4	6	-9	-3	0

			y_1	x_2	x_3
1	f	0	0	0	-1
	x_1	3	-4	2	0
	y_2	-4	32	-4	0
	y_3	$-3\frac{1}{2}$	4	$-1\frac{1}{2}$	-1
	y_4	33	-36	15	0

			y_1	y_2	x_3
2	f	0	0	0	-1
	x_1	1	12	$\frac{1}{2}$	0
	x_2	1	-8	$-\frac{1}{4}$	0
	y_3	-2	-8	$-\frac{3}{8}$	-1
	y_4	18	84	$3\frac{3}{4}$	0

			y_3	y_2	x_3
3	f	0	0	0	-1
	x_1	-2	$1\frac{1}{2}$	$-\frac{1}{16}$	$-1\frac{1}{2}$
	x_2	3	-1	$\frac{1}{8}$	1
	y_1	$\frac{1}{4}$	$-\frac{1}{8}$	$\frac{3}{64}$	$\frac{1}{8}$
	y_4	-3	$10\frac{1}{2}$	$-\frac{3}{16}$	$-10\frac{1}{2}$

			y_3	y_4	x_3
4	f	0	0	0	-1
	x_1	-1	-2	$-\frac{1}{3}$	2
	x_2	1	6	$\frac{2}{3}$	-6
	y_1	$-\frac{1}{2}$	$2\frac{1}{2}$	$\frac{1}{4}$	$-2\frac{1}{2}$
	y_2	16	-56	$-5\frac{1}{3}$	56

Table 5 (*continued*)

Tableau	Basic variables	Values basic variables	x_1	y_4	x_3
5	f	0	0	0	-1
	y_3	$\frac{1}{2}$	$-\frac{1}{2}$	$\frac{1}{6}$	-1
	x_2	-2	3	$-\frac{1}{3}$	0
	y_1	$-1\frac{3}{4}$	$1\frac{1}{4}$	$-\frac{1}{6}$	0
	y_2	44	-28	4	0

			x_1	y_1	x_3
1a	f	0	0	0	-1
	x_2	$1\frac{1}{2}$	$\frac{1}{2}$	-2	0
	y_2	2	2	24	0
	y_3	$-1\frac{1}{4}$	$\frac{3}{4}$	1	-1
	y_4	$10\frac{1}{2}$	$-7\frac{1}{2}$	-6	0

			x_1	y_1	y_3
2a	f	$1\frac{1}{4}$	$-\frac{3}{4}$	-1	-1
	x_2	$1\frac{1}{2}$	$\frac{1}{2}$	-2	0
	y_2	2	2	24	0
	x_3	$1\frac{1}{4}$	$-\frac{3}{4}$	-1	-1
	y_4	$10\frac{1}{2}$	$-7\frac{1}{2}$	-6	0

decrease in the objective function while preserving existing dual feasibility. Of the two resulting pivots, that one is chosen which gives the greatest change in the objective function.

This method has obviously the greatest change variants of the simplex method and the dual method as special cases. If the starting solution is neither primally or dually feasible, the method does not necessarily terminate, but for the problems used in [30] it did. In view of the fact that the symmetric method is more general than the greatest change variant of the simplex method, it could be more efficient than the latter method; instead, it was found to be less efficient in the computational experiments in [46].

4.6. Dual variables and Lagrangean multipliers

Let us again consider the typical maximization problem: Maximize

$$f = p'x \tag{99}$$

subject to

$$Ax \leq b, \tag{100}$$

$$x \geq 0. \tag{101}$$

A is an $m \times n$ matrix.

Let us now apply the classical maximization technique using the Lagrangean multipliers. The difficulty is what to do with the inequalities. If they are ignored it is obvious that an infinite solution is found if any element of p is nonzero. The inequalities can be classified according to whether they are binding for the optimal solution or not. An inequality is called *binding* for a solution if this solution satisfies this inequality with an equality sign; for the nonnegativity constraints in (101) this means that a constraint is binding if the variable concerned is zero in the optimal solution. Hence, if we knew which of the nonnegativity constraints are binding, the classical maximization technique could be applied.

Let us for a moment write all inequalities of (100) and (101) as equations. Then the following Lagrangean expression may be written:

$$\phi = p'x - u'(Ax - b) - v'(-x); \tag{102}$$

u is a vector of Lagrangean multipliers belonging to the constraints of (100) and v is the vector of Lagrangean multipliers belonging to the constraints of (101). For the constraints which are satisfied with a strict inequality, the corresponding multipliers can be taken as zero. Taking derivatives of ϕ with respect to the x-variables and putting these at zero, we find

$$\partial \phi / \partial x = p - A'u + v = 0 \tag{103}$$

or

$$A'u - v = p. \tag{104}$$

These are the constraints of the dual problem. After putting the u- and v-variables connected with strict inequalities at zero, the system (104) may be solved for the remaining variables. The primal variables are solved directly from the equalities in (100) and (101).

From (104) we may conclude that the dual variables may be interpreted as Lagrangean multipliers connected with inequality constraints. The u-variables are connected with the constraints (100) or, which is the same, with the y-variables and the v-variables are connected with constraints $x \geq 0$. The property of complementary slackness may also be explained using Lagrangean multipliers. If a nonnegativity constraint is binding, its variable has a value zero, but it will have a multiplier which may have nonzero values; on the other hand, if a constraint is not taken as a binding, the variable concerned may take a nonzero value, but there is no multiplier.

In the case in which the original problem has equalities instead of inequalities in (100), none of the u-variables should be put at zero, so that all constraints are included in the Lagrangean (102) for any solution.

The linear programming problem could be solved using the Lagrangean multiplier technique by computing the solutions belonging to all possible combinations of binding constraints and taking the feasible solution with the highest value of the objective function. Such a method could be called a *complete enumeration method*. The number of solutions generated in this manner would be very high, so that it would be useless for all problems but those of the smallest size. It is the virtue of the simplex method and related methods that it computes only a small number of basic solutions, eliminating at each step a great number of other basic solutions, which are in the simplex method those with a lower value of the objective function.

4.7. Dual upper and lower bounds

Let us consider the typical maximization problem: Maximize f

$$f = p'x,$$
$$Ax \le b,$$
$$x \ge 0, \tag{105}$$

and its dual problem: Minimize g

$$g = b'u,$$
$$A'u \ge p,$$
$$u \ge 0. \tag{106}$$

The meaning of lower and upper bounds $\underline{x} \le x \le \bar{x}$ of the primal problem is obvious and methods to deal with these constraints in a convenient way in the simplex method are well known. Now let us add the lower and upper bounds:

$$\underline{u} \le u \le \bar{u} \tag{107}$$

to the dual problem. The first question is whether any meaning can be attached to such constraints and the second is whether these constraints and their equivalents in the primal problem can be conveniently handled in the simplex and the dual method.

The dual problem with the additional constraints can be written as: Maximize g

$$g = b'u,$$
$$A'u \geq p,$$
$$u \geq \underline{u},$$
$$-u \geq -\bar{u},$$
$$u \geq 0. \tag{108}$$

The corresponding primal problem must be: Maximize f

$$f = p'x + \underline{u}'y - \bar{u}'z,$$
$$Ax + y - z \leq b,$$
$$x, y, z \geq 0. \tag{109}$$

The y- and z-variables correspond with the lower and upper bounds in the dual problem. The y-variables can be interpreted as the amounts of the resources not used by the firm in the production process, but sold at prices given by the vector $\underline{u}$; the z-variables can be interpreted as the amounts of the resources bought at prices given by the vector $\bar{u}$. Variables like these should occur in many practical cases.

Since the y- and z-variables correspond with upper bounds in the dual problem, it should be possible to develop a technique which does not require an explicit introduction of these variables in the tableaux. We should distinguish two cases, firstly the case in which the simplex method is used, secondly the case in which the dual method is used for the solution of the primal problem. The latter case must be the more familiar one, since the application of the dual method to the primal problem is equivalent to an application of the simplex method to the dual problem. Hence we shall first treat this case.

For an application of the dual method, a typical minimization problem is suitable. The typical minimization problem is: Minimize f

$$f = c'x,$$
$$Ax \geq b,$$
$$x \geq 0. \tag{110}$$

The dual problem is: Maximize g

$$g = b'u,$$
$$A'u \le c,$$
$$u \ge 0. \tag{111}$$

To this problem the dual constraints are added:

$$\underline{u} \le u \le \bar{u}. \tag{112}$$

The corresponding primal problem is: Minimize f

$$f = c'x - \underline{u}'y + \bar{u}'z,$$
$$Ax - y + z \ge b,$$
$$x, y, z \ge 0. \tag{113}$$

This problem can be interpreted as a cattlefeed problem in which also amounts of pure protein, etc., can be sold (the y-variables) and bought (the z-variables).

Let us consider the following cattlefeed problem: Minimize f

$$f = 24x_1 + 30x_2 + 40x_3 + 50x_4 + \tfrac{3}{4}z_1 + z_2,$$
$$12x_1 + 12x_2 + 40x_3 + 60x_4 + z_1 \ge 20,$$
$$2x_1 + 6x_2 + 12x_3 + 2x_4 + z_2 > 5,$$
$$x_1, x_2, x_3, x_4, z_1, z_2 \ge 0. \tag{114}$$

This means that pure protein can be bought for $\tfrac{3}{4}$ monetary units per unit of protein and pure fat for 1. For this problem the dual method can be used, but we want to do this without explicitly having the z-variables in the tableau; this corresponds to an application of the simplex method to the dual problem without explicitly having the constraints $u_1 \le \tfrac{3}{4}$, $u_2 \le 1$ in the tableau. One way to find the method would be to set up the corresponding dual problem, apply the simplex method with the upper bounds technique to u and make the corresponding steps in the primal tableau. However, it is somewhat more interesting to find out the corresponding method directly from the primal tableau.

The first three rows of tableau 0 of table 6 give the set-up tableau with a solution which is optimal in the narrow sense but not feasible. As is usual in the dual method, the basic variable with the most negative value leaves the basis; in this case it is y_1. Now the variable to enter the basis should be selected in such a way that the optimality in the narrow sense is preserved. This means that none of the nonbasic y-variables should get

Table 6

The cattlefeed problem with dual upper bounds

Tableau	Basic variables	Value basic variables	x_1	x_2	x_3	x_4
0	f	0	-24	-30	-40	-50
	Ratio		2	$2\frac{1}{2}$	1	$\frac{5}{6}$
	f^*	15	-15	-21	-10	-5
	Ratio		$7\frac{1}{2}$	$3\frac{1}{2}$	$\frac{5}{6}$	$2\frac{1}{2}$
	$y_1\|y_1^*$	-20	-12	-12	-40	-60
	y_2	-5	-2	-6	-12	-2
			x_1	x_2	y_2	x_4
1	f	$19\frac{1}{6}$	$-13\frac{1}{3}$	-16	$-\frac{5}{6}$	$-3\frac{2}{3}$
	y_1^*	$-3\frac{1}{3}$	$-8\frac{2}{3}$	8	$-3\frac{1}{3}$	$-53\frac{1}{3}$
	x_3	$\frac{5}{12}$	$\frac{1}{6}$	$\frac{1}{2}$	$-\frac{1}{12}$	$\frac{1}{6}$

shadow prices exceeding the prices at which the pure nutrients could be bought, because if the shadow price exceeds this buying price, it is better to buy the pure nutrient instead of the raw materials. Hence, in addition to taking the usual ratios we should also pay attention to the elements in the f-row in columns of nonbasic y-variables. In tableau 0 there are no nonbasic y-variables, so that we only consider the usual ratios which are found in the additional row. The minimum ratio is $\frac{5}{6}$ in the x_4-column. But this $\frac{5}{6}$ would be the minimum premium we have to put on protein in order to allow an ingredient, in this case the third, to fulfil the protein requirement. Now $\frac{5}{6} > \frac{3}{4}$, so that it is better to buy the pure protein at $\frac{3}{4}$. Hence we buy the 20 units required protein, but in order to account for that, we change y_1 into y_1^* and subtract $\frac{3}{4}$ times its row from the f-row to obtain the row f^*. All elements in the f^*-row should remain nonnegative. The value of the objective function is $\frac{3}{4} \times 20 = 15$.

In this technique starred y-variables give minus the amount of pure nutrient which is bought; they should be negative. If starred y-variables are positive, this means that a negative quantity of pure nutrient is being bought, which means an unfeasibility which should be undone by the dual method in the following steps. Note that in the step in which y_1^*

replaced y_1, no transformation of the tableau was necessary. If the variable z_1 had been explicitly included in the tableau, z_1 would have entered the basis replacing y_1.

Next we go on to tackle the fat deficiency; y_2 should leave the basis. Taking the usual ratios we find that the minimum one is $\frac{5}{6}$ for x_3. Hence $\frac{5}{6}$ should be the shadow price of fat. This is less than the price per pure fat which is 1. Since there are no nonbasic y-variables, there is no possibility that their shadow price exceeds their upper bounds. Hence x_3 should enter the basis. Pivoting on -12, tableau 1 is found, which is feasible and hence optimal.

It is not difficult to devise explicit rules for the dual method with dual upper bounds; they are of course entirely analogous to the rules of the simplex method with primal upper bounds.

So far we did not consider dual lower bounds, which can be interpreted as selling prices for pure nutrients in the cattlefeed problem. Let us assume that these are $\frac{1}{4}$ for protein and $\frac{1}{2}$ for fat. This means that the cost of barley is not 24 but $24 - \frac{1}{4}$, $12 - \frac{1}{2} \times 2 = 20$, and so on. The value of the objective function in the initial tableau is not 0 but $\frac{1}{4} \times 20 + \frac{1}{2} \times 5 = 7\frac{1}{2}$. The same results may be obtained by subtracting $\frac{1}{4}$ times the y_1 row and $\frac{1}{2}$ times the y_2 row from the f-row. We may then proceed as indicated before, but should express upper bounds as deviations from their lower bounds.

We shall now consider the more interesting case of the simplex method with dual upper bounds. As an example we take the following production planning problem. Maximize f

$$f = 3x_1 + x_2 + 2x_3 + 2x_4 - \tfrac{3}{4}z_1 - \tfrac{3}{4}z_2 - \tfrac{3}{4}z_3$$
$$4x_1 + 3x_2 + 2x_3 \qquad - z_1 \leq 100,$$
$$x_1 \qquad + x_3 + 2x_4 - z_2 \leq 40,$$
$$2x_1 + 4x_2 + x_3 + 2x_4 - z_3 \leq 60. \qquad (115)$$

The z-variables can be interpreted as additional amounts of the three resources, which can be bought for 0.75 per unit. We should develop for this problem a variant of the simplex method which does not explicitly introduce the z-variables.

The rows 1, 3, 4, and 5 of tableau 0 of table 7 give the set-up tableau of the problem without the z-variables. The simplex method is initiated by taking x_1 into the basis. y_1 is the leaving basic variable. If y_1 is leaving the basis its shadow price will be $\frac{25}{32}$. This exceeds the price at which addi-

Table 7

The simplex method with dual upper bounds

Tableau	Basic variables	Values basic variables	x_1	x_2	x_3	x_4
0	f	0	$-3\frac{1}{8}$	-1	-2	-2
	f^*	75	$-\frac{1}{8}$	$1\frac{1}{4}$	$-\frac{1}{2}$	-2
	$y_1\mid y_1^*$	100	4	3	2	0
	y_2	40	1	0	1	2
	y_3	60	2	4	1	2

Tableau	Basic variables	Values basic variables	y_3	x_2	x_3	x_4
1	f	$78\frac{3}{4}$	$\frac{1}{16}$	$1\frac{1}{2}$	$-\frac{7}{16}$	$-1\frac{7}{8}$
	y_1^*	-20	-2	-5	0	-4
	y_2	10	$-\frac{1}{2}$	-2	$\frac{1}{2}$	1
	x_1	30	$\frac{1}{2}$	2	$\frac{1}{2}$	1

Tableau	Basic variables	Values basic variables	y_3	x_2	x_3	$y_1^*\mid y_1$
2	f	$88\frac{1}{8}$	1	$3\frac{27}{32}$	$-\frac{7}{16}$	$-\frac{15}{32}\mid\frac{9}{32}$
	f^*	$91\frac{7}{8}$	$\frac{1}{4}$	$1\frac{13}{32}$	$-\frac{1}{16}$	$\frac{15}{32}$
	x_4	5	$\frac{1}{2}$	$1\frac{1}{4}$	0	$-\frac{1}{4}$
	$y_2\mid y_2^*$	5	-1	$-3\frac{1}{4}$	$\frac{1}{2}$	$\frac{1}{4}$
	x_1	25	0	$\frac{3}{4}$	$\frac{1}{2}$	$\frac{1}{4}$

Tableau	Basic variables	Values basic variables	y_3	x_2	x_1	y_1
3	f	95	$\frac{1}{4}$	$1\frac{1}{2}$	$\frac{1}{8}$	$\frac{1}{2}$
	x_4	5	$\frac{1}{2}$	$1\frac{1}{4}$	0	$-\frac{1}{4}$
	y_2^*	-20	-1	-4	-1	0
	x_3	50	0	$1\frac{1}{2}$	2	$\frac{1}{2}$

tional amounts of the first resource can be bought, which is $\frac{3}{4}$ or $\frac{24}{32}$. Hence we may give the first resource a shadow price of $\frac{3}{4}$ and let y_1 become negative, which means that additional quantities of this resource are being bought. Since the first resource costs $\frac{3}{4}$ per unit, we add $\frac{3}{4}$ times the row of y_1 to the f-row, which results in the f^*-row. To y_1 a star is attached. Now the ratio in the y_1^*-row can be ignored since y_1^* may become negative, which is interpreted as buying the resource. The next smallest ratio is that in y_3. If x_1 replaces y_3, the shadow price of the third resource will become $\frac{1}{16}$, which is smaller than its buying price $\frac{3}{4}$; hence we do not buy additional quantities of the third resource and x_1 replaces y_3 in the normal way. Tableau 1 results.

Proceeding with the simplex method, we select x_4 as the new basic variable. When selecting the leaving basic variable, we should take care that y_1^* does not become positive. In starred rows, ratios should therefore be taken for negative elements in the column of the new basic variable. The minimum ratio is found in the row of y^*. A formal pivot operation takes place, and tableau 2 results with y_1^* as a nonbasic variable. Now the shadow price of y_1^* was increased in tableau 0 from 0 to $\frac{3}{4}$. In order to undo this, we should subtract $\frac{3}{4}$ from the shadow price, which means that $\frac{3}{4}$ should be added to the element in the f-row. Accordingly $-\frac{15}{32}$

Table 8

The dual method with primal upper bounds

Tableau	Basic variables	Values basic variables				
		(1)	(2)	u_1^*	u_2	u_3
0	f	0	75	-100	-40	-60
	v_1	$-3\frac{1}{8}$	$-\frac{1}{8}$	-4	-1	-2
	v_2	-1	$1\frac{1}{4}$	-3	0	-4
	v_3	-2	$\frac{1}{2}$	-2	-1	-1
	v_4	-2	-2	0	-2	-2
				u_1^*	u_2	v_1
1	f	$78\frac{3}{4}$		20	-10	-30
	u_3	$\frac{1}{16}$		2	$\frac{1}{2}$	$-\frac{1}{2}$
	v_2	$1\frac{1}{2}$		5	2	-2
	v_3	$-\frac{7}{16}$		0	$-\frac{1}{2}$	$-\frac{1}{2}$
	v_4	$1\frac{7}{8}$		4	-1	-1
				v_4	$u_2\vert u_2^*$	v_1
2	f	$88\frac{1}{8}$	$91\frac{7}{8}$	-5	-5	-25
	u_3	1	$\frac{1}{4}$	$-\frac{1}{2}$	1	0
	v_2	$3\frac{27}{32}$	$1\frac{13}{32}$	$-1\frac{1}{4}$	$3\frac{1}{4}$	$-\frac{3}{4}$
	v_3	$-\frac{7}{16}$	$\frac{1}{16}$	0	$-\frac{1}{2}$	$-\frac{1}{2}$
	$u_1^*\vert u_1$	$-\frac{15}{32}\,\frac{9}{32}$	$\frac{15}{32}$	$\frac{1}{4}$	$-\frac{1}{4}$	$-\frac{1}{4}$
				v_4	u_2^*	v_3
3	f	95		-5	20	-50
	u_3	$\frac{1}{4}$		$-\frac{1}{2}$	1	0
	v_2	$1\frac{1}{2}$		$-1\frac{1}{4}$	4	$-1\frac{1}{2}$
	v_1	$\frac{1}{8}$		0	1	-2
	u_1	$\frac{1}{2}$		$\frac{1}{4}$	0	$-\frac{1}{2}$

is changed into $\frac{9}{32}$ and the star in y_1^* is deleted. The next variable to enter the basis is x_3. The variable to leave the basis is y_2, but it is found that its shadow price in the new tableau, which would be 1, exceeds the buying price $\frac{3}{4}$. Hence y_2 is changed into y_2^*, $\frac{3}{4}$ times the row of y_2, is added to the f-row, which results in the f^*-row. The next smallest ratio is in the row of x_1, which has no dual upper bound. Tableau 3 gives the optimal solution, which implies that 20 units of the second resource are bought. It may be checked that the value of the objective function is correct:

$$5 \times 2 + 50 \times 2 - 20 \times \tfrac{3}{4} = 95.$$

Table 9

The simplex method with dual lower and upper bounds

Tableau	Basic variables	Values basic variables	x_1	x_2	x_3	x_4
	f	0	$-3\frac{1}{8}$	-1	-2	-2
	f^*	50	$-1\frac{3}{8}$	$\frac{3}{4}$	-1	-1
0	y_1	100	4	3	2	0
	y_2	40	1	0	1	2
	y_3	60	2	4	1	2
			y_1	x_2	x_3	x_4
	f	$84\frac{3}{8}$	$\frac{11}{32}$	$1\frac{25}{32}$	$-\frac{5}{16}$	-1
1	x_1	25	$\frac{1}{4}$	$\frac{3}{4}$	$\frac{1}{2}$	0
	y_2	15	$\frac{1}{4}$	$-\frac{3}{4}$	$\frac{1}{2}$	2
	y_3	10	$-\frac{1}{2}$	$2\frac{1}{2}$	0	2
			y_1	x_2	x_3	y_3
	f	$89\frac{3}{8}$	$3\frac{1}{32}$	$\frac{3}{32}$	$-\frac{5}{16}$	$\frac{1}{2}$
	f^*	$91\frac{7}{8}$	$\frac{7}{32}$	$1\frac{13}{32}$	$-\frac{1}{16}$	0
2	x_1	25	$\frac{1}{4}$	$\frac{3}{4}$	$\frac{1}{2}$	0
	y_2	5	$\frac{1}{4}$	$-3\frac{1}{4}$	$\frac{1}{2}$	-1
	x_4	5	$-\frac{1}{4}$	$1\frac{1}{4}$	0	$\frac{1}{2}$
			y_1	x_2	x_1	y_3
	f	95	$\frac{1}{4}$	$1\frac{1}{2}$	$\frac{1}{8}$	0
3	x_3	50	$\frac{1}{2}$	$1\frac{1}{2}$	2	0
	y_2^*	-20	0	-4	-1	-1
	x_4	5	$-\frac{1}{4}$	$1\frac{1}{4}$	0	$\frac{1}{2}$

It is interesting to compare this with the corresponding method of the dual problem, which amounts to the dual method with primal upper bounds. Table 8 gives these solutions. In tableau 0, v_1 leaves the basis and u_1 should enter it. But if u_1 replaces v_1 as basic variables, u_1 exceeds its upper bound $\frac{3}{4}$. Hence u_1 is put at its upper bound, so that this variable is no longer a candidate for the new basic variable. Then u_3 is found as the new basic variable which now replaces v_1 without complications. In tableau 1 v_4 should leave the basis and it is found that u_1^* is the new basic variable; there are no complications since the value of u_1^* is $-\frac{15}{32}$, which is less than $-\frac{3}{4}$. After pivoting tableau 2 is found in which we change over from u_1^* to u_1. Now v_3 is the leaving basic variable. u_2 should be the new basic variable, but its value would be $\frac{7}{8}$, which exceeds $\frac{3}{4}$, so that u_2 is put at its upper bound. v_1 is then the new basic variable. Tableau 3 results, which is the optimal tableau.

We shall now deal with the case of the simplex method with dual lower and upper bounds. The same maximization problem is used, but now we have the lower bounds $\underline{u}_1 = \frac{1}{4}$, $\underline{u}_2 = \frac{1}{4}$, $\underline{u}_3 = \frac{1}{4}$. This means that any spare capacity can be sold for $\frac{1}{4}$ per unit. Hence, we should never use slack variables but always use the resource selling variables instead. In the set-up tableau this amounts to adding to each of the elements in the f-row the sum of the other elements times $\frac{1}{4}$, in the f^*-row in tableau 2 of table 9. Now the simplex procedure starts as in the other case, but the upper bounds of the shadow prices of the y-variables are not $\frac{3}{4}$ but $\frac{1}{2}$. The final solution of the problem is the same as before.

CHAPTER 5

SENSITIVITY ANALYSIS AND PARAMETRIC PROGRAMMING

5.1. Sensitivity analysis of the objective function

An obvious question arising after a linear programming problem has been solved is: what happens to the solution of the problem when there is a change in coefficients? This question will be answered in this chapter. We consider different kinds of changes. Firstly, we may consider changes around the given values of the coefficients which are not large enough to affect the feasibility or the optimality of the basic solution. The analysis of these changes is called *sensitivity analysis*. Secondly, we may consider changes in the coefficient in a given range which changes the set of basic variables of the optimal solution; this is called *parametric programming*.

Sensitivity analysis and parametric programming may further be distinguished according to which coefficients of the problem change. Firstly, changes in the objective function coefficients may be considered, secondly, those of the constant terms of the constraints and finally those of variables in the constraints. Altogether, six cases may be distinguished.

The first case is the sensitivity analysis of the coefficients in the objective function. Let us take the objective function of the example used for explanation of the simplex method:

$$f = 3x_1 + x_2 + \tfrac{1}{2}x_3 + 2x_4 + 2x_5. \tag{1}$$

For each of the coefficients the question may be asked: what is the effect of a change in this coefficient on the optimal solution and between which boundaries may these values vary without changing the set of variables in the basis?

Let us consider a general linear programming problem of the form: Maximize

$$f = p'x$$

subject to

$$Ax \le b,$$

$$x \ge 0,$$

and let the first q x-variables and the last $m-q$ y-variables be basic in the

125

optimal solution. The variables and the constraints can always be re-arranged in such a manner that this is true. A, p, b, x and y may then be partitioned accordingly, see tableau I of table 1. The optimal solution is then generated by block-pivoting on A_{11}, which results in tableau II.

In this tableau, the value of f and the coefficients in the f-row for nonbasic variables depend on the elements of p. For the value of f we have

$$f = p^{1\prime} A_{11}^{-1} b^1 = p_1 \hat{x}_1 + p_2 \hat{x}_2 + \ldots + p_q \hat{x}_q, \tag{2}$$

where $\hat{x}_1, \ldots, \hat{x}_q$ indicate the values of the basic x-variables in the optimal solution. The interpretation is obvious.

Let us now consider changes in p^2, the coefficients of x-variables which are nonbasic in the optimal solution. From the expression

$$-p^{2\prime} + p^{1\prime} A_{11}^{-1} A_{12}$$

we note that each coefficient occurs only in its own column. Consider the first variable in x^2, which is x_{q+1}. Let us denote the first column of $A_{11}^{-1} A_{12}$ by

$$\begin{bmatrix} \bar{a}_{1,q+1} \\ \cdot \\ \cdot \\ \bar{a}_{q,q+1} \end{bmatrix}$$

The f-element of x_{q+1} can then be written as

$$\bar{a}_{0,q+1} = -p_{q+1} + p_1 \bar{a}_{1,q+1} + p_2 \bar{a}_{2,q+1} + \ldots + p_q \bar{a}_{q,q+1}. \tag{3}$$

This element should be nonnegative for a maximization problem. If a

Table 1

Set-up tableau and optimal tableau

Tableau	Basic variables	Value basic variables	x^1	x^2	f	y^1	y^2
I	f	0	$-p^{1\prime}$	$-p^{2\prime}$	1	0	0
	y^1	b^1	A_{11}	A_{12}	0	I	0
	y^2	b^2	A_{21}	A_{22}	0	0	I
II	f	$p^{1\prime} A_{11}^{-1} b^1$	0	$-p^{2\prime} + p^{1\prime} A_{11}^{-1} A_{12}$	1	$p^{1\prime} A_{11}^{-1}$	0
	x^1	$A_{11}^{-1} b^1$	I	$A_{11}^{-1} A_{12}$	0	A_{11}^{-1}	0
	y^2	$b^2 - A_{21} A_{11}^{-1} b^1$	0	$A_{22} - A_{21} A_{11}^{-1} A_{12}$	0	$-A_{21} A_{11}^{-1}$	I

change Δp_{q+1} in p_{q+1} is considered, we should have, in order that the solution remains optimal

$$- (p_{q+1} + \Delta p_{q+1}) + p_1 \bar{a}_{1,q+1} + \ldots + p_q \bar{a}_{q,q+1} \geq 0$$

or

$$\Delta p_{q+1} \leq \bar{a}_{0,q+1}. \tag{4}$$

Hence only an increase of p_{q+1} above the corresponding element in the f-row will make the current solution nonoptimal.

For example, let us consider the coefficient of x_2 in the production planning example of chapter 2, which was 1. In the optimal tableau, x_2 is nonbasic with a coefficient in the f-row of $\frac{1}{2}$. The optimal solution will remain the same for $\Delta p_2 \leq \frac{1}{2}$ or $p_2 + \Delta p_2 \leq 1\frac{1}{2}$, so that the price of product 2 should not increase over $1\frac{1}{2}$.

Let us now find out for which changes of coefficients of basic variables the optimal solution remains the same. The vector p^1 occurs in all coefficients of the f-row. Consider again the coefficient in the f-row for x_{q+1} in the optimal tableau, which is given by (3). Let us have instead of p_1, $p_1 + \Delta p_1$. For optimality we should have

$$- p_{q+1} + (p_1 + \Delta p_1)\bar{a}_{1,q+1} + p_2 \bar{a}_{2,q+1} + \ldots + p_q \bar{a}_{q,q+1} \geq 0 \tag{5}$$

or

$$\Delta p_1 \bar{a}_{1,q+1} \geq - \bar{a}_{0,q+1}. \tag{6}$$

From all nonbasic variables similar constraints are obtained, the general form of which is

$$\Delta p_1 \bar{a}_{1j} \geq - \bar{a}_{0j}. \tag{7}$$

Consider now the cases in which $\bar{a}_{1j} > 0$. We then find from (7)

$$\Delta p_1 \geq - \bar{a}_{0j} / \bar{a}_{1j}. \tag{8}$$

An effective lower bound is then given by

$$\Delta p_1 \geq \underset{j}{\text{Max}} \, (- \bar{a}_{0j} / \bar{a}_{1j}) \text{ for } \bar{a}_{1j} > 0. \tag{9}$$

For $\bar{a}_{1j} < 0$, we derive from (7)

$$\Delta p_1 \leq - \bar{a}_{0j} / \bar{a}_{1j}, \tag{10}$$

so that an effective upper bound is

$$\Delta p_1 \leq \underset{j}{\text{Min}} \, (- \bar{a}_{0j} / \bar{a}_{1j}) \text{ for } \bar{a}_{1j} < 0. \tag{11}$$

(9) and (11) may be combined as follows

$$\text{Max}_j \, (-\bar{a}_{0j}/\bar{a}_{1j}|\bar{a}_{1j} > 0) \le \Delta p_1 \le \text{Min}_j \, (-\bar{a}_{0j}/\bar{a}_{1j}|\bar{a}_{1j} < 0), \qquad (12)$$

which may be written as

$$p_1 - \text{Min}_j \, (\bar{a}_{0j}/\bar{a}_{1j}|\bar{a}_{1j} > 0) \le p_1 + \Delta p_1 \le p_1 + \text{Min}_j \, (\bar{a}_{0j}/-\bar{a}_{1j}|\bar{a}_{1j} < 0).$$

$$(13)$$

For example for tableau 3 of table 3 in chapter 2, the following bounds are found for p_1:

$$3 - \text{Min} \, (\tfrac{1}{2}/4, \, 0/1) \le p_1 + \Delta p_1 \le 3 + \text{Min} \, (1/1),$$

or

$$3 \le p_1 + \Delta p_1 \le 4.$$

The solution of tableau 3 is therefore optimal only for values of p_1 between 3 and 4.

Note that changes in the coefficients of basic variables between these bounds, although they do not change the values of basic variables, change the opportunity costs of the constraints and the reduced revenues of nonbasic x-variables.

5.2. Sensitivity analysis of constant terms

In this section the consequences of changes in the elements of the vector b are analysed. As in the analysis of the objective function, the tableau of the optimal solution can be expressed in terms of partitioned vectors and matrices, see tableau II of table 1. From this, it is observed that changes in b only affect the values of basic variables.

For the value of f we find

$$f = p^{1\prime} A_{11}^{-1} b^1 = \hat{u}_1 b_1 + \hat{u}_2 b_2 + \ldots + \hat{u}_q b_q; \qquad (14)$$

$\hat{u}_1, \ldots, \hat{u}_q$ are the values of the dual variables for the first q constraints in the optimal solution. From this we observe that f is only dependent on constant terms of binding constraints and that the coefficients are the opportunity costs.

Consider now the values of the slack variables y^2:

$$y^2 = b^2 - A_{21} A_{11}^{-1} b^1. \qquad (15)$$

The coefficients of b^2 only occur in the corresponding slack variable. For y_{q+1} we may write

$$y_{q+1} = \bar{a}_{q+1,0} = b_{q+1} - \bar{a}_{q+1,1}b_1 - \dots - \bar{a}_{q+1,q}b_q. \tag{16}$$

$\bar{a}_{q+1,0}$ is by definition the value of y_{q+1} in tableau II and the row vector

$$[- \bar{a}_{q+1,1} - \bar{a}_{q+1,2} \dots - \bar{a}_{q+1,q}]$$

gives the elements of the first row of the matrix $- A_{21}A_{11}^{-1}$.

Since the value of y_{q+1} should be nonnegative, we have for any change in b_{q+1}, indicated by Δb_{q+1}:

$$b_{q+1} + \Delta b_{q+1} - \bar{a}_{q+1,1}b_1 - \dots - \bar{a}_{q+1,q}b_q \geq 0 \tag{17}$$

or

$$\Delta b_{q+1} \geq - \bar{a}_{q+1,0}. \tag{18}$$

This means that the supply of resources for non-binding constraints should not be decreased below the surplus amount in the optimal solution.

Changes in constant terms of binding constraints do affect the values of all basic variables as is indicated by the expression for basic variables $A_{11}^{-1}b^1$ and $b_2 - A_{21}A_{11}^{-1}b^1$. Let us write these as

$$\begin{bmatrix} A_{11}^{-1}b^1 \\ b^2 - A_{21}A_{11}^{-1}b^1 \end{bmatrix} = \begin{bmatrix} \bar{a}_{10} \\ \cdot \\ \bar{a}_{q0} \\ \bar{a}_{q+1,0} \\ \cdot \\ \bar{a}_{m0} \end{bmatrix}, \tag{19}$$

and let us write the elements in the columns of y^1 as

$$\begin{bmatrix} A_{11}^{-1} \\ - A_{21}A_{11}^{-1} \end{bmatrix} = \begin{bmatrix} \bar{a}_{11} & \cdots & \bar{a}_{1q} \\ \cdot & \cdots & \cdot \\ \cdot & \cdots & \cdot \\ \cdot & \cdots & \cdot \\ \bar{a}_{m1} & \cdots & \bar{a}_{mq} \end{bmatrix} \tag{20}$$

The value of, say x_1, can be written as

$$\bar{a}_{10} = \bar{a}_{11}b_1 + \bar{a}_{12}b_2 + \dots + \bar{a}_{1q}b_q. \tag{21}$$

Consider now a change in b_1 of Δb_1; the value of x_1 which should be nonnegative, is then

$$\bar{a}_{11}(b_1 + \Delta b_1) + \bar{a}_{12}b_2 + \dots + \bar{a}_{1q}b_q \geq 0, \tag{22}$$

which results in

$$\bar{a}_{11}\Delta b_1 \geq - \bar{a}_{10}. \tag{23}$$

For the basic variable in the ith row we obtain similarly

$$\bar{a}_{i1}\Delta b_1 \geq - \bar{a}_{i0}. \tag{24}$$

From this, effective upper and lower bounds for Δb_1 are obtained as follows. For $\bar{a}_{i1} > 0$, (24) may be written as

$$\Delta b_1 \geq - \bar{a}_{i0}/\bar{a}_{i1}, \tag{25}$$

so that the effective lower bound is

$$\Delta b_1 \geq \underset{i}{\text{Max}} \, (- \bar{a}_{i0}/\bar{a}_{i1} | a_{i1} > 0). \tag{26}$$

For $\bar{a}_{i1} < 0$, (24) may be written as

$$\Delta b_1 \leq - \bar{a}_{i0}/\bar{a}_{i1}, \tag{27}$$

so that the effective upper bound is

$$\Delta b_1 \leq \underset{i}{\text{Min}} \, (- \bar{a}_{i0}/\bar{a}_{i1} | \bar{a}_{i1} < 0). \tag{28}$$

(26) and (28) may be combined as

$$b_1 - \underset{i}{\text{Max}} \, (\bar{a}_{i0}/\bar{a}_{i1} | \bar{a}_{i0} > 0) \leq b_1 + \Delta b_1 \leq b_1 + \underset{i}{\text{Min}} \, (\bar{a}_{i0}/- \bar{a}_{i1} | \bar{a}_{i1} < 0). \tag{29}$$

For example, if lower and upper bounds of capacity on the first machine which x_1, x_4, and x_5 are basic are required, (29) is applied as follows

$$100 - 10/\tfrac{1}{2} \leq 100 + \Delta b_1 \leq 100 + 5/\tfrac{1}{4}$$

or

$$80 \leq 100 + \Delta b_1 \leq 120,$$

so that it must be concluded that the basis of the optimal solution remains unaltered for capacity on the first machine varying from 80 to 120. Note that the *values* of the basic variables change; the dual variables do not.

5.3. Programming with a parametric objective function

Suppose we wish to find optimal solutions for the production planning problem for all possible net revenues of the first product, the net revenues of other products remaining the same. In this case the objective function is

$$f = p_1 x_1 + x_2 + \tfrac{1}{2}x_3 + 2x_4 + 2x_5, \tag{30}$$

which may be written as

$$f = 0x_1 + x_2 + \tfrac{1}{2}x_3 + 2x_4 + 2x_5$$
$$+ \lambda (x_1 + 0x_2 + 0x_3 + 0x_4 + 0x_5), \tag{31}$$

with λ instead of p_1.

It may also be interesting to know the effect of the level of certain kinds of variable costs on the optimal production program; as an example the wage level may be taken. Let in the production planning problem the labor costs per unit of product be $1, \tfrac{1}{2}, \tfrac{1}{4}, \tfrac{3}{4}$ and $\tfrac{3}{4}$ for products 1–5, assuming that the wage level is 1. The objective function for a general wage level λ is

$$f = 4x_1 + 1\tfrac{1}{2}x_2 + \tfrac{3}{4}x_3 + 2\tfrac{3}{4}x_4 + 2\tfrac{3}{4}x_5$$
$$- \lambda (x_1 + \tfrac{1}{2}x_2 + \tfrac{1}{4}x_3 + \tfrac{3}{4}x_4 + \tfrac{3}{4}x_5). \tag{32}$$

The objective functions (31) and (32) are examples of a *parametric objective function* with a parameter λ. A general objective function with a parameter λ may be written as

$$f = p_1 x_1 + p_2 x_2 + \ldots + p_n x_n + \lambda (q_1 x_1 + q_2 x_2 + \ldots + q_n x_n). \tag{33}$$

In the general parametric problem, optimal solutions are desired for all values of λ between given upper and lower bounds λ_l and λ_u; for instance it may be that λ is varied between 0 and ∞. This general parametric objective function may be put into the simplex tableau format by using an extra row for the parametric terms; this row is indicated by f_λ, whereas the coefficients without λ are given in a row indicated by f_c; the first two rows of the set-up tableau then become

$$
\begin{array}{cccccc}
f_c & 0 & -p_1 & -p_2 & \cdots & -p_n \\
f_\lambda & 0 & -q_1 & -q_2 & \cdots & -q_n.
\end{array}
$$

These rows and the corresponding rows in subsequent tableaux should be read as follows:

$$0 + 0\lambda = (-p_1 - q_1\lambda)x_1 + (-p_2 - q_2\lambda)x_2 + \ldots + (-p_n - q_n\lambda)x_n + f. \quad (34)$$

Suppose that we have a tableau with an optimal solution for $\lambda = \lambda^*$; λ^* is a given value of λ. Such a tableau can be represented as in table 2. In this tableau all nonbasic variables are indicated as x-variables and all basic variables as y-variables. Since the solution is optimal in the wide sense, it is feasible, so that $b_1, \ldots, b_m \geq 0$. The coefficient of x_j in the row of objective function is $p_j + q_j\lambda$. Assuming that we have a maximization problem, we have, since the solution is optimal for $\lambda = \lambda^*$,

$$p_j + q_j\lambda^* \geq 0, \qquad j = 1, \ldots, n. \quad (35)$$

The value of the objective function is $p_0 + q_0\lambda^*$. Figures 1a, 1b, and 1c give the values of $p_j + q_j\lambda$ for three cases: $q_j > 0$, $q_j = 0$, and $q_j < 0$. As is obvious from figure 1c, $p_j + q_j\lambda$ is decreasing for increasing λ if $q_j < 0$; we have $p_j + q_j\lambda = 0$ for $\lambda = -p_j/q_j$. For $\lambda > -p_j/q_j$, negative values of $p_j + q_j\lambda$ are found, so that for these values of λ the solution is no longer optimal, since the jth column has a negative coefficient in the row of objective function. If there is only one nonbasic variable with a negative q_j, say x_1, the maximum value of λ for which the present solution is optimal is $\bar{\lambda} = -p_1/q_1$. If there are more nonbasic variables with negative q's, for instance x_1, x_2 and x_3, we should compute $\lambda_1 = -p_1/q_1$, $\lambda_2 = -p_2/q_2$, $\lambda_3 = -p_3/q_3$ and take the minimum of λ_1, λ_2 and λ_3 (see figure 2).

If $q_j > 0$, $p_j + q_j\lambda$ is decreasing for decreasing values of λ (see figure 1a). This means that $p_j + q_j\lambda$ will become 0 for some value of $\lambda \leq \lambda^*$, namely for $\lambda = -p_j/q_j$. If there is only one nonbasic variable x_j with a

Table 2

A general tableau for a problem with a parametric objective function

Basic variables	Values basic variables	x_1	x_2	.	x_n
f_c	p_0	p_1	p_2	.	p_n
f_λ	q_0	q_1	q_2	.	q_n
y_1	b_1	a_{11}	a_{12}	.	a_{1n}
.	.	.	.	.	.
y_m	b_m	a_{m1}	a_{m2}	.	a_{mn}

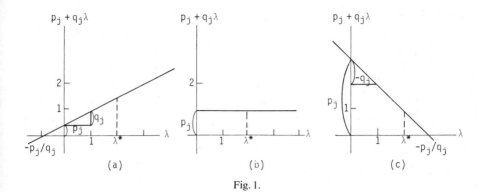

Fig. 1.

positive q_i, then $-p_i/q_i$ will be the lower bound of values of λ for which the variables with positive q's, say x_4, x_5, and x_6, we should compute $\lambda_4 = -p_4/q_4$, $\lambda_5 = -p_5/q_5$ and $\lambda_6 = -p_6/q_6$ and take as the lower bound $\underline{\lambda}$ the

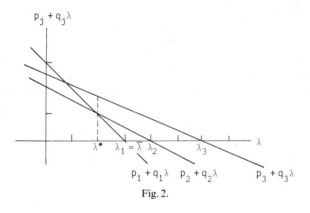

Fig. 2.

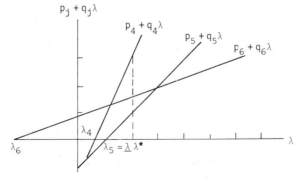

Fig. 3.

maximum of λ_4, λ_5, and λ_6 (see figure 3). Finally we have the cases in which $q_j = 0$. Since $p_j + q_j\lambda = 0$, it follows that $p_j \geq 0$; no upper or lower bound of λ are found (see figure 1b).

In this manner upper and lower bounds of λ can be determined for which the present solution is optimal. Figure 4 combines the cases in which $q_j < 0$ with cases $q_j > 0$, adding $p_7 + q_7\lambda$ with $q_7 = 0$. It is obvious that if for $\lambda = \lambda^*$ the solution is optimal, it follows that $\underline{\lambda} \leq \bar{\lambda}$. If the solution is optimal for no values of λ, the lower and upper bounds of λ can still be determined as before, but in this case it is found that $\underline{\lambda} > \bar{\lambda}$ (see figure 5). If it is not known for which values of λ a tableau is optimal, these values can be determined, if $\underline{\lambda} \leq \bar{\lambda}$; these values are given by $\underline{\lambda} \leq \lambda \leq \bar{\lambda}$. We may state formally: If

$$\bar{\lambda} = \operatorname*{Min}_{j}(- p_j/q_j \,|\, q_j < 0),$$

$$\underline{\lambda} = \operatorname*{Max}_{j}(- p_j/q_j \,|\, q_j > 0); \tag{36}$$

and if $\underline{\lambda} \leq \bar{\lambda}$ and $p_j \geq 0$ for j with $q_j = 0$, then the solution is optimal for $\underline{\lambda} \leq \lambda \leq \bar{\lambda}$; otherwise the solution is optimal for no value of λ.

In parametric programming optimal solutions must be found for all values of λ between a given lower bound λ_l and a given upper bound λ_u. The procedure can be started in two ways. For a given tableau, for instance the set-up tableau, the values of λ are determined for which the solution is optimal. If these values exist and are within the range λ_l and λ_u, λ may be varied to generate solutions for other intervals within this range.

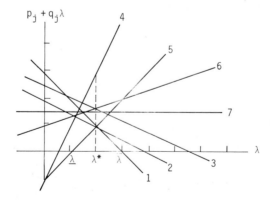

Fig. 4.

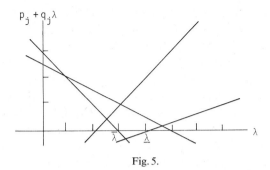

Fig. 5.

If these values do not exist or are not within the range λ_l to λ_u, the second way to start the procedure should be employed. An optimal solution for some value of λ between λ_l and λ_u, for instance for $\lambda = \lambda_l$, is generated by adding in a separate row, the f_c-row and λ_l times the f_λ-row. The new row is used as the objective function row; an optimal solution should then be generated by, for instance, the simplex method. After this, the f_c- and f_λ-rows are used to find $\underline{\lambda}$ and $\bar{\lambda}$ for the solution concerned; the parametric procedure then may be started.

If optimal tableaux and solutions are desired for $\lambda > \bar{\lambda}$, the nonbasic variable corresponding with $\bar{\lambda}$ in (36) should be introduced into the basis, since this is the first variable having a negative coefficient in the f-equation for an increasing value of λ. The variable to leave the basis is selected as in the simplex method. In the new tableau, new upper bounds and lower bounds $\underline{\lambda}^*$ and $\bar{\lambda}^*$ can be determined using (36). It will be proved below that the lower bound of the new tableau is equal to the upper bound of the previous one, and that its upper bound is not less than its lower bound, so that the new tableau is optimal for at least one value of λ. In the same manner a new tableau may be generated which is optimal for higher values of λ, and so on, until a tableau is found with $\underline{\lambda} < \lambda_u \leq \bar{\lambda}$ or, if $\lambda_u = \infty$, one having no column with $q_j < 0$; this tableau is optimal for an interval of λ having no upper bound.

On the other hand, if optimal tableaux and solutions are desired for $\lambda < \underline{\lambda}$, the nonbasic variable connected with $\underline{\lambda}$ in (36) should become basic, since this is the first variable having a negative coefficient for a decreasing value of λ; the leaving basic variable is again determined as in the simplex method. In the resulting tableau, upper and lower bounds for λ may be determined using (36); this upper bound must be the same as the lower bound of the initial tableau and the lower bound must be at most equal to this, so that this tableau is optimal for at least one value of λ. From this tableau, another one may be generated by introducing into the basis the variable connected with the lower bound, after which a tableau results which is optimal for a new interval of λ, which is lower than the previous one. Finally a solution must be found which has non-negative elements in the f_c-row; this solution is optimal for $\lambda = 0$. If solutions also should be found for negative values of λ, the procedure should be applied until all elements in the f_λ-row are nonpositive. The resulting solution is optimal for $\lambda \to -\infty$. It can be proved that for any range of λ, all optimal solutions can be generated in a finite number of iterations.

5.4. Numerical example

Let us consider the following variant of the production planning problem treated before. Maximize, for $0 \le \lambda \le \infty$,

$$f = (4 - \lambda)x_1 + (1\tfrac{1}{2} - \tfrac{1}{2}\lambda)x_2 + (2\tfrac{3}{4} - \tfrac{3}{4}\lambda)x_3 + (3\tfrac{1}{2} - \tfrac{3}{4}\lambda)x_4 \qquad (37)$$

subject to

$$4x_1 + 3x_2 + 2x_3 \qquad\quad \le 100,$$
$$x_1 \qquad + x_3 + 2x_4 \le 40,$$
$$2x_1 + 4x_2 + x_3 + 2x_4 \le 60, \qquad (38)$$

and

$$x_1, x_2, x_3, x_4 \ge 0. \qquad (39)$$

λ can be interpreted as labor costs per unit of labor or as any other kind of variable costs; in case of an interpretation as labor costs, to which we shall adhere in the following, the coefficients of λ in the factors between brackets in (37) can be interpreted as the quantity of labor used per unit of the product concerned. The total quantity of labor used in the production program is given by

$$x_1 + \tfrac{1}{2}x_2 + \tfrac{3}{4}x_3 + \tfrac{3}{4}x_4. \qquad (40)$$

The problem may be put into a simplex tableau format with the coefficients of the objective function equation separated into two rows; f_c is the row for the coefficients without λ and f_λ is the row of coefficients which should be multiplied by λ. The result is tableau 0 of table 3.

Since none of the coefficients in the f_λ-row, the q_j's in our general nota-

Table 3

Numerical example for programming with a parametric objective function

Tableau	Basic variables	Values basic variables	Ratio	x_1	x_2	x_3	x_4
	f_c	0		-4	$-1\frac{1}{2}$	$-2\frac{3}{4}$	$-3\frac{1}{2}$
	f_λ	0		1	$\frac{1}{2}$	$\frac{3}{4}$	$\frac{3}{4}$
0	Ratio			4	3	$3\frac{2}{3}$	$\boxed{4\frac{2}{3}}$
	y_1	100		4	3	2	0
	y_2	40	20	1	0	1	2
	y_3	60	30	2	4	1	2
				x_1	x_2	x_3	y_2
	f_c	70		$-2\frac{1}{4}$	$-1\frac{1}{2}$	-1	$1\frac{3}{4}$
	f_λ	-15		$\frac{5}{8}$	$\frac{1}{2}$	$\frac{3}{8}$	$-\frac{3}{8}$
1	Ratio			$\boxed{3\frac{3}{5}}$	3	$2\frac{2}{3}$	$(4\frac{2}{3})$
	y_1	100	25	4	3	2	0
	x_4	20	40	$\frac{1}{2}$	0	$\frac{1}{2}$	$\frac{1}{2}$
	y_3	20	20	$\underline{1}$	4	0	-1
				y_3	x_2	x_3	y_2
	f_c	115		$2\frac{1}{4}$	$7\frac{1}{2}$	-1	$-\frac{1}{2}$
	f_λ	$-27\frac{1}{2}$		$-\frac{5}{8}$	-2	$\frac{3}{8}$	$\frac{1}{4}$
2	Ratio			$(3\frac{3}{5})$		$\boxed{2\frac{2}{3}}$	2
	y_1	20	10	-4	-13	2	4
	x_4	10	20	$-\frac{1}{2}$	-2	$\frac{1}{2}$	1
	x_1	20		1	4	0	-1
				y_3	x_2	y_1	y_2
	f_c	125		$\frac{1}{4}$	1	$\frac{1}{2}$	$1\frac{1}{2}$
	f_λ	$-31\frac{1}{4}$		$\frac{1}{8}$	$\frac{7}{16}$	$-\frac{3}{16}$	$-\frac{1}{2}$
3	Ratio			$\boxed{-2}$	$-2\frac{2}{7}$	$(2\frac{2}{3})$	
	x_3	10		-2	$-6\frac{1}{2}$	$\frac{1}{2}$	2
	x_4	5		$\frac{1}{2}$	$1\frac{1}{4}$	$-\frac{1}{4}$	0
	x_1	20		1	4	0	-1

tion, are negative, there is no upper limit for the value of λ for which this solution is optimal. On the other hand, all coefficients are positive, so that to find $\underline{\lambda}$ for this tableau, all ratios are compared:

$$\underline{\lambda} = \text{Max}(4/1,\ 1\tfrac{1}{2}/\tfrac{1}{2},\ 2\tfrac{3}{4}/\tfrac{3}{4},\ 3\tfrac{1}{2}/\tfrac{1}{4}) = \text{Max}(4,\ 3,\ 3\tfrac{2}{3},\ 4\tfrac{2}{3}) = 4\tfrac{2}{3}.$$

The ratios are given in an additional row and a square is drawn around the lower bound $4\tfrac{2}{3}$, as we shall do in subsequent tableaux; upper bounds for a tableau are indicated by parentheses. The solution of tableau 0 is optimal for $4\tfrac{2}{3} \leq \lambda \leq +\infty$.

A solution which is optimal for $\lambda < 4\tfrac{2}{3}$ is found by taking into the basis the variable connected with $4\tfrac{2}{3}$, which is x_4; as in the simplex method it is found that y_2 leaves the basis, so that after transformation tableau 1 results. As indicated before, the upper bound for λ for this solution is equal to the lower bound of the previous one.

The lower bound for λ is found by comparing ratios in columns with positive elements in the f_λ-row:

$$\underline{\lambda} = \text{Max}(2\tfrac{1}{4}/\tfrac{5}{8},\ 1\tfrac{1}{2}/\tfrac{1}{2},\ 1/\tfrac{3}{8}) = \text{Max}(3\tfrac{3}{5},\ 3,\ 2\tfrac{2}{3}) = 3\tfrac{3}{5}.$$

Hence the solutions of tableau 1 are optimal for $3\tfrac{3}{5} \leq \lambda \leq 4\tfrac{2}{3}$. The variable connected with $3\tfrac{3}{5}$, which is x_1, is taken into the basis; the leaving basic variable turns out to be y_3. Tableau 2 results, which is found to be optimal for $2\tfrac{2}{3} \leq \lambda \leq 3\tfrac{3}{5}$. Then tableau 3 is found which is optimal for $-2 \leq \lambda \leq 2\tfrac{2}{3}$. Since optimal solutions are required for $0 \leq \lambda \leq \infty$, we stop.

We could also have started by generating from the set-up tableau first a solution which is optimal for $\lambda = 0$, after which optimal solutions for increasing values of λ are found. This is done in table 4. Tableau 00 gives the set-up tableau. Since our starting-point is $\lambda = 0$, the f_λ-row does not contribute anything to the objective function, so that the f_c-row may be used instead of an additional row consisting of the f_c-row plus λ_l times the f_λ-row. After three iterations of the simplex method, the optimal solution for $\lambda = 0$ is found in tableau 0. In order to find the upper bound for λ, we compare the ratios for columns with negative elements in the f_λ-rows:

$$\bar{\lambda} = \text{Min}(\tfrac{1}{2}/\tfrac{3}{16},\ 1\tfrac{1}{2}/\tfrac{1}{2}) = \text{Min}(2\tfrac{2}{3},\ 3) = 2\tfrac{2}{3}.$$

The variable connected with $2\tfrac{2}{3}$ is y_1, so that this variable enters the basis; tableau 1 results. The lower bound for λ in this tableau is of course $2\tfrac{2}{3}$. The upper bound is found to be $3\tfrac{3}{5}$. The following two tableaux are found in the same fashion. In tableau 3, there are no negative elements in the

Table 4

Example for programming with parametric objective function with starting-point $\lambda = 0$

Tableau	B.v.	V.b.v.	Ratio	x_1	x_2	x_3	x_4
	f_c	0		-4	$-1\frac{1}{2}$	$-2\frac{3}{4}$	$-3\frac{1}{2}$
	f_λ	0		1	$\frac{1}{2}$	$\frac{3}{4}$	$\frac{3}{4}$
00	y_1	100	25	4	3	2	0
	y_2	40	40	1	0	1	2
	y_3	60	30	2	4	1	2
				y_1	x_2	y_2	y_3
	f_c	125		$\frac{1}{2}$	1	$1\frac{1}{2}$	$\frac{1}{4}$
	f_λ	$-31\frac{1}{4}$		$-\frac{3}{16}$	$\frac{7}{16}$	$-\frac{1}{2}$	$\frac{1}{8}$
0	Ratio			$(2\frac{2}{3})$		3	
	x_1	20		0	4	-1	1
	x_3	10	20	$\frac{1}{2}$	$-6\frac{1}{2}$	2	-2
	x_4	5		$-\frac{1}{4}$	$1\frac{1}{4}$	0	$\frac{1}{2}$
				x_3	x_2	y_2	y_3
	f_c	115		-1	$7\frac{1}{2}$	$-\frac{1}{2}$	$2\frac{1}{4}$
	f_λ	$-27\frac{1}{2}$		$\frac{3}{8}$	-2	$\frac{1}{4}$	$-\frac{5}{8}$
1	Ratio			$[2\frac{2}{3}]$	$3\frac{3}{4}$		$(3\frac{3}{5})$
	x_1	20	20	0	4	-1	1
	y_1	20		2	-13	4	-4
	x_4	10		$\frac{1}{2}$	-2	1	$-\frac{1}{2}$
				x_3	x_2	y_2	x_1
	f_c	70		-1	$-1\frac{1}{2}$	$1\frac{3}{4}$	$-2\frac{1}{4}$
	f_λ	-15		$\frac{3}{8}$	$\frac{1}{2}$	$-\frac{3}{8}$	$\frac{5}{8}$
2	Ratio					$(4\frac{2}{3})$	$[3\frac{3}{5}]$
	y_3	20		0	4	-1	1
	y_1	100		2	3	0	4
	x_4	20	40	$\frac{1}{2}$	0	$\frac{1}{2}$	$\frac{1}{2}$
				x_3	x_2	x_4	x_1
	f_c	0		$-2\frac{3}{4}$	$-1\frac{1}{2}$	$-3\frac{1}{2}$	-4
	f_λ	0		$\frac{3}{4}$	$\frac{1}{2}$	$\frac{3}{4}$	1
3	Ratio					$[4\frac{2}{3}]$	
	y_3	60		1	4	2	2
	y_1	100		2	3	0	4
	y_2	40		1	0	2	1

f_λ-row, so that its solution has no upper bound. Comparing table 4 with table 3, we find that tableaux 0–3 of the first table are the same as tableaux 0–3 of the last table in reverse order.

From this example it is obvious that the use of a given tableau as a starting-point by determining its lower and upper bounds $\underline{\lambda}$ and $\bar{\lambda}$, has, if $\underline{\lambda} \le \bar{\lambda}$, the advantage that no unnecessary tableaux are generated as can be the case if first an optimal solution is generated for $\lambda = 0$ or its given lower or upper bound $\underline{\lambda}$ and $\bar{\lambda}$. On the other hand, if for the given tableau, $\underline{\lambda} > \bar{\lambda}$, so that its solution is not optimal for any value of λ, the last method, which is always applicable, must be used. Furthermore, if for the given tableau we have $\lambda_l < \underline{\lambda} < \bar{\lambda} < \lambda_u$, so that the interval of λ for which the given solution is optimal lies within the bounds between which λ is varied in the problem, λ must be varied first upwards and then downwards, which slightly complicates the method.

Table 5 gives a summary of the results, in which the solution numbers of table 3 are used. For each tableau the lower and upper bounds of λ and the values of basic variables are given. In two separate columns denoted λ by $f(\underline{\lambda})$ and $f(\bar{\lambda})$ the values of f for the lower and upper bounds of λ are given. It is obvious that the values of f decrease with increasing λ, since increasing costs can only lead to worse results if optimal decisions are taken. Figure 6 gives a graphic representation of the dependence of f on λ.

It is also interesting to look at the total amount of labor used, which is given by the f_λ-column with a minus sign. This amount decreases from $31\frac{1}{4}$ for $\lambda = 0$ to zero for $\lambda \ge 4\frac{2}{3}$. Figure 7 gives a graphic representation of this. Total labor income which is equal to the amount of labor used times λ is represented in figure 8. It should be noted that this is a curve which increases for values of λ within a lower and upper bound for a certain solution; however, if a solution changes, there is a downwards jump. The slope of the increasing sections decreases as the amount of labor used

Table 5

Summary of results

Solution	$\underline{\lambda}$	$\bar{\lambda}$	f_c	f_λ	$f(\underline{\lambda})$	$f(\bar{\lambda})$	x_1	x_2	x_3	x_4	y_1	y_2	y_3
3	(0)	$2\frac{2}{3}$	125	$-31\frac{1}{4}$	125	$41\frac{2}{3}$	20	0	10	5	0	0	0
2	$2\frac{2}{3}$	$3\frac{3}{5}$	115	$-27\frac{1}{2}$	$41\frac{2}{3}$	16	20	0	0	10	20	0	0
1	$3\frac{3}{5}$	$4\frac{2}{3}$	70	-15	16	0	0	0	0	20	100	0	20
0	$4\frac{2}{3}$	∞	0	0	0	0	0	0	0	0	100	40	60

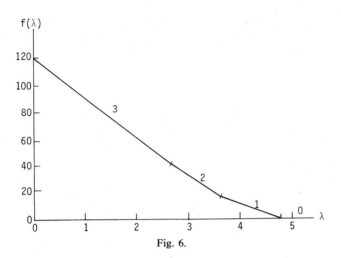

Fig. 6.

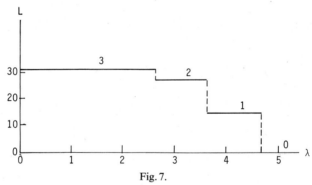

Fig. 7.

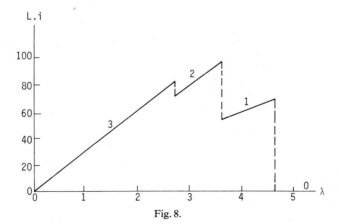

Fig. 8.

decreases. It is obvious that total labor income has many local maxima, but since it is a one-dimensional problem in λ, it is not difficult to find the global maximum.

5.5. Some proofs

We shall prove that for two successive tableaux, which are optimal for an increasing value of λ, the upper bound of λ of the first tableau coincides with the lower bound of λ for the second one, and that the upper bound of this last tableau is not smaller than this bound. Indicating the bound of the first tableau without and the second one with stars, we find that this amounts to

$$\bar{\lambda} = \underline{\lambda}^* \le \bar{\lambda}^*. \tag{41}$$

Table 6 gives a representation of the relevant elements of the two tableaux. The f_c- and f_λ-rows are modified in such a manner that $\lambda - \bar{\lambda}$ is used instead of λ. It is assumed that the variable connected with $\bar{\lambda}$ is x_k, so that

$$p_k + q_k \bar{\lambda} = p_k + q_k(-p_k/q_k) = 0. \tag{42}$$

Further, since $\bar{\lambda}$ is an upper bound of λ for an optimal solution, the coefficients of all other nonbasic variables in the objective function should be nonnegative for $\lambda = \bar{\lambda}$, so that we have for the coefficient in the jth column

$$p_j + q_j \bar{\lambda} \ge 0. \tag{43}$$

Let x_r be the leaving basic variable, so that a_{rk} is the pivot of the transformation; the result is tableau II. Since $p_k + q_k \bar{\lambda} = 0$, all elements in

Table 6

Some relevant parts of successive simplex tableaux for parametric programming

Tabl.	Basic variables	Values basic variables	x_j	x_k	x_r
I	$f_c + \bar{\lambda} f_\lambda$ $f_\lambda(\lambda - \bar{\lambda})$ x_r	$b_c + \bar{\lambda} b_\lambda$ $b_\lambda(\lambda - \bar{\lambda})$ b_r	$p_j + q_j \bar{\lambda} \ge 0$ $q_j(\lambda - \bar{\lambda})$ a_{rj}	$p_k + q_k \bar{\lambda} = 0$ $q_k(\lambda - \bar{\lambda})$ a_{rk}	0 0 1
II	$f_c + \bar{\lambda} f_\lambda$ $f_\lambda(\lambda - \bar{\lambda})$ x_k	$b_c + \bar{\lambda} b_\lambda$ $(b_\lambda - q_k a_{rk}^{-1} b_r)(\lambda - \bar{\lambda})$ $a_{rk}^{-1} b_r$	$p_j + q_j \bar{\lambda} \ge 0$ $(q_j - q_k a_{rk}^{-1} a_{rj})(\lambda - \bar{\lambda})$ $a_{rk}^{-1} a_{rj}$	0 0 1	0 $-q_k a_{rk}^{-1}(\lambda - \bar{\lambda})$ a_{rk}^{-1}

the $(f_c + f_\lambda \bar{\lambda})$-row are unchanged. The lower bound of λ for which this tableau is optimal is found as

$$\underline{\lambda}^* - \bar{\lambda} = \underset{j}{\text{Max}} \left(\frac{-(p_j + q_j \bar{\lambda})}{q_j - q_k a_{rk}^{-1} a_{rj}}, \frac{0}{-q_k a_{rk}^{-1}} \right) \tag{44}$$

for positive denominators of the ratios. Since $q_k < 0$ because x_k was the new basic variable in the last tableau and $a_{rk} > 0$ because it was the pivot, $-q_k a_{rk}^{-1} > 0$, so that the ratio $0/-q_k a_{rk}^{-1} = 0$ should be included. Of the other ratios, the numerator is nonpositive, so that the maximum must be 0, which means $\underline{\lambda}^* = \bar{\lambda}$.

For the upper bound of λ in the new tableau we have

$$\bar{\lambda}^* - \bar{\lambda} = \underset{j}{\text{Min}} \left(\frac{-(p_j + q_j \bar{\lambda})}{q_j - q_k a_{rk}^{-1} a_{rj}}, \frac{0}{-q_k a_{rk}^{-1}} \right) \tag{45}$$

for negative denominators. Since $-q_k a_{rk}^{-1} > 0$, the ratio $0/-q_k a_{rk}^{-1}$ should not be included. Of the other ratios the denominators are negative and the numerators nonpositive, so that they all are nonnegative. Hence $\bar{\lambda}^* \geq \bar{\lambda}$.

It is interesting to consider under which conditions $\bar{\lambda}^* = \bar{\lambda}$. If x_l is the new basic variable in tableau II, we would have

$$p_l + q_l \bar{\lambda} = \bar{\lambda}^* - \bar{\lambda} = 0, \tag{46}$$

which implies

$$-p_l/q_l = \bar{\lambda} = -p_k/q_k \qquad or \qquad p_l = q_l = 0. \tag{47}$$

If $q_l < 0$, there must have been a tie in the selection of the new basic variable in tableau I, but it may also occur that $q_l > 0$, so that the variable connected with the column l is not considered as a new basic variable in tableau I; the last case can occur since the condition

$$q_l - q_k a_{rk}^{-1} a_{rl} < 0. \tag{48}$$

merely implies

$$a_{rl} > (q_l a_{rk})/q_k. \tag{49}$$

If the problem is perturbed as in the dual method, both cases of (47) cannot occur. Hence for dually perturbed problems we have strictly increasing upper bounds of λ in successive tableaux. Since each of the successive tableaux in parametric programming has its own interval of λ, no tableau can recur, and since there is only a finite number of combination of basic variables, a finite number of tableaux and solutions are found which are optimal for a certain interval of λ. The first of these intervals has no lower bound but has some finite upper bound $\bar{\lambda}_1$; the last of these intervals has a lower bound λ_p, but no upper bound.

Let us now consider the objective function in two successive tableaux. In tableau I it is

$$f = b_c + \bar{\lambda} b_\lambda + b_\lambda(\lambda - \bar{\lambda}) \tag{50}$$

and in tableau II:

$$f = b_c + \bar{\lambda} b_\lambda + (b_\lambda - q_k a_{rk}^{-1} b_r)(\lambda - \bar{\lambda}). \tag{51}$$

This means that if optimal solutions are considered for $\lambda < \bar{\lambda}$, we have

$$df/d\lambda = b_\lambda, \tag{52}$$

and for $\lambda > \bar{\lambda}$

$$df/d\lambda = b_\lambda - q_k a_{rk}^{-1} b_r \geq b_\lambda; \tag{53}$$

the last relation follows from $q_k < 0$, $a_{rk} > 0$ and $b_r \geq 0$. Hence $df/d\lambda$ is nondecreasing. For $\lambda = \bar{\lambda}$, the left-hand derivative is (52), but the right-hand derivative is (53).

From this, the conclusion may be drawn that f as a function of λ is piece-wise linear; the slope of the pieces increases or remains the same. Figure 6 is an example of this. In this figure, the slope is never positive. In other cases, the slope may become positive, as in figure 9. Note the typical form of the function, which in a later chapter will be indicated as convex. In most practical cases, f will only have downward-going pieces ending with a horizontal piece (as in figure 6) or it will start with a horizontal piece and have further upward-going pieces. If in parametric programming no leaving basic variable can be found, then $f \to \infty$ for $\lambda = \bar{\lambda}$. Such a case is indicated by the dashed line in figure 9.

For a minimization problem it can easily be shown that $f(\lambda)$ is piece-wise linear with slopes which decrease or stay the same. The general

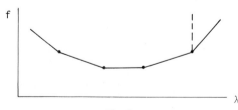

Fig. 9.

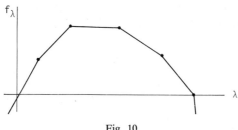

Fig. 10.

shape is then that of figure 10. The slope of this function is called concave. In most cases either the downward- or the upward-going part is
absent.

5.6. Parametric constant terms

In this section we consider the case in which the constant terms of the
constraints are dependent on the parameter λ. The problem to be solved
may be formulated as follows. Maximize, for a given range of values of
λ, the objective function

$$f = p_1 x_1 + p_2 x_2 + ... + p_n x_n \tag{54}$$

subject to

$$a_{11} x_1 + a_{12} x_2 + ... + a_{1n} x_n \leq b_1 + a_1 \lambda,$$
$$a_{21} x_1 + a_{22} x_2 + ... + a_{2n} x_n \leq b_2 + a_2 \lambda,$$

$$...$$

$$a_{m1} x_1 + a_{m2} x_2 + ... + a_{mn} x_n \leq b_m + a_m \lambda, \tag{55}$$

and

$$x_1, x_2, ..., x_n \geq 0. \tag{56}$$

In matrix notation, this problem may be stated as: Maximize

$$f = p'x \tag{57}$$

subject to

$$Ax \leq b + \lambda a, \tag{58}$$

$$x \geq 0. \tag{59}$$

This problem could be solved by first solving its dual problem, which
is: Minimize

$$g = (b + \lambda a)' u \tag{60}$$

subject to

$$A'u \geq p, \tag{61}$$

$$u \geq 0. \tag{62}$$

From the solutions of this problem, which has a parametric objective function, the solutions of the primal problem may be found. However, it is much more convenient to find the solutions of this problem in a direct manner.

Instead of the two rows which were used in the parametric variation of the objective function, two columns are used for the values of basic variables which now are linear expressions in λ, one for the coefficients without λ, which is called the constant-term column, and one for the λ-coefficients which is called the λ-term column. In the set-up tableau, the constant-term column consists of the vector b and the λ-term column of the vector a; the elements of both columns in the f-row are both zero. As in programming with a parametric objective function, the procedure may be started with a given tableau (e.g. the set-up tableau) or a given value of λ.

Let us first consider the case in which optimal solutions for $0 \leq \lambda \leq \infty$ are required. If the set-up tableau is not optimal for any value of λ, we may try to find the optimal solution for $\lambda = 0$. This is done by the simplex method (or any other method), ignoring the λ-term column, since $\lambda = 0$, but transforming this column all the same. The resulting tableau may be represented as in table 7, where i indicates a typical row and j a typical column. Since the solution is optimal, all b's as well as all p's, are nonnegative.

Table 7

Representation of a tableau for programming with a parametric constant term

	Values bas. var.		Nonbasic variables				
	c-term	λ-term					
f	b_0	a_0	p_1	$\cdot$	p_j	$\cdot$	p_n
	b_1	a_1	a_{11}	$\cdot$	a_{1j}	$\cdot$	a_{1n}
	$\cdot$	$\cdot$	$\cdot$	$\cdot$	$\cdot$		$\cdot$
Basic var.	b_i	a_i	a_{i1}	$\cdot$	a_{ij}	$\cdot$	a_{in}
	$\cdot$	$\cdot$	$\cdot$	$\cdot$	$\cdot$		$\cdot$
	b_m	a_m	a_{m1}	$\cdot$	a_{mj}	$\cdot$	a_{mn}

A solution optimal in the wide sense having been obtained for $\lambda = 0$, an increase in λ is considered. The value of the basic variable in the ith row of the current tableau can be written as

$$b_i + a_i\lambda.$$

For $a_i \geq 0$, the value of this variable increases or remains the same, but for $a_i < 0$, this value decreases and becomes negative eventually; it becomes zero for a value of λ which is found from

$$b_1 + a_1\lambda = 0$$

or

$$\lambda = b_i/- a_i. \tag{63}$$

But other basic variables with negative a's may already be negative for this value of λ if the value of λ given in (63) for these variables is lower. Hence we should determine an upper bound for λ for which the values of all basic variables $b_i + a_i\lambda$ are nonnegative; this upper bound is determined by

$$\bar{\lambda} = \text{Min}\,(b_i/- a_i | a_i < 0). \tag{64}$$

The solution of the current tableau is therefore feasible and hence optimal in the wide sense for $0 \leq \lambda \leq \bar{\lambda}$.

Let us assume that this critical value is connected with only one basic variable, for example the one in the rth row; let us denote this variable by x_r. The value of x_r is given by

$$b_r + a_r\lambda = b_r + a_r\bar{\lambda} + a_r(\lambda - \bar{\lambda}) = b_r + a_r(b_r/- a_r) + a_r(\lambda - \bar{\lambda}) = a_r(\lambda - \bar{\lambda}).$$

For $\lambda > \bar{\lambda}$, x_r becomes negative, so that the solution is no longer feasible. In order to maintain feasibility, x_r should leave the basis. A nonbasic variable should replace x_r in the basis in such a manner that the resulting solution remains feasible for $\lambda > \bar{\lambda}$ and that it keeps the property of optimality in the narrow sense. This is the same situation as in the dual method, where a negative basic variable was leaving the basis and a new basic variable was selected in such a manner that the solution is optimal in the narrow sense. The new basic variable should therefore be connected with

$$\text{Min}_j\,(- p_j/- a_{rj} | a_{rj} < 0). \tag{65}$$

Let the new basic variable be x_k in the kth column. The negative element

a_{rk} is then the pivot of the transformation. The value of x_k in the new tableau is

$$\frac{b_r + a_r\lambda}{a_{rk}} = \frac{a_r}{a_{rk}}(\lambda - \bar\lambda),\tag{66}$$

which indeed is nonnegative for $\lambda > \bar\lambda$.

For the new solution, a critical value can be determined as in (64). Let this be $\bar\lambda^*$. This solution is then feasible and hence optimal in the wide sense for $\bar\lambda = \underline\lambda^* \leq \lambda \leq \bar\lambda^*$. A leaving basic variable and a new basic variable are then determined as described, and so on. The procedure can terminate in two ways. Firstly, when a solution is found having no negative a_i's; this solution is optimal in the wide sense for all values of λ not less than the critical value of λ of the previous solution. Secondly, when no new basic variable can be found because the elements in the row of the leaving basic variables and in the columns of all nonbasic variables are nonnegative. This means that for values of λ higher than the upper bound no feasible solution exists. It is easy to adapt the procedure for ranges of values of λ other than $0 \leq \lambda \leq \infty$.

It can be proved that for any tableau $\bar\lambda = \underline\lambda^* \leq \bar\lambda^*$, where $\underline\lambda^*$ is defined as

$$\underset{i}{\text{Max}}\,(-b_i/a_i\,|\,a_i > 0);\tag{67}$$

further that a_0 is nonincreasing and that b_0 is nondecreasing for $\lambda > 0$. Further it may be proved that the procedure terminates in a finite number of iterations for suitably perturbed problems. These statements may be proved either directly as in section 5.5 or by generating the corresponding tableau of the dual problem which has a parametric objective function and drawing on the results of section 5.5.

As an example of application of this parametric procedure a variant of the production planning problem used in section 5.4 will be solved. The problem is as follows. Maximize for $0 \leq \lambda \leq \infty$ the function

$$f = 4x_1 + 1\tfrac{1}{2}x_2 + 2\tfrac{3}{4}x_3 + 3\tfrac{1}{2}x_4\tag{68}$$

subject to

$$4x_1 + 3x_2 + 2x_3 \qquad \leq 100,$$
$$x_1 \qquad + x_3 + 2x_4 \leq 40,$$
$$2x_1 + 4x_2 + x_3 + 2x_4 \leq 60,\tag{69}$$
$$x_1 + \tfrac{1}{2}x_2 + \tfrac{3}{4}x_3 + \tfrac{3}{4}x_4 \leq \lambda,\tag{70}$$
$$x_1, x_2, x_3, x_4 \geq 0.\tag{71}$$

Note that this problem is related to the one used in section 5.2 in a special manner; the relationships between the two problems will be further explored in chapter 6.

This problem may be put into a simplex tableau format as in tableau 00 of table 8. The solution of this tableau is not optimal in the narrow sense, but it is feasible for any nonnegative value of λ. Let us first generate an optimal solution for $\lambda = 0$ by the simplex method. Normally x_1, the variable with the most negative coefficient in the f-row, would enter the basis, after which other iterations follow, but it can be shown that in cases like

Table 8

Example of programming with parametric constant terms

Tableau	Basic var.	Value basic var. c-term	Value basic var. λ-term	Ratio	x_1	x_2	x_3	x_4
	f	0	0		-4	$-1\frac{1}{2}$	$-2\frac{3}{4}$	$-3\frac{1}{2}$
	Ratio				4	3	$3\frac{2}{3}$	$4\frac{2}{3}$
00	y_1	100	0		4	3	2	0
	y_2	40	0		1	0	1	2
	y_3	60	0		2	4	1	2
	y_λ	0	1	0	1	$\frac{1}{2}$	$\frac{3}{4}$	$\frac{3}{4}$
					x_1	x_2	x_3	y_λ
	f	0	$4\frac{2}{3}$		$\frac{2}{3}$	$\frac{5}{6}$	$\frac{3}{4}$	$4\frac{2}{3}$
	Ratio				$\frac{2}{5}$	$\frac{5}{8}$	$\frac{3}{4}$	$1\frac{1}{4}$
0	y_1	100	0		4	3	2	0
	y_2	40	$-2\frac{2}{3}$	$\boxed{15}$	$-1\frac{1}{3}$	$-1\frac{1}{3}$	-1	$-2\frac{2}{3}$
	y_3	60	$-2\frac{2}{3}$	$22\frac{1}{2}$	$-\frac{2}{3}$	$2\frac{2}{3}$	-1	$-2\frac{2}{3}$
	x_4	0	$1\frac{1}{3}$		$1\frac{1}{3}$	$\frac{2}{3}$	1	$1\frac{1}{3}$
					y_2	x_2	x_3	y_λ
	f	16	$3\frac{3}{5}$		$\frac{2}{5}$	$\frac{3}{10}$	$\frac{7}{20}$	$3\frac{3}{5}$
	Ratio				1	$\frac{7}{12}$		$2\frac{1}{4}$
1	y_1	196	$-6\frac{2}{5}$	$30\frac{5}{8}$	$2\frac{2}{5}$	$-\frac{1}{5}$	$-\frac{2}{5}$	$-6\frac{2}{5}$
	x_1	-24	$1\frac{3}{5}$	(15)	$-\frac{3}{5}$	$\frac{4}{5}$	$\frac{3}{5}$	$1\frac{3}{5}$
	y_3	44	$-1\frac{3}{5}$	$\boxed{27\frac{1}{2}}$	$-\frac{2}{5}$	$3\frac{1}{5}$	$-\frac{3}{5}$	$-1\frac{3}{5}$
	x_4	32	$-\frac{4}{5}$	40	$\frac{4}{5}$	$-\frac{2}{5}$	$\frac{1}{5}$	$-\frac{4}{5}$

Table 8　(*continued*)

Tableau	Basic var.	Value basic var. c-term	Value basic var. λ-term	Ratio	y_2	x_2	y_3	y_λ
	f	$41\frac{2}{3}$	$2\frac{2}{3}$		$\frac{1}{6}$	$2\frac{1}{6}$	$\frac{7}{12}$	$2\frac{2}{3}$
	Ratio				$\frac{13}{14}$	$\frac{7}{8}$	$\frac{1}{8}$	
	y_1	$116\frac{2}{3}$	$-5\frac{1}{3}$	$\boxed{31\frac{1}{4}}$	$2\frac{2}{3}$	$-2\frac{1}{3}$	$-\frac{2}{3}$	$-5\frac{1}{3}$
2	x_1	20	0		-1	4	1	0
	x_3	$-73\frac{1}{3}$	$2\frac{2}{3}$	$(27\frac{1}{2})$	$\frac{2}{3}$	$-5\frac{1}{3}$	$-1\frac{2}{3}$	$2\frac{2}{3}$
	x_4	$46\frac{2}{3}$	$-1\frac{1}{3}$	35	$\frac{2}{3}$	$\frac{2}{3}$	$\frac{1}{3}$	$-1\frac{1}{3}$
				Ratio	y_2	x_2	y_3	y_1
	f	125	0		$1\frac{1}{2}$	1	$\frac{1}{4}$	$\frac{1}{2}$
	y_λ	$-31\frac{1}{4}$	1	$(31\frac{1}{4})$	$-\frac{1}{2}$	$\frac{7}{16}$	$\frac{1}{8}$	$-\frac{3}{16}$
3	x_1	20	0		-1	4	1	0
	x_3	10	0		2	$-6\frac{1}{2}$	-2	$\frac{1}{2}$
	x_4	5	0		0	$1\frac{1}{4}$	$\frac{1}{2}$	$-\frac{1}{4}$

this one (with all elements positive in a row with a zero basic variable) the optimal solution is found in one step by selecting the new basic variable connected with

$$\text{Max}_{j}\ (-p_i/a_{r_i}|p_i < 0), \qquad (72)$$

where p_i is the element in the objective function row in the jth column and a_{r_i} the element in the row with the zero basic variable in the same column; we assumed $a_{r_i} > 0$ for all j. Hence in tableau 00, x_4 enters the basis and y_λ leaves it. The result is tableau 0, which is optimal in the narrow sense and feasible for $\lambda = 0$.

Starting the parametric procedure, we find that y_2 becomes zero for $\lambda = 15$ and y_3 becomes zero for $\lambda = 22\frac{1}{2}$. The solution of tableau 0 is therefore optimal in the wide sense for $0 \le \lambda \le 15$. This solution is

$$f = 4\frac{2}{3}\lambda, \ y_1 = 100, \ y_2 = 40 - 2\frac{2}{3}\lambda, \ y_3 = 60 - 2\frac{2}{3}\lambda, \ x_4 = 1\frac{1}{3}\lambda.$$

Since the variable connected with $\lambda = 15$ is y_2, this variable should leave the basis. The new basic variable is found as in the dual method by taking the minimum of the ratios:

$$(\tfrac{2}{3}/1\tfrac{2}{3}, \ \tfrac{5}{6}/1\tfrac{1}{3}, \ \tfrac{3}{4}/1, \ 4\tfrac{2}{3}/2\tfrac{2}{3}) = (\tfrac{2}{5}, \ \tfrac{5}{8}, \ \tfrac{3}{4}, \ 1\tfrac{3}{4}).$$

The minimum is $\frac{2}{5}$ and is connected with x_1, so that this variable should enter the basis.

The result is tableau 1. The upper bound of λ is found in the usual manner and turns out to be $27\frac{1}{2}$, so that this solution is optimal for $15 \le \lambda \le 27\frac{1}{2}$. The basic variable connected with this upper bound is y_3, so that this variable should leave the basis. The new basic variable is determined as in the dual method and is found to be x_3. After transformation tableau 2 results. The solution of this tableau turns out to be optimal in the wide sense for $27\frac{1}{2} \le \lambda \le 31\frac{1}{4}$. y_1 now leaves the basis and y_λ is the new basic variable. Tableau 3 is optimal for $31\frac{1}{4} \le \lambda \le \infty$. The solution of this tableau is

$$f = 125, x_1 = 20, x_3 = 10, x_4 = 5, y_\lambda = -31\frac{1}{4} + \lambda.$$

The solution, apart from y_λ, is no longer dependent on λ; if the available quantity of resource corresponding with constraint (70) exceeds $31\frac{1}{4}$, the excess amount is not used. Note that as λ increases, the successive values of the c-term of f increase; furthermore, the dual variable connected with constraint (70) decreases from $4\frac{2}{3}$ in tableau 0 to 0 in tableau 3; this is in accordance with the statement made before that in table 7 b_0 is nondecreasing and a_0 is nonincreasing. The consequence is that in a maximization problem f^λ will have the form of figure 10 (concave) and, in a minimization problem, that of figure 9 (convex).

5.7. Parametric variations of constraint coefficients

Here we shall treat parametric variations in the elements of the matrix A of the general problem. The general problem is in this case as follows. Maximize f,

$$f = p'x,$$
$$(A + \lambda B)x \le b,$$
$$x \ge 0. \tag{73}$$

Let us consider a solution in which the first q x-variables have replaced the first q slack variables. Partitioning the matrices A and B and the vectors p and b accordingly, we find for the values of the primal variables

$$(A_{11} + \lambda B_{11})^{-1}b^1, \; b^2 - (A_{21} + \lambda B_{21})(A_{11} + \lambda B_{11})^{-1}b^1 \tag{74}$$

and for the values of dual variables

$$p^{1\prime}(A_{11} + \lambda B_{11})^{-1}, - p^{1\prime}(A_{11} + \lambda B_{11})^{-1}(A_{12} + \lambda B_{12}). \qquad (75)$$

For $B_{11} \neq 0$, these expressions are nonlinear in λ because they contain $(A_{11} + \lambda B_{11})^{-1}$. It may be proved that in this case each tableau-element can be expressed as a ratio of two polynomials. It is not impossible to deal with such a parametric programming problem, but the necessary computations are rather complex even in small problems.

We shall limit ourselves to treat an example in which only one coefficient of A varies. Let us consider once more the production planning problem but let us suppose, in order to obtain a sufficiently general example, that the available capacity on machine 3 is not 60, but 70. For the same matrix of coefficients in the constraints as used before, the set-up tableau is as given in tableau 0 of table 9. After an application of the simplex method, the optimal solution is found in tableau 2.

Now separate parametric variations of each of the coefficients of A are considered. Suppose we consider variations of a_{ij} around a given value; then x_j is the corresponding nonbasic variable in the set-up tableau and y_i the corresponding basic variable in the set-up tableau. The following four cases may apply for the optimal tableau for the given value of a_{ij}:

(i) x_j is nonbasic and y_i is basic (as in the set-up tableau),
(ii) x_j and y_i are both nonbasic,
(iii) x_j and y_i are both basic,
(iv) x_j is basic and y_i is nonbasic.

First case (i) is considered. As an example we shall treat a variation of the coefficient of x_2 in the row of y_3 in tableau 0 of table 9. This coefficient is 4, but we want to find optimal solutions for $4 + \lambda$ instead; λ may take positive as well as negative values. Since in the optimal tableau y_3 is basic and x_2 is nonbasic, as they were in the set-up tableau, the only thing that happened to the element in the y_3-row and the x_2-column is that some multiples of other elements in the x_2-column were added to it. So if λ is added to it in tableau 0, it is not changed in tableau 2. Other elements of tableau 2 do not contain λ, since the row of y_3 was not used in the iterations for pivoting. Hence having $4 + \lambda$ instead of 4 in the row of y_3 and the column of x_2 results in having $3\frac{1}{4} + \lambda$ for the same element of tableau 2 and no other changes. Since this change does not affect the feasibility nor the optimality in the narrow sense of the solution, the change of this coefficient does not affect the optimal solution at all. For the typical maximization problem it can be said that changes of coeffi-

Table 9

Variation of a constraint coefficient

Tableau	Basic variables	Value basic variables	x_1	x_2	x_3	x_4
0	f	0	-3	-1	-2	$-2\frac{3}{4}$
	y_1	100	$4(+\beta)$	$3(+\mu)$	2	0
	y_2	40	$1(+\gamma)$	0	1	2
	y_3	70	$2(+\alpha)$	$4(+\lambda)$	1	2

			y_1	x_2	x_3	x_4
1	f	75	$\frac{3}{4}$	$1\frac{1}{4}$	$-\frac{1}{2}$	$-2\frac{3}{4}$
	x_1	25	$\frac{1}{4}$	$\frac{3}{4}$	$\frac{1}{2}$	0
	y_2	15	$-\frac{1}{4}$	$-\frac{3}{4}$	$\frac{1}{2}$	2
	y_3	20	$-\frac{1}{2}$	$2\frac{1}{2}$	0	2

			y_1	x_2	x_3	y_2
2	f	$95\frac{5}{8}$	$\frac{13}{32}$	$\frac{7}{32}$	$\frac{3}{16}$	$1\frac{3}{8}$
	x_1	25	$\frac{1}{4}$	$\frac{3}{4}$	$\frac{1}{2}$	0
	x_4	$7\frac{1}{2}$	$-\frac{1}{8}$	$-\frac{3}{8}$	$\frac{1}{4}$	$\frac{1}{2}$
	y_3	5	$-\frac{1}{4}$	$3\frac{1}{4}(+\lambda)$	$-\frac{1}{2}$	-1

			x_1	x_2	x_3	y_2
3	f	55	$-1\frac{5}{8}$	-1	$-\frac{5}{8}$	$1\frac{3}{8}$
	y_1	100	$4(+\beta)$	$3(+\mu)$	2	0
	x_4	20	$\frac{1}{2}$	0	$\frac{1}{2}$	$\frac{1}{2}$
	y_3	30	$1(+\alpha)$	4	0	-1

			y_1	x_2	x_3	y_2
4	f	$95\frac{5}{8}$	$\frac{13}{32}$	$\frac{7}{32}+\frac{13}{32}\mu$	$\frac{3}{16}$	$1\frac{3}{8}$
	x_1	25	$\frac{1}{4}$	$\frac{3}{4}+\frac{1}{4}\mu$	$\frac{1}{2}$	0
	x_4	$7\frac{1}{2}$	$-\frac{1}{8}$	$-\frac{3}{8}-\frac{1}{8}\mu$	$\frac{1}{4}$	$\frac{1}{2}$
	y_3	5	$-\frac{1}{4}$	$3\frac{1}{4}-\frac{1}{4}\mu$	$-\frac{1}{2}$	-1

			y_1	x_2	x_3	y_2
5	f	$95\frac{5}{8}$	$\frac{13}{32}$	$\frac{7}{32}$	$\frac{3}{16}$	$1\frac{3}{8}$
	x_1	25	$\frac{1}{4}$	$\frac{3}{4}$	$\frac{1}{2}$	0
	x_4	$7\frac{1}{2}$	$-\frac{1}{8}$	$-\frac{3}{8}$	$\frac{1}{4}$	$\frac{1}{2}$
	y_3	$5-25\alpha$	$-\frac{1}{4}-\frac{1}{4}\alpha$	$3\frac{1}{4}-\frac{3}{4}\alpha$	$-\frac{1}{2}-\frac{1}{2}\alpha$	-1

			y_1	x_2	x_3	y_2
6 × $(1+\beta/4)^{-1}$	f	$95\frac{5}{8}+13\frac{3}{4}\beta$	$\frac{13}{32}$	$\frac{7}{32}-\frac{1}{4}\beta$	$\frac{3}{16}-\frac{5}{32}\beta$	$1\frac{3}{8}+\frac{11}{32}\beta$
	x_1	25	$\frac{1}{4}$	$\frac{3}{4}$	$\frac{1}{2}$	0
	x_4	$7\frac{1}{2}+5\beta$	$-\frac{1}{8}$	$-\frac{3}{8}$	$\frac{1}{4}+\frac{1}{8}\beta$	$\frac{1}{2}+\frac{1}{8}\beta$
	y_3	$5+7\frac{1}{2}\beta$	$-\frac{1}{4}$	$3\frac{1}{4}+\beta$	$-\frac{1}{2}$	$-1-\frac{1}{4}\beta$

cients of x-variables which are nonbasic in the optimal solution in rows of slack variables which are basic in the optimal solution do not affect the optimal solution at all.

Next an example of case (ii) is considered. As an example, the coefficient of x_2 in the y_1-row of tableau 0 is changed from 3 into $3 + \mu$. The elements of the optimal tableau, tableau 2, should be expressed in terms of μ. This is done by first generating a tableau for which case (i) applies. In this tableau the dependence on μ is restricted to just one element, to which μ should be added. From this modified tableau, the tableau of the optimal solution is generated, which explicitly gives its dependence on u. Hence in tableau 2, y_1 is introduced into the basis. For the leaving basic variable any basic variable may be taken; let us choose x_1. Tableau 3 results. Since for this tableau case (i) applies, the element in the column of x_2 and the row of y_1 is changed into $3 + \mu$. The tableau of the optimal solution is found by introducing into the basis x_1, which replaces y_1. The result is tableau 4, of which only the column of x_2 depends on μ. Note that the coefficients of μ in the x_2-column are the same as in the y_1 column, which is explained by the fact that in tableau 3 the column of y_1 (deleted) and the column of coefficients of μ in the column of x_2 are identical unit vectors. Tableau 4 may therefore be immediately be inferred from tableau 2.

The element in the first row in the x_2-column is $\frac{1}{32}(7 + 13\mu)$; this is nonnegative for $\mu \geq -\frac{7}{13}$. Hence tableau 4 is optimal for a coefficient of x_2 in the first constraint which is not less than $2\frac{6}{13}$. For $\mu < -\frac{7}{13}$, x_2 should be introduced into the basis. The variable to leave the basis can be found by taking ratios as in the simplex method for $\mu = -\frac{7}{13}$. These ratios are

$$\frac{25}{(3 - 7/13)/4} = 40\frac{5}{8}, \qquad \frac{5}{(13 + 7/13)/4} = 1\frac{21}{44}.$$

The minimum ratio is the last one, so that y_3 leaves the basis.

The next tableau has as its elements ratios of linear expressions in μ. All elements can be computed but here we shall only compute the values of primal and dual variables in order to know what the next solution is and for which values of μ it is optimal in the wide sense. We find

$$f = 95\frac{5}{8} - \frac{5}{8}(7 + 13\mu)/(13 - \mu),$$

$$x_1 = 25 - \frac{3 + \mu}{4} \frac{4}{13 - \mu} 5 = \frac{30(10\frac{1}{3} - \mu)}{13 - \mu},$$

$$x_4 = 7\frac{1}{2} - \frac{3 + \mu}{8} \frac{4}{13 - \mu} 5 = \frac{5(21 - \mu)}{13 - \mu},$$

$$x_2 = 20/(13 - \mu).$$

Hence the solution is feasible for any negative value of μ. For the value of the dual variables corresponding with y_1, y_3, x_3, and y_2 we find

$$u_1 = \frac{11}{2(13 - \mu)}, \qquad u_3 = \frac{-7 - 13\mu}{8(13 - \mu)}, \qquad v_3 = \frac{23 + 5\mu}{8(13 - \mu)}, \qquad u_2 = \frac{75 - \mu}{4(13 - \mu)}.$$

The solution is optimal for $-4\frac{3}{5} \leq \mu \leq -\frac{7}{13}$. Since negative coefficients of x_2 in the first constraint are not likely, we may stop the variation of μ.

As an example of case (iii), a variation of the coefficient of x_1 in the third constraint is considered; hence we add α to its value 2 in tableau 0. In order to generate from tableau 2 a tableau in which x_1 is nonbasic, one of the nonbasic variables of tableau 2 should be introduced into the basis; let us take y_1. The result is tableau 3. In this tableau, α can be added to the element in the row of y_3 and the column of x_1; this is the only element changing with α. Now the same iteration is performed in the opposite direction, so that y_1 replaces x_1 as a basic variable. The result is tableau 5, which is the same as tableau 2 apart from the y_3-row. The constant coefficients of this row are the same as in tableau 2, the α-coefficients are equal to minus the first row of tableau 2, so that tableau 5 may be inferred from tableau 2. From tableau 5 it is observed that the current optimal solution remains feasible for $\alpha \leq \frac{1}{5}$. For $\alpha > \frac{1}{5}$, y_3 is negative. The solution $\alpha > \frac{1}{5}$ can be found by determining the new basic variable as in the dual method for $\alpha = \frac{1}{5}$. We find

$$\frac{\frac{13}{32}}{(1 + \frac{1}{5})/4} = 1\frac{17}{48}, \qquad \frac{\frac{3}{16}}{(1 + \frac{1}{5})/2} = \frac{5}{16}, \qquad \frac{1\frac{3}{8}}{1} = 1\frac{3}{8}.$$

Hence x_3 should replace y_3 as basic variable; the pivot is $-\frac{1}{2} - \frac{1}{2}\alpha$.

As before only the primal and dual basic variables of the next solution are determined. We find

$$x_1 = \frac{30}{1 + \alpha}, \qquad x_4 = \frac{10 - 5\alpha}{1 + \alpha}, \qquad x_3 = \frac{-10 + 50\alpha}{1 + \alpha},$$

$$f = 95\frac{5}{8} + \frac{15 - 75\alpha}{8(1 + \alpha)}, \qquad u_1 = \frac{10}{32}, \qquad v_2 = \frac{8 - \alpha}{16(1 + \alpha)},$$

$$u_3 = \frac{3}{8(1 + \alpha)}, \qquad u_2 = \frac{8 + 11\alpha}{8(1 + \alpha)}.$$

Hence the solution is optimal in the wide sense for $\frac{1}{5} \leq \alpha \leq 2$. For values of α higher than 2, x_4 should leave the basis. The new basic variable is determined as in the dual method, taking $\alpha = 2$. In this manner optimal solutions for all ranges of values of α are generated.

Finally we turn to case (iv). An example of this case is the coefficient of x_1 in the first constraint which is 4, but which is changed into $4 + \beta$. In order to determine the elements of the optimal tableau as functions of β, x_1 should leave the basis and y_1 should enter it. Pivoting on the element $\frac{1}{4}$, we find tableau 3. β is added to the element in the row of y_1 and the column of x_1, after which x_1 is made basic again, replacing y_1. Tableau 6 results. It is found that this tableau may be easily derived from tableaux 2 and 3, since in the numerator of the elements the terms without β are the same as those in tableau 2; in the row of x_1 and the column of y_1 the β-terms are missing, but for other elements the β-terms are equal to $\frac{1}{4}$ times the corresponding elements of tableau 3. Tableau 6 is optimal in the wide sense for $-\frac{2}{3} \le \beta \le \frac{7}{8}$. Solutions for higher and lower values of β can be found in a similar manner.

Some complications arise if the element that should be used as a pivot is zero. This is the case if the coefficient of x_1 in the second constraint is varied. Since the element of tableau 2 in the row of x_1 and the column of y_2 is 0, it cannot be used as a pivot. In this case it is not possible to make x_1 nonbasic and y_2 basic in one iteration, but x_1 may be made nonbasic by introducing y_1 into the basis, which results in tableau 3, after which another iteration follows in which y_2 enters the basis and x_4 or y_3 leaves it. After having added γ to the element in the row of y_2 and the column of x_1, two iterations are made in order to find the solution of tableau 2 so that it depends on γ.

Another complication arises if an upper or lower bound of the parameter is found in the denominator of the element; for instance it would be $\beta = -4$ in tableau 6. In that case we should go back to tableau 3 and find the optimal solution for $\beta = -4$.

From the above it may be concluded that programming with parametric coefficients of the constraints is not very difficult, but that it leads to some computational complications.

CHAPTER 6

PARAMETRIC METHODS FOR LINEAR PROGRAMMING

6.1. The primal-dual method

Parametric programming may be used not only for finding solutions to problems in which some coefficients of a linear programming problem are dependent on a parameter; it may also be used in methods other than the simplex and the dual method for the solution of ordinary linear programming problems. In these cases the problem is reformulated by the inclusion of a parameter either in the coefficients of the objective function or in those of the constant terms, or both, after which an optimal solution is generated having certain properties; this solution is then the optimal solution of the original problem.

The most well-known parametric method for linear programming is the primal-dual method. This method as it was originally described is rather complicated and uses alternately the primal and the dual problem. In this section an interpretation of this method as a parametric method will be given, which makes the method much simpler to understand and to implement. Section 6.2 gives the original interpretation of the primal-dual method.

Related with any problem with a parametric objective function there is a problem with parametric constant terms and vice versa; this relationship is called parametric equivalence, which is explained in section 6.3. Parametric equivalence of problems gives rise to parametric equivalence of methods, so that any parametric method for linear programming has a method which is parametrically equivalent to it.

Another concept used is that of dual equivalence of methods, which means that any linear programming method has a method which is its dual equivalent. Section 6.4 develops the equivalent dual method of the primal-dual method. Section 6.5 discusses methods which are either the dual equivalent or the parametric equivalent, or both, of the primal-dual method. Hence the primal-dual method is one of four related parametric methods for linear programming.

Sections 6.6 and 6.7 deal with two other parametric methods for linear programming, the gradient method and the capacity method, and their related methods.

Section 6.8 deals with parametric programming with a parameter occurring both in the coefficients of the objective function and those of the constant term and develops parametric methods which make use of this.

Let us consider the following problem. Minimize

$$f = c'x \tag{1}$$

subject to

$$Ax = b,$$
$$x \geq 0. \tag{2}$$

It is assumed that the elements of b and c are nonnegative and that there exists at least one feasible solution to the constraints (2). If in some problems b has some negative elements, they may be made nonnegative by a multiplication of the constraint by -1. The assumption that the elements of c are nonnegative is nontrivial. Note that the typical minimization problem can be written in the form (1)–(2). In this section a method for solving this problem will be treated. This method is called the *primal-dual method*; it was discovered by Dantzig, Ford and Fulkerson.[1] The name of the method is related to an interpretation which was originally given to it, and which can be found in section 6.2. Here an interpretation of this method as a parametric method as formulated by van de Panne and Whinston[27] is given.

Let us consider the following related problem which we shall call the *extended problem*. Minimize

$$f^* = c'x + \lambda e'z \tag{3}$$

subject to

$$Ax + z = b,$$
$$x, z \geq 0; \tag{4}$$

z is a vector of m artificial variables, e is a vector of m unit elements, and λ is a variable parameter. The artificial variables are added only to equations which have no basic variable with a nonnegative value. Here we assume that none of the equations has such a variable.

For a sufficiently high value of λ, the solution of the extended problem must be the same as that of the original problem because such a value of λ will prevent the z-variables from having nonzero values. This is the

---

[1] See [7].

basic idea of the two-phase simplex method for linear programming, in which the extended problem with λ having a very high value is solved instead of the original problem. This amounts to minimizing first $e'z$ since the terms in λ are dominant, which leads to a feasible solution of the original problem; the original objective function is then used to find the optimal solution.

The primal-dual method also uses the extended problem, but instead of solving the problem for a high value of λ, we solve the problem first for $\lambda = 0$, after which λ is increased parametrically. The initial basic feasible solution $z = b$ is for $\lambda = 0$ also an optimal solution, since all elements of c are assumed to be nonnegative. After that, parametric linear programming is used to trace the optimal solutions of the extended problem for increasing values of λ. The solution of $\lambda \to \infty$ must be the solution of the original problem, if there is such a solution.

As an example, we take the problem used by Dantzig[6] for the primal-dual method. Minimize

$$f = x_1 + 4x_2 + 8x_3 + 8x_4 + 23x_5 \tag{5}$$

subject to

$$x_1 + 4x_2 - 5x_3 + 7x_4 - 4x_5 = 8,$$
$$- 4x_2 + 4x_3 - 4x_4 + 4x_5 = 2,$$
$$x_2 - 3x_3 + 4x_4 - 2x_5 = 2,$$
$$x_1, x_2, x_3, x_4, x_5 \geq 0. \tag{6}$$

In the formulation of the extended problem $\lambda(z_1 + z_2 + z_3)$ is added to the objective function and z_1, z_2 and z_3 are added to the left side of the respective equations of (6). The initial basic solution is then $z_1 = 8$, $z_2 = 2$ and $z_3 = 2$ and the corresponding initial tableau is obtained by subtracting λ times the sum of the equality constraints of the extended problem from its objective function. This objective function becomes

$$f^* = x_1 + 4x_2 + 8x_3 + 8x_4 + 23x_5$$
$$+ \lambda(12 - x_1 - x_2 + 4x_3 - 7x_4 + 2x_5). \tag{7}$$

Putting the λ-terms in a separate row, the initial tableau as given in tableau 0 of table 1 is obtained; the value of the terms without λ is indicated by f_c, that of the terms with λ by f_λ. By adding to the f_c-row λ times the f_λ-row for specific values of λ, a row is obtained which represents the objective function for specific values of λ.

In tableau 0 of table 1, the specific value of λ is first taken to be 0. For

Table 1

An application of the primal-dual method

Tableau	Basic variables	Values of basic variables	Nonbasic variables				
			x_1	x_2	x_3	x_4	x_5
0	f_c	0	-1	-4	-8	-8	-23
	f_λ	12	1	1	-4	7	-2
	z_1	8	$\underline{1}$	4	-5	7	-4
	z_2	2	$\overline{0}$	-4	4	-4	4
	z_3	2	0	1	-3	4	-2

Tableau	Basic variables	Values of basic variables		x_2	x_3	x_4	x_5
1	f_c	8		0	-13	-1	-27
	f_λ	4		-3	1	0	2
	x_1	8		4	5	7	-4
	z_2	2		-4	$\underline{4}$	-4	4
	z_3	2		1	$-\overline{3}$	4	-2

Tableau	Basic variables	Values of basic variables		x_2		x_4	x_5
2	f_c	$14\tfrac{1}{2}$		-13		-14	-14
	f_λ	$3\tfrac{1}{2}$		-2		1	1
	x_1	$10\tfrac{1}{2}$		-1		2	1
	x_3	$\tfrac{1}{2}$		-1		-1	1
	z_3	$3\tfrac{1}{2}$		-2		$\underline{1}$	1

Tableau	Basic variables	Values of basic variables		x_2			x_5
3	f_c	$63\tfrac{1}{2}$		-41			0
	f_λ	0		0			0
	x_1	$3\tfrac{1}{2}$		3			-1
	x_3	4		-3			$\underline{2}$
	x_4	$3\tfrac{1}{2}$		-2			1

Tableau	Basic variables	Values of basic variables		x_2	x_3		
4	f	$63\tfrac{1}{2}$		-41	0		
	x_1	$5\tfrac{1}{2}$		$1\tfrac{1}{2}$	$\tfrac{1}{2}$		
	x_5	2		$-1\tfrac{1}{2}$	$\tfrac{1}{2}$		
	x_4	$1\tfrac{1}{2}$		$-\tfrac{1}{2}$	$-\tfrac{1}{2}$		

this value, the initial solution is also an optimal solution of the extended problem, since all coefficients in the f_c-row are negative. Next, consider for what range of λ the present solution is an optimal one. Its upper bound is determined by

$$\underset{j}{\text{Min}} \, (- p_j / q_j | q_j > 0), \qquad (8)$$

where p_j stands for the element in the f_c-row and in the jth column and q_j for the element in the f_λ-row and in the same column. In tableau 0, it turns out that the highest value of λ for which the solution is optimal is 1, because for that value the coefficient of x_1 in the objective function becomes zero. According to the usual parametric procedure, x_1 enters the basis and z_1 leaves it. Tableau 0 is then transformed into tableau 1. Note that the value of f_λ has decreased. The column of z_1 is deleted.

The solution of tableau 1 is optimal for $\lambda = 1$. An upper bound on λ for which this solution is optimal is found by applying (8) again; this upper bound turns out to be $\lambda = 13$. x_3 must then enter the basis and z_2 leaves it; the column of z_2 is deleted in the resulting tableau.

In tableau 2, it turns out that the minimum in (8) is not unique, because the coefficients of x_4 and x_5 in the f-equation both vanish for $\lambda = 14$. This is a degenerate solution but it causes no difficulty because, whichever variable enters into the basis, the value of f_λ decreases. Hence, either variable may enter the basis. Choosing x_4, we find that z_3 must leave the basis. The solution of the resulting tableau is found to be the optimal solution for $\lambda \to \infty$. This optimal solution is not unique because the coefficient of f in the column of x_5 is zero. The corresponding alternative extreme-point optimal solution is generated in tableau 4.

6.2. The original formulation of the primal-dual method

The original formulation of the primal-dual method by Dantzig, Ford and Fulkerson, which also can be found in various textbooks, is rather different from the formulation in the previous section. Both formulations are equivalent in the sense that they lead to the same sequence of solutions. First the original formulation will be indicated, after which the equivalence with the parametric formulation of section 6.1 will be shown.

The primal-dual method starts with a feasible solution to the dual of the original problem. The general form of this dual problem is: Maximize

$$b'u \tag{9}$$

subject to

$$A'u \le c, \tag{10}$$

where u is a vector of m nonconstrained variables. Introducing a vector v of n slack variables, the constraints (10) can be written as

$$A'u + v = c,$$
$$v \ge 0. \tag{11}$$

Because it was assumed that the elements of c are nonnegative, an initial feasible solution is $v = c$, $u = 0$. However, this solution is not likely to be an optimal one, since all or some elements of b may be positive.

We also consider the primal feasibility problem. Minimize

$$e'z \tag{12}$$

subject to

$$Ax + z = b,$$
$$x, z \ge 0, \tag{13}$$

and its dual: Maximize

$$b'u^* \tag{14}$$

subject to

$$A'u^* \le 0, \tag{15}$$

$$u^* \le e. \tag{16}$$

Constraint (15) can be written as

$$A'u^* + v^* = 0,$$
$$v^* \ge 0. \tag{17}$$

The primal-dual method is based on the following ideas. An initial solution of the dual problem (9)–(11) is available. Suppose there are some vectors u^*, v^* which give a positive value of the objective function of the dual feasibility problem; for this latter solution, $v^* \ge 0$ need not be satisfied. If the first solution is $\bar{u}$, $\bar{v}$, and the second $\bar{u}^*$, $\bar{v}^*$, then the solution

$$\bar{u} + k\bar{u}^*, \qquad \bar{v} + k\bar{v}^* \tag{18}$$

must give a higher value of the objective function of the dual problem for $k > 0$, since $b'\bar{u}^*$ was assumed to be positive. However, $\bar{v}^*$ was not necessarily positive, so that for some value of $k \geq 0$, $\bar{v} + k\bar{v}^*$ might become negative. Hence we determine

$$\bar{k} = \underset{i}{\text{Min}} \, (\bar{v}_i / - \bar{v}_i^* | \bar{v}_i^* < 0). \tag{19}$$

If this value of k is used in (18), the objective function of the dual is increased as much as possible without making its solution unfeasible. After this, a new solution to the dual feasibility problem is generated and added to the solution of the dual as before. The dual objective function is increased until no improvement is possible because the objective function of the dual feasibility problem has become zero. The optimal solution of the dual problem has been obtained.

The solutions of the dual feasibility problem are obtained via the primal feasibility problem; the dual variables then appear in the row of the objective function. The method usually starts with a solution of the dual problem $v = c$, $u = 0$, which is a feasible solution. For the primal feasibility problem, the solution $z = b$, $x = 0$ is taken. The corresponding basic solution of the dual feasibility problem is

$$u^* = e, \qquad v^* = -A'u^* = -A'e; \tag{20}$$

the value of its objective function is $b'e$, which is positive if b has at least one positive element. Adding a multiple k determined by (19) of the solution (20) to that of the dual problem, we find that the objective function of the dual problem is increased for $k \neq 0$; $k = 0$ can only occur if some basic variables are zero for corresponding negative v^*-variables. This can only occur in the first iteration, because there are some additional requirements for solutions of the feasibility problem in later iterations. These are, that given the improved solution of the dual problem, a *restricted* primal unfeasibility problem is solved, in which the variables to enter the basis are restricted to those which have zero corresponding variables in the dual problem. These will be the basic variables and the variable connected with $\bar{k}$ in (19); if the minimum was not unique, then all the variables connected with this minimum are included plus possible other variables with zero v-variables. The optimal solution to this restricted primal feasibility problem is one with v^*-variables of the corresponding dual which are zero. Hence, when we next add the solution of the corresponding dual feasibility problem to the dual solution, k is nonzero. After this, another restricted primal feasibility problem is solved and so on until the objective function of the feasibility problem has become zero; in that case the optimal solution of the dual problem has been found, and also the solution of the original problem.

Let us now compare the original formulation with the parametric approach using the numerical example presented in the previous section. Tableau 0 of table 1 gives the initial solution of the primal feasibility problem; the f_λ-row gives the value of its objective function and the values of the basic variables of the corresponding dual solution. Hence we have

$$v_1^* = -1, \qquad v_2^* = -1, \qquad v_3^* = 4, \qquad v_4^* = -7, \qquad v_5^* = 2.$$

The f_c-row now gives the values of the objective function of the original dual problem and its corresponding solution. The solution of the dual problem is

$$v_1 = 1, \qquad v_2 = 4, \qquad v_3 = 8, \qquad v_4 = 8, \qquad v_5 = 23.$$

Now k times the f_λ-row is added to the f_c-row, thus increasing the objective function of the dual from 0 to $12k$. The maximum value of k turns out to be 1; the f-row for $\lambda = 1$ gives the improved value of the objective function of the dual as well as the corresponding solution of the dual. According to the parametric procedure, x_1 must enter the basis. The same is true for the primal-dual method, since in the restricted primal feasibility problem only x_1 and the basic variables may be in the basis. In the resulting transformation both methods transform the rows of basic variables and the f_λ-row in the same way; the original formulation has no f_c-row and it does not transform the present solution of the dual problem. The parametric procedure transforms the f-rows but their sum for $\lambda = 1$ does not change since the element in the x_1-column is zero.

In the next tableau the original formulation adds k times the f_λ-row to the f-row for $\lambda = 1$. $\bar{k}$ is then found to be 12 and is connected with x_3. The parametric procedure adds λ times the f_λ-row to the f_c-row and finds $\lambda = 13$, connected with x_3. The result, the sum of the f_c- and f_λ-rows for $\lambda = 13$, is in both cases the same. As is easily seen, k is equal to the increase in λ.

Each cycle in the original formulation corresponds with a particular value of λ in the parametric approach. The restricted columns of the primal feasibility problem are the same as the columns which have the same ratio λ of elements in the f_c- and f_λ-row. Usually the optimal solution to the restricted primal feasibility problem will be obtained in one iteration, but it is possible that it takes more iterations. This can be the case when the maximum from which k or λ is found is not unique. The adjusted dual solution then contains more than one zero apart from the basic variables. An example of this can be found in tableau 2, where both x_4 and x_5 are connected with the minimum in (19). x_4 and x_5 are then both columns of the restricted primal feasibility problem and if x_5 is chosen as a basic variable instead of x_4, it takes two iterations to obtain the solution of the restricted problem. Again there is no substantial difference with the parametric procedure.

Hence it may be concluded that both procedures are equivalent and differ only by having k in the original formulation which is equal to the increment of λ used in the parametric method. The two methods may, therefore, be seen as two alternative interpretations of the same algorithm.

6.3. Parametric equivalence

The problems solved in sections 5.4 and 5.6 are very similar. In the problem of section 5.4 we have as objective function

$$f = 4x_1 + 1\tfrac{1}{2}x_2 + 2\tfrac{3}{4}x_3 + 3\tfrac{1}{2}x_4 - \lambda(x_1 + \tfrac{1}{2}x_2 + \tfrac{3}{4}x_3 + \tfrac{3}{4}x_4). \tag{21}$$

The problem of section 5.6 has the same objective function but without the λ-terms; instead it has the additional constraint

$$x_1 + \tfrac{1}{2}x_2 + \tfrac{3}{4}x_3 + \tfrac{3}{4}x_4 \le \mu. \tag{22}$$

The set-up tableaux of both problems (see tables 3 and 8 of chapter 5) have the same elements, with, however, different interpretations; the f_λ-row of tableau 00 of table 3 is part of the objective function and the y_μ-row of tableau 00 of table 8 is that of a constraint. The first problem can be interpreted as a variation of the cost of a resource of which an abundant supply is available; the second is interpreted as a variation of the available supply of a resource which does not cost anything. It will be shown that both variations generate solutions which are equivalent in a certain sense.

Let us consider the following problem with a general parametric objective function; this problem is called problem A: Maximize

$$f = (p - \lambda q)'x \tag{23}$$

subject to

$$Ax \le b,$$

$$x \ge 0. \tag{24}$$

The related problem with a parametric constraint which we call problem B, is as follows. Maximize

$$f = p'x \tag{25}$$

subject to

$$Ax \le b,$$

$$q'x \le \mu,$$

$$x \ge 0. \tag{26}$$

The set-up tableau for both problems is the same, apart from the following differences (see tableau I of table 2), in which the elements b_λ and b_0 should be put at zero. For problem A there is a row f_λ, the elements of which are interpreted as the coefficients of the objective function to be multiplied with λ. For problem B the same row is interpreted as belonging to a parametric constraint with slack variable y_μ. In problem B there is an extra μ-column which contains the part of the values of basic variables to be multiplied by μ; this column has a unit in the row of y_μ and zero's elsewhere.

Now suppose that for problem A a basic solution has been found which is optimal in the wide sense for $\underline{\lambda} \le \lambda \le \bar{\lambda}$. The tableau of such a solution can be represented as in the set-up tableau given in tableau I, with the basic variables being denoted as y-variables and the nonbasic

Table 2

Corresponding parametric tableaux

Tableau	Basic variables	c-term	(μ-term)	x_1	x_2	.	x_n
I	f_c	b_0	0	$-p_1$	$-p_2$	.	$-p_n$
	$f_\lambda(y_\mu)$	b_λ	1	q_1	q_2		q_n
	y_1	b_1	0	a_{11}	a_{12}	.	a_{1n}
	y_2	b_2	0	a_{21}	a_{22}	.	a_{2n}
	.	.	.	.	.	.	.
	y_m	b_m	0	a_{m1}	a_{m2}	.	a_{mn}
				y_μ	x_2	.	x_n
II	f	$b_0 + p_1 q_1^{-1} b_\lambda$	$p_1 q_1^{-1}$	$p_1 q_1^{-1}$	$-p_2 + p_1 q_1^{-1} q_2$	.	$-p_n + p_1 q_1^{-1} q_n$
	x_1	$q_1^{-1} b_\lambda$	q_1^{-1}	q_1^{-1}	$q_1^{-1} q_2$	.	$q_1^{-1} q_n$
	y_1	$b_1 - a_{11} q_1^{-1} b_\lambda$	$-a_{11} q_1^{-1}$	$-a_{11} q_1^{-1}$	$a_{12} - a_{11} q_1^{-1} q_2$	.	$a_{1n} - a_{11} q_1^{-1} q_n$
	y_2	$b_2 - a_{21} q_1^{-1} b_\lambda$	$-a_{21} q_1^{-1}$	$-a_{21} q_1^{-1}$	$a_{22} - a_{21} q_1^{-1} q_2$	.	$a_{2n} - a_{21} q_1^{-1} q_n$
	.	.	.	.	.	.	.
	y_m	$b_m - a_{m1} q_1^{-1} b_1$	$-a_{m1} q_1^{-1}$	$-a_{m1} q_1^{-1}$	$a_{m2} - a_{m1} q_1^{-1} q_2$	.	$a_{mn} - a_{m1} q_1^{-1} q_n$

ones as x-variables; the elements b_0 and b_λ will in general be different from zero. Since the solution is feasible, the elements $b_1, \ldots, b_m$ are nonnegative; furthermore $-p_j + q_j \lambda \geq 0$ for $\underline{\lambda} \leq \lambda \leq \bar{\lambda}$, for each j. Let us assume that x_1 is connected with $\underline{\lambda}$, so that, when λ is varied upwards, x_1 has been the leaving basic variable in the previous iteration; this implies $\underline{\lambda} = p_1/q_1$ with $q_1 > 0$.

From this tableau, a corresponding tableau for problem B will be generated. The f_λ-row is interpreted as the row of the slack variable y_μ and the μ-term column is used. Let us introduce into the basis the variable connected with $\underline{\lambda}$, which according to our assumption is x_1, and let us take as the leaving basic variable y_μ. Hence q_1 is the pivot of the transformation which results in tableau II.

The element b_λ in tableau II can be interpreted as minus the quantity of the source used; hence $b \leq 0$. Let us consider the values of basic variables of tableau II for $\mu = -b_\lambda$; we find

$$f = b_0 + p_1 q_1^{-1} b_\lambda \quad + \quad q_1^{-1} \mu = 0,$$
$$x_1 = q_1^{-1} b_\lambda \quad\quad\quad + \; p_1 q_1^{-1} \mu = b_0.$$
$$y_1 = b_1 - a_{11} q_1^{-1} b_\lambda - a_{11} q_1^{-1} \mu = b_1,$$

$$\cdot$$
$$\cdot$$
$$\cdot$$

$$y_m = b_m - a_{m1} q_1^{-1} b_\lambda - a_{m1} q_1^{-1} \mu = b_m, \tag{27}$$

By assumption, the lower bound for μ for which this solution is feasible is found from the value of x_1:

$$q_1^{-1} b_\lambda + q_1^{-1} \mu = 0 \quad\quad \text{or} \quad\quad \mu = -b_\lambda;$$

for this value of μ all other basic variables are nonnegative, see (27). The elements in the f-row for the nonbasic x-variables may be written as

$$-p_j + p_1 q_1^{-1} q_j = -p_j + \underline{\lambda} q_j, j = 2, \ldots, n.$$

Since the solution of tableau I was optimal for $\underline{\lambda} \leq \lambda \leq \bar{\lambda}$, these elements are nonnegative; hence the solution of tableau II is optimal in the narrow sense. It is feasible for $-b_\lambda \leq \mu \leq \bar{\mu}$; the upper bound $\bar{\mu}$ may be determined from the tableau.

It may be concluded that tableaux I and II give corresponding solutions of problems A and B; the values of the basic variables and those of dual variables corresponding with nonbasic variables are the same for the lower bounds of the parameters λ and μ in each problem. For higher values of λ the dual variables of problem A change linearly until the optimality requirement causes a variable to enter the basis. For higher values of μ the primal variables of problem B change linearly until the feasibility requirement causes a variable to leave the basis.

If problems A and B have corresponding tableaux, we should find a sequence of such tableaux for increasing values of λ and decreasing values of μ and vice versa. This means that the same variables should enter or leave the basis. This is indeed the case and it can be shown as follows. First it should be noted that, since problem B has an additional constraint, tableau II has one additional basic variable compared with tableau I, namely x_1, which has a value zero for $\mu = -b_\lambda = \underline{\mu}$. Let us consider increasing values of λ and decreasing values of μ. Since x_1 in tableau I is connected with $\underline{\lambda}$, it has left the basis in the previous iteration. If in tableau II μ is decreased it means that the variable connected with μ should leave the basis. Hence the leaving basic variable is the same for both problems, but it leaves the basis in problem B one iteration later than in problem A. In tableau I the new basic variable is determined by

$$\operatorname*{Min}_{j}(p_i/q_i \,|\, q_i < 0). \tag{28}$$

In tableau II it is determined by

$$\operatorname*{Min}_{j}\left(\frac{-p_i + p_1 q_1^{-1} q_i}{-q_1^{-1} q_i} \,\middle|\, q_1^{-1} q_i < 0\right). \tag{29}$$

Since $q_1^{-1} > 0$, this reduces to

$$\operatorname*{Min}_{j}((p_i/q_i)q_1 - p_1 \,|\, q_i < 0). \tag{30}$$

The variable connected with (30) must be the same as the variable connected with (28); hence the same variable enters the basis in both tableaux.

These relationships can conveniently be checked by comparing table 3 with table 8 of chapter 5. Table 3 indicates the basic variables in the successive iterations. If λ is decreased and μ increased, then the tableaux of table 8 always contain an additional basic variable, which is the one which in table 3 will be basic in the next iteration. Table 4 gives a complete summary of the results of the equivalent problems. For problem A and table 3 of chapter 5 the first two rows apply, while for problem

Table 3

Comparison of iterations for equivalent parametric problems

Tableau	00	0	1	2	3
	λ		$4\frac{2}{3}$	$3\frac{3}{5}$	$2\frac{2}{3}$
Table 3		y_1	y_1	y_1	x_1
		y_2	y_3	x_1	x_3
		y_3	x_4	x_4	x_4
	μ		15	$27\frac{1}{2}$	$31\frac{1}{4}$
Table 8	y_1	y_1	y_1	y_1	y_λ
	y_2	y_2	y_3	x_1	x_1
	y_3	y_3	x_1	x_3	x_3
	y_λ	x_4	x_4	x_4	x_4

B and table 8 of chapter 5 the last two rows should be used. For problem A, which is the problem with a parametric objective function, the primal variables change at critical values of λ only, while the dual variables change linearly with λ. Hence, in the second column the values of the primal variables are given for tableau 0, in the fourth column the corresponding values for tableau 1, and so on. For the dual variables, values can only be given for specific values of λ. The third column gives these values for $\lambda = 4\frac{2}{3}$; they can be obtained either from tableau 0 or tableau 1. $f(A)$ indicates the value of f for the critical value of λ.

As far as the values of primal and dual variables other than f are concerned, the same results can be stated for the equivalent problem having parametric constant terms, problem B, if the column headings are changed, see the last two rows. Now the values of the dual variables change at critical values of μ, but the values of the primal variables vary linearly with μ. The third column now gives the values of the dual variables for tableau 0 for table 8, whereas the second column gives the values of primal variables of the same solution for $\mu = 0$, and the fourth column gives these values for $\mu = 15$.

Figure 1 illustrates for the example the relation between λ and the amount of the resource used, which is denoted by m (in tableau I of table 2 indicated by $-b_\lambda$), see the dashed lines. The full lines give the relation between μ and the dual variable of the parametric constraint which is denoted by l for the example of table 8. Note that, as they should, m and

Table 4
Summary of results for equivalent parametric problems

Problem A, table 3	Tableau	0		1		2		3	
	λ		$4\frac{2}{3}$		$3\frac{3}{5}$		$2\frac{2}{3}$		0
	$-f_\lambda$	0		15		$27\frac{1}{2}$		$31\frac{1}{4}$	
	$f(A)$	0		16		$41\frac{2}{3}$		125	
	x_1	−		−		20		20	
	v_1		$\frac{2}{3}$		0		−		−
	x_2	−		−		−		−	
	v_2		$\frac{5}{6}$		$\frac{3}{10}$		$2\frac{1}{6}$		1
	x_3	−		−		−		10	
	v_3		$\frac{3}{4}$		$\frac{7}{20}$		0		−
	x_4	−		20		10		5	
	v_4	0		−		−		−	
	y_1	100		100		20		−	
	u_1		−		−		−		$\frac{1}{2}$
	y_2	40		−		−		−	
	u_2		−		$\frac{2}{5}$		$\frac{1}{6}$		$1\frac{1}{2}$
	y_3	60		20		−		−	
	u_3		−		−		$\frac{7}{12}$		$\frac{1}{4}$
	$f(B)$	0		70		115		125	
	f_μ		$4\frac{2}{3}$		$3\frac{3}{5}$		$2\frac{2}{3}$		0
Problem B, table 8	Tableau		0		1		2		3
	μ	0		15		$27\frac{1}{2}$		$31\frac{1}{4}$	

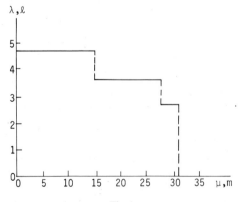

Fig. 1.

l decrease for increasing λ and μ. It can easily be imagined that in the case of many products and one scarce resource, the curve of figure 1 would be a nearly continuous downwards sloping curve, which is the demand curve for a resource of production in an economy with given prices.

Since for a problem with general parametric objective function an equivalent problem can be formulated with a parametric constraint it may be asked whether a problem with general parametric constant terms also has an equivalent problem. Since the dual of a problem with parametric constant terms is a problem with a parametric objective function, the equivalent parametric problem of this dual problem can be used to find the equivalent parametric problem of the original problem.

Let us consider the following general problem with parametric constant terms. Maximize for $0 \leq \lambda \leq \infty$

$$f = p'x \tag{31}$$

subject to

$$Ax \leq b + a\lambda,$$
$$x \geq 0. \tag{32}$$

Its dual problem is: Minimize

$$g = (b + a\lambda)'u \tag{33}$$

subject to

$$A'u \geq p,$$
$$u \geq 0. \tag{34}$$

For this dual problem an equivalent parametric problem can be formulated; this turns out to be: Minimize

$$g = b'u \tag{35}$$

subject to

$$A'u \geq p,$$
$$-a'u \geq -\mu,$$
$$u \geq 0. \tag{36}$$

The dual of this problem is: Maximize

$$f = p'x - \mu x_{\mu} \tag{37}$$

subject to

$$Ax - ax_\mu \le b, \tag{38}$$
$$x, x_\mu \ge 0.$$

This problem is equivalent to that of (31)–(32). The details of the relationship between the two problems and between corresponding solutions are left as an exercise.

The problem (31)–(32) can be viewed as a production planning problem with a parametric growth in resources, or as a problem with an activity at a level λ which varies autonomously and which adds a_i units of resource i for all i.

In the equivalent parametric problem (37)–(38) there is a resource-producing activity with a level x_μ which can be freely varied, but which carries a cost of μ per unit which is parametrically varied.

Since the problem (37)–(38) is a problem with a parametric objective function, the equivalent parametric problem of this problem may be considered. It turns out to be: Maximize

$$f = p'x \tag{39}$$

subject to

$$Ax - ax_\mu \le b,$$
$$x_\mu \le \lambda,$$
$$x, x_\mu \ge 0. \tag{40}$$

This problem is formally different from that of (31)–(32), so that parametric equivalence is not symmetric. It is easily seen, however, that the solutions of both problems are the same apart from x_μ, which only occurs in the problem (39)–(40).

It may be observed that the problems (31)–(32), (37)–(38) and (39)–(40) have an increasing order of complexity; the second problem has an additional variable compared with the first one and the third problem has an additional constraint compared with the second problem. Hence taking the equivalent parametric problem always leads to slightly larger problems.

6.4. The dual equivalent of the primal-dual method

Consider the typical minimization problem. Minimize

$$f = c'x \tag{41}$$

subject to

$$Ax \geq b,$$

$$x \geq 0. \tag{42}$$

We assume that $c \geq 0$ and $b \geq 0$. This problem cannot be solved by the simplex method in one phase. The dual problem is: Maximize

$$g = b'u \tag{43}$$

subject to

$$A'u \leq c,$$

$$u \geq 0. \tag{44}$$

This problem can be solved by the simplex method in one phase, using the initial basic feasible solution $v = c$.

Earlier it was demonstrated that the dual method applied to the primal problem leads to successive solutions of the primal problem which correspond to the successive solutions of an application of the simplex method to the dual problem. Hence there is a certain relationship between the simplex method and the dual method, which may be given the name of *dual equivalence.*

More formally, it may be stated that one method is the dual equivalent of another method if the first method applied to the primal problem gives a sequence of solutions which corresponds to the sequence of solutions generated by an application of the second method to the dual problem. Hence the dual method is the dual equivalent of the simplex method and the simplex method is the dual equivalent of the dual method. Since the dual of the dual is the original problem, if a method is dually equivalent with a second method, then that second method is dually equivalent to the first method.

The usefulness of a dually equivalent method is that the primal problem may not satisfy the conditions for application of a certain method, but its dual may satisfy these; in that case, the dually equivalent method may be used. For instance, the simplex method may not be applied to the typical minimization problem because the basic solution of the set-up tableau is not feasible. However, its dual problem has a basic feasible solution, so that the simplex method may be applied to it, which means that the primal problem satisfies the conditions for the dual equivalent of the simplex method, which is the dual method.

These ideas will now be applied to the primal-dual method. This method can only be applied to problems for which the solution of the set-up tableau is optimal in the narrow sense. It can therefore not be applied to typical maximization problems formulated as: Maximize

$$f = p'x \tag{45}$$

subject to

$$Ax \le b,$$
$$x \ge 0, \tag{46}$$

because the initial basic solution $y = b$ is not optimal in the narrow sense. However, its dual problem is: Minimize

$$g = b'u \tag{47}$$

subject to

$$A'u \ge p,$$
$$u \ge 0. \tag{48}$$

The initial basic solution of this problem is $v = -p$, which is not feasible, but which is optimal in the narrow sense, so that the primal-dual method may be applied to it. The dual equivalent of the primal-dual method is then the method which generates for the primal problem successive solutions which correspond to the successive solutions of an application of the primal-dual method to the dual problem.

Let us now consider the dual equivalent of the primal-dual method. In the primal-dual method, the typical minimization problem is solved by solving the following parametric problem for $\lambda \to \infty$: Minimize

$$f = c'x + \lambda e'z \tag{49}$$

subject to

$$Ax - y + z = b,$$
$$y, z \ge 0. \tag{50}$$

The elements of c should be nonnegative.

The dual of a typical minimization is a typical maximization problem and can therefore not be solved by the primal-dual method. However, the dual of a typical maximization problem is a typical minimization problem: Minimize

$$g = b'u \tag{51}$$

subject to

$$A'u \ge p,$$
$$u \ge 0. \tag{52}$$

For this problem the extended problem can be formulated, which is: Minimize

$$g^* = b'u + \lambda e'w \qquad (53)$$

subject to

$$A'u - v + w = p,$$
$$u, v, w \geq 0. \qquad (54)$$

For notational reasons w is used instead of z. The solution for $\lambda \to \infty$ then gives the optimal solution of the dual problem.

The dual problem of this extended dual problem is: Maximize

$$f^* = p'x \qquad (55)$$

subject to

$$Ax \leq b,$$
$$x \leq \lambda e,$$
$$x \geq 0. \qquad (56)$$

Instead of additional variables, we have additional constraints. These constraints give for each variable x_i a parametric upper bound $x_i \leq \lambda$. Just as the z-variables in the primal-dual method are called artificial, we may call these additional constraints artificial. For $\lambda \to \infty$ these constraints are not binding unless there is an infinite optimal solution; hence the optimal solution of the problem with the additional constraints for $\lambda \to \infty$ and that of the original problem must be the same, except for an infinite optimal solution.

Let us now consider the construction of the initial tableau for the dual equivalent of the primal-dual method. This tableau can easily be constructed by comparison with the construction of the initial tableau for an application of the primal-dual method to the dual problem. This last construction is done in table 5. The set-up gives the equations of the problem (53)–(54) with v as the vector of basic variables. Then w, the vector of artificial variables is made basic instead of v, which results in the initial tableau. This tableau has an optimal solution for $\lambda = 0$, so that the parametric variation may start from here.

Table 6 gives the set-up tableau for the equivalent dual problem (55)–(56). The z's are the slack variables of the artificial constraints. In table 5, the w-variables replaced the v-variables, so that in table 6 the x-variables (corresponding to the v-variables) should replace the z-variables (corresponding to the w-variables). After block-pivoting on the underlined unit matrix, the initial tableau of table 6 results. This tableau

Table 5

Set-up and initial tableau for an application of the primal-dual method to the dual problem

	Basic variables	Val. basic var.	u	v	w	g^*
Set-up tableau	g_c^*	0	$-b'$	0	0	1
	g_λ^*	0	0	0	$-e'$	1
	v	$-p$	$-A'$	I	$-I$	0
Initial tableau	g_c^*	0	$-b'$	0	0	1
	g_λ^*	$e'p$	$e'A'$	$-e'$	0	1
	w	p	A'	$-I$	I	0

is optimal in the narrow sense ($p \geq 0$) and is feasible for $\lambda = 0$ ($b \geq 0$). After this, the parametric variation of λ may start. The solutions which are obtained in this manner must be the corresponding solutions found by an application of the primal-dual method in table 5, since for the same value of λ each problem is the dual of the other problem.

As an example for this method, let us consider the following simple problem. Maximize

$$f = 2x_1 + 3x_2 \tag{57}$$

subject to

$$-x_1 + x_2 \leq 1\tfrac{1}{2},$$
$$-x_1 + 2x_2 \leq 4,$$
$$x_1 + x_2 \leq 6\tfrac{1}{2},$$
$$x_1, x_2 \geq 0. \tag{58}$$

Table 6

Set-up and initial tableau for the dual equivalent of the primal-dual method

	Basic variables	Val. bas. var.		x	y	z	f^*
		c-term	λ-term				
Set-up tableau	f^*	0	0	$-p'$	0	0	1
	y	b	0	A	I	0	0
	z	0	e	I	0	I	0
Initial tableau	f^*	0	$p'e$	0	0	p'	1
	y	b	$-Ae$	0	I	$-A$	0
	x	0	e	I	0	I	0

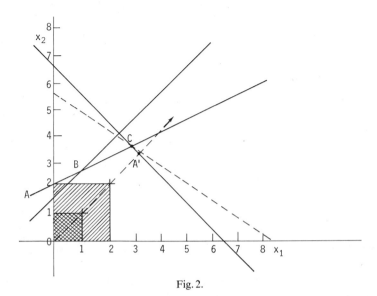

Fig. 2.

Figure 2 gives a geometrical interpretation of this problem. The dashed line through C is a line of equal values of the objective function. To this problem the following constraints are added:

$$x_1 \leq \lambda, \tag{59}$$

$$x_2 \leq \lambda. \tag{60}$$

For $\lambda = 1$, the feasible region is doubly shaded, for $\lambda = 2$, it is shaded once. It is obvious that for $\lambda = 0$, the optimal solution is $x_1 = 0, x_2 = 0$. An increasing value of λ gives higher values of x_1 and x_2; the solution follows the $45°$ line until the point A' is reached. After that, the solution moves along $A'C$, the third constraint of (58); this means that the constraint $x_1 \leq \lambda$ has become nonbinding. In C the second constraint becomes binding for the solution replacing $x_2 \leq \lambda$. Any increase in λ now leaves the optimal solution of the problem including (59) and (60) unchanged; hence we must have found the optimal solution to the problem without these constraints.

Table 7 gives the tableaux for an application of the method. Tableau 0 is the initial tableau constructed in accordance with the initial tableau of table 6. For $\lambda = 3\frac{1}{4}$, y_3 becomes zero, which means that the third constraint is binding; this is in A' in figure 2. z_1 is found to be the new basic variable, so that (59) is not binding any more. The solution of tableau 1 is feasible for $3\frac{1}{4} \leq \lambda \leq 3\frac{1}{2}$. For $\lambda = 3\frac{1}{2}$, y_2 becomes zero and should therefore

Table 7

An application of the dual equivalent of the primal-dual method

Tableau	Basic variables	Val. bas. var.		Ratio	z_1	z_2
		c-term	λ-term			
	f	0	5		2	3
	y_1	$1\frac{1}{2}$	0		1	-1
0	y_2	4	-1	4	1	-2
	y_3	$6\frac{1}{2}$	-2	$3\frac{1}{4}$	-1	-1
	$[x_1$	0	1		1	0]
	$[x_2$	0	1		0	1]
					y_3	z_2
	f	13	1		2	1
	y_1	8	-2	4	1	-2
1	y_2	$10\frac{1}{2}$	-3	$3\frac{1}{2}$	1	-3
	$(z_1$	$-6\frac{1}{2}$	2		-1	1)
	x_1	$6\frac{1}{2}$	-1	$6\frac{1}{2}$	1	-1
	$[x_2$	0	1		0	1]
					y_3	y_2
	f	$16\frac{1}{2}$	0		$2\frac{1}{3}$	$\frac{1}{3}$
	y_1	1	0		$\frac{1}{3}$	$-\frac{2}{3}$
2	$(z_2$	$-3\frac{1}{2}$	1		$-\frac{1}{3}$	$-\frac{1}{3}$)
	$(z_1$	-3	1		$-\frac{2}{3}$	$\frac{1}{3}$)
	x_1	3	0		$\frac{2}{3}$	$-\frac{1}{3}$
	x_2	$3\frac{1}{2}$	0		$\frac{1}{3}$	$\frac{1}{3}$

leave the basis; z_2 is found to be the new basic variable. In tableau 2 the optimal solution of the original problem has been found because the additional constraints are nonbinding for $\lambda \to \infty$. Note that in tableaux 1 and 2 the rows of z-variables can be deleted; furthermore, the rows of additional constraints need not be written down because they are trivial. This means that compared with the tableau for the simplex method only one additional λ-term column is required.

The method described requires that an initial feasible solution is available since in this case it can always be formulated as the typical maximization problem (45)–(46) with $b \geq 0$. If not all the elements of p are positive, we should only add additional constraints for the x-variables that have positive coefficients. In the initial tableau only those x-variables should become basic that have additional constraints.

Since the method requires the same initial solution as the simplex method, it can be compared with that method. Geometrically, the dual equivalent of the primal-dual method seems to have advantages, since, as in figure 2, it goes *into* the feasible region, in contrast to the simplex method which proceeds along the edges of the feasible region; in figure 2 it can be seen that the simplex method requires 3 iterations, going from the origin via A and B to C.

But figures are sometimes deceptive. In many cases the number of variables, denoted by n, greatly exceeds the number of constraints, denoted by m. In the case of n variables with positive coefficients in the objective function the number of additional constraints is also n. This means that the dual equivalent of the primal-dual method will require at least n iterations since only if all z-variables have become basic, the additional constraints are not binding any more. Since it is known that the simplex method requires usually $m-3m$ iterations to find the optimal solution, the simplex method should be preferred to the dual equivalent of the primal-dual method, except for problems with few variables with positive coefficients in the objective function; in the last kind of problems the dual equivalent of the primal-dual method may have an advantage.

6.5. The parametric and dual equivalents of the primal-dual method

The extended problem (13)–(14) for the primal-dual method may be written as follows: Maximize

$$f = -c'x - \lambda e'z \tag{61}$$

subject to

$$Ax + z = b, \tag{62}$$

$$z \geq 0. \tag{63}$$

This is a parametric programming problem with a parametric objective function. Then there must be a problem which is parametrically equivalent. Using the relationship between problems A and B in section 6.3, we find for this parametric equivalent: Maximize

$$f = -c'x \tag{64}$$

subject to

$$Ax + z = b, \tag{65}$$

$$e'z \leq \mu, \tag{66}$$

$$x, z \geq 0. \tag{67}$$

The objective function can be written as: Minimize

$$f = c'x. \tag{68}$$

Since the problems (61)–(63) and (65)–(68) are parametrically equivalent, the successive solutions found for an increasing value of λ in (61)–(63) and for a decreasing value of μ in (65)–(68) should be the same for critical values of the parameter. Hence the optimal solution to the original problem must be found by solving the problem (65)–(68) first for $\mu \to \infty$ and then for decreasing values of μ until $\mu = 0$, at which point the solution of the original problem should have been reached.

The set-up tableau for this problem can be found in tableau I of table 8. In this tableau, the z-variables cannot be considered as basic variables since they have nonzero elements in the z_μ-row. After block-pivoting on the unit-matrix tableau II is found, which provides a solution which is optimal in the wide sense for $\mu \geq e'b$. For lower values of μ, z_μ will leave the basis and one of the x-variables will become basic.

Table 9 gives an application of this method to the problem used in section 6.1.

Note that a method which is the parametric equivalent of another method generates the same solution as that method and hence has no advantage over that method except that we may have the parameter in the constant terms instead of the objective function or vice versa. The importance of parametric equivalence is that it enables us to recognize that seemingly unrelated methods are essentially the same. On the other

Table 8

Set-up and initial tableau for equivalent parametric method of
the primal-dual method

Tableau	Basic variables	Values basic variables			
		c-term	μ-term	x	z
I	f	0	0	$-c'$	0
	z_μ	0	1	0	e'
		b	0	A	I
II	f	0	0	$-c'$	0
	z_μ	$-e'b$	1	$-e'A$	0
	z	b	0	A	I

Table 9

Application of equivalent parametric method of the primal-dual method

Tableau	Basic variables	Val. bas. var. c-term	Val. bas. var. λ-term	x_1	x_2	x_3	x_4	x_5
	f	0	0	-1	-4	-8	-8	-23
	z_μ	-12	1	-1	-1	4	-7	2
0	z_1	8	0	1	4	-5	7	-4
	z_2	2	0	0	-4	4	-4	4
	z_3	2	0	0	1	-3	4	-2
				z_μ	x_2	x_3	x_4	x_5
	f	12	-1	-1	-3	-12	-1	-25
	x_1	12	-1	-1	1	-4	7	-2
1	z_1	-4	1	1	3	$\underline{-1}$	0	-2
	z_2	2	0	0	-4	4	-4	4
	z_3	2	0	0	1	-3	4	-2
				z_μ	x_2		x_4	x_5
	f	60	-13	-13	-39		-1	-1
	x_1	28	-5	-5	-11		7	6
2	x_3	4	-1	-1	-3		0	2
	z_2	-14	4	4	8		$\underline{-4}$	-4
	z_3	14	-3	-3	-8		4	4
				z_μ	x_2			x_5
	f	$63\frac{1}{2}$	-14	-14	-41			0
	x_1	$3\frac{1}{2}$	2	2	3			-1
3	x_3	4	-1	-1	-3			2
	x_4	$3\frac{1}{2}$	-1	-1	-2			1
	z_3	0	1	$\underline{1}$	0			0
					x_2			x_5
	f	$63\frac{1}{2}$	0		-41			0
	x_1	$3\frac{1}{2}$	0		3			-1
4	x_3	4	0		-3			2
	x_4	$3\frac{1}{2}$	0		-2			1
	z_μ	0	1		0			0

hand, the concept of dually equivalent methods is more useful because it leads to methods which are basically different.

In previous sections, the primal-dual method was taken as point of departure, from which the dual equivalent and the parametric equivalent of the primal-dual method were developed. These two methods can be used to develop dually and parametrically equivalent methods.

Let us first consider the dual equivalent of the dual equivalent of the primal-dual method. Since the dual of the dual is the original problem, the dual equivalent indicated is simply the primal-dual method.

Next, consider the parametric equivalent of the dual equivalent of the primal-dual method. The extended problem for the dual equivalent of the primal-dual method is (see (55) and (56)): Maximize

$$f = p'x \tag{69}$$

subject to

$$Ax \le b, \tag{70}$$

$$x \le \lambda e, \tag{71}$$

$$x \ge 0. \tag{72}$$

The optimal solution of the original problem is found by first taking the optimal solution for $\lambda = 0$ and then varying λ parametrically until $\lambda \to \infty$.

The equivalent parametric problem is: Maximize

$$f = p'x - \mu z_\mu \tag{73}$$

subject to

$$Ax \le b, \tag{74}$$

$$x - ez_\mu \le 0, \tag{75}$$

$$x, z_\mu \ge 0. \tag{76}$$

Table 10

Set-up and initial tableau for the parametric-dual equivalent of the primal-dual method

	Basic variables	V.b.v.	x	y	z_μ	z
Set-up tableau	f_c	0	$-p'$	0	0	0
	f_μ	0	0	0	1	0
	y	b	A	I	0	0
	z	0	$\underline{I}$	0	$-e$	I
			x	y	z_μ	z
Initial tableau	f_c	0	0	0	$-p'e$	p'
	f_μ	0	0	0	1	0
	y	b	0	I	Ae	$-A$
	x	0	I	0	$-e$	I

Adding vectors of slack variables y and z in the inequalities of (74) and (75), the set-up tableau is as given in table 10. The initial tableau is found by replacing the z-variables by the x-variables as basic variables. The result is the initial tableau of table 10. This tableau is optimal for $p \geq 0$, $b \geq 0$ and $\mu \to \infty$. μ is parametrically varied in a downwards direction until a solution is found for $\mu = 0$. This must be the optimal solution, since for $\mu = 0$ the additional constraints (75) are not binding for the solution of the original problem.

Table 11 gives an application to the problem (54)–(58). Tableau 0 gives the initial tableau constructed in accordance with the initial tableau of table 10. In the first iteration z_μ enters the basis and y_3 leaves it, in the

Table 11

An application of the parametric-dual equivalent of the primal-dual method

Tableau	Basic variables	Values of basic variables	z_1	z_2	z_μ
0	f_c	0	2	3	-5
	f_μ	0	0	0	1
	y_1	$1\frac{1}{2}$	1	-1	0
	y_2	4	1	-2	1
	y_3	$6\frac{1}{2}$	-1	-1	2
	x_1	0	1	0	-1
	x_2	0	0	1	-1

			z_1	z_2	y_3
1	f_c	$16\frac{1}{4}$	$-\frac{1}{2}$	$\frac{1}{2}$	$2\frac{1}{2}$
	f_μ	$-3\frac{1}{4}$	$\frac{1}{2}$	$\frac{1}{2}$	$-\frac{1}{2}$
	y_1	$1\frac{1}{2}$	1	-1	0
	y_2	$\frac{3}{4}$	$1\frac{1}{2}$	$-1\frac{1}{2}$	$-\frac{1}{2}$
	z_μ	$3\frac{1}{4}$	$-\frac{1}{2}$	$-\frac{1}{2}$	$\frac{1}{2}$
	x_1	$3\frac{1}{4}$	$\frac{1}{2}$	$-\frac{1}{2}$	$\frac{1}{2}$
	x_2	$3\frac{1}{4}$	$-\frac{1}{2}$	$\frac{1}{2}$	$\frac{1}{2}$

			y_2	z_2	y_3
2	f_c	$16\frac{1}{2}$	$\frac{1}{3}$	0	$2\frac{1}{3}$
	f_μ	$-3\frac{1}{2}$	$-\frac{1}{3}$	1	$-\frac{1}{3}$
	y_1	1	$-\frac{2}{3}$	0	$\frac{1}{3}$
	z_1	$\frac{1}{2}$	$\frac{2}{3}$	-1	$-\frac{1}{3}$
	z_μ	$3\frac{1}{2}$	$\frac{1}{3}$	-1	$\frac{1}{3}$
	x_1	3	$-\frac{1}{3}$	0	$\frac{2}{3}$
	x_2	$3\frac{1}{2}$	$\frac{1}{3}$	0	$\frac{1}{3}$

second iteration z_1 enters the basis and y_2 leaves it. In tableau 2 the optimal solution for $\mu = 0$ and therefore the optimal solution of the original problem has been reached. Note that the tableaux of tables 7 and 11 correspond with each other in the manner described before. As in table 7, there is considerable redundancy in table 11.

Consider now methods derived from the parametric equivalent of the primal-dual method. Firstly, we know that the parametrically equivalent problem of a parametrical equivalent of an original problem is a problem which is the same as the original problem apart from being nominally more complicated.

For the dual equivalent of the parametric equivalent of the primal-dual method, consider the typical maximization problem: Maximize

$$f = p'x \tag{77}$$

subject to

$$Ax \le b,$$
$$x \ge 0. \tag{78}$$

Its dual problem is: Minimize

$$g = b'u \tag{79}$$

subject to

$$A'u \ge p,$$
$$u \ge 0. \tag{80}$$

The extended problem for an application of the parametric equivalent of the primal-dual method to this problem is formulated as follows (see (65)–(68)): Minimize

$$g = b'u \tag{81}$$

subject to

$$A'u + w \ge p, \tag{82}$$

$$e'w \le \mu, \tag{83}$$

$$u, w \ge 0. \tag{84}$$

In the dual equivalent of this method, the dual of this problem is solved, which is: Maximize

$$f = p'x - \mu z_\mu \tag{85}$$

subject to

$$Ax \le b, \tag{86}$$

$$x - ez_\mu \leq 0, \tag{87}$$

$$x, z_\mu \geq 0. \tag{88}$$

This problem is identical with the problem (73)–(76), which was the parametric equivalent of the dual equivalent of the primal-dual method. From this it may be concluded that the parametric equivalent of the dual equivalent of the primal-dual method is identical with the dual equivalent of the parametric equivalent of that method. For this reason we may call this method the parametric-dual equivalent of the primal-dual method.

We may try to derive additional methods by taking parametric or dual equivalents of this method, but it is rather obvious that no new methods are derived, since dual equivalence is symmetric and so is parametric equivalence apart from a trivial enlargement of the problem.

The results can be summarized in the following way. We have considered four related methods for linear programming the primal-dual method, its dual equivalent, its parametric equivalent and its parametric-dual equivalent. They use parametric programming as a means to obtain the optimal solution. Two methods require a basic feasible solution (the dual equivalent and the parametric-dual equivalent and two methods require an initial basic solution which is optimal in the narrow sense (the primal-dual method and its parametric equivalent).

6.6. The gradient method and its equivalent methods

The dual equivalent of the primal-dual method turned out to be a method which can be used for the typical maximization problem. The solution of this method moves initially into the feasible region along a 45°-line until a constraint is met; after that the solutions move along this constraint, until another constraint is met; then the solutions move along this constraint, and so on. It could be objected that moving along 45°-line is somewhat arbitrary; this amounts to the objection that all x-variables with positive coefficients have the same additional constraint $x_j \leq \lambda$. Let us consider the following simple problem. Maximize f for

$$f = 5x_1 + x_2,$$
$$x_1 + 2x_2 \leq 6,$$
$$3x_1 + 2x_2 \leq 12,$$
$$x_1, x_2 \geq 0. \tag{89}$$

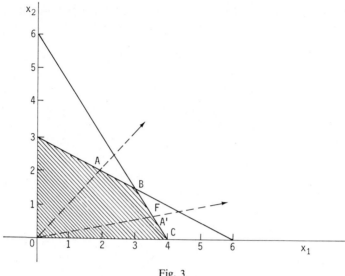

Fig. 3.

In the dual equivalent of the primal dual method the constraints $x_1 \leq \lambda$, $x_2 \leq \lambda$ would be added. Figure 3 gives a geometric representation of this problem. The solutions would move from the origin to A, then to B and then to the optimal solution C. It is obvious that moving along $0A$ does not give the highest increase in the objective function per unit of movement. If we wish to have the highest increase per unit of movement, we should move along OA'. The direction OA' is found as the *gradient* of the objective function and is defined as

$$\begin{bmatrix} \partial f / \partial x_1 \\ \partial f / \partial x_2 \end{bmatrix} = \begin{bmatrix} 5 \\ 1 \end{bmatrix}. \tag{90}$$

Instead of adding the constraints $x_1 \leq \lambda$, $x_2 \leq \lambda$, we should add the constraints

$$x_1 \leq 5\lambda$$

$$x_2 \leq \lambda. \tag{91}$$

In this case a parametric increase of λ would lead to solutions moving from 0 to A' and then to C. Though this method, which may be called the *gradient method*, does not necessarily lead to fewer iterations than the dual equivalent of the primal-dual method, it may be considered more attractive because the movement of solutions is in the optimal direction.

The gradient method for the typical maximization problem may be outlined as follows. Let the problem be: Maximize f for

$$f = p'x,$$
$$Ax \leq b,$$
$$x \leq 0. \tag{92}$$

All elements of b are assumed to be nonnegative. To this problem we add the constraints

$$x_j \leq \lambda p_j, \tag{93}$$

for all x_j with $p_j > 0$. As in the dual equivalent of the primal-dual method, an initial solution which is optimal for $\lambda = 0$ has the slack variables of the second set of relations in (92) and the x-variables in (93) as basic variables; the initial tableau is the same as in the dual equivalent of the primal-dual method if e is replaced by p. Table 12 gives an application of this method to the problem used in the previous section. In this case we add the constraints

$$x_1 \leq 2\lambda,$$
$$x_2 \leq 3\lambda. \tag{94}$$

Figure 4 gives a geometrical representation. The dashed line through A

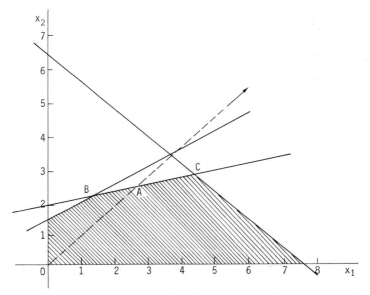

Fig. 4.

Table 12

Application of the gradient method to an example

Tableau	Basic var.	Val. bas. var.		z_1	z_2
		c-term	λ-term		
	f	0	13	2	3
	y_1	$1\frac{1}{2}$	-1	1	-1
0	y_2	4	-4	1	-2
	y_3	$6\frac{1}{2}$	-5	-1	-1
	x_1	0	2	1	0
	x_2	0	3	0	1

				z_1	y_2
	f	6	7	$3\frac{1}{2}$	$1\frac{1}{2}$
	y_1	$-\frac{1}{2}$	1	$\frac{1}{2}$	$-\frac{1}{2}$
1	z_2	-2	2	$-\frac{1}{2}$	$-\frac{1}{2}$
	y_3	$4\frac{1}{2}$	-3	$-1\frac{1}{2}$	$-\frac{1}{2}$
	x_1	0	2	1	0
	x_2	2	1	$\frac{1}{2}$	$\frac{1}{2}$

				y_3	y_2
	f	$16\frac{1}{2}$	0	$2\frac{1}{3}$	$\frac{1}{3}$
	y_1	1	0	$\frac{1}{3}$	$-\frac{2}{3}$
2	z_2	$-3\frac{1}{2}$	3	$-\frac{1}{3}$	$-\frac{1}{3}$
	z_1	-3	2	$-\frac{2}{3}$	$\frac{1}{3}$
	x_1	3	0	$\frac{2}{3}$	$-\frac{1}{3}$
	x_2	$3\frac{1}{2}$	0	$\frac{1}{3}$	$\frac{1}{3}$

is the gradient of the objective function. The solution moves from the origin to A, where the constraints $-x_1 + 2x_2 \leq 4$ becomes binding; the constraint $x_2 \leq 3\lambda$ ceases to be binding. The solution then moves to C, where the constraint $x_1 + x_2 \leq 6\frac{1}{2}$ becomes binding and the constraints $x_1 \leq 2\lambda$ ceases to be binding. Since the additional constraints are not binding anymore, C must be the optimal solution of the original problem.

The initial tableau of an application of the gradient method to this problem is tableau 0 of table 12. This tableau is optimal for $\lambda = 0$. Varying λ upwards we find that y_2 leaves the basis and further that z_2 should enter it. The result is tableau 1, which is optimal for $1 \leq \lambda \leq 1\frac{1}{2}$. y_3 leaves the basis and z_1 enters it. Tableau 2 results, which contains the optimal solution because all z-variables are basic.

It should be noted that the added constraints $x_j \leq \lambda p_j$ need not lead to additional rows in the tableau since as long as the x-variables of such a constraint is nonbasic, the corresponding row does not change and as soon as a z-variable has entered the basis, its row may be deleted since an added constraint which has become nonbinding is not needed any more. Hence, compared with the simplex method, the gradient method has tableaux which have only one additional column, namely the λ-term column.

The same objections may be raised against the gradient method as were raised against the dual equivalent of the primal-dual method. Each constraint added implies at least one iteration necessary for its removal. This means that unless the number of additional constraints is less than the number of proper constraints, the gradient method is not likely to be more effective than the simplex method.

The treatment of the parametric equivalent of the gradient method is left as an exercise. We now turn to the dual equivalent of the gradient method. Since the dual equivalent of the primal-dual method and the gradient method only differ by having the vector e in the first method and p in the second method, the dual equivalent of the gradient method should be the same as the primal-dual method if e is replaced by b. This may be checked as follows. The gradient method proceeds from the problem: Maximize f where

$$f = p'x,$$
$$Ax \leq b,$$
$$x \leq \lambda p,$$
$$x \geq 0. \tag{95}$$

Its dual problem is: Minimize g where

$$g = b'u + \lambda p'w$$
$$A'u + w \geq p,$$
$$u, w \geq 0. \tag{96}$$

This problem has an optimal solution for $\lambda = 0$ with the w-variables as basic variables. λ is varied upwards until the optimal solution has been found for $\lambda \to \infty$.

In the dual equivalent of the gradient method there is a primal problem of the same structure as (96). As point of departure we have a typical minimization problem: Minimize f where

$$f = c'x,$$
$$Ax \ge b,$$
$$x \ge 0. \tag{97}$$

Introducing a vector of artificial variables z in the constraints and in the objective function, we find

$$f = c'x + \lambda b'z,$$
$$Ax + z \ge b,$$
$$z \ge 0. \tag{98}$$

The artificial variables are now weighted by their values in the initial solution instead of being given the same coefficient. As in the primal-dual method an optimal solution for $\lambda = 0$ is found with all z-variables as basic variables. λ then is parametrically increased until a solution is found which is optimal for $\lambda \to \infty$. The initial tableau for the dual equivalent of

Table 13

An application of the dual equivalent of the gradient method

Tableau	Bas. var.	V. b. v.	x_1	x_2	x_3	x_4	y_1	y_2
0	f_c	0	-40	-77	-39	-75	0	0
	f_λ	26	50	119	64	125	-1	-5
	z_1	1	10	14	4	0	-1	0
	z_2	5	8	21	12	$\underline{25}$	0	-1
	Ratio		$\frac{4}{5}$	$\frac{77}{119}$	$\frac{39}{64}$	$\frac{3}{5}$		
			x_1	x_2	x_3		y_1	y_2
1	f_c	15	-16	-14	-3		0	-3
	f_λ	1	10	14	4		-1	0
	z_1	1	10	14	4		-1	0
	x_4	$\frac{1}{5}$	$\frac{8}{25}$	$\frac{21}{25}$	$\frac{12}{25}$		0	$-\frac{1}{25}$
	Ratio		$1\frac{3}{5}$	1	$\frac{3}{4}$			
			x_1	x_2			y_1	y_2
2	f_c	$15\frac{3}{4}$	$-8\frac{1}{2}$	$-3\frac{1}{2}$			$-\frac{3}{4}$	-3
	f_λ	0	0	0			0	0
	x_3	$\frac{1}{4}$	$2\frac{1}{2}$	$3\frac{1}{2}$			$-\frac{1}{4}$	0
	x_4	$\frac{2}{25}$	$-\frac{22}{25}$	$-\frac{21}{25}$			$\frac{3}{25}$	$-\frac{1}{25}$

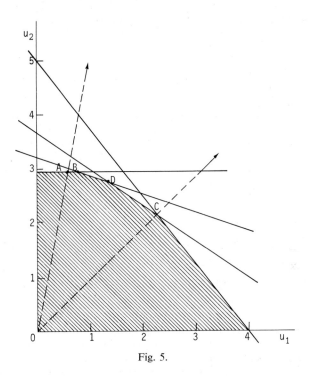

Fig. 5.

the gradient method is the same as that of the primal-dual method except for the vector e which is replaced by b.

As an example let us solve the following problem by the dual equivalent of the gradient method. Minimize f where

$$f = 40x_1 + 77x_2 + 39x_3 + 75x_4$$
$$10x_1 + 14x_2 + 4x_3 \qquad \geq 1,$$
$$8x_1 + 21x_2 + 12x_3 + 25x_4 \geq 5,$$
$$x_1, x_2, x_3, x_4 \geq 0. \qquad (99)$$

Tableau 0 of table 13 gives the initial tableau. In two iterations the optimal solution is found. Figure 5 gives a geometric representation of the problem in terms of the dual variables u_1 and u_2. It is seen that the solution moves from the origin along the gradient to A and then to the optimal solution B. If the primal-dual method had been used, the solution in terms of the dual variables would have moved along the 45°-line of C, then to D and then to B. Of course, the dual equivalent of the gradient method does not necessarily have fewer iterations than the primal-dual method, but in terms of the space of dual variables it is more likely to move into the right direction.

It should be noted that the dual equivalent of the gradient method does not have the disadvantages of the gradient method since the number of variables added is at most as large as the number of constraints. On the other hand, the method requires an initial solution which is optimal in the narrow sense (a nonnegative vector c in a minimization problem), which may not always be available. The treatment of the parametric equivalent of the dual equivalent of the gradient method is left as an exercise.

6.7. The capacity method and its equivalent methods

Let us consider the typical maximization problem: Maximize f where

$$f = p'x,$$
$$Ax \leq b,$$
$$x \geq 0. \tag{100}$$

It is assumed that the elements of b are nonnegative. Let us add to this problem the additional constraint

$$e'x \leq \lambda. \tag{101}$$

First an optimal solution for $\lambda = 0$ is found; after that λ is parametrically varied until it approaches infinity. The resulting solution should be the optimal solution of the original problem. This method has been proposed by Houthakker[57] for rather restricted quadratic programming problems. Houthakker called this method the *capacity method*. It was proved later that this method was applicable to any convex quadratic programming problem and hence to linear programming problems; furthermore a reformulation of the method could be given in terms of parametric programming.[2]

The initial solution is easily found. Let y be the vector of the slack variables of the constraints of (100) and y_λ the slack variable of (101). The set-up tableau has the y-variables and y_λ as basic variables. The initial solution which is optimal for $\lambda = 0$ is found by introducing into the

---

[2] See VAN DE PANNE and WHINSTON[72].

basis the x-variable with the largest p-element. The parametric increase in λ then follows until for $\lambda \to \infty$ the optimal solution of problem (100) is found.

Table 14 gives an application of the capacity method to the following problem. Maximize f where

$$f = 3x_1 + 4x_2,$$
$$-x_1 + 2x_2 \le 2,$$
$$x_1 - x_2 \le 1,$$
$$x_1, x_2 \ge 0. \tag{102}$$

After the capacity constraint

$$x_1 + x_2 \le \lambda \tag{103}$$

is added, tableau 0 can be written down. After the introduction of the variable with the most negative coefficient in the f-row, which turns out to be x_2, it is found that y_λ leaves the basis. The normal parametric procedure is followed until y_λ enters the basis again. This means that the optimal solution has been reached. Note that the columns of y_λ, and the λ-term column are always the same. The capacity method requires one additional row compared with the tableaux of the simplex method; it also requires an initial iteration in order to obtain an initial optimal solution.

Table 14

An application of the capacity method

Tableau	Basic var.	Values of basic var. c-term	λ-term	x_1	x_2	Tableau	Basic var.	Values of basic var. c-term	λ-term	y_1	y_λ
0	f	0	0	−3	−4	2	f	$\frac{2}{3}$	$3\frac{1}{3}$	$\frac{1}{3}$	$3\frac{1}{3}$
	y_1	2	0	−1	2		x_1	$-\frac{2}{3}$	$\frac{2}{3}$	$-\frac{1}{3}$	$\frac{2}{3}$
	y_2	1	0	1	−1		y_2	$2\frac{1}{3}$	$-\frac{1}{3}$	$\frac{2}{3}$	$-\frac{1}{3}$
	y_λ	0	1	1	1		x_2	$\frac{2}{3}$	$\frac{1}{3}$	$\frac{1}{3}$	$\frac{1}{3}$
				x_1	y_λ					y_1	y_2
1	f	0	4	1	4	3	f	24	0	7	10
	y_1	2	−2	−3	−2		x_1	4	0	1	2
	y_2	1	1	2	1		y_λ	−7	1	−2	−3
	x_2	0	1	1	1		x_2	3	0	1	1

The capacity method is obviously related to the dual equivalent of the primal-dual method. In the last method we added a constraint $x_j \leq \lambda$ for each x_j with $p_j > 0$. In the capacity method we have a joint constraint $\Sigma \, x_j \leq \lambda$ for these variables.

The equivalent parametric method uses the following extended problem.[3] Maximize f where

$$f = (p - \mu e)'x,$$
$$Ax \leq b,$$
$$x \geq 0. \tag{104}$$

For $\mu \to \infty$ the initial solution with the slack variables y as basic variables is optimal. μ is varied in a downwards direction until an optimal solution for $\mu = 0$ is found, which is the optimal solution of the original problem (100).

This method could also have been obtained by the following reasoning. The set-up tableau of the typical maximization problem has feasibility but not optimality in the narrow sense, since each variable x_j has a coefficient in the f-row of $-p_j$, which is generally negative. To each coefficient $-p_j$, which is negative, a parametric term μ may be added, so that we obtain $-p_j + \mu$. For large values of μ, the solution is optimal in the wide sense. μ is then varied downwards until an optimal solution for $\mu = 0$ is obtained, which must be the optimal solution of the original problem.

It can be proved that this method (and hence also the capacity method) is equivalent to a variant of the simplex method in which not necessarily the nonbasic variable with the most negative element in the f-row enters the basis. In the initial tableau we have for the f_c and f_μ-row:

$$f_c \quad 0 \quad -p_1 \quad -p_2 \quad \cdot \quad -p_n$$
$$f_\mu \quad 0 \quad 1 \quad 1 \quad \cdot \quad 1$$

It is obvious that this solution is optimal for $\mu = \infty$. The new basic variable is determined from the ratios

$$\text{Max} \, (p_1/1, p_2/1, ..., p_n/1). \tag{105}$$

Let in any subsequent tableau the f_c-row and f_μ-row be as follows

---

[3] The following problem has the problem (100)–(101) as its parametric equivalent. Taking the parametric equivalent of (100)–(101) results in a slightly larger problem. It will be remembered that parametric equivalence is not quite symmetric.

$$
\begin{array}{cccccc}
f_c & p_0 & p_1 & p_2 & \cdot & p_n \\
f_\mu & q_0 & q_1 & q_2 & \cdot & q_n.
\end{array}
$$

Since μ is varied in a downwards direction, we should determine

$$
\max_{j=1,\ldots,n} \ (-p_j/q_j \,|\, q_j > 0) = -p_k/q_k. \tag{106}
$$

Since $-p_k/q_k$ is a lower bound of μ, it should be positive, so that p_k is negative. As in the simplex method, a nonbasic variable with a negative element in the f-row is selected. The leaving basic variable is selected as in the simplex method. Hence, the parametric equivalent to the capacity method, and therefore also the capacity method itself, can be interpreted as a special case of the simplex method.

As in the gradient method and its dual equivalent we may use the idea of having p instead of e in the capacity method and its parametric equivalent. The extended problem for the latter method would be as follows. Maximize f where

$$
f = (p - \mu p)'x,
$$
$$
Ax \le b,
$$
$$
x \ge 0. \tag{107}
$$

The initial tableau in partitioned form would be as in table 15. The solution of this tableau is optimal for $\mu \ge 1$ if all elements of p are nonnegative. Any x-variable now may enter the basis replacing a y-varible; we would obtain a general tableau as is given in table 15.

Table 15

Initial and general tableaux for modified parametric equivalent of the capacity method

	Bas. var.	V.b.v.	x^1	x^2
Initial tableau	f_c	0	$-p^{1\prime}$	$-p^{2\prime}$
	f_μ	0	$p^{1\prime}$	$p^{2\prime}$
	y^1	b^1	A_{11}	A_{12}
	y^2	b^2	A_{21}	A_{22}
			y^1	x^2
General tableau	f_c	$p^{1\prime}b^1$	$p^{1\prime}A_{11}^{-1}$	$-p^{2\prime} + p^{1\prime}A_{11}^{-1}A_{12}$
	f_μ	$-p^{1\prime}b^1$	$-p^{1\prime}A_{11}^{-1}$	$p^{2\prime} - p^{1\prime}A_{11}^{-1}A_{12}$
	x^1	$-A_{11}^{-1}b^1$	A_{11}^{-1}	$A_{11}^{-1}A_{12}$
	y^2	$b^2 - A_{21}A_{11}^{-1}b^1$	$-A_{21}A_{11}^{-1}$	$A_{22} - A_{21}A_{11}^{-1}A_{12}$

The solution of this tableau is optimal for $\mu = 1$. In order to operate the parametric procedure, any nonbasic variable is taken into the basis having a negative element in the f-row. This is exactly the same as in the simplex method, which is not surprising at all, since the objective function in (107) can be written as $(1 - \mu)p'x$. But then there is no reason to add the term $-\mu p'x$ to the objective function. For the same reason it can be argued that the capacity method and its parametric equivalent have no advantage over the simplex method, but that they have the disadvantage of an additional column or row, and in the case of the capacity method, one additional iteration.

Let us consider the dual equivalent of the capacity method. The typical minimization problem is: Minimize f where

$$f = c'x$$
$$Ax \geq b,$$
$$x \geq 0. \tag{108}$$

The vectors b and c are supposed to have nonnegative elements; only the last assumption is nontrivial. The dual problem is: Maximize g where

$$g = b'u,$$
$$A'u \leq c,$$
$$u \geq 0. \tag{109}$$

To this problem the constraint is added:

$$e'u \leq \lambda. \tag{110}$$

The primal problem having (109) and (110) as its dual is: Maximize f where

$$f = c'x + \lambda z$$
$$Ax + ze \geq b,$$
$$x, z \geq 0. \tag{111}$$

z is an artificial variable, just as (110) can be called an artificial constraint.

This method is obviously related to the two-phase method, in which only one artificial variable was introduced, which was called the fudge-method. In the set-up tableau all slack variables are basic. The initial tableau is obtained by taking z into the basis replacing the y-variable corresponding with the largest element of b. The resulting solution is optimal for $\lambda = 0$. The optimal solution of the original problem is found by an application of parametric programming until a solution is found for $\lambda \to \infty$.

The parametric equivalent[4] of the dual equivalent of the capacity method is: Minimize f where

$$f = c'x,$$
$$Ax \geq b - \mu e,$$
$$x \geq 0. \tag{112}$$

μ is varied from ∞ in a downwards direction. It can be shown that this method, as well as the dual equivalent of the capacity method amounts to a special case of the dual method; it seems to have no advantages over the dual method. If the vector e is replaced by b, a method which is equivalent with the dual method results.

This method may be thought of as one in which, if the set-up tableau has optimality in the narrow sense but unfeasibilities in the form of negative values of basic variables, a positive parameter μ is added to this value. The solution is then optimal for larger values of μ, which is varied downwards parametrically until it reaches zero. Artificial variables can be treated in the same way if the constant term of the equalities from which they arise is given a negative value by multiplying the equation by -1.

6.8. Self-dual parametric methods

In this section we shall deal with methods for problems which are neither typical maximization problems nor typical minimization problems, that is, problems having a set-up tableau with a solution which is neither optimal in the narrow sense nor feasible. After reordering, such problems can be written as follows: Maximize

$$f = p^{1\prime}x^1 - p^{2\prime}x^2 \tag{113}$$

subject to

$$A_{11}x^1 + A_{12}x^2 \leq b^1, \tag{114}$$

$$A_{21}x^1 + A_{22}x^2 = b^2, \tag{115}$$

$$A_{31}x^1 + A_{32}x^2 \geq b^3, \tag{116}$$

$$x^1, x^2 \geq 0, \tag{117}$$

---

[4] Again the method of which the original method is the parametric equivalent is used.

with p^1, p^2, b^1, b^2, $b^3 \geq 0$. For reasons of symmetry between the primal and dual problem, a vector of x-variables with no nonnegativity constraints could have been included, but since these occur very rarely, this has not been done.

The first approach to this problem is to deal with the unfeasibilities as in the primal-dual method and to deal with the non-optimalities as in its dual equivalent. This results in the following extended problem: Maximize

$$f = p^{1\prime}x^1 - p^{2\prime}x^2 - \lambda(e'z^1 + e'z^2) \tag{118}$$

subject to

$$A_{11}x^1 + A_{12}x^2 + y^1 = b^1, \tag{119}$$

$$A_{21}x^1 + A_{22}x^2 + z^1 = b^2. \tag{120}$$

$$A_{31}x^1 + A_{32}x^2 - y^2 + z^2 = b^3, \tag{121}$$

$$x^1 \leq \mu e, \tag{122}$$

$$x^1, x^2, y^1, y^2, z^1, z^2 \geq 0. \tag{123}$$

Note that the artificial variables introduced have obtained parametric penalty terms in the objective function and that for variables which are non-optimal in the set-up tableau, parametric constraints are added. The set-up and initial tableau are then as given in table 16.

The solution of the initial tableau is optimal for $\lambda = 0$ and $\mu = 0$. If we put $\mu = \lambda$, optimal solutions may be found for increasing values of λ. Upper bounds for λ may be found either in the objective function or in the values of basic variables. In the first case we proceed as in parametric programming with a parametric objective function, in the second as in parametric programming with parametric constant terms. The solution for $\lambda \to \infty$ must be the optimal solution of the original problem.

It is also possible to vary first λ upwards and after that, vary μ upwards, or vice-versa. Another possibility is to put $\mu = c\lambda$, where c is some positive constant.

Parametric equivalence may be used to obtain a problem having instead of the artificial constraints (122) the following constraints

$$x^1 - ez_\mu \leq 0, \tag{124}$$

and which has the objective function

$$f = p^{1\prime}x^1 - p^{2\prime}x^2 - \lambda(e'z^1 + e'z^2) - \lambda^* z_\mu. \tag{125}$$

Table 16
Set-up and initial tableaux for the self-dual primal-dual method

| Tableau | Basic variables | Values of basic variables | | x^1 | x^2 | y^1 | y^2 | y^3 | z^1 | z^2 |
		c-term	μ-term							
0	f	0	0	$-p^{1'}$	$p^{2'}$	0	0	0	0	0
	f_Λ	0	0	0	0	0	0	0	e'	e'
	y^1	b^1	0	A_{11}	A_{12}	I	0	0	0	0
	y^2	b^2	0	A_{21}	A_{22}	0	$-I$	0	$\dfrac{I}{0}$	$\dfrac{I}{0}$
	y^3	b^3	0	A_{31}	A_{32}	0	0	I	0	0
		0	e	$\underline{I}$	0	0	0	0	0	0
I	f	0	$p^{1'}e$	0	$p^{2'}$	0	0	$p^{1'}$	0	0
	f_Λ	$-e'b^1-e'b^2$	$e'A_{21}e+e'A_{31}e$	0	$-e'A_{22}-e'A_{32}$	0	e'	$e'A_{21}+e'A_{31}$	0	0
	y^1	b^1	$-A_{11}e$	0	A_{12}	I	0	$-A_{11}$	0	0
	z^1	b^2	$-A_{21}e$	0	A_{22}	0	0	$-A_{21}$	I	I
	z^2	b^3	$-A_{31}e$	0	A_{32}	0	$-I$	$-A_{31}$	0	I
	x^1	0	e	I	0	0	0	I	0	0

Note that there are now two parameters, λ, which is to be varied from 0 to ∞ and λ^*, which is to be varied from ∞ to 0. This can be handled by first varying λ upwards keeping λ^* at ∞ and then varying λ^* downwards, keeping λ at ∞, or the other way around. An alternative is to substitute $\lambda^* = 1/\lambda$, but this has the disadvantage of coefficients which are no longer a linear function of λ.

It is also possible to use parametric equivalence to obtain a problem having parametric constant terms only, but also in this case we end up with values of basic variables which are a function of two parameters to be varied in the opposite direction.

Instead of using a combination of the primal-dual method and its dual equivalent we may use a combination of the gradient method and its dual equivalent; this method can be called the self-dual gradient method. This method amounts to the following modification of the objective function and the artificial constraint (123) for the extended problem:

$$f = p^{1\prime}x^1 - p^{2\prime}x^2 - \lambda(b^{1\prime}z^1 + b^{2\prime}z^2), \tag{126}$$

$$x^1 \le \mu p^1. \tag{127}$$

The corresponding changes in the initial tableau can be easily found.

Finally the self-dual capacity method may be formulated. Let the following problem be considered: Maximize

$$f = p^{1\prime}x^1 - p^{2\prime}x^2 \tag{128}$$

subject to

$$A_{11}x^1 + A_{12}x^2 \le b^1,$$
$$A_{21}x^1 + A_{22}x^2 \ge b^2,$$
$$x^1, x^2 \ge 0. \tag{129}$$

It is assumed that variables and constraints are ordered in such a way that p^1, p^2, b^1 and $b^2 \ge 0$.

The extended problem is then as follows: Maximize

$$f = (p^1 - \lambda e)'x^1 - p^{2\prime}x^2 \tag{130}$$

subject to

$$A_{11}x^1 + A_{12}x^2 + y^1 = b^1$$
$$- A_{21}x^1 - A_{22}x^2 + y^2 = - b^2 + \mu e,$$
$$x^1, x^2, y^1, y^2 \ge 0. \tag{131}$$

The solution of the initial tableau with y^1 and y^2 as basic variables is optimal in the wide sense for large values of λ and μ. λ and μ are then varied downwards until both are equal to zero.

The problem (128)–(129) did not include equality constraints; if these occur they should be eliminated first by selecting a suitable basic variable for each equation.

The self-dual capacity method was proposed by Dantzig,[5] who called it the *self-dual parametric method*. It is closely related to the symmetric method[6], since the capacity method can be considered as a variant of the simplex method and the dual equivalent of the capacity method as a variant of the dual method. However, convergence in a finite number of iterations can be proved for the self-dual capacity method, which is not true for the symmetric method.

As for the self-dual primal-dual method and the self-dual gradient method, parametric equivalence can be used to obtain an extended problem with a parametric objective function only or parametric constant terms only.

---

[5] See DANTZIG [6].
[6] See section 4.5.

CHAPTER 7

POST-OPTIMALITY ANALYSIS, THE REVERSE SIMPLEX
METHOD AND RELATED METHODS

7.1. Post-optimality analysis via the reverse simplex method

The solution found by linear programming is a unique one, or, if not, there are only a few extreme-point solutions. However, in many cases the optimal solution or solutions are not the only ones of interest, since the decisionmaker will usually also be interested in near-optimal solutions. This is so, because frequently it will be impossible to specify exactly the constraints of the problem; furthermore, the objective function does not always reflect precisely the preferences of the decisionmaker.

In these cases the linear programming formulation of a problem can be said to give the most important objectives and constraints of the problem; other factors may and frequently do remain outside the formulation, either because it is too difficult to incorporate them or because the decisionmaker is only vaguely aware of them. The only way to deal with these factors is to face the decisionmaker with alternative decisions and to leave it to him to select the alternative he prefers. The difficulty then is that the number of all conceivable alternatives is so large that the decisionmaker cannot evaluate them.

This suggests the following approach. First the optimal solution for the linear programming formulation is determined. Any additional binding constraints imposed on the solution will always lower the objective function and so will any other objectives which the decisionmaker has in mind besides the main objective of maximizing profits. However, these factors will only be able to decrease the objective function of the linear programming problem by a limited amount. The optimal solution for the decisionmaker is therefore one which satisfies the constraints of the linear programming problem and which has a value of the objective function differing only by a limited amount from the optimal solution of the programming problem. The decisionmaker may therefore find his optimal solution by considering all solutions which satisfy these requirements. There will, of course, be an infinity of these solutions, but it suffices to give only the extreme-point solutions.

202

The extreme-point solutions can be determined by linear programming using a method in which the amount by which the value of the objective may differ from that of the optimal value in the programming formulation is varied from 0 to an amount the decisionmaker does not want to lose in any case. This method can be said to use the simplex method in the reverse direction and is therefore called the *reverse simplex method*. In this section this approach will be presented and applied to a simple example. A different approach, based on the theory of graphs which follows from a proposal of Charnes and Cooper[1] for a problem of a similar kind, is given in section 7.2.

Other approaches which are not discussed here have been given by Balinski[1], Burdet[2], and Maňas and Nedoma[22]. A practical application was discussed by Moroney and Dimond[23].

Though the reverse simplex method was designed for post-optimality analysis, it can be used as a general method for finding all extreme points of a system of inequalities, as is shown in section 7.3. The dual equivalent of the reverse simplex method, which is the reverse dual method, is developed in section 7.4. In this method solutions are generated which are optimal in the narrow sense but not necessarily feasible.

The linear programming problem can be formulated as follows. Maximize f:

$$f(x) = p'x,$$
$$Ax \le b,$$
$$x \ge 0. \tag{1}$$

Let the value of the objective function for the optimal solution[2] x^0 be f^0. It is assumed that the optimal solution is unique. The difference of f^0 and the value of the objective function for some solution is called the *loss* of that solution; any feasible solution which is not optimal has a positive loss.

Suppose that the maximum loss allowed is k. This means that only solutions are considered which satisfy, in addition to (1), the inequality

$$f(x) = p'x \ge f^0 - k. \tag{2}$$

The problem is now to generate all extreme-point solutions satisfying (1) and (2). For $k = 0$, we find, of course, the optimal solution x^0. For in-

---

[1] See CHARNES and COOPER[4], pp. 308–310.
[2] Throughout this chapter, any superindex is to be interpreted as such and not as an exponent; except in some cases where ϵ's are involved.

creasing values of k an increasing number of extreme-point solutions must be found; for some large value of k the extreme-point solutions of (1) and (2) are in any case the same as those of (1) if these are finite. In practical problems the value of k will be relatively small, but even in these cases the number of extreme-point solutions of (1) and (2) will usually be large.

The idea on which the reverse simplex method is based is quite simple. In the normal simplex method variables are introduced into the basis which increase the value of the objective function until the optimal solution is found. In the reverse simplex method the optimal solution is taken as a starting-point for the generation of the extreme-point solution of (1) and (2). In this and in subsequent solutions variables are introduced into the basis which decrease the value of the objective function; the maximal loss k sets a lower bound to this decrease.

The main difference between the normal simplex method and the reverse simplex method is that in the normal simplex method only one variable is introduced into the basis; in the reverse simplex method all variables which decrease the objective function are sooner or later introduced into the basis. Corresponding with the normal simplex method is the effect of iterations on the objective function: in the normal simplex method the objective function increases in each iteration, while in the reverse simplex method it decreases with each iteration, provided there is no degeneracy.

Let us consider an iteration in the normal simplex method. Tableau I of table 1 gives a representation of the relevant parts of a simplex tableau. x_j is the variable entering the basis since $p_j < 0$. x_r leaves the basis since it is connected with

$$\underset{i}{\text{Min}}\,(b_i/a_{ij}\,|\,a_{ij} > 0) = b_r/a_{rj}. \tag{3}$$

Tableau I then is transformed into tableau II with a_{rj} as the pivot element of the transformation. In tableau II we find that the value of the objective function has increased by $-p_j b_r/a_{rj}$; furthermore, we find that the element in the column of x_r and in the last row, $-p_j/a_{rj}$, is positive.

Let us now consider introducing x_r again into the basis in tableau II. Since $-p_j/a_{rj}$ is positive, the objective function decreases. The variable to leave the basis is found by comparing the ratios

$$\frac{b_1 - (a_{1j}b_r)/a_{rj}}{-a_{1j}/a_{rj}}, \,...,\, \frac{b_r/a_{rj}}{1/a_{rj}}, \,...,\, \frac{b_m - (a_{mj}b_r)/a_{rj}}{-a_{mj}/a_{rj}} \tag{4}$$

Table 1

Successive tableaux for a simplex iteration

Tableau	Value basic variable	x_j	x_r
I	b_0	p_j	0
	b_1	a_{1j}	0
	.	.	.
	b_r	a_{rj}	1
	.	.	.
	b_m	a_{mj}	0
II	$b_0 - (p_j b_r)/a_{rj}$	0	$-p_j/a_{rj}$
	$b_1 - (a_{1j}b_r)/a_{rj}$	0	$-a_{1j}/a_{rj}$
	.	.	.
	b_r/a_{rj}	1	$1/a_{rj}$
	.	.	.
	$b_m - (a_{mj}b_r)/a_{rj}$	0	$-a_{mj}/a_{rj}$

for positive denominators; these ratios reduce to

$$b_r + \frac{b_1 a_{rj}}{-a_{1j}}, \ ..., \ b_r, \ ..., \ b_r + \frac{b_m a_{rj}}{-a_{mj}} \tag{5}$$

for negative $a_{1j}, ..., a_{mj}$. The minimum ratio is obviously b_r, so that x_r is the leaving basic variable. The resulting tableau, which is obtained by using $1/a_{rj}$ as a pivot, is then again tableau I. Hence we have proved that for each iteration of the simplex method there is an iteration reversing the simplex iteration by introducing a nonbasic variable with a positive element in the last row into the basis and determining the leaving basic variable in the usual manner.

We now consider the case in which p_j is positive and x_j is introduced into the basis; this means that the value of the objective function will decrease. The variable to leave the basis is determined as usual and turns out to be x_r; tableau I is therefore transformed into tableau II with a_{rj} as the pivot. The objective function has decreased by $-(p_j b_r)/a_{rj}$ and for the element of tableau II in the last row and in the column of x_r we find $-p_j/a_{rj}$, which is negative.

The normal simplex method may now be applied to tableau II by introducing x_r into the basis because $-p_j/a_{rj}$ is negative. Comparing the ratios as in (4) and (5), we find that x_j leaves the basis and tableau I is obtained. We therefore have proved that to each iteration of the reverse simplex method there corresponds an iteration of the normal simplex method.

The reverse simplex method will now be described in more detail. For our purposes it is useful to express the objective function as a deviation from the minimally required value of the objective function, $f^0 - k$. This means that instead of f, we use $f - (f^0 - k)$, which amounts to using the slack variable which can be introduced in (2), which will be called y_k.

The reverse simplex method uses simplex tableaux which are quite the same as those of the normal simplex method. Table 2 gives a representation of the simplex tableau at iteration s, in which the tableau is rearranged in such a way that the columns of basic variables are arranged in a unit matrix.

Table 2

Representation of simplex tableau at the sth iteration of the reverse simplex method

Val. bas. var.	Nonbasic variables				y_k	Basic variables				
b^s	p_1^s	.	p_j^s	.	1	0	.	0	.	0
b_1^s	a_{11}^s	.	a_{1j}^s	.	0	1	.	0	.	0
.	.	.	.	.	.	.	.	.	.	.
b_i^s	a_{i1}^s	.	a_{ij}^s	.	0	0	.	1	.	0
.	.	.	.	.	.	.	.	.	.	.
b_m^s	a_{m1}^s	.	a_{mj}^s	.	0	0	.	0	.	1

Table 2 can also serve as a representation of the starting tableau in which case $s = 0$. The starting tableau is the tableau of the optimal solution of the original problem in which the last element of the first column, the value of the objective function for the optimal solution f^0, has been replaced by k, so that $b^0 = k$. This means that in this and the following tableaux the value of the objective function is measured as a deviation from $f^0 - k$.

For the solution of the starting tableau we have $b_i^0 > 0$ for each i and $p_j^0 > 0$ for each j; if we had not assumed that the optimal solution is unique, the weak inequality sign would apply. Introduction of a nonbasic variable x_j into the basis at a level θ leads to a loss of $p_j^0 \theta$, which is positive. The variable to leave the basis is determined, just as in the simplex method, as the one corresponding with

$$\theta_j^0 = \underset{i}{\text{Min}}\{b_i^0 / a_{ij}^0 | a_{ij}^0 > 0\} = b_r^0 / a_{rj}^0, \tag{6}$$

where r is the index of the leaving basic variable. The value of y_k corresponding with the new extreme-point solution, which is denoted as k_{0j}, is

$$k_{0j} = k - p_j^0(b_r^0/a_{rj}^0). \tag{7}$$

k_{0j} is determined for each j. Because the extreme-points are ranked in order of decreasing values of y_k, the largest k_{0j} is selected. Since the value of y_k cannot be negative, the value of y_k for the second solution is determined as

$$k_1 = \underset{j}{\text{Max}}\,(k_{0j}, 0) = k_{0c}; \tag{8}$$

we assume that this minimum is unique.

If $k_1 \neq 0$, the next tableau is then generated by introducing the corresponding nonbasic variable x_c into the basis, the leaving basic variable being determined by (6). The case $k_1 = 0$ will be dealt with below. If in (6) for some j, $a_{ij}^0 \leq 0$ for all i, no corresponding extreme-point solution can be found in the usual manner, because in this case x_c can be increased indefinitely; the loss then also increases indefinitely. However, the maximal loss k resulting in $y_k = 0$ provides a bound for the solution, so that this case is similar to the case $k_1 = 0$, and can be dealt with in the same manner.

We have now found an extreme-point solution with $y_k = k_1$. All other extreme-point solutions we shall generate have smaller values of y_k. There are no other extreme-point feasible solutions with y_k between k and k_1, since if there was such a solution, the optimal solution which is our starting solution, could be generated from this solution in one or more steps of the ordinary simplex method. If it could be generated in one step, then it should be possible to perform the corresponding iteration in the reverse direction, as shown before. If it could be generated in more steps, the last solution before the optimal one has a larger value than k_1; from this solution the optimal solution could be generated in one step and it should then be possible to make the corresponding iteration in the reverse direction.

Having found a second extreme-point solution, we are now looking for another one ranking next in order of y_k. First, there are the other extreme-point solutions which can be reached from the first tableau. Secondly, there are the extreme-point solutions which can be reached from the second tableau. The latter solutions are determined by introducing into the basis the nonbasic variables x_j having[3] $p_j^1 > 0$; a solution obtained by introducing x_j with $p_j^1 < 0$ will have $y_k > k_1$ and must in this case be the optimal solution. It is not possible that an extreme-point solution with y_k between k and k_1 is reached, since x_j with $p_j^1 > 0$ is introduced into the basis, which decreases the value of the objective function, so that the loss must be greater than k_1. We therefore determine in the second tableau for each j with $p_j^1 > 0$,

$$\theta_j^1 = \underset{i}{\text{Min}} \{(b_i^1/a_{ij}^1)|a_{ij}^1 > 0\} = b_r^1/a_{rj}^1. \tag{9}$$

The value of y_k resulting from introducing x_j into the basis is therefore

$$k_{1j} = k_1 - p_j^1 (b_r^1/a_{rj}^1), \tag{10}$$

The extreme-point solution ranking next in order of decreasing values of y_k is then the one corresponding with

$$k_2 = \underset{j}{\text{Max}} (k_{0j}, k_{1j}|k_{0j} < k_1, p_j^1 > 0); \tag{11}$$

again the minimum is assumed to be unique. If the minimum corresponds with one of the k_{0j}'s, the next solution is generated from tableau 0, if it corresponds with one of the k_{1j}, it is generated from tableau 1; for the case $k_2 = 0$, see below.

A general iteration, say iteration s, may now be described as follows. First, determine for each j with $p_j^s > 0$.

$$\underset{i}{\text{Min}} \{(b_i^s/a_{ij}^s)|a_{ij}^s > 0\} = b_r^s/a_{rj}^s \tag{12}$$

and

$$k_{sj} = k_s - p_j^s(b_r^s/a_{rj}^s). \tag{13}$$

After this, determine

$$k_{s+1} = \underset{j}{\text{Max}} (k_{1j}, k_{2j}, ..., k_{sj}|0 < k_{1j}, ..., k_{s-1,j} < k, 0). \tag{14}$$

The next tableau is then generated from the one corresponding with the maximum.

In the case there is no positive $k_{1j}, ..., k_{sj}$ left, all other extreme-points which can be generated from the tableaux which have been generated so far have a loss exceeding the maximum loss k. The remaining extreme points are situated between extreme points of the original problem with a loss less than k and those with a loss more than k, in such a way that the

---

[3] Again, only the strict inequalities are considered.

loss is exactly k. These extreme points may be found by introducing non-basic variables with positive p_j^s, which have not yet been introduced in the tableaux generated so far; these variables take such values that the corresponding solutions with a loss of exactly k. This means that a nonbasic variable x_j in tableau s with $p_j^s > 0$ and $k_{sj} < 0$ is given a value

$$x_j = k_s / p_j^s; \tag{15}$$

the corresponding values of the basic variables in the tableau are

$$x_i = b_i^s - a_{ij}^s x_j. \tag{16}$$

In the tableaux these solutions can be obtained by pivoting on the p_j^s's, so that y_k leaves the basis. Because no further iterations take place it is not necessary to generate all elements of the simplex tableaux of these solutions; only the values of variables and possibly the elements of the first row are needed.

That all extreme-point solutions satisfy (1) and (2) are generated by this procedure can be proved as follows. From any extreme-point solution of (1) and (2) it is possible to reach the optimal solution in a finite number of steps by the simplex method. The reverse simplex method proceeds from the optimal solution in the reverse direction, investigating every extreme-point solution downwards; if there is a path going upwards, we are sure of tracing this when investigating every path going downwards.

To illustrate the reverse simplex method, a small example will be used. It involves a production planning problem with 4 products and 2 constraints. The problem is as follows. Maximize f:

$$f = 3x_1 + 4x_2 + 5x_3 + 6x_4,$$

$$x_1 + x_2 + x_3 + x_4 \le 18,$$

$$2x_3 + 3x_4 \le 6,$$

$$x_1, x_2, x_3, x_4 \ge 0.$$

The initial tableau for an application of the normal simplex method is the tableau 00 of table 3. The tableau of the optimal solution is reached after a few iterations and is given in tableau 0; the corresponding value of the objective function is 76 and the maximum loss is assumed to be 20.

Starting the procedure, we replace 76 in tableau 0 by 20. Considering

Table 3

An application of the reverse simplex method

Tableau	Basic variables	Values bas. var.	x_1	x_2	x_3	x_4
00	f	0	-3	-4	-5	-6
	y_1	18	1	1	1	1
	y_2	6	0	0	2	3
			x_1	y_1	x_3	y_2
0	f/y_k	76|20	1	4	$\frac{1}{3}$	$\frac{2}{3}$
	x_2	16	1	1	$\frac{1}{3}$	$-\frac{1}{3}$
	x_4	2	0	0	$\frac{2}{3}$	$\frac{1}{3}$
	k_{0j}		4	-44	19	16
			x_1	y_1	x_4	y_2
1	y_k	19	1	4	$-\frac{1}{2}$	$\frac{1}{2}$
	x_2	15	1	1	$-\frac{1}{2}$	$-\frac{1}{2}$
	x_3	3	0	0	$1\frac{1}{2}$	$\frac{1}{2}$
	k_{1j}		4	-41	–	16
			x_1	y_1	x_3	x_4
2	y_k	16	1	4	-1	-2
	x_2	18	1	1	1	1
	y_2	6	0	0	2	3
	k_{2j}		-2	-56	–	–
			x_2	y_1	x_3	y_2
3	y_k	4	-1	3	0	1
	x_1	16	1	1	$\frac{1}{3}$	$-\frac{1}{3}$
	x_4	2	0	0	$\frac{2}{3}$	$\frac{1}{3}$
	k_{3j}		–	-44	4	-2
			x_2	y_1	x_3	y_2
4	y_k	4	-1	3	0	1
	x_1	15	1	1	0	$-\frac{1}{2}$
	x_3	3	0	0	1	$\frac{1}{2}$
	k_{4j}		–	-41	–	-2

the introduction of x_1 into the basis, we find

$$\theta_1^0 = 16$$

and

$$k_{01} = 20 - 1 \times 16 = 4.$$

The eventual pivot element $a_{rj}^1 = 1$ is underlined and $k_{01} = 4$ is entered in an additional row. For x_3, y_1 and y_2, we find in a similar manner[4]

$$k_{03} = 19, \qquad k_{05} = -44, \qquad k_{06} = 16.$$

The extreme-point with the maximum value of y_k is found by taking k_1 as the maximum element of the row k_{0j}; this is 19, the element in the x_3-column, so that x_3 enters the basis. Tableau 1 is found by pivoting on the element $\frac{2}{3}$. For the value of y_k we find 19, which corresponds with k_1.

In tableau 1, the k_{1j} are to be determined; this is done for the nonbasic variables with a positive entry in the y_k-row, so that x_4 is not considered. For x_1, for instance, we find

$$\theta_1^1 = 15$$

and

$$k_{11} = k_1 - p_1^1 \theta_1^1 = 19 - 1 \times 15 = 4;$$

it is a coincidence that $k_{01} = k_{11}$. k_{15} and k_{16} are found in the same way. Determining the maximum of the relevant k_{0j} and k_{1j}, we find

$$k_2 = k_{06} = 16.$$

This maximum is not unique since also $k_{16} = 16$, but this causes no difficulties. Tableau 2 is generated by introducing y_2 in the basis of tableau 0 and pivoting in the element $\frac{1}{3}$.

The tableaux 3 and 4 are generated in the same way. There is a slight complication since a number of k_{nj}'s may have the same value, which indicates that solutions may be obtained which have the same value of the objective function. In some cases these solutions are different, for example for k_{01} and k_{11}, but in other cases they are the same, such as for k_{11} and k_{33}. However, it is easy to avoid generating the same solution two or more times by disregarding the k_{nj}'s, which lead to solutions which have already been generated.

---

[4] The y-variables are considered as additional x-variables so that y_1 is given the index 5 and y_2 the index 6.

In section 7.3 we will deal with a perturbation method which results in each different solution having a different value of the objective function, so that for such a perturbed objective function k_{01} and k_{11} can no longer be equal. If this perturbation is applied, then if there are k_{sj}'s with the same value, only one of the corresponding solutions should be generated; the other k_{sj}'s of this value will necessarily lead to the same value.

In tableau 4 the maximum of the relevant k_{sj}'s turns out to be negative. The nonbasic variables with negative k_{sj} are then introduced in the basis.

Table 4

Basic variables	Values of basic variables	y_1
y_k	20	$\frac{4}{1}$
x_2	16	
x_4	2	0
		y_k
y_1	5	$\frac{1}{4}$
x_2	11	$-\frac{1}{4}$
x_4	2	0

Table 5

List of extreme-point near-optimal solutions

Solution number	Value objective function	x_1	x_2	x_3	x_4	y_1	y_2
0	76	–	16	–	2	–	–
1	75	–	15	3	–	–	–
2	72	–	18	–	–	–	6
3	60	16	–	–	2	–	–
4	60	15	–	3	–	–	–
5	56	–	11	–	2	5	–
6	56	–	$10\frac{1}{4}$	3	–	$4\frac{3}{4}$	–
7	56	16	2	–	–	–	6
8	56	–	14	–	–	4	6
9	56	$14\frac{2}{3}$	–	–	2	$1\frac{1}{3}$	–
10	56	$17\frac{1}{3}$	–	–	$\frac{2}{3}$	–	4
11	56	$13\frac{2}{3}$	–	3	–	$1\frac{1}{3}$	–
12	56	17	–	1	–	–	4

The solutions are then found by pivoting on the positive p_j^{s}'s. For instance, introducing y_1 into the basis in tableau 0 in such a way that the loss is 20, we have to pivot on the element 4. The relevant parts of the two successive tableaux are then as in table 4. All extreme-point solutions which can be found in this manner are given in table 5, nos 5–12.

Note that the total number of extreme-point solutions, 13, is fairly high for a problem of such a small size. The maximum loss, 20, is relatively high, but even if it had been chosen much lower, say between 4 and 16, there would have been 9 extreme-point solutions. The decisionmaker's task of selecting the preferred solution is therefore not likely to be a trivial one.

7.2. A method based on graph theory

An important difference between the normal simplex method and the reverse simplex method is that in the first method only the current tableau, or some parts of it, are necessary to perform an iteration, while in the reverse simplex method, as it was presented here, any of the previous tableaux may have to be used. When using computers for larger problems, the memory space required to store all previous tableaux, the corresponding inverses or the columns of the product form of the inverse may exhaust the available space. An alternative for the reverse simplex method is to store only the $k_{1j}, k_{2j}, ..., k_{3j}$ and the numbers of the basic variables in the corresponding solutions and to generate each tableau when necessary. Any tableau problem can be generated from the optimal solution or any other solution in at most m steps.

In view of this, it is useful to consider an alternative method in which iterations use, apart from some other information only the preceding tableau, so that the product form of the inverse can be used as in the normal simplex method. This means that we want to consider a method in which the successive solutions are adjacent extreme points as in the simplex method. This method was proposed by Charnes and Cooper[4].

We may proceed as follows. The starting-point is the tableau of the optimal solution. In this tableau f^0 is replaced by k, so that $f(x)$ is measured from its lower bound $f^0 - k$. We now want to generate all extreme-point feasible solutions in which $f(x)$, measured as indicated, is nonnegative. This can be done by generating all basic feasible solutions of the equation system given by the tableau in which $f(x)$ is now considered as an ordinary basic variable. Any of the nonbasic variables may now be introduced into the basis; the leaving basic variable is determined as in the simplex method by comparing positive ratios for all rows, *including the f-row*. In the resulting tableau we have found a new extreme-point feasible solution. In this tableau we may introduce any of the nonbasic variables into the basis, determining the leaving basic variable as before. The resulting tableau may be treated in the same manner, and so on.

The difficulty is that we may find the same solution again and again without being sure that all extreme-point feasible solutions are generated. Since the only element of choice in each iteration is that of the choice of the nonbasic variable which is to enter the basis, the problem is to find a rule for the new basic variable at each iteration which results in generating in a sequence of iterations all extreme-point feasible solutions.

As an initial example let us consider an even smaller problem than that used before. The problem is to maximize f:

$$f = 5x_1 + 6x_2,$$
$$2x_1 + 3x_2 \leq 6,$$
$$x_1, x_2 > 0. \tag{17}$$

The maximum loss allowed is 5.

The set-up tableau is then as given in tableau 0 of table 6; tableau 1 gives the optimal solution, in which the value of the objective function, 15, is replaced by 5. Introducing x_2, the first nonbasic variable into the basis, we find that x_1 leaves the basis. If in tableau 2 the first nonbasic variable x_1, is introduced into the basis, we find solution 1 again. If x_3 is selected as a nonbasic variable, f leaves the basis; the resulting tableau is tableau 3. Introducing now x_1, into the basis we find tableau 4. Introducing then f into the basis, tableau 1 is found again. There exist no other extreme point feasible solutions apart from those of tableaux 1–4, as can be ascertained by introducing other nonbasic variables in each of the tableaux.

In larger problems it would not have been so easy to find a sequence of adjacent basic feasible solutions containing every basic feasible solution. To find such a sequence, it must be realized that the various solutions can be arranged in a network representation. From the tableaux of table 6 we find that from solution 1 we may reach solutions 2 and 4, from solution 2 we may reach solutions 1 and 3, from solution 3 we may reach solutions 2

Table 6
Generation of feasible extreme points for a small example

Tableau	Basic variables	Val. basic variables	x_1	x_2
0	f	0	-5	-6
	x_3	6	$\underline{2}$	3
			x_3	x_2
1	f	15|5	$2\frac{1}{2}$	$1\frac{1}{2}$
	x_1	3	$\frac{1}{2}$	$1\frac{1}{2}$
			x_1	x_3
2	f	2	-1	2
	x_2	2	$\frac{2}{3}$	$\frac{1}{3}$
			x_1	f
3	x_3	1	$-\frac{1}{2}$	$\frac{1}{2}$
	x_2	$1\frac{2}{3}$	$\frac{5}{6}$	$-\frac{1}{6}$
			f	x_2
4	x_3	2	$\frac{2}{5}$	$\frac{3}{5}$
	x_1	2	$-\frac{1}{5}$	$\underline{1\frac{1}{5}}$

and 4, and from solution 4 we may reach solutions 1 and 3. These relations are given in the network in figure 1, in which the nodes correspond with the solutions. From this network it can be concluded that finding all feasible solutions is equivalent to finding a path through the network such that all nodes are visited at least once. In graph theory this is called a *Hamiltonian path*.

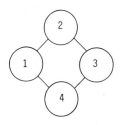

Fig. 1.

Even in such a small network it is easily possible to have a path which misses a node; for example the path 1–2–1–4–1 does not include node 3. For larger programming problems, the network is much more complicated; for instance, in the problem used in section 7.1 each node is connected by 4 arcs, since in each tableau there are 4 nonbasic variables. For such problems it is much less easy to find a path visiting all nodes.

We may, of course, try out the following simple procedure: at each node, go to another node which has not been visited before. In the example of figure 1 this method works and generates all nodes.

Let us apply this method to the example used in section 7.1. The solutions successively generated are given in table 7. For the moment the

Table 7
Simplex tableaux for an application of the Tarry method

Tableau	Basic variable	Values basic variables	x_1	x_3	x_5	x_6
0	f	20	1	$\frac{1}{3}$	4	$\frac{2}{3}$
	x_2	16	1	$\frac{1}{3}$	1	$-\frac{1}{3}$
	x_4	2	0	$\frac{2}{3}$	0	$\frac{1}{2}$

Tableau	Basic variable	Values basic variables	x_2	x_3	x_5	x_6
1	f	4	-1	0	3	1
	x_1	16	1	$\frac{1}{3}$	1	$-\frac{1}{3}$
	x_4	2	0	$\frac{2}{3}$	0	$\frac{1}{3}$

Tableau	Basic variable	Values basic variables	x_2	x_4	x_5	x_6
2	f	4	-1	0	3	$1\frac{1}{2}$
	x_1	15	1	$-\frac{1}{2}$	1	$-\frac{1}{2}$
	x_3	3	0	$1\frac{1}{2}$	0	$\frac{1}{2}$

Tableau	Basic variable	Values basic variables	x_1	x_4	x_5	x_6
3	f	19	1	$-\frac{1}{2}$	4	$\frac{1}{2}$
	x_2	15	1	$-\frac{1}{2}$	1	$-\frac{1}{2}$
	x_3	3	0	$1\frac{1}{2}$	0	$\frac{1}{2}$

Tableau	Basic variable	Values basic variables	x_1	x_4	x_6	f
6	x_5	$4\frac{3}{4}$	$\frac{1}{4}$	$-\frac{1}{8}$	$\frac{1}{8}$	$\frac{1}{4}$
	x_2	$10\frac{1}{4}$	$\frac{3}{4}$	$-\frac{3}{8}$	$-\frac{5}{8}$	$-\frac{1}{4}$
	x_3	3	0	$-1\frac{1}{2}$	$\frac{1}{2}$	0

Table 7 (*continued*)

Tableau	Basic variable	Values basic variables	x_2	x_4	x_6	f
7	x_5	$1\frac{1}{3}$	$-\frac{1}{3}$	0	$\frac{1}{3}$	$\frac{1}{3}$
	x_1	$13\frac{2}{3}$	$1\frac{1}{3}$	$-\frac{1}{2}$	$-\frac{5}{6}$	$-\frac{1}{3}$
	x_3	3	0	$1\frac{1}{2}$	$\frac{1}{2}$	0

Tableau	Basic variable	Values basic variables	x_2	x_3	x_6	f
8	x_5	$1\frac{1}{3}$	$-\frac{1}{3}$	0	$\frac{1}{3}$	$\frac{1}{3}$
	x_1	$14\frac{2}{3}$	$1\frac{1}{3}$	$\frac{1}{3}$	$-\frac{2}{3}$	$-\frac{1}{3}$
	x_4	2	0	$\frac{2}{3}$	$\frac{1}{3}$	0

Tableau	Basic variable	Values basic variables	x_1	x_3	x_6	f
9	x_5	5	$\frac{1}{4}$	$\frac{1}{12}$	$\frac{1}{6}$	$\frac{1}{4}$
	x_2	11	$\frac{3}{4}$	$\frac{1}{4}$	$-\frac{1}{2}$	$-\frac{1}{4}$
	x_4	2	0	$\frac{2}{3}$	$\frac{1}{3}$	0

Tableau	Basic variable	Values basic variables	x_1	x_3	x_4	f
12	x_6	6	0	2	3	0
	x_2	14	$\frac{3}{4}$	$1\frac{1}{4}$	$1\frac{1}{2}$	$-\frac{1}{4}$
	x_5	4	$\frac{1}{4}$	$-\frac{1}{4}$	$-\frac{1}{2}$	$\frac{1}{4}$

Tableau	Basic variable	Values basic variables	x_3	x_4	x_5	f
13	x_6	6	2	3	0	0
	x_1	16	-1	-2	4	1
	x_2	2	2	3	-3	-1

Tableau	Basic variable	Values basic variables	x_2	x_4	x_5	f
14	x_6	4	-1	0	3	1
	x_1	17	$\frac{1}{2}$	$-\frac{1}{2}$	$2\frac{1}{2}$	$\frac{1}{2}$
	x_3	1	$\frac{1}{2}$	$1\frac{1}{2}$	$-1\frac{1}{2}$	$-\frac{1}{2}$

Tableau	Basic variable	Values basic variables	x_2	x_3	x_5	f
15	x_6	4	-1	0	$\frac{3}{2}$	1
	x_1	$17\frac{1}{3}$	$\frac{2}{3}$	$\frac{1}{3}$	$\frac{3}{2}$	$-\frac{1}{3}$
	x_4	$\frac{2}{3}$	$\frac{1}{3}$	$\frac{2}{3}$	-1	$-\frac{1}{3}$

Tableau	Basic variable	Values basic variables	x_1	x_3	x_4	x_5
32	f	16	-1	-1	-2	$\frac{4}{1}$
	x_2	18	1	1	1	$\frac{1}{1}$
	x_6	6	0	2	3	0

exact numbering of the solutions is irrelevant. Each solution in this table is generated from its predecessor, but from solution 6 no new solution can be generated. However, there exists another solution, namely solution 32, which has not been generated yet. Hence it is not always possible to generate consecutively all solutions by this method.

However, this method may still be useful by first applying it and after no new solutions can be generated from the last one check if there are some other solutions which can be generated from other, previous solutions. Any such solution can be used as a starting point for a chain of new solutions. In our example there would be only one new solution. This "jumping" to another solution can be repeated as often as necessary.

The following rule may be used to avoid isolating nodes. In each node, go to an adjacent unvisited node with the maximum accessibility, where accessibility is defined as the number of times this node could have been visited from an adjacent node which has been generated before. If this is applied to the example, the following path is found: 1, 2, 3, 32, 12, 6, 9, 8, 7, 14, 13, 15. Hence all nodes have been visited. In this case this procedure works, but there is no guarantee that it works in all cases.

Charnes and Cooper[4] have proposed to use in problems like these a method devised by Tarry for finding a path through a network such that each node is visited at least once. This method and its application to our present problem will now be explained.

The problem of finding a path through a network such that all nodes are visited is a well-known problem in graph theory and can be solved in various ways. Following Charnes and Cooper, we shall consider the method devised by Tarry. This method requires that at each node it is known through which arc it was first reached; further it must be known whether an arc has already been traversed in a certain direction. The rule of Tarry's method is then as follows: *At any node, go to another node along an arc which has not yet been traversed in that direction, but do not go along the arc that was used to reach the first-mentioned node for the first time unless there is no other choice.*

If this rule is followed, every arc will be traversed twice, once in one direction and once in the opposite direction. Each node will be visited at least once. Applying this rule to the network of figure 1, we find the following. Solution 1 is used as a starting-point; we can go to solutions 2 and 4, but choose to go to solution 2. In solution 2 we can in principle go to solutions 1 and 3, but we have to go to solution 3, since solution 1 was used to reach solution 2 for the first time. For similar reasons we go to solutions 3, 4, and 1. In solution 1, we cannot go to solution 2, since we went already once in that direction. Though solution 1 was reached first coming from solution 4, there is no other choice, so that we return to solution 4. In solution 4, we go back to solution 3, then to solution 2, then to solution 1, all for the same reasons. Every arc has been traversed twice and every solution has been visited.

The Tarry method will now be applied to the example used in section 7.1. The initial solution is the optimal solution of the linear programming problem; it is given in tableau 0 of table 7. When the Tarry method allows this, we shall always introduce into the basis the nonbasic variable with the smaller number; for this purpose f is considered to be x_7. The information required for an application of the Tarry method is given in table 8. Since solutions reappear many times, they are indicated in table 8 by the number of the iteration in which they first occur. Each solution is indicated by parentheses in the columns of basic variables; the spaces in the columns of nonbasic variables are used for the number of the solution which is obtained after the nonbasic variable concerned is introduced into the basis; the number of the iteration is indicated within brackets.

In tableau 0 of table 7, x_1 is introduced into the basis; we therefore enter 1 as the number of the iteration in the column of x_1 in the row of solution 0 in table 8. From tableau 0 we find that x_2 leaves the basis; the resulting solution which is indicated as solution 1 is given in tableau 1.

Table 8
Information for an application of the Tarry method

Sol no.	x_1	x_2	x_3	x_4	x_5	x_6	f
0	1	–	3(5)	–	9(36)	32(38)	–
1	–	0(52*)	2	–	8(21)	15(23)	–
2	–	3	–	1(51*)	7(28)	14(30)	–
3	2(50*)	–	–	0(4)	6	32(34)	–
6	7	–	–	9(11)	–	12(41)	3(49*)
7	–	6(48*)	–	8	–	14(26)	2(29)
8	–	9	7(47*)	–	–	15(19)	1(22)
9	8(46*)	–	6(10)	–	–	12	0(37)
12	13	–	6(40)	9(45*)	–	–	32(42)
13	–	–	14	15(17)	12(44*)	–	32
14	–	13(31*)	–	15	7(25)	–	2(27)
15	–	13(16)	14(24*)	–	8(18)	–	1(20)
32	13(43*)	–	3(33)	0(35)	12(39)	–	–

This solution is then also entered in table 8 with strokes in the columns of x_1, x_4, x_5.

To indicate the solution from which solution 1 was reached for the first time, a star is placed in the column of x_2, since by introducing this variable into the basis solution 0 is generated. In solution 1, we cannot introduce x_2 into the basis unless there is no other possibility, so that x_3 is introduced into the basis. Since this is the second iteration, a 2 is put in the column of x_3 in the row of solution 1 in table 8. From tableau 1 it is found that x_4 leaves the basis. After the transformation, tableau 2 containing solution 2 is obtained, which is entered in table 8 with strokes in the columns of x_1, x_3, and f and a star in the column of x_4. Solution 3 is generated in the same manner.

In solution 3, x_4 is introduced into the basis, yielding solution 0 again. Hence solution 4 is equivalent with solution 0. In order to save space, we do not reproduce the tableau of this solution again, but if a computer was used, this tableau would be generated from tableau 3. In table 8 the nonbasic variable with the next blank space in the row of solution 0 is selected as the new basic variable; this happens to be x_3, so that the number of the next iteration, 5, is entered in the space within brackets. The resulting solution happens to be the same as solution 3, which is also indicated in the column of x_3. In this solution x_4 is introduced into the basis. Solution 6 turns out to be a solution which has not been generated before, so that it is entered in table 8.

All other iterations are performed in the same manner. If in some row of table 8 no blank space is left except the starred one, then the corresponding variable of this starred space enters the basis; the solution concerned will not be generated again.

After 52 iterations we return to solution 0; every solution has then been generated four times, which is the number of nonbasic variables in each solution. Note that the last new solution was generated in the 32nd iteration, so that the last 20 iterations only served to prove that there is no other solution. In fact, the first 15 iterations, which is 30% of all iterations gave all solutions except one. In other problems, however, things may turn out differently.

The number of iterations in the Tarry method is equal to the number of extreme-point feasible solutions times the number of nonbasic variables in a tableau. In linear programming with inequality constraints the number of nonbasic variables is equal to the number of x-variables. Thus, if there are n variables in a problem with inequality constraints, every extreme-point feasible solution has to be generated n times for this method. Since the number of variables in most cases exceeds the number of constraints, the reverse simplex method in the version in which each previous tableau is regenerated from the optimal tableau will usually require fewer computations.

Another advantage which the reverse simplex method has over the Tarry method is that it is easier in the first method to vary the value of k, since all solutions are ranked according to increasing loss. An increase in k will leave all solutions having a smaller loss than k the same, while those having a loss of exactly k can be easily adapted. In the Tarry method or any other adjacent extreme-point method, the entire procedure would have to be started all over again.

7.3. The reverse simplex method for the generation of extreme points of a system of inequalities

The reverse simplex method can also be used in cases in which all extreme-point solutions of a system of equations and inequalities are desired. We may for example have the general system

$$Ax + y = b,$$
$$x, y \geq 0; \tag{18}$$

b is assumed to be nonnegative.

We could try to introduce each of the nonbasic variables in any tableau that is generated and note each basic solution, but we would never be sure when to terminate; the same solutions can be generated time and again. In order to prevent cycling, an objective function may be introduced. This could be just any objective function. An optimal solution is generated for this system, after which the reverse simplex method may be applied. If the following equation for the objective function is added

$$e'x + f = 0, \tag{19}$$

the system has an optimal solution $y = b$, $f = 0$; the reverse simplex method can be started at once.

The addition of an objective function in (19) has the disadvantage that an extra row is required; furthermore it may happen that in any application of the reverse simplex method a number of the elements in the f-row is zero, which means that cycling of solutions is possible. These objections do not apply[5] to the objective function equation

$$\epsilon'x + f = 0, \tag{20}$$

where ϵ' is the vector

$$\epsilon' = [\epsilon, \epsilon^2, \epsilon^3, ..., \epsilon^n].$$

ϵ is considered to be a very small number.[6] The equations of (18) and (20) may be put into tableau format, see tableau 0 of table 9. The coefficients of $\epsilon, \epsilon^2, \epsilon^3, ..., \epsilon^n$ in the equation of the objective function are put into separate rows. If none of the elements of b is zero, x_n will be the first variable to enter the basis since

$$\epsilon_n \{\operatorname*{Min}_i (b_i/a_{in} | a_{in} > 0)\} < \epsilon_j \{\operatorname*{Min}_i (b_i/a_{ij} | a_{ij} > 0)\}, j \not\in n. \tag{21}$$

Table 9

The reverse simplex method with lexicographic ordering

Tableau	Basic variables	Values basic variables	x_1	.	x_n
	$f(\epsilon_1)$	0	1	.	0
	.	.	.	.	.
0	$f(\epsilon_n)$	0	0	.	1
	y_1	b_1	a_{11}	.	a_{1n}
	.	.	.	.	.
	y_m	b_m	a_{m1}	.	a_{mn}
			x_1	.	y_1
	$f(\epsilon_1)$	0	1	.	0
	.	0	.	.	.
1	$f(\epsilon_n)$	$-a_{1n}^{-1}b_1$	$-a_{1n}^{-1}a_{11}$	.	$-a_{1n}^{-1}$
	x_n	$a_{1n}^{-1}b_1$	$a_{1n}^{-1}a_{11}$	.	a_{1n}^{-1}
	.	.	.	.	.
	y_m	$b_m - a_{mn}a_{1n}^{-1}b_1$	$a_{m1} - a_{mn}a_{1n}^{-1}a_{11}$	.	$-a_{mn}a_{1n}^{-1}$

If (19) has been added as a constraint, there is at least one positive a_{ij} in each column j. Let y_1 be the leaving basic variable, so that a_{1n} is the pivot. In the next tableau it is observed the $f(\epsilon_n)$-row is equal to minus the x_n-row. It may be proved that any row $f(\epsilon_j)$ is equal to minus the

[5] Nondegeneracy of solutions is assumed; otherwise the normal perturbation technique may be applied, so that there is a double lexicographic ordering.

[6] The superindices are in this case exponents.

x_j-row if x_j is a basic variable; if x_j is nonbasic, the $f(\epsilon_j)$-row remains as it is in tableau 0. This means that it is not necessary to carry along the $f(\epsilon)$-rows, since all required information is available in the other rows, see the x_n-row of tableau 1 of table 9.

The lexicographic variant of the reverse simplex method will be used to generate all extreme-point solutions satisfying the following constraints

$$x_1 + x_2 + x_3 + x_4 \le 18,$$
$$2x_3 + 3x_4 \le 6,$$
$$x_1, x_2, x_3, x_4 \ge 0. \tag{22}$$

Tableau 0 of table 10 is the set-up tableau which can be used as an optimal tableau if the equation of the objective function is added to it:

$$0 = \epsilon x_1 + \epsilon^2 x_2 + \epsilon^3 x_3 + \epsilon^4 x_4 + f. \tag{23}$$

The coefficients of these equations should be ordered in rows according to decreasing exponents of ϵ, so that we get table 11. Since terms in ϵ with a lower exponent always dominate those in ϵ with a higher exponent, except when the first-mentioned terms have a zero coefficient, we should only bother to write down the nonzero terms with the lowest exponent of ϵ. In tableau 0 of table 10 the losses of solutions to be generated from this tableau are found in the usual manner. The smallest loss is $2\epsilon^4$, which occurs in the column of x_4. Hence x_4 is selected as the new basic variable and y_2 is the leaving basic variable. Tableau 1 results.

Table 10

Application of the lexicographic reverse simplex method

Tableau	Basic variables	Values basic variables	x_1	x_2	x_3	x_4
0	y_1	18	1	1	1	1
	y_2	6	0	0	2	3
	k_{0j}		18ϵ	$18\epsilon^2$	$3\epsilon^3$	$2\epsilon^4$
			x_1	x_2	x_3	y_2
1	y_1	16	1	1	$\frac{1}{3}$	$-\frac{1}{3}$
	x_4	2	0	0	$\frac{2}{3}$	$\frac{1}{3}$
	k_{1j}		16ϵ	$16\epsilon^2$	$3\epsilon^3$	—

Table 10   (*continued*)

Tableau	Basic variables	Values basic variables	x_1	x_2	y_2	x_4
2	y_1	15	1	1	$-\frac{1}{2}$	$-\frac{1}{2}$
	x_3	3	0	0	$\frac{1}{2}$	$1\frac{1}{2}$
	k_{2j}		15ϵ	$15\epsilon^2$	—	—

			x_1	y_1	y_2	x_4
3	x_2	15	1	1	$-\frac{1}{2}$	$-\frac{1}{2}$
	x_3	3	0	0	$\frac{1}{2}$	$1\frac{1}{2}$
	k_{3j}		15ϵ	—	$18\epsilon^2$	$16\epsilon^2$

			x_1	y_1	x_3	y_2
4	x_2	16	1	1	$\frac{1}{3}$	$-\frac{1}{3}$
	x_4	2	0	0	$\frac{2}{3}$	$\frac{1}{3}$
	k_{4j}		16ϵ	—	—	$18\epsilon^2$

			x_1	y_1	x_3	x_4
5	x_2	18	1	1	1	1
	y_2	6	0	0	2	3
	k_{5j}		18ϵ	—	—	—

			x_2	y_1	y_2	x_4
6	x_1	15	1	1	$-\frac{1}{2}$	$-\frac{1}{2}$
	x_3	3	0	0	$\frac{1}{2}$	$1\frac{1}{2}$
	k_{6j}		—	—	18ϵ	16ϵ

			y_1	x_2	x_3	y_2
7	x_1	16	1	1	$\frac{1}{3}$	$-\frac{1}{3}$
	x_4	2	0	0	$\frac{2}{3}$	$-\frac{1}{3}$
	k_{7j}		—	—	—	18ϵ

			y_1	x_2	x_3	x_4
8	x_1	18	1	1	1	1
	y_2	6	0	0	2	3
	k_{8j}		—	—	—	—

Table 11

Values bas. var.	x_1	x_2	x_3	x_4
0	1	0	0	0
0	0	1	0	0
0	0	0	1	0
0	0	0	0	1

In this tableau, the deleted rows of the objective function are the same except that the fourth row should be replaced by minus the row of x_4. These rows are reproduced in table 12.

Table 12

V.b.v.	x_1	x_2	x_3	y_2
0	1	0	0	0
0	0	1	0	0
0	0	0	1	0
-2	0	0	$-\frac{2}{3}$	$-\frac{1}{3}$
$-2\epsilon^4$	ϵ	ϵ^2	ϵ^3	$-\frac{1}{3}\epsilon^4$

In the last row the other rows are added after multiplication by the corresponding power of ϵ; dominated terms are neglected. Since the coefficients in the column of x_4 is negative, y_2 should not be made basic, because this will lead to solutions obtained before. The k_{1j} in tableau 1 are obtained as usual; again dominated terms are neglected.

x_3 happens to have in tableau 0 and in tableau 1 a loss of $3\epsilon^3$, so that there is a tie, but it turns out that the same solution is generated in both cases, namely a solution with y_1 and x_3 as basic variables. Hence this solution may be generated either from tableau 0 or tableau 1. The result is in both cases tableau 2. The rows of the objective function which are constructed from this tableau are given in table 13.

Table 13

V.b.v.	x_1	x_2	y_2	x_4
0	1	0	0	0
0	0	1	0	0
-3	0	0	$-\frac{1}{2}$	$-1\frac{1}{2}$
0	0	0	0	1
$-3\epsilon^3$	ϵ	ϵ^2	$-\frac{1}{2}\epsilon^3$	$-1\frac{1}{2}\epsilon^3$

The last row is found by multiplication of the rows by the appropriate powers of ϵ, addition of the rows and deletion of dominated terms. Since the elements of y_2 and x_4 in the f-row are negative, their introduction as basic variables should not be considered. The losses of solutions obtained by having x_1 and x_2 as basic variables are determined as usual.

The smallest loss of solutions to be generated is now $15\epsilon^2$, which is found if x_2 is introduced as basic variable in tableau 2. The result is tableau 3. The rows of the objective function for this tableau are shown in table 14. The losses k_{3j} follow from this.

Table 14

V.b.v.	x_1	y_1	y_2	x_4
0	1	0	0	0
-15	-1	-1	$\frac{1}{2}$	$\frac{1}{2}$
-3	0	0	$-\frac{1}{2}$	$-1\frac{1}{2}$
0	0	0	0	1
$-15\epsilon^2$	ϵ	$-\epsilon^2$	$\frac{1}{2}\epsilon^2$	$\frac{1}{2}\epsilon^2$

The next smallest loss is $16\epsilon^2$, which is found if x_2 is introduced as basic variable in tableau 1 or in x_4 is introduced as basic variable in tableau 3. The same solution results in both cases. All other tableaux are found in the same manner. After tableau 8 has been generated, there is no other solution to be generated, which means that all extreme-point solutions of the constraints (22) have been found.

The lexicographic variant of the reverse simplex method may be used in parametric programming with a parameter in the objective function; if more than one coefficient in the f-row becomes zero for a certain value of λ, all solutions that are optimal for that value of λ may be generated by adding ϵ-terms to the objective function for the zero coefficients in the f-row, after which the method may be applied.

7.4. The reverse dual method

In view of the equivalence between primal and dual methods, it seems rather obvious that if there is a reverse simplex method, it is possible to devise a reverse dual method. In the case of the reverse simplex method, all extreme points of the feasible region, that is all extreme-point solutions of a system of linear inequalities, were generated in decreasing order of the value of the objective function. This means that primal feasibility was maintained, but optimality in the narrow sense, which implies nonnegativity of the dual variables, was no longer required. In the reverse dual method solutions are generated which are optimal in the narrow sense, so that all dual variables are nonnegative, but feasibility of a solution is not required.

This means that we are interested in all solutions to the problem which are efficient, with possible shortages of scarce resources and negative quantities to be produced in the production planning example. If there is a shortage in scarce resources, we may consider problems with larger amounts of scarce resources. Negative production, however, usually does not have an interpretation. For this reason, the reverse dual method does not have such a direct interpretation as the reverse simplex method had. Of course, the reverse dual method may always be interpreted as a method for finding all extreme-point solutions for the "outsider" of the interpretation of the dual problem.

It turns out that the dual method can be used as a subroutine in methods for more complicated problems. The reverse dual method and its lexicographic variant will be explained using a few examples. The theoretical considerations on which it is based are similar to those of the reverse simplex method.

Let us consider the typical maximization problem which may be interpreted as a production planning problem and which was used before in the exposition of parametric programming. Maximize f:

$$f = 3x_1 + x_2 + 2x_3 + 2\tfrac{3}{4}x_4,$$
$$4x_1 + 3x_2 + 2x_3 \qquad \leq 100,$$
$$x_1 \qquad + x_3 + 2x_4 \leq 20,$$
$$2x_1 + 4x_2 + x_3 + 2x_4 \leq 30,$$
$$x_1, x_2, x_3, x_4 \geq 0. \tag{24}$$

Tableau 00 of table 15 in the set-up tableau and the initial tableau for an application of the simplex method which after two iterations gives the optimal solution in tableau 0.

In the reverse dual method the value of the objective function is increased by allowing unfeasibilities to occur; so that instead of suboptimal solutions we may speak about post-optimal solutions. Since the objective function increases, the word *gain* instead of loss may be used. Successive solutions ordered according to increasing gains will be gener-

POST-OPTIMALITY ANALYSIS

Table 15
An application of the reverse dual method

Tableau	Basic variables	Values bas. variables	Gain	x_1	x_2	x_3	x_4
				x_1	x_2	x_3	x_4
00	f	0		-3	-1	-2	$-2\frac{3}{4}$
	y_1	100		4	3	2	0
	y_2	20		1	0	1	2
	y_3	30		$\underline{2}$	4	1	2
				y_3	x_2	y_2	x_4
0	$f\|f^*$	50/0		1	3	1	$1\frac{1}{4}$
	y_1	40	$12\frac{1}{2}$	-2	-5	0	$\underline{-4}$
	x_3	10	$7\frac{1}{2}$	-1	$\underline{-4}$	2	2
	x_1	10	10	1	4	$\underline{-1}$	0
				y_3	x_3	y_2	x_4
1	f^*	$7\frac{1}{2}$		$\frac{1}{4}$	$\frac{3}{4}$	$2\frac{1}{2}$	3
	y_1	$27\frac{1}{2}$	$16\frac{2}{3}$	$-\frac{3}{4}$	$-1\frac{1}{4}$	$-2\frac{1}{2}$	$-6\frac{1}{2}$
	x_2	$-2\frac{1}{2}$	—	$\frac{1}{4}$	$-\frac{1}{4}$	$-\frac{1}{2}$	$-\frac{1}{2}$
	x_1	20	∞	0	1	1	2
				y_3	x_2	x_1	x_4
2	f^*	10		2	7	1	$1\frac{1}{4}$
	y_1	40	$22\frac{1}{2}$	-2	-5	0	$\underline{-4}$
	x_3	30	∞	1	4	2	2
	y_2	-10	—	-1	-4	$\underline{-1}$	0
				y_3	x_2	y_2	y_1
3	f^*	$12\frac{1}{2}$		$\frac{3}{8}$	$1\frac{7}{16}$	1	$\frac{5}{16}$
	x_4	-10	—	$\frac{1}{2}$	$1\frac{1}{4}$	0	$-\frac{1}{4}$
	x_3	30	$18\frac{1}{8}$	-2	$-6\frac{1}{2}$	2	$\frac{1}{2}$
	x_1	10	$22\frac{1}{2}$	1	4	$\underline{-1}$	0
				y_1	x_3	y_2	x_4
4	f^*	$16\frac{2}{3}$		$\frac{1}{3}$	$\frac{1}{3}$	$1\frac{2}{3}$	$\frac{5}{6}$
	y_3	$-36\frac{2}{3}$	—	$-1\frac{1}{3}$	$1\frac{2}{3}$	$3\frac{1}{3}$	$8\frac{2}{3}$
	x_2	$6\frac{2}{3}$	$18\frac{3}{4}$	$\frac{1}{3}$	$-\frac{2}{3}$	$-1\frac{1}{3}$	$-2\frac{2}{3}$
	x_1	20	∞	0	1	1	2

ated. The objective function will be measured from its optimal value, which is 50, so that the value of f in the tableau, 50 is replaced by 0.

Each basic variable with a nonnegative value is considered a leaving basic variable. For each such variable the corresponding new basic variable is determined as in the dual method and the gain of the new solutions is given in an additional column. Hence for $y_1 = 40$ in tableau 0, the new basic variable is found from

$$\text{Min } (\tfrac{1}{2}, \tfrac{3}{5}, 1\tfrac{1}{4}/4) = \tfrac{5}{16}.$$

and the gain is $\tfrac{5}{16} \times 40 = 12\tfrac{1}{2}$. In the same manner the gains $7\tfrac{1}{2}$ and 10 are found for the basic variables x_3 and x_1. The minimum gain is $7\tfrac{1}{2}$. Hence the solution resulting from having x_3 as leaving basic variable and x_2 as new basic variable should be generated first, which is done by a transformation of the tableau with -4 as a pivot. The result is tableau 1. In its solution x_2 has a negative value, which is in most cases difficult to interpret.

For this tableau the gain should be generated which is done by first taking y_1 as the leaving basic variable. We determine

$$\text{Min } (\tfrac{1}{4}/\tfrac{3}{4}, \tfrac{3}{4}/1\tfrac{1}{4}, 2\tfrac{1}{2}/2\tfrac{1}{2}, 3/6\tfrac{1}{2}) = \tfrac{1}{3},$$

so that the gain is

$$\tfrac{1}{3} \times 27\tfrac{1}{2} + 7\tfrac{1}{2} = 16\tfrac{2}{3}.$$

x_2, which has a negative value in tableau 1, should not be considered a leaving basic variable because if a positive pivot is chosen in its row, the new basic variable, in this case y_3, would have in the next tableau a negative coefficient in the f-row, so that the resulting solution is not optimal. If a negative pivot is chosen, the resulting solution, if it is optimal in the narrow sense, would have a lower value of the objective function and should therefore have been generated before, for similar reasons as used in the corresponding argument in the reverse simplex method. Hence, for basic variables with negative values there is no new solution with a higher gain.

For the last basic variable, x_1, we have essentially the same situation; it cannot be replaced by one of the nonbasic variables without decreasing the objective function. In this row the gain is denoted as ∞, because one of the nonbasic variables can be given the value ∞. This case differs from the previous one, since there the negative value was the reason why no new solution with a higher gain could be found; for x_1 the reason is that there are no negative elements in its row.

Proceeding from tableaux 0 and 1, we select the next solution with the least gain, which is found by replacing x_1 in tableau 0 as a basic variable by y_2. The result is tableau 2, of which the gains are determined as described before. The next solution with the least gain is found from tableau 0 if x_4 replaces y_1, which results in tableau 3. The next solution has a gain of $16\frac{2}{3}$, which is found from tableau 1 if y_3 replaces y_1. We could go on in the same manner until all finite extreme points are generated.

Note that in this case it would not be meaningful to add a constraint in terms of the maximum gain on the problem, since primal feasibility is not required in the solutions. In the corresponding dual problem such a constraint would be quite appropriate and solutions corresponding to such a constraint would be easily generated.

In the lexicographic variant of the reverse dual method, the values of basic variables in the optimal tableau, tableau 0, which are 40, 10, and 10, are replaced by ϵ, ϵ^2 and ϵ^3, so that these values may be written as the product

$$
\begin{bmatrix} f^* \\ y_1 \\ x_3 \\ x_1 \end{bmatrix} = \begin{bmatrix} 0 & 0 & 0 \\ 1 & 0 & 0 \\ 0 & 1 & 0 \\ 0 & 0 & 1 \end{bmatrix} \begin{bmatrix} \epsilon \\ \epsilon^2 \\ \epsilon^3 \end{bmatrix}.
$$

Since the columns of the unit matrix are the same as those of the basic variables they do not have to be written down. The gains of tableau 0 are now $\frac{5}{16}\epsilon$, $1\frac{1}{3}\epsilon^2$, and $\frac{3}{8}\epsilon^3$, of which the third is the least so that the next solution is found by replacing x_1 by x_2. The result would be tableau 2, with the following values of basic variables

$$
\begin{bmatrix} f^* \\ y_1 \\ x_2 \\ y_2 \end{bmatrix} = \begin{bmatrix} 0 & 0 & 2 \\ 1 & 0 & -2 \\ 0 & 1 & 1 \\ 0 & 0 & -1 \end{bmatrix} \begin{bmatrix} \epsilon \\ \epsilon^2 \\ \epsilon^3 \end{bmatrix} = \begin{bmatrix} 2\epsilon^3 \\ \epsilon - 2\epsilon^3 \\ \epsilon^2 + \epsilon^3 \\ -\epsilon^3 \end{bmatrix}.
$$

The corresponding gains are found as follows. For y_1:

$$
(\epsilon - 2\epsilon^3)\tfrac{3}{8} + 2\epsilon^3 \sim \tfrac{3}{8}\epsilon.
$$

For x_3: ∞, since there are no negative elements in this row. For y_2: no gain, since the value of the basic variable is $-\epsilon^3$, which is negative. Hence the gains to be considered are $\frac{5}{16}\epsilon$, $1\frac{1}{3}\epsilon^2$, and $\frac{3}{8}\epsilon$, of which the second is the smallest, so that in tableau 0 x_2 should replace x_3, which results in tableau 1. The values of basic variables are found to be

$$
\begin{bmatrix} f^* \\ y_1 \\ x_2 \\ x_1 \end{bmatrix} = \begin{bmatrix} 0 & 2\frac{1}{2} & 0 \\ 1 & -2\frac{1}{2} & 0 \\ 0 & -\frac{1}{2} & 0 \\ 0 & 1 & 1 \end{bmatrix} \begin{bmatrix} \epsilon \\ \epsilon^2 \\ \epsilon^3 \end{bmatrix} = \begin{bmatrix} 2\frac{1}{2}\epsilon^2 \\ \epsilon - 2\frac{1}{2}\epsilon^2 \\ -\frac{1}{2}\epsilon^2 \\ \epsilon^2 - \epsilon^3 \end{bmatrix}.
$$

From this new gains can be found, and so on. The same tableaux and solutions should be found as in the normal reverse dual method, but the order in which they are found will be different.

CHAPTER 8

QUADRATIC FORMS

8.1. Introduction

A *homogeneous quadratic form* in n variables may be represented as follows:

$$f(x_1, x_2, \ldots, x_n) = c_{11}x_1^2 + c_{22}x_2^2 + \ldots + c_{nn}x_n^2$$
$$+ 2c_{12}x_1x_2 + \ldots + 2c_{n-1,n}x_{n-1}x_n. \tag{1}$$

In matrix formulation this form is given by

$$f(x) = x'Cx \tag{2}$$

with

$$C = \begin{bmatrix} c_{11} & c_{12} & \cdot & c_{1n} \\ c_{12} & c_{22} & \cdot & c_{2n} \\ \cdot & \cdot & \cdot & \cdot \\ c_{1n} & c_{2n} & \cdot & c_{nn} \end{bmatrix};$$

C is called the matrix of the quadratic form.

A *nonhomogeneous quadratic form* also contains linear and constant terms; it may therefore be represented as

$$f(x_1, \ldots, x_n) = c_0 + c_1x_1 + \ldots + c_nx_n + c_{11}x_1^2 + \ldots$$
$$+ c_{nn}x_n^2 + 2c_{12}x_1x_2 + \ldots + 2c_{n-1,n}x_{n-1}x_n; \tag{3}$$

the corresponding matrix formulation is

$$f(x) = c_0 + c'x + x'Cx, \tag{4}$$

with c a vector of n elements,

$$c = \begin{bmatrix} c_1 \\ c_2 \\ \cdot \\ \cdot \\ \cdot \\ c_n \end{bmatrix}.$$

A homogeneous quadratic form is said to be *positive definite* if it takes only positive values for any set of values of the x-variable which are not all zero; if all x-variables are zero, the quadratic form will of course be zero. A quadratic form is said to be *positive semi-definite*, if the form can only take positive or zero values for any set of values of the x-variables. A quadratic form $f(x_1, \ldots, x_n)$ is *negative definite* or *negative semi-definite* if $-f(x_1, \ldots, x_n)$ is positive definite or positive semi-definite, respectively. It is said to be *indefinite* if the form takes positive values for certain values of the x-variables and negative values for other values of the x-variables. From these definitions it follows that a quadratic form is either positive (negative) (semi-)definite or indefinite.

The simplest example of a positive definite quadratic form is the function

$$f(x_1, x_2) = x_1^2 + x_2^2 = [x_1 \quad x_2] \begin{bmatrix} 1 & 0 \\ 0 & 1 \end{bmatrix} \begin{bmatrix} x_1 \\ x_2 \end{bmatrix} = x'x. \tag{5}$$

It is obvious that this function cannot take negative values, since it is a sum of squares; further it is zero only for $x_1 = 0$ and $x_2 = 0$. A graphical representation for constant values of the function is given in figure 1; it consists of circles centered around the origin.

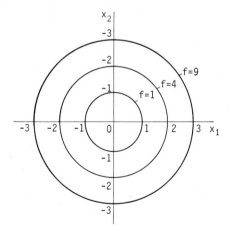

Fig. 1.

The function

$$f(x_1, x_2) = x_1^2 = [x_1 \quad x_2] \begin{bmatrix} 1 & 0 \\ 0 & 0 \end{bmatrix} \begin{bmatrix} x_1 \\ x_2 \end{bmatrix} \tag{6}$$

is an example of a positive semi-definite quadratic form; see figure 2. Examples of negative definite and negative semi-definite quadratic forms can easily be found by multiplying the expressions in (5) and (6) by -1.

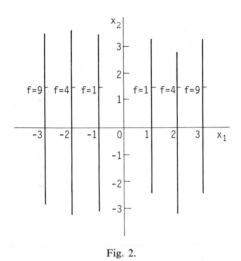

Fig. 2.

The simplest example of an indefinite quadratic form is given by

$$f(x_1, x_2) = x_1^2 - x_2^2; \tag{7}$$

a graphical representation can be found in figure 3. It is obvious that this function may take both positive and negative values, since x_1 may have a nonzero value and x_2 a zero value, so that the function is positive, while it is also possible to keep x_1 zero and give x_2 a nonzero value, so that the function is negative.

An example of a positive definite quadratic form in n variables is

$$f(x) = x_1^2 + x_2^2 + \ldots + x_n^2 = x'x. \tag{8}$$

An example of a negative definite quadratic form is given by

$$f(x) = -x_1^2 - x_2^2 - \ldots - x_n^2 = -x'x. \tag{9}$$

An example of an indefinite quadratic form is

$$f(x) = x_1^2 + \ldots + x_q^2 - x_{q+1}^2 - \ldots - x_n^2 = x'Jx, \tag{10}$$

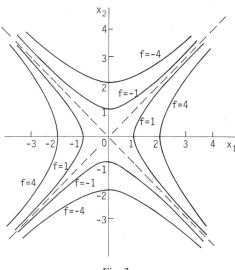

Fig. 3.

with

$$
J = \begin{bmatrix}
1 & \cdot & 0 & 0 & \cdot & 0 \\
\cdot & \cdot & \cdot & \cdot & \cdot & \cdot \\
0 & \cdot & 1 & 0 & \cdot & 0 \\
0 & \cdot & 0 & -1 & \cdot & 0 \\
\cdot & \cdot & \cdot & \cdot & \cdot & \cdot \\
0 & \cdot & 0 & 0 & \cdot & -1
\end{bmatrix}.
$$

In (8) the matrix of the quadratic form is a unit matrix, in (9) a negative unit matrix, and in (10) a matrix with diagonal elements 1 and -1.

Any quadratic form which has a diagonal matrix can immediately be classified according to any of the five classes. If the diagonal elements of the matrix are all positive, the form is positive definite; if they are positive or zero, the form is positive semi-definite; if they are all negative, the form is negative definite; if they are all negative or zero, the form is negative semi-definite. In the remaining case, in which both positive and negative diagonal elements occur, the form is indefinite.

Quadratic forms with a matrix which is not diagonal are not so easily classified, but it will be shown that these quadratic forms can be transformed into a quadratic form with a diagonal matrix, after which their classification is straightforward.

One of the most important properties of homogeneous positive definite and positive semi-definite quadratic forms is that they have a minimum value, namely zero. In the case of a positive definite quadratic form this value can only be reached if all variables are zero; for a positive semi-definite form this is not necessary. A homogeneous negative definite or semi-definite form has a maximum value zero, which is obtained in the definite case for all variables equal to zero. Indefinite quadratic forms, on the other hand, can have any positive or negative values. For nonhomogeneous quadratic forms similar results hold, but these will be given later.

8.2. Reduction to diagonal form

Any quadratic form can be represented as a quadratic form with a symmetric matrix. A quadratic form $x'Cx$, in which C is an asymmetric matrix, may be represented as $x'\bar{C}x$ with

$$\bar{C} = \tfrac{1}{2}(C + C'); \tag{11}$$

$\bar{C}$ is obviously symmetric. In the following we shall only consider symmetric quadratic forms.

A general quadratic form may be reduced to a quadratic form with a diagonal matrix using the familiar process of "completing the square", which is connected with the name of Lagrange. Let us consider the general homogeneous quadratic form in n variables:

$$x'Cx = c_{11}x_1^2 + 2c_{12}x_1x_2 + \ldots + 2c_{1n}x_1x_n$$
$$+ c_{22}x_2^2 + \ldots + c_{nn}x_n^2. \tag{12}$$

Assume that $c_{11} \neq 0$; then (12) may be written as

$$c_{11}\{x_1^2 + 2c_{11}^{-1}(c_{12}x_2 + \ldots + c_{1n}x_n)x_1 + c_{11}^{-2}(c_{12}x_2 + \ldots + c_{1n}x_n)^2\}$$
$$- c_{11}^{-1}(c_{12}x_2 + \ldots + c_{1n}x_n)^2 + c_{22}x_2^2 + \ldots + c_{nn}x_n^2, \tag{13}$$

or

$$c_{11}(x_1 + c_{11}^{-1}c_{12}x_2 + \ldots + c_{11}^{-1}c_{1n}x_n)^2 + (c_{22} - c_{11}^{-1}c_{12}^2)x_2^2$$
$$+ 2(c_{23} - c_{11}^{-1}c_{12}c_{13})x_2x_3 + \ldots + (c_{nn} - c_{11}^{-1}c_{1n}^2)x_n^2. \tag{14}$$

This is a quadratic form with one quadratic term in a new variable that may be defined as

$$y_1 = x_1 + c_{11}^{-1}c_{12}x_2 + \ldots + c_{11}^{-1}c_{1n}x_n, \tag{15}$$

and a general quadratic form involving all x-variables except x_1. On this quadratic form, the device of "completing the square" may be applied for x_2, and so on, until a quadratic form in y-variables is obtained having a diagonal matrix.

Expression (14) is a quadratic form with variables $y_1, x_2, ..., x_n$, which was obtained from the original quadratic form $x'Cx$ by using y_1, as given by (15) as a variable instead of x_1. This quadratic form may be written in matrix formulation as

$$[y_1 \quad x_2 \quad \cdot \quad x_n] \begin{bmatrix} c_{11} & 0 & \cdot & 0 \\ 0 & c_{22}-c_{11}^{-1}c_{12}^2 & \cdot & c_{2n}-c_{11}^{-1}c_{12}c_{1n} \\ \cdot & \cdot & \cdot & \cdot \\ 0 & c_{2n}-c_{11}^{-1}c_{12}c_{1n} & \cdot & c_{nn}-c_{11}^{-1}c_{1n}^2 \end{bmatrix} \begin{bmatrix} y_1 \\ x_2 \\ \cdot \\ x_n \end{bmatrix}$$

$$= x^{*\prime}C^*x^*. \quad (16)$$

From (15), the relationship between x and x^* may be deduced, which is

$$x^* = \begin{bmatrix} y_1 \\ x_2 \\ \cdot \\ x_n \end{bmatrix} = \begin{bmatrix} 1 & c_{11}^{-1}c_{12} & \cdot & c_{11}^{-1}c_{1n} \\ 0 & 1 & \cdot & 0 \\ \cdot & \cdot & \cdot & \cdot \\ 0 & 0 & \cdot & 1 \end{bmatrix} \begin{bmatrix} x_1 \\ x_2 \\ \cdot \\ x_n \end{bmatrix}; \quad (17)$$

this may be written as

$$x^* = T_1 x. \quad (18)$$

Substituting this in the quadratic form (16) we find

$$x^{*\prime}C^*x^* = x' T_1'C^*T_1 x, \quad (19)$$

which should be equal to the original quadratic form $x'Cx$; hence

$$T_1'C^*T_1 = C. \quad (20)$$

or, since T_1 is nonsingular,

$$C^* = T_1'^{-1}CT_1^{-1}. \quad (21)$$

From (17) we find

$$T_1^{-1} = \begin{bmatrix} 1 & -c_{11}^{-1}c_{12} & \cdot & -c_{11}^{-1}c_{1n} \\ 0 & 1 & \cdot & 0 \\ \cdot & \cdot & \cdot & \cdot \\ 0 & 0 & \cdot & 1 \end{bmatrix}; \quad (22)$$

from this follows

$$T_1'^{-1} = \begin{bmatrix} 1 & 0 & \cdot & 0 \\ -c_{11}^{-1}c_{12} & 1 & \cdot & 0 \\ & & \cdot & \\ & \cdot & \cdot & \cdot \\ -c_{11}^{-1}c_{1n} & 0 & \cdot & 1 \end{bmatrix}.$$

(23)

The resulting quadratic form and its matrix are given by (16) and (21). Let us introduce the following notation for this matrix:

$$C^* = \begin{bmatrix} c_{11} & 0 & 0 & \cdot & 0 \\ 0 & c_{22}^* & c_{23}^* & \cdot & c_{2n}^* \\ 0 & c_{23}^* & c_{33}^* & \cdot & c_{3n}^* \\ \cdot & \cdot & \cdot & \cdot & \cdot \\ 0 & c_{2n}^* & c_{3n}^* & \cdot & c_{nn}^* \end{bmatrix}.$$

Now the process of "completing the square" may be applied to this quadratic form for the variable x_2 instead of for x_1. If $c_{22} \neq 0$, a quadratic form is obtained in the variables

$$x^{**} = \begin{bmatrix} y_1 \\ y_2 \\ x_3 \\ \cdot \\ x_n \end{bmatrix} = \begin{bmatrix} 1 & 0 & 0 & \cdot & 0 \\ 0 & 1 & c_{22}^{*-1}c_{23}^* & \cdot & c_{22}^{*-1}c_{2n}^* \\ 0 & 0 & 1 & \cdot & 0 \\ \cdot & \cdot & \cdot & \cdot & \cdot \\ 0 & 0 & 0 & \cdot & 1 \end{bmatrix} \begin{bmatrix} y_1 \\ x_2 \\ x_3 \\ \cdot \\ x_n \end{bmatrix}.$$

(25)

The resulting quadratic form is

$$x^{**\prime}C^{**}x^{**} = x^{**\prime}T_2'^{-1}C^*T_2^{-1}x^{**},$$

(26)

with

$$T_2'^{-1} = \begin{bmatrix} 1 & 0 & 0 & \cdot & 0 \\ 0 & 1 & 0 & \cdot & 0 \\ 0 & -c_{22}^{*-1}c_{23}^* & 1 & \cdot & 0 \\ \cdot & \cdot & & \cdot & \cdot \\ 0 & -c_{22}^{*-1}c_{2n}^* & 0 & \cdot & 1 \end{bmatrix}.$$

(27)

The process is continued until finally a quadratic form is obtained with a diagonal matrix:

$$y'Dy = [y_1 \quad y_2 \quad \cdot \quad y_n] \begin{bmatrix} d_1 & 0 & \cdot & 0 \\ 0 & d_2 & \cdot & 0 \\ \cdot & \cdot & \cdot & \cdot \\ 0 & 0 & \cdot & d_n \end{bmatrix} \begin{bmatrix} y_1 \\ y_2 \\ \cdot \\ y_n \end{bmatrix}$$

(28)

with $d_1 = c_{11}$, $d_2 = c_{22}^*$, and so on. The complication which arises if a diagonal element to be used for the reduction is zero will be treated further on. The reduction to diagonal form may be represented as

$$x'Cx = x^{*'}T_1'^{-1}CT_1^{-1}x^* = x^{**'}T_2'^{-1}T_1'^{-1}CT_1^{-1}T_2^{-1}x^{**} = \ldots$$

$$= y'T'^{-1}CT^{-1}y = y'Dy. \tag{29}$$

with

$$T = T_{n-1} \ldots T_2T_1. \tag{30}$$

The relationship between y and x is

$$y = Tx. \tag{31}$$

The matrix T is easily found from its factors as given by (30) as

$$T = \begin{bmatrix} 1 & c_{11}^{-1}c_{12} & c_{11}^{-1}c_{12} & \cdot & c_{11}^{-1}c_{1n} \\ 0 & 1 & c_{22}^{*-1}c_{23}^* & \cdot & c_{22}^{*-1}c_{2n}^* \\ 0 & 0 & 1 & \cdot & \cdot \\ \cdot & \cdot & \cdot & \cdot & \cdot \\ 0 & 0 & 0 & \cdot & 1 \end{bmatrix}. \tag{32}$$

As a numerical example, let us take the quadratic form

$$[x_1 \quad x_2 \quad x_3] \begin{bmatrix} 5 & 2 & 4 \\ 2 & 2 & 1 \\ 4 & 1 & 7 \end{bmatrix} \begin{bmatrix} x_1 \\ x_2 \\ x_3 \end{bmatrix}, \tag{33}$$

which is reduced as

$$[y_1 \quad x_2 \quad x_3] \begin{bmatrix} 5 & 0 & 0 \\ 0 & 1\tfrac{1}{5} & -\tfrac{3}{5} \\ 0 & -\tfrac{3}{5} & 3\tfrac{4}{5} \end{bmatrix} \begin{bmatrix} y_1 \\ x_2 \\ x_3 \end{bmatrix} = [y_1 \quad y_2 \quad x_3] \begin{bmatrix} 5 & 0 & 0 \\ 0 & 1\tfrac{1}{5} & 0 \\ 0 & 0 & 3\tfrac{1}{2} \end{bmatrix} \begin{bmatrix} y_1 \\ y_2 \\ x_3 \end{bmatrix}, \tag{34}$$

which has the desired form. For T we have

$$T = T_2T_1 = \begin{bmatrix} 1 & 0 & 0 \\ 0 & 1 & -\tfrac{1}{2} \\ 0 & 0 & 1 \end{bmatrix} \begin{bmatrix} 1 & \tfrac{2}{5} & \tfrac{4}{5} \\ 0 & 1 & 0 \\ 0 & 0 & 1 \end{bmatrix} = \begin{bmatrix} 1 & \tfrac{2}{5} & \tfrac{4}{5} \\ 0 & 1 & -\tfrac{1}{2} \\ 0 & 0 & 1 \end{bmatrix}. \tag{35}$$

Hence

$$y = \begin{bmatrix} y_1 \\ y_2 \\ y_3 \end{bmatrix} = \begin{bmatrix} x_1 + \tfrac{2}{5}x_2 + \tfrac{4}{5}x_3 \\ x_2 - \tfrac{1}{2}x_3 \\ x_3 \end{bmatrix}. \tag{36}$$

If all diagonal elements of the matrix D are positive, the quadratic form in D and D itself and also the original quadratic form and its matrix

as well as the intermediate quadratic forms and their matrices are positive definite, since their form will be positive for all nonzero values of the
y-variables which applies also, the matrix T and its factors being nonsingular, to all nonzero values of the x-variables. If one or more of the
d's is zero but the remaining ones positive, the quadratic forms and their
matrices are positive semi-definite; if all d's are negative, the term negative definite applies and if one or more d's are zero but the other ones
negative, the term negative semi-definite applies. In the remaining cases,
in which both positive and negative d's occur, the forms and their matrices are indefinite.

So far, it was tacitly assumed that in the reduction to diagonal form, the
coefficient of the quadratic term to be treated next, is nonzero, or if it was
zero, then all other remaining terms were zero. If such a coefficient is
zero, but there is another quadratic term with a nonzero coefficient, we
may simply interchange the variables concerned. Hence the case remains
in which a partly reduced quadratic form has no quadratic terms left, but
only bilinear terms of the type $2c_{ij}x_ix_j$. Such a quadratic form is expressed
as

$$c_{ij}[x_i \quad x_j]\begin{bmatrix} 0 & 1 \\ 1 & 0 \end{bmatrix}\begin{bmatrix} x_i \\ x_j \end{bmatrix}. \tag{37}$$

Since

$$\tfrac{1}{2}(x_i + x_j)^2 - \tfrac{1}{2}(x_i - x_j)^2 = 2x_ix_j, \tag{38}$$

we may put

$$\begin{bmatrix} y_i \\ y_j \end{bmatrix} = \begin{bmatrix} x_i + x_j \\ x_i - x_j \end{bmatrix} = \begin{bmatrix} 1 & 1 \\ 1 & -1 \end{bmatrix}\begin{bmatrix} x_i \\ x_j \end{bmatrix}, \tag{39}$$

so that we have

$$\begin{bmatrix} x_i \\ x_j \end{bmatrix} = \begin{bmatrix} \tfrac{1}{2} & \tfrac{1}{2} \\ \tfrac{1}{2} & -\tfrac{1}{2} \end{bmatrix}\begin{bmatrix} y_i \\ y_j \end{bmatrix}. \tag{40}$$

Substituting this into (37), we find

$$[y_i \quad y_j]\begin{bmatrix} \tfrac{1}{2}c_{ij} & 0 \\ 0 & -\tfrac{1}{2}c_{ij} \end{bmatrix}\begin{bmatrix} y_i \\ y_j \end{bmatrix}. \tag{41}$$

A transformation of the type (40) leading to (41) is called an *off-
diagonal transformation*. As (41) shows, it leads to a positive and a
negative diagonal element, which means that the quadratic form concerned must be indefinite. If there were other bilinear terms in x_i or x_j, for
instance a term $c_{ik}x_ix_k$, it may be transformed into

$$\tfrac{1}{2}c_{ik}y_ix_k + \tfrac{1}{2}c_{ik}y_jx_k,$$

which can be removed by a normal diagonal transformation for the variables y_i and y_j.

As a numerical example, let us reduce the quadratic form

$$[x_1 \quad x_2 \quad x_3 \quad x_4] \begin{bmatrix} 25 & 10 & 20 & 15 \\ 10 & 4 & 5 & 10 \\ 20 & 5 & 16 & 10 \\ 15 & 10 & 10 & 7 \end{bmatrix} \begin{bmatrix} x_1 \\ x_2 \\ x_3 \\ x_4 \end{bmatrix}. \tag{42}$$

The first transformation involves a premultiplication of the matrix by

$$T_1'^{-1} = \begin{bmatrix} 1 & 0 & 0 & 0 \\ -\frac{2}{5} & 1 & 0 & 0 \\ -\frac{4}{5} & 0 & 1 & 0 \\ -\frac{3}{5} & 0 & 0 & 1 \end{bmatrix}. \tag{43}$$

and a postmultiplication by its transpose. The result is:

$$[y_1 \quad x_2 \quad x_3 \quad x_4] \begin{bmatrix} 25 & 0 & 0 & 0 \\ 0 & 0 & -3 & 4 \\ 0 & -3 & 0 & -2 \\ 0 & 4 & -2 & 0 \end{bmatrix} \begin{bmatrix} y_1 \\ x_2 \\ x_3 \\ x_4 \end{bmatrix}, \tag{44}$$

with

$$\begin{bmatrix} y_1 \\ x_2 \\ x_3 \\ x_4 \end{bmatrix} = \begin{bmatrix} 1 & \frac{2}{5} & \frac{4}{5} & \frac{3}{5} \\ 0 & 1 & 0 & 0 \\ 0 & 0 & 1 & 0 \\ 0 & 0 & 0 & 1 \end{bmatrix} \begin{bmatrix} x_1 \\ x_2 \\ x_3 \\ x_4 \end{bmatrix}. \tag{45}$$

We find that the matrix in (44) has no further nonzero diagonal elements, while some off-diagonal elements are nonzero. Hence we have to make a nondiagonal transformation. We take

$$\begin{bmatrix} y_1 \\ x_2^* \\ x_3^* \\ x_4 \end{bmatrix} = \begin{bmatrix} 1 & 0 & 0 & 0 \\ 0 & 1 & 1 & 0 \\ 0 & 1 & -1 & 0 \\ 0 & 0 & 0 & 1 \end{bmatrix} \begin{bmatrix} y_1 \\ x_2 \\ x_3 \\ x_4 \end{bmatrix}, \tag{46}$$

so that

$$\begin{bmatrix} y_1 \\ x_2 \\ x_3 \\ x_4 \end{bmatrix} = \begin{bmatrix} 1 & 0 & 0 & 0 \\ 0 & \frac{1}{2} & \frac{1}{2} & 0 \\ 0 & \frac{1}{2} & -\frac{1}{2} & 0 \\ 0 & 0 & 0 & 1 \end{bmatrix} \begin{bmatrix} y_1 \\ x_2^* \\ x_3^* \\ x_4 \end{bmatrix}. \tag{47}$$

Hence the matrix of (44) should be postmultiplied by the matrix of (47) and premultiplied by its transpose. The quadratic form then becomes

$$
[y_1 \quad x_2^* \quad x_3^* \quad x_4]
\begin{bmatrix}
25 & 0 & 0 & 0 \\
0 & -1\frac{1}{2} & 0 & 1 \\
0 & 0 & 1\frac{1}{2} & 3 \\
0 & 1 & 3 & 0
\end{bmatrix}
\begin{bmatrix}
y_1 \\ x_2^* \\ x_3^* \\ x_4
\end{bmatrix}.
\tag{48}
$$

We may now proceed in the usual fashion. We obtain the following quadratic forms:

$$
[y_1 \quad y_2 \quad x_3^* \quad x_4]
\begin{bmatrix}
25 & 0 & 0 & 0 \\
0 & -1\frac{1}{2} & 0 & 0 \\
0 & 0 & 1\frac{1}{2} & 3 \\
0 & 0 & 3 & \frac{2}{3}
\end{bmatrix}
\begin{bmatrix}
y_1 \\ y_2 \\ x_3^* \\ x_4
\end{bmatrix} =
\tag{49}
$$

$$
[y_1 \quad y_2 \quad y_3 \quad x_4]
\begin{bmatrix}
25 & 0 & 0 & 0 \\
0 & -1\frac{1}{2} & 0 & 0 \\
0 & 0 & 1\frac{1}{2} & 0 \\
0 & 0 & 0 & -3\frac{1}{3}
\end{bmatrix}
\begin{bmatrix}
y_1 \\ y_2 \\ y_3 \\ x_4
\end{bmatrix}
\tag{50}
$$

The last quadratic form has a diagonal matrix.

8.3. Completing the square and pivoting

We shall now consider in some more detail the matrix multiplications belonging to one transformation of completing the square. The first transformation corresponds with a premultiplication of C by $T_1'^{-1}$ and a postmultiplication by T_1^{-1}. We find

$$
T_1'^{-1} C =
\begin{bmatrix}
1 & 0 & \cdot & 0 \\
-c_{11}^{-1}c_{12} & 1 & \cdot & 0 \\
\cdot & & \cdot & \\
-c_{11}^{-1}c_{1n} & 0 & \cdot & 1
\end{bmatrix}
\begin{bmatrix}
c_{11} & c_{12} & \cdot & c_{1n} \\
c_{12} & c_{22} & \cdot & c_{2n} \\
\cdot & & \cdot & \\
c_{1n} & c_{2n} & \cdot & c_{nn}
\end{bmatrix} =
$$

$$
\begin{bmatrix}
c_{11} & c_{12} & \cdot & c_{1n} \\
0 & c_{22} - c_{11}^{-1}c_{12}^2 & \cdot & c_{2n} - c_{11}^{-1}c_{12}c_{1n} \\
\cdot & \cdot & \cdot & \cdot \\
0 & c_{2n} - c_{11}^{-1}c_{12}c_{1n} & \cdot & c_{nn} - c_{11}^{-1}c_{1n}^2
\end{bmatrix}.
\tag{51}
$$

After postmultiplication by T_1^{-1}, the matrix C^* given in (16) is obtained. This postmultiplication only results in changing all elements, except the first, of the first row of $T_1'^{-1}C$ into zero; the elements of other rows are unchanged.

The premultiplication of C by $T_1'^{-1}$ yields a result similar to that of a pivoting operation on C with c_{11} as a pivot. In the latter case, C is premultiplied by the inverse of the matrix S_1,

$$
S_1 = \begin{bmatrix} c_{11} & 0 & \cdot & 0 \\ c_{12} & 1 & \cdot & 0 \\ \cdot & & \cdot & \cdot \\ c_{1n} & 0 & \cdot & 1 \end{bmatrix},
$$

which results in

$$
S_1^{-1}C = \begin{bmatrix} c_{11}^{-1} & 0 & \cdot & 0 \\ -c_{11}^{-1}c_{12} & 1 & \cdot & 0 \\ \cdot & & \cdot & \cdot \\ -c_{11}^{-1}c_{1n} & 0 & \cdot & 1 \end{bmatrix} \begin{bmatrix} c_{11} & c_{21} & \cdot & c_{1n} \\ c_{12} & c_{22} & \cdot & c_{2n} \\ \cdot & \cdot & & \cdot \\ c_{1n} & c_{2n} & \cdot & c_{nn} \end{bmatrix} =
$$

$$
\begin{bmatrix} 1 & c_{11}^{-1}c_{12} & \cdot & c_{11}^{-1}c_{1n} \\ 0 & c_{22}-c_{11}^{-1}c_{12}^2 & \cdot & c_{2n}-c_{11}^{-1}c_{12}c_{1n} \\ \cdot & & \cdot & \cdot \\ 0 & c_{2n}-c_{11}^{-1}c_{12}c_{1n} & \cdot & c_{nn}-c_{11}^{-1}c_{1n}^2 \end{bmatrix}. \tag{52}
$$

Apart from the first row, the results are the same as in (51) and (16). Another pivoting operation can be applied to the last matrix of (52) with $c_{22} - c_{11}^{-1}c_{12}^2$ as a pivot; the resulting matrix is then the same as C^{**} apart from the first two rows. Similar pivoting operations can be applied using appropriate diagonal elements of the successive resulting matrices, until no possible nonzero diagonal pivots remain. It will be clear that the successive pivots will be equal to the d's of the diagonal form. After a number of diagonal transformations, which can be represented as pivot transformations, the diagonal elements of the remaining rows and columns are all zero, while some off-diagonal elements are nonzero, off-diagonal transformations must be used; these are different from pivot-transformations. In such cases, the quadratic form and its matrix are indefinite.

Earlier[1] it was proved that the pivots of the successive pivot operations are equal to the ratio of the corresponding principal determinant to the previous principal determinant:

$$
d_i = D_i/D_{i-1}, \tag{53}
$$

where d_i is the ith diagonal element after $i-1$ pivot operations and D_i is the ith principal determinant of the original matrix. Using this relation,

---

[1] See chapter 1, eq. (99).

we may obtain the conditions for positive and negative definite matrices as they are conventionally stated, according to which a matrix is positive definite if successive principal determinants are positive, and negative definite if successive principal determinants have an alternating sign, the first being negative.

The formulation of positive and negative (semi-)definite matrices in terms of pivots of pivot operations is important, because it is much more efficient than the computation of the successive principal determinants; further it has the advantage that in the determination of maximum or minimum values of quadratic forms these pivot operations are performed anyway, as will be shown later.

For the sake of completeness, another form of "completing the square" which is slightly closer to pivoting operations should be mentioned. If in (14) the expression within brackets in the first term is multiplied by c_{11}, which is compensated by a division of the coefficient of this term by c_{11}^2, one obtains for the first term

$$c_{11}^{-1}(c_{11}x_1 + c_{12}x_2 + ... + c_{1n}x_n)^2. \tag{54}$$

The equivalents of y_1, T_1' and $T_1'^{-1}$, to be indicated with a bar, then are

$$\bar{y}_1 = c_{11}x_1 + c_{12}x_2 + ... + c_{1n}x_n, \tag{55}$$

$$\bar{T}_1' = \begin{bmatrix} c_{11} & 0 & \cdot & 0 \\ c_{12} & 1 & \cdot & 0 \\ \cdot & & \cdot & \cdot \\ c_{1n} & 0 & \cdot & 1 \end{bmatrix}, \tag{56}$$

and

$$\bar{T}_1'^{-1} = \begin{bmatrix} c_{11}^{-1} & 0 & \cdot & 0 \\ -c_{11}^{-1}c_{12} & 1 & \cdot & 0 \\ \cdot & & \cdot & \cdot \\ -c_{11}^{-1}c_{1n} & 0 & \cdot & 1 \end{bmatrix}. \tag{57}$$

Hence the matrix by which C is premultiplied is the same in the first pivoting operation and in the first operation of completing the square; in the latter case, however, a postmultiplication by $\bar{T}_1'^{-1}$ follows, resulting in

$$\bar{C}^* = \bar{T}_1'^{-1} C \bar{T}_1'^{-1} = \begin{bmatrix} c_{11}^{-1} & 0 & \cdot & 0 \\ 0 & c_{22} - c_{11}^{-1}c_{12}^2 & \cdot & c_{2n} - c_{11}^{-1}c_{12}c_{1n} \\ \cdot & & \cdot & \\ 0 & c_{2n} - c_{11}^{-1}c_{12}c_{1n} & \cdot & c_{nn} - c_{11}^{-1}c_{1n}^2 \end{bmatrix}. \tag{58}$$

In the following operations the matrix of the pivoting operation and that of this version of Lagrange's procedure are not quite the same, since the latter operation will have zeroes in the pivot column above the pivot element; for instance for the second operation we would have

$$
\bar{T}_2^{-1} = \begin{bmatrix}
1 & 0 & 0 & \cdot & 0 \\
0 & c_{22}^{*-1} & 0 & \cdot & 0 \\
0 & -c_{22}^{*-1}c_{23}^{*} & 1 & \cdot & 0 \\
\cdot & \cdot & & \cdot & \cdot \\
0 & -c_{22}^{*-1}c_{2n}^{*} & 0 & \cdot & 1
\end{bmatrix}
\tag{59}
$$

In the matrix of the pivoting operation the element $(1,2)$ would have been $-c_{22}^{*-1}c_{11}^{-1}c_{12}$. This alternative version of Lagrange's procedure leads to a diagonal matrix with diagonal elements being the reciprocal of the corresponding diagonal elements, as far as they are nonzero, of the diagonal matrix obtained by the first version.

8.4. Diagonalization of nonhomogeneous quadratic forms

A general nonhomogeneous quadratic form can be represented as

$$
f(x) = c_0 + 2c'x + x'Cx,
\tag{60}
$$

where c_0 is a given constant, c is a column vector of n elements and C is a symmetric matrix. Let us consider the maximization or minimization of $f(x)$. The conventional treatment of this problem is to formulate the first-order and second-order conditions for a maximum or minimum of $f(x)$. The first-order conditions for both a maximum and a minimum are

$$
\frac{\partial f(x)}{\partial x} = 2c + 2Cx = 0,
\tag{61}
$$

and the second-order conditions for a minimum are

$$
D_i > 0, \quad i = 1, ..., n,
\tag{62}
$$

where D_i is the ith principal minor of C. For a maximum the second-order conditions are

$$
(-1)^i D_i > 0, \quad i = 1, ..., n.
\tag{63}
$$

These second-order conditions could be derived from a diagonalization of the matrix C.

If the conventional approach is followed, the equations of the first-order conditions are used to compute solutions, after which the second-order conditions are used to verify whether the solutions are maxima, minima, or neither of these. In any case, first-order and second-order conditions are derived and used separately.

In the following, two methods are given for finding the maximum or minimum of a quadratic form; in both methods computation of the optimal solution and verification of the nature of the solution are combined. The first method is based on the diagonalization of the quadratic form and is explained in this section, the second method is based on direct maximalization and will be treated in the next section.

A nonhomogeneous quadratic form can be considered a homogeneous one with an additional variable x_{n+1}, which can only take the value 1. Hence (60) may be written as

$$f(x) = [x \quad 1] \begin{bmatrix} C & c \\ c' & c_0 \end{bmatrix} \begin{bmatrix} x \\ 1 \end{bmatrix}. \tag{64}$$

This quadratic form may be reduced to a quadratic form with a diagonal matrix but first the n ordinary x-variables should be treated, so that transformations involving x_{n+1} are performed only when the possibilities of transformation (diagonal and off-diagonal ones) involving other variables have been exhausted. Then there are two possibilities. Either a quadratic form with a diagonal matrix is obtained without a transformation involving $x_{n+1} = 1$, or a quadratic form of the following type is found:

$$d_1 y_1^2 + ... + d_k y_k^2 + 2d_{0,k+1} x_{k+1} + ... + 2d_{0n} x_n + d_{n+1}, \tag{65}$$

with

$$d_{0,k+1}, ..., d_{0n} \neq 0.$$

In the first case, a quadratic form of the following structure is obtained:

$$[y_1 \quad y_2 \quad y_n \quad 1] \begin{bmatrix} d_1 & 0 & \cdot & 0 & 0 \\ 0 & d_2 & \cdot & 0 & 0 \\ \cdot & \cdot & \cdot & & \cdot \\ 0 & 0 & \cdot & d_n & 0 \\ 0 & 0 & \cdot & 0 & d_{n+1} \end{bmatrix} \begin{bmatrix} y_1 \\ y_2 \\ \cdot \\ y_n \\ 1 \end{bmatrix} =$$

$$d_1 y_1^2 + d_2 y_2^2 + ... + d_n y_n^2 + d_{n+1}. \tag{66}$$

This quadratic form has a minimum d_{n+1} for $y_1 = ... = y_n = 0$ if d_1, $d_2, ..., d_n$ are nonnegative, which corresponds to a positive (semi-)definite matrix C. The values of x for $y = 0$ are found from the relation (31)

$$\begin{bmatrix} y \\ 1 \end{bmatrix} = \begin{bmatrix} 0 \\ 1 \end{bmatrix} = T \begin{bmatrix} x \\ 1 \end{bmatrix} = T_n T_{n-1} ... T_2 T_1 \begin{bmatrix} x \\ 1 \end{bmatrix}, \tag{67}$$

so that

$$\begin{bmatrix} x \\ 1 \end{bmatrix} = T_1^{-1} T_2^{-1} \dots T_{n-1}^{-1} T_n^{-1} \begin{bmatrix} 0 \\ 1 \end{bmatrix}, \tag{68}$$

or

$$[x' \quad 1] = [0 \quad 1] T_n'^{-1} T_{n-1}'^{-1} \dots T_2'^{-1} T_1'^{-1} \tag{69}$$

As an example, let us apply this method to the quadratic form with the following coefficients

$$c_0 = 0, \quad c = \begin{bmatrix} 9 \\ 8 \\ 11 \\ 10 \end{bmatrix}, \quad C = \begin{bmatrix} -4 & 2 & -2 & 0 \\ 2 & -2 & -1 & 2 \\ -2 & -1 & -6 & 5 \\ 0 & 2 & 5 & -9 \end{bmatrix}.$$

The matrix of the extended quadratic form, indicated by C_0, is

$$C_0 = \begin{bmatrix} -4 & 2 & -2 & 0 & 9 \\ 2 & -2 & -1 & 2 & 8 \\ -2 & -1 & -6 & 5 & 11 \\ 0 & 2 & 5 & -9 & 10 \\ 9 & 8 & 11 & 10 & 0 \end{bmatrix}.$$

The successive transformed matrices of the quadratic form are indicated by C_1, C_2, C_3, and C_4. We find

$$T_1'^{-1} = \begin{bmatrix} 1 & 0 & 0 & 0 & 0 \\ -\frac{1}{2} & 1 & 0 & 0 & 0 \\ \frac{1}{2} & 0 & 1 & 0 & 0 \\ 0 & 0 & 0 & 1 & 0 \\ 2\frac{1}{4} & 0 & 0 & 0 & 1 \end{bmatrix}, \quad C_1 = \begin{bmatrix} -4 & 0 & 0 & 0 & 0 \\ 0 & -1 & -2 & 2 & 12\frac{1}{2} \\ 0 & -2 & -5 & 5 & 6\frac{1}{2} \\ 0 & 2 & 5 & -9 & 10 \\ 0 & 12\frac{1}{2} & 6\frac{1}{2} & 10 & 20\frac{1}{2} \end{bmatrix},$$

$$T_2'^{-1} = \begin{bmatrix} 1 & 0 & 0 & 0 & 0 \\ 0 & 1 & 0 & 0 & 0 \\ 0 & -2 & 1 & 0 & 0 \\ 0 & 2 & 0 & 1 & 0 \\ 0 & 12\frac{1}{2} & 0 & 0 & 1 \end{bmatrix}, \quad C_2 = \begin{bmatrix} -4 & 0 & 0 & 0 & 0 \\ 0 & -1 & 0 & 0 & 0 \\ 0 & 0 & -1 & 1 & -18\frac{1}{2} \\ 0 & 0 & 1 & -5 & 35 \\ 0 & 0 & -18\frac{1}{2} & 35 & 176\frac{1}{2} \end{bmatrix},$$

$$T_3'^{-1} = \begin{bmatrix} 1 & 0 & 0 & 0 & 0 \\ 0 & 1 & 0 & 0 & 0 \\ 0 & 0 & 1 & 0 & 0 \\ 0 & 0 & 1 & 1 & 0 \\ 0 & 0 & -18\frac{1}{2} & 0 & 1 \end{bmatrix}, \quad C_3 = \begin{bmatrix} -4 & 0 & 0 & 0 & 0 \\ 0 & -1 & 0 & 0 & 0 \\ 0 & 0 & -1 & 0 & 0 \\ 0 & 0 & 0 & -4 & 16\frac{1}{2} \\ 0 & 0 & 0 & 16\frac{1}{2} & 518\frac{3}{4} \end{bmatrix},$$

$$T_4'^{-1} = \begin{bmatrix} 1 & 0 & 0 & 0 & 0 \\ 0 & 1 & 0 & 0 & 0 \\ 0 & 0 & 1 & 0 & 0 \\ 0 & 0 & 0 & 1 & 0 \\ 0 & 0 & 0 & 4\frac{1}{8} & 1 \end{bmatrix}, \quad C_4 = \begin{bmatrix} -4 & 0 & 0 & 0 & 0 \\ 0 & -1 & 0 & 0 & 0 \\ 0 & 0 & -1 & 0 & 0 \\ 0 & 0 & 0 & -4 & 0 \\ 0 & 0 & 0 & 0 & 586\frac{13}{16} \end{bmatrix}.$$

Since the diagonal elements of C_4 in all rows except the last are negative, the quadratic form has a maximum of $586\frac{13}{16}$. This maximum is found according to (69) for

$$x' = [4\tfrac{1}{8} \quad -14\tfrac{3}{8} \quad 49\tfrac{1}{2} \quad 32\tfrac{3}{16}].$$

If it happens at any transformation that a row and column of the matrix of the quadratic form consists entirely of zeroes, that row and column and its corresponding variable may be deleted.

In the case no further transformation can be performed without involving x_{n+1}, a quadratic form of the following type is obtained:

$$[y_1 \; \cdot \; y_k \; x_{k+1} \; \cdot \; x_n \; 1]
\begin{bmatrix}
d_1 & \cdot & 0 & 0 & \cdot & 0 & 0 \\
\cdot & & \cdot & \cdot & & \cdot & \cdot \\
0 & \cdot & d_k & 0 & \cdot & 0 & 0 \\
0 & \cdot & 0 & 0 & \cdot & 0 & d_{0,k+1} \\
\cdot & & \cdot & \cdot & & \cdot & \cdot \\
0 & \cdot & 0 & 0 & \cdot & 0 & d_{0n} \\
0 & \cdot & 0 & d_{0,k+1} & \cdot & d_{0n} & d_{n+1}
\end{bmatrix}
\begin{bmatrix}
y_1 \\ \cdot \\ y_k \\ x_{k+1} \\ \cdot \\ x_n \\ 1
\end{bmatrix} \cdot$$

(70)

This means that we should diagonalize a linear function. Let us diagonalize the function

$$2x = [x \quad 1]\begin{bmatrix} 0 & 1 \\ 1 & 0 \end{bmatrix}\begin{bmatrix} x \\ 1 \end{bmatrix}.$$

(71)

An off-diagonal transformation is necessary; the result is

$$2x = [x+1 \quad x-1]\begin{bmatrix} \tfrac{1}{2} & 0 \\ 0 & -\tfrac{1}{2} \end{bmatrix}\begin{bmatrix} x+1 \\ x-1 \end{bmatrix}.$$

(72)

This is an indefinite quadratic form. All remaining linear terms can be treated in this manner. We conclude that if a quadratic form is linear in one or more variables, it is indefinite and it has no maximum or minimum value. The latter statement is, of course, obvious without any diagonalization.

8.5. Maximization of a quadratic form

In the previous section it was shown that a quadratic form may be transformed into one with a diagonal matrix; in case the elements of the diagonal matrix all have the same sign, a maximum or a minimum of this quadratic form may be found from the transformation matrix. In this

section a more direct approach towards maximization and minimization
will be followed. It should be noted that this approach is self-contained in
the sense that it does not rely on results obtained before. The resulting
method is closely related to linear programming methods and forms the
basis of quadratic programming methods. The notation used in this section is the same as the one to be used in quadratic programming.

We consider the general nonhomogeneous quadratic form

$$f(x) = c_0 + c'x - \tfrac{1}{2}x'Cx. \tag{73}$$

Let us define the vector $-v$ as the vector of derivatives of $f(x)$ with
respect to the x-variables:

$$\frac{\partial f(x)}{\partial x} = -v = c - Cx. \tag{74}$$

After substitution for Cx in (73) according to (74) we find

$$f = c_0 + c'x - \tfrac{1}{2}x'(c + v) = c_0 + \tfrac{1}{2}c'x - \tfrac{1}{2}v'x. \tag{75}$$

In order to avoid the factor $\tfrac{1}{2}$, the function $F = 2f(x)$ may be considered,
so that we have instead of (75)

$$F = 2c_0 + c'x - v'x. \tag{76}$$

This is a nonlinear expression for the quadratic form, but it is linear for
solutions such that $-v'x = 0$. If the term $-v'x$ is deleted in (76) this
expression may be combined with the equations (74) to form the following equation system:

$$2c_0 = -c'x + F,$$
$$-c = -Cx + v. \tag{77}$$

This is a system in canonical form with F and the v-variables as basic
variables. Note that the terms in $-v'x$ are zero for the basic solution
since all x-variables are nonbasic and hence zero. This system may be put
into tableau-form (see table 1). This may be called the set-up tableau for
the method. Note that the entire tableau is symmetric.

Table 1

Basic variables	Values basic variables	x
F	$2c_0$	$-c'$
v	$-c$	$-C$

First the method is explained by means of an example of application; after that a more general treatment follows. Let the quadratic form to be maximized be of the type (73) with

$$x = \begin{bmatrix} x_1 \\ x_2 \\ x_3 \\ x_4 \end{bmatrix}, \qquad c_0 = 0, \qquad c = \begin{bmatrix} 9 \\ 8 \\ 11 \\ 10 \end{bmatrix}, \qquad C = \begin{bmatrix} 4 & -2 & 2 & 0 \\ -2 & 2 & 1 & -2 \\ 2 & 1 & 6 & -5 \\ 0 & -2 & -5 & 9 \end{bmatrix}.$$

These data are used to form a tableau as for (77); the result is tableau 0 of table 2, in which the values of basic variables are given in the last column instead of the first and the F-row is the last instead of the first row.

The F-row is interpreted as the equation

$$0 = -9x_1 - 8x_2 - 11x_3 - 10x_4 + F, \tag{78}$$

but this equation is valid only if the deleted bilinear terms $-v'x$ are zero. The equation for F which is valid for all values of the x- and v-variables is

$$F = 9x_1 + 8x_2 + 11x_3 + 10x_4 - v_1x_1 - v_2x_2 - v_3x_3 - v_4x_4. \tag{79}$$

Let us consider a change in the nonbasic variable x_1, keeping all other x-variables at zero. We then obtain for F

$$F = 9x_1 - v_1x_1. \tag{80}$$

But v_1 changes if x_1 changes; the first row of tableau 0 gives the equation

$$-9 = -4x_1 + 2x_2 - 2x_3 + 0x_4 + v_1, \tag{81}$$

so that

$$v_1 = -9 + 4x_1. \tag{82}$$

Substituting this into (80) we find

$$F = 18x_1 - 4x_1^2. \tag{83}$$

Hence we find that F increases for an increase in x_1, but at a decreasing rate. The maximum value of F is found for

$$\partial F/\partial x_1 = 18 - 8x_1 = 0, \tag{84}$$

or $x_1 = 2\frac{1}{4}$. The corresponding value of F is $20\frac{1}{4}$. The values of the other v-variables can easily be determined from the value of x_1 and the equations implied by their rows.

Table 2

Example of quadratic maximization in tableau form

Tableau	Basic variables	x_1	x_2	x_3	x_4	Values basic variables
0	v_1	$\underline{-4}$	2	-2	0	-9
	v_2	$\overline{2}$	-2	-1	2	-8
	v_3	-2	-1	-6	5	-11
	v_4	0	2	5	-9	-10
	F	-9	-8	-11	-10	0
		v_1	x_2	x_3	x_4	
1	x_1	$-\frac{1}{4}$	$-\frac{1}{2}$	$\frac{1}{2}$	0	$2\frac{1}{4}$
	v_2	$\frac{1}{2}$	-1	-2	2	$-12\frac{1}{2}$
	v_3	$-\frac{1}{2}$	$\underline{-2}$	-5	5	$-6\frac{1}{2}$
	v_4	0	2	5	-9	-10
	F	$-2\frac{1}{4}$	$-12\frac{1}{2}$	$-6\frac{1}{2}$	-10	$20\frac{1}{4}$
		v_1	v_2	x_3	x_4	
2	x_1	$-\frac{1}{2}$	$-\frac{1}{2}$	$1\frac{1}{2}$	-1	$8\frac{1}{2}$
	x_2	$-\frac{1}{2}$	-1	2	-2	$12\frac{1}{2}$
	v_3	$-1\frac{1}{2}$	-2	$\underline{-1}$	1	$18\frac{1}{6}$
	v_4	1	2	1	-5	-35
	F	$-8\frac{1}{2}$	$-12\frac{1}{2}$	$18\frac{1}{2}$	-35	$176\frac{1}{2}$
		v_1	v_2	v_3	x_4	
3	x_1	$-2\frac{3}{4}$	$-3\frac{1}{2}$	$1\frac{1}{2}$	$\frac{1}{2}$	$36\frac{1}{4}$
	x_2	$-3\frac{1}{2}$	-5	2	0	$49\frac{1}{2}$
	x_3	$-1\frac{1}{2}$	2	-1	-1	$-18\frac{1}{2}$
	v_4	$-\frac{1}{2}$	0	1	$\underline{-4}$	$-16\frac{1}{2}$
	F	$-36\frac{1}{4}$	$-49\frac{1}{2}$	$18\frac{1}{2}$	$-16\frac{1}{2}$	$518\frac{3}{4}$
		v_1	v_2	v_3	v_4	
4	x_1	$-2\frac{13}{16}$	$-3\frac{1}{2}$	$1\frac{5}{8}$	$\frac{1}{8}$	$32\frac{3}{16}$
	x_2	$-3\frac{1}{2}$	-5	2	0	$49\frac{1}{2}$
	x_3	$1\frac{5}{8}$	2	$-1\frac{1}{4}$	$-\frac{1}{4}$	$-14\frac{3}{8}$
	x_4	$\frac{1}{8}$	0	$-\frac{1}{4}$	$-\frac{1}{4}$	$4\frac{1}{8}$
	F	$-32\frac{3}{16}$	$-49\frac{1}{2}$	$14\frac{3}{8}$	$-4\frac{1}{8}$	$586\frac{13}{16}$

The same results are obtained by a transformation of tableau 0 using the element -4 as a pivot. The result is tableau 1. In this tableau v_1 is nonbasic and therefore zero; x_1 is basic and may be nonzero. Note that for the basic solution of the new tableau we have $-v'x = 0$.

If the coefficient -4 of the quadratic term in (83) had been nonnegative, F could have been increased indefinitely by an indefinite increase in x_1. But if this coefficient had been positive, say 4, a *minimum* for this part of the quadratic form would have been found, since we would have

$$F = 18x_1 + 4x_1^2. \tag{85}$$

In that case a decrease in x_1 would lead to a decrease in F, but the quadratic term causes the function to have a minimum which is found from

$$\partial F/\partial x_1 = 18 + 8x_1 = 0, \text{ or } x_1 = -2\tfrac{1}{4}. \tag{86}$$

We now proceed from tableau 1 of table 2. It should be noted that this tableau has some symmetry properties which also occur in subsequent tableaus. The matrices above and to the left and below and to the right of the dashed lines are symmetric; the two remaining matrices are skew-symmetric. Let us now find out how a change in the nonbasic variable x_2 affects F. As in (80) we find

$$F = 20\tfrac{1}{4} + 12\tfrac{1}{2}x_2 - v_2x_2, \tag{87}$$

and as in (82),

$$v_2 = -12\tfrac{1}{2} + x_2. \tag{88}$$

After substitution of (88) in (87) we find

$$F = 20\tfrac{1}{4} + 25x_2 - x_2^2. \tag{89}$$

The maximal value of F is found for

$$\partial F/\partial x_2 = 25 - 2x_2 \tag{90}$$

or $x_2 = 12\tfrac{1}{2}$; the maximal value of F is $176\tfrac{1}{2}$. The same values are obtained after pivoting on the principal element in the second row of tableau 1; the result is tableau 2.

It is obvious that a similar result would have been obtained if we had pivoted on the principal elements in the third or fourth row, introducing x_3 and x_4 into the basis. More interesting is the result of a nonzero value of v_1. We find as before

$$F = 20\tfrac{1}{4} + 2\tfrac{1}{4}v_1 - v_1x_1, \tag{91}$$

and

$$x_1 = 2\tfrac{1}{4} + \tfrac{1}{4}v_1. \tag{92}$$

Substituting (92) into (91), we have

$$f = 20\tfrac{1}{4} + 2\tfrac{1}{4}v_1 - v_1(2\tfrac{1}{4} + \tfrac{1}{4}v_1) = 20\tfrac{1}{4} - \tfrac{1}{4}v_1^2. \tag{93}$$

Hence any nonzero value of v_1 decreases F.

In tableau 2, x_3 is introduced into the basis, replacing v_3; this results in tableau 3. In this tableau x_4 replaces v_4, which results in tableau 4. The solution of this tableau is

$$x_1 = 32\tfrac{3}{16}, \qquad x_2 = 49\tfrac{1}{2}, \qquad x_3 = -14\tfrac{3}{8}, \qquad x_4 = 4\tfrac{1}{8}, \qquad F = 586\tfrac{13}{16}.$$

We should prove that this is the optimal solution. We may write for F:

$$F = 586\tfrac{13}{16} + [32\tfrac{3}{16} \quad 49\tfrac{1}{2} - 14\tfrac{3}{8} \quad 4\tfrac{1}{8}] \begin{bmatrix} v_1 \\ v_2 \\ v_3 \\ v_4 \end{bmatrix} - v'x. \tag{94}$$

According to the first four rows of tableau 4 we have

$$x = \begin{bmatrix} 32\tfrac{3}{16} \\ 49\tfrac{1}{2} \\ -14\tfrac{3}{8} \\ 4\tfrac{1}{8} \end{bmatrix} - \begin{bmatrix} -2\tfrac{13}{16} & -3\tfrac{1}{2} & 1\tfrac{5}{8} & \tfrac{1}{8} \\ -3\tfrac{1}{2} & -5 & 2 & 0 \\ 1\tfrac{5}{8} & 2 & -1\tfrac{1}{4} & -\tfrac{1}{4} \\ \tfrac{1}{8} & 0 & -\tfrac{1}{4} & -\tfrac{1}{4} \end{bmatrix} \begin{bmatrix} v_1 \\ v_2 \\ v_3 \\ v_4 \end{bmatrix}, \tag{95}$$

which may be written as

$$x = \hat{x} - (-R)v, \tag{96}$$

so that

$$x - \hat{x} = Rv. \tag{97}$$

From the tableau follows that $-R = -C^{-1}$; hence (97) may be written as

$$v = C(x - \hat{x}). \tag{98}$$

We may now write for F

$$F = 586\tfrac{13}{16} + \hat{x}'v - v'x = 586\tfrac{13}{16} - (x - \hat{x})'v =$$
$$= 586\tfrac{13}{16} - (x - \hat{x})'C(x - \hat{x}). \tag{99}$$

In section 8.3 it was shown that the diagonal elements of a diagonalized matrix are equal to the successive principal pivots. Hence F may be written as

$$F = 586\tfrac{13}{16} + (x - \hat{x})'T'(-D)T(x - \hat{x}), \tag{100}$$

where $-D$ is a diagonal matrix with the successive pivots as elements. Since then pivots were all negative, $x = \hat{x}$ must be the optimal solution.

A more general treatment of the method may be given by considering a general tableau which must have the form of table 3.

Table 3

	v^1	x^2	Values basic variables
x^1	$-R$	P	r
v^2	$-P'$	$-Q$	q
F	$-r'$	q	q_0

This is the tableau obtained after k iterations; x^1 contains the variables $x_1, ..., x_k$, v^2 contains the variables $v_{k+1}, ..., v_n$, v^1 contains the variables $v_1, ..., v_k$ and x_2 the variables $x_{k+1}, ..., x_n$. $-R$ is a $k \times k$ matrix, $-Q$ is a $(n-k) \times (n-k)$ matrix, P is a $k \times (n-k)$ matrix. A special case is $k = 0$, which is the set-up tableau in which all x-variables are nonbasic, $-Q = -C$, $q = c$, and R, P, and r do not exist; another special case is $k = n$, so that all x-variables are basic and Q, P, and q do not exist.

Such a general quadratic tableau has symmetry properties implied in its representation. We wish to show that if in such a tableau a principal pivot of $-Q$ is used, another quadratic tableau is obtained having the same symmetry properties. This may be proved by simply performing such a transformation on a general standard tableau. In table 4 this is done for the case $k = 1$ and $n = 3$. It is obvious that tableau II has the required symmetry properties. It is interesting to find out what happens with the diagonal elements of $-R$ and $-Q$ in a transformation. For the transformed diagonal elements of $-R$ we find

$$-r_{11} - p_{11}^2 q_{11}^{-1}.$$

Hence, if the principal pivots are all negative, the elements if $-R$ are nonincreasing. For the transformed diagonal element of $-Q$ we find

$$-q_{22} + q_{21}q_{12}q_{11}^{-1};$$

since $q_{12} = q_{21}$, the diagonal elements if $-Q$ are nondecreasing. Therefore, if a diagonal element of $-Q$ is nonnegative, it cannot become negative.

Table 4

Transformation of a quadratic tableau with principal pivot

Tableau	Basic variables	v_1	x_2	x_3	Values basic variables
I	x_1	$-r_{11}$	p_{11}	p_{12}	r_1
	v_2	$-p_{11}$	$-q_{11}$	$-q_{12}$	q_1
	v_3	$-p_{12}$	$-q_{21}$	$-q_{22}$	q_2
	F	$-r_1$	q_1	q_2	q_0

	Basic variables	v_1	v_2	x_3	Values basic variables
II	x_1	$-r_{11}-p_{11}^2 q_{11}^{-1}$	$q_{11}^{-1}p_{11}$	$p_{12}-p_{11}q_{11}^{-1}q_{12}$	$r_1+p_{11}q_{11}^{-1}q_1$
	x_2	$q_{11}^{-1}p_{11}$	$-q_{11}^{-1}$	$q_{11}^{-1}q_{12}$	$-q_{11}^{-1}q_1$
	v_3	$-p_{12}+q_{21}q_{11}^{-1}p_{11}$	$-q_{21}q_{11}^{-1}$	$-q_{22}+q_{21}q_{11}^{-1}q_{12}$	$q_2-q_{21}q_{11}^{-1}q_1$
	F	$-r_1+q_1q_{11}^{-1}p_{11}$	$q_1q_{11}^{-1}$	$q_2-q_1q_{11}^{-1}q_{12}$	$q_0+q_1^2q_{11}^{-1}$

It is obvious that if $-Q$ has a positive diagonal element, the quadratic form has no finite maximum. If some diagonal elements of $-Q$ are zero, this does not necessarily mean that the quadratic form has no finite maximum. If $-q_{ii}=0$ and $q_i \neq 0$, there is no finite maximum, but if $q_i = 0$, there may be a finite maximum, since a change of the corresponding x-variable was not affecting the value of F. In this case other diagonal elements of $-Q$ which are negative should be used first. If finally a matrix $-Q$ and a vector q emerge which consist of zero elements, a finite maximum has been reached; if $-Q$ or q contains a nonzero element, the quadratic form has no finite maximum.

8.6. Maximizing quadratic forms subject to linear equalities

Now the problem of maximization of quadratic forms subject to linear equations will be considered. Minimization of a quadratic form is, of course, equivalent to maximization of minus this form. This problem can be formulated as follows. Maximize

$$f(x) = c_0 + p'x - \tfrac{1}{2}x'Cx \qquad (101)$$

subject to

$$Ax = b. \tag{102}$$

x and p are column vectors of n-elements, b is a column vector of m-elements, C is an $n \times n$ matrix, and A an $m \times n$ matrix. For the vector of coefficients of the linear term in the objective function p is used rather than c in order to conform with the notation for quadratic programming problems. Note that the x-variables are still unrestricted as to sign.

An obvious way to solve this problem is to find a basic solution for (102). If for instance, A and x may be partitioned as

$$A = [A_1 \quad A_2], \qquad x = \begin{bmatrix} x^1 \\ x^2 \end{bmatrix},$$

where A_1 is an $m \times m$ nonsingular matrix and x^1 an m-element vector, the system (102) may be written as

$$x^1 = A_1^{-1}b - A_1^{-1}A_2x^2. \tag{103}$$

This may be substituted in (101) so that an expression in the variables in x^2 is obtained. An unrestricted quadratic form is obtained which may be maximized as indicated before.

The following approach does exactly the same, but has the advantage that all operations are transformations of a tableau. Furthermore, some transformations will be indicated which are useful in quadratic programming.

Let us add to the left side of (102) the vector of artificial variables y, so that we have

$$Ax + y = b. \tag{104}$$

For a feasible solution of the problem y should be a zero vector. For the problem (101) and (104), the following Lagrangean expression can be formed:

$$\phi = c_0 + p'x - \tfrac{1}{2}xCx - u'(Ax + y - b), \tag{105}$$

where u is an m-element vector of Lagrangean multipliers. Note that $\phi = f$ for $Ax + y = b$. The derivatives of ϕ with respect to the x-variables are defined as the elements of the vector $-v$:

$$\partial\phi/\partial x = -v = p - Cx - A'u. \tag{106}$$

Let us substitute for Cx in (101) according to (106); this results in

$$f = c_0 + \tfrac{1}{2}p'x - \tfrac{1}{2}v'x + \tfrac{1}{2}u'Ax. \tag{107}$$

After substitution for Ax according to (104) the following expression for f is found:

$$f = c_0 + \tfrac{1}{2}p'x + \tfrac{1}{2}b'u - \tfrac{1}{2}v'x - \tfrac{1}{2}u'y. \tag{108}$$

Instead of f, $F = 2f$ may be considered, so that the factor $\tfrac{1}{2}$ disappears:

$$F = 2c_0 + p'x + b'u - v'x - u'y. \tag{109}$$

If the bilinear terms of this equation are deleted, it may be written, together with the equations of (104) and (106) as the equation system

$$2c_0 = -p'x - b'u + F,$$
$$-p = - Cx - A'u + v,$$
$$b = Ax + y. \tag{110}$$

This is a canonical equation system with F, the v-variables and the y-variables as basic variables. It may be put in tableau-form as in table 5, which is called the set-up tableau for the problem.

Table 5

Set-up tableau for a quadratic maximization problem

Basic variables	Values basic variables	x	u
F	$2c_0$	$-p'$	$-b'$
v	$-p$	$-C$	$-A'$
y	b	A	0

In this tableau we can distinguish primal variables and their corresponding dual variables; the corresponding dual variables of the x-variables are the v-variables and the corresponding dual variables of the y-variables are the u-variables. In the maximization problem first a feasible solution should be found, which means that the y-variables, which are artificial variables, should become nonbasic. The y-variables should be replaced by their corresponding u-variables, which are the Lagrangean multipliers of the constraints. But it is not possible to replace the y-variables directly by the u-variables since the corresponding pivots are all zero. Hence the y-variables should first be replaced by x-variables; after that, the u-variables replace the dual variables of the x-variables concerned which are v-variables. After all y-variables have become non-basic in this fashion, the proper optimization may start.

For an example of an application of this method, the following problem is used. Maximize

$$f = 3x_1 + 2x_2 + x_3 - 1\tfrac{1}{2}x_1^2 - x_2^2 - \tfrac{1}{6}x_3^2$$

subject to

$$x_1 + 2x_2 + x_3 = 4.$$

The set-up tableau is tableau 0 of table 6; the column "values of basic variables" is given as the last column in order to demonstrate certain symmetry properties in subsequent tableaux. In the first iteration y should be replaced by some x-variable with a nonzero element in the y-row, say x_1; then in a following iteration the corresponding variable of y, which is u, should replace the corresponding variable of x_1, which is v_1. If the element in the row of v_1 and the column of x_1 is nonzero, we may just as well pivot first on this element and afterwards introduce into the basis, replacing y, since the necessary pivot element then will have become nonzero. This has the advantage that the tableau obtained after the first iteration has desirable symmetry properties. The underlined element -3 is therefore used as a pivot for the first transformation; in the resulting tableau, this underlined element $-\tfrac{1}{3}$ is used for the second transformation, which results in tableau 2.

This tableau does not contain any artificial variables, so that the maximization may start. Note that this tableau has symmetry properties indicated by the dashed lines. The y-column and the u-row may be deleted in this tableau, but they should be retained if the value of u, the Lagrangean multiplier, is desired for the optimal solution. If the equation

$$x_1 + 2x_2 + x_3 = 4$$

had been used to eliminate x_1 from the objective function, the following quadratic form in x_2 and x_3 would have been obtained:

$$-12 + 20x_2 + 10x_3 - 7x_2^2 - 1\tfrac{2}{3}x_3^2 - 6x_2x_3.$$

The coefficients of this quadratic form can be found in the row of v_2, v_3, and F and the columns of x_2, x_3, and Values basic variables of tableau 2.

As explained in the previous section, the dependence of F on x_2 can be determined as follows:

$$F = -24 + 2 \times 20x_2 - 14x_2^2.$$

Table 6

The maximization of a constrained quadratic form

Tableau	Basic variables	x_1	x_2	x_3	u	Values basic variables
	v_1	$\underline{-3}$	0	0	-1	-3
	v_2	$\underline{0}$	-2	0	-2	-2
0	v_3	0	0	$-\frac{1}{3}$	-1	-1
	y	1	2	1	0	4
	F	$\underline{-3}$	-2	-1	-4	0

Tableau	Basic variables	v_1	x_2	x_3	u	
	x_1	$-\frac{1}{3}$	0	0	$\frac{1}{3}$	1
	v_2	0	-2	0	-2	-2
1	v_3	0	0	$-\frac{1}{3}$	-1	-1
	y	$\frac{1}{3}$	2	1	$-\frac{1}{3}$	3
	F	-1	-2	-1	$\underline{-3}$	3

Tableau	Basic variables	v_1	x_2	x_3	y	
	x_1	0	2	1	1	4
	v	-2	$\underline{-14}$	-6	-6	-20
2	v_3	-1	-6	$-3\frac{1}{3}$	-3	-10
	u	-1	-6	-3	-3	-9
	F	-4	-20	-10	-9	-24

Tableau	Basic variables	v_1	v_2	x_3	y	
	x_1	$-\frac{2}{7}$	$\frac{1}{7}$	$\frac{1}{7}$	$\frac{1}{7}$	$1\frac{1}{7}$
	x_2	$\frac{1}{7}$	$-\frac{1}{14}$	$\frac{3}{7}$	$\frac{3}{7}$	$1\frac{3}{7}$
3	v_3	$-\frac{1}{7}$	$-\frac{3}{7}$	$\underline{-\frac{16}{21}}$	$-\frac{3}{7}$	$-1\frac{3}{7}$
	u	$-\frac{1}{7}$	$-\frac{3}{7}$	$-\frac{3}{7}$	$-\frac{3}{7}$	$-\frac{3}{7}$
	F	$-1\frac{1}{7}$	$-1\frac{3}{7}$	$-1\frac{3}{7}$	$-\frac{3}{7}$	$4\frac{4}{7}$

Tableau	Basic variables	v_1	v_2	v_3	y	
	x_1	$-\frac{5}{16}$	$\frac{1}{16}$	$\frac{3}{16}$	$\frac{1}{16}$	$\frac{7}{8}$
	x_2	$\frac{1}{16}$	$-\frac{5}{16}$	$\frac{3}{16}$	$\frac{3}{16}$	$\frac{5}{8}$
4	x_3	$\frac{3}{16}$	$\frac{9}{16}$	$-1\frac{5}{16}$	$\frac{9}{16}$	$1\frac{7}{8}$
	u	$-\frac{1}{16}$	$-\frac{3}{16}$	$-\frac{9}{16}$	$-\frac{3}{16}$	$\frac{3}{8}$
	F	$-\frac{7}{8}$	$-\frac{5}{8}$	$-1\frac{7}{8}$	$\frac{3}{8}$	$7\frac{1}{4}$

The maximum value of F is obtained for $x_2 = \frac{20}{14} = 1\frac{3}{7}$. The same solution is obtained after pivoting on the element -14. In tableau 3, F may be written as the following function of x_3:

$$F = 4\frac{4}{7} + 2 \times 1\frac{3}{7}x_3 - \frac{16}{21}x_3^2.$$

The maximum of F is found for

$$x_3 = 1\frac{3}{7}/\frac{16}{21} = 1\frac{7}{8}.$$

The same value is obtained after pivoting on the element $-\frac{16}{21}$, which results in tableau 4. In this tableau all x-variables have entered the basis; y is not allowed to enter the basis, since it is an artificial variable. Since the pivots used from the start of the maximization, tableau 2, onwards have been negative, the optimal solution must have been obtained.

If the objective function has a finite maximum without the constraints imposed, it is possible to reverse the sequence of feasibility and maximization. In this case, proceeding from the set-up tableau, first all x-variables replace their corresponding v-variables; then the artificial y-variables are replaced by their corresponding u-variables. However, it may happen that the unconstrained objective function has no finite maximum while the constrained objective function has a finite maximum. For example, consider the problem in which the objective function can be partitioned as

$$\begin{bmatrix} p^1 \\ p^2 \end{bmatrix}' \begin{bmatrix} x^1 \\ x^2 \end{bmatrix} - \frac{1}{2} \begin{bmatrix} x^1 \\ x^2 \end{bmatrix}' \begin{bmatrix} 0 & 0 \\ 0 & C_{22} \end{bmatrix} \begin{bmatrix} x^1 \\ x^2 \end{bmatrix}$$

with the constraints

$$x^1 + Ax^2 = b.$$

If some of the elements of p^1 are nonzero, the unconstrained objective function has no finite maximum while the constrained objective function may have one. Hence it is advisable, at least for problems with equality constraints, to find first a feasible solution to the constraints.

Just as for unconstrained maximization, the method allows us to find the maximum solution or to demonstrate that there is no finite maximum. It should be noted that the conventional analytic approach is much more complicated.

CHAPTER 9

THE SIMPLEX AND THE DUAL METHOD FOR QUADRATIC PROGRAMMING

9.1. Introduction

A quadratic programming problem may be formulated as follows. Maximize

$$f = p'x - \tfrac{1}{2}x'Cx \tag{1}$$

subject to

$$Ax \le b, \tag{2}$$

$$x \ge 0. \tag{3}$$

It is assumed that C is a positive semi-definite matrix. In this case the problem is called a *convex quadratic programming problem*; if C is not positive semi-definite, it is a *general quadratic programming problem*. Since in the last type of problems a number of local optima may occur, these problems are more difficult to solve if the number of variables and constraints is not quite small. In convex quadratic programming problems, a local optimum is always a global optimum since the function f in (1) is concave for C positive semi-definite, as will be proved later on.

Since the objective function is no longer linear, the derivatives of f with respect to the x-variables change if these variables change:

$$df/dx = p - Cx. \tag{4}$$

Hence the determination of the values of the dual variables is not as easy as in linear programming, but the dual variables may be determined as follows. Let us introduce the vector of slack variables y in (2). The constraints are then

$$Ax + y = b, \tag{5}$$

$$x, y \ge 0. \tag{6}$$

The following Lagrangean expression may be formulated for the quadratic programming problem:

$$\phi = p'x - \tfrac{1}{2}x'Cx - u'(Ax + y - b); \tag{7}$$

u is a vector of Lagrangean multipliers. Note that $\phi = f$ for all solutions x, y satisfying (5).

The derivatives of ϕ with respect to x, which are defined as the elements of the vector $-v$, are

$$d\phi/dx = -v = p - Cx - A'u. \tag{8}$$

Taking the derivatives of ϕ with respect to y, we find

$$d\phi/dy = -u. \tag{9}$$

The minus-sign in the definition of the u- and v-variables will prove to be convenient. It should be noted that (8), apart from the term $-Cx$, is the same as the constraints of the dual problem in linear programming, since it may be written as

$$A'u - v = p - Cx. \tag{10}$$

f may be expressed in terms of the primal variables x and y and the dual variables u and v in the following manner. Let us substitute in (1) for Cx according to (8); the result is

$$f = p'x - \tfrac{1}{2}x'(p + v - A'u) = \tfrac{1}{2}p'x - \tfrac{1}{2}v'x + \tfrac{1}{2}u'Ax.$$

After substitution for Ax according to (5) we find

$$f = \tfrac{1}{2}p'x + \tfrac{1}{2}b'u - \tfrac{1}{2}v'x - \tfrac{1}{2}u'y. \tag{11}$$

Instead of f, $F = 2f$ may be considered, so that the coefficients $\tfrac{1}{2}$ vanish:

$$F = p'x + b'u - v'x - u'y. \tag{12}$$

This is a nonlinear expression, but the nonlinear terms are zero for any solution of (5) and (8) in which, if a primal variable is nonzero, then its corresponding dual variable is zero and vice versa.

The simplex method for quadratic programming uses the equation systems (5) and (8) and also eq. (12) in which the nonlinear terms have been deleted. Hence we have the following system:

$$0 = -p'x - b'u + F,$$
$$-p = -Cx - A'u \quad\quad + v,$$
$$b = \quad Ax \quad\quad\quad\quad + y. \tag{13}$$

Note that this equation system is linear, which facilitates the solution of quadratic programming problems to a great extent. These equations can be put into a tableau-format so that the set-up tableau for a quadratic programming problem results; see table 1.

Table 1

Set-up tableau for a quadratic programming
problem

Basic variables	Values basic variables	Nonbasic variables	
		x	u
F	0	$-p'$	$-b'$
v	$-p$	$-C$	$-A'$
y	b	A	0

The v- and y-variables and F are basic in this tableau and the x- and u-variables are nonbasic. The equation system (13) and its corresponding tableau may be transformed in the usual manner, so that other systems and tableaux are generated with different basic variables. Note that in the set-up tableau if a primal variable is basic, then its corresponding dual variable is nonbasic and vice versa. Any tableau for which this is true is called a *standard tableau* and its corresponding solution is a *standard solution*. Hence for any standard solution, the bilinear terms in (11) and (12) vanish, so that the value of F in this problem is equal to twice the objective function. A tableau in which a primal variable and its corresponding dual variable are both basic, or in which there are more such pairs of corresponding primal and dual basic variables is called a *nonstandard tableau* and its corresponding solution is called a *nonstandard solution*.

The above set-up tableau has properties of symmetry and skew-symmetry, which are typical for these tableaux. Note that elements in the rows of F and the dual variables and in the columns of Values of basic variables and the corresponding primal variables form the symmetric matrix

$$\begin{bmatrix} 0 & -p' \\ -p & -C \end{bmatrix}.$$

Furthermore, the elements in the rows of the primal variables and in the columns of the corresponding dual variables form the symmetric matrix 0. Finally, there is skew-symmetry for the elements both in rows and columns of primal variables (the column values of basic variables is counted as a primal variable) and the elements both in rows and columns of dual variables (F is counted as a dual variable); the matrices are

$$[b \quad A] \qquad \text{and} \qquad \begin{bmatrix} -b' \\ -A' \end{bmatrix}.$$

We shall prove that these symmetry and skew-symmetry properties are valid for any standard tableau. Furthermore it will be proved that in any standard tableau the matrix of elements in rows of primal variables and in columns of the corresponding dual variables (excluded F and V.b.v.) is negative semi-definite; the same is true for the matrix of elements in rows of dual variables and columns of primal variables. Hence a general standard tableau may be represented as in table 2.

Table 2

A general standard tableau

Basic variables	Values basic variables	x	u
F	q_0	q'	$-r'$
v	q	$-Q$	$-P'$
y	r	P	$-R$

v is in this tableau the vector of all dual basic variables, y the vector of all primal basic variables, and x and u are the vectors of primal and dual nonbasic variables. The matrices $-Q$ and $-R$ are negative semi-definite. The set-up tableau is a particular case of a general standard tableau in which $-R = 0$.

A standard tableau can be transformed into any other standard tableau by one or more applications of any of the following steps:

(1) replace a dual basic variable by its corresponding primal variable,

(2) replace a primal basic variable by its corresponding dual variable,

(3) replace a primal basic variable by another primal variable and replace the corresponding dual variable of the last primal variable by the corresponding dual variable of the first primal variable.

We shall show that each of these steps preserves the properties of a standard tableau. Tableau I of table 3 is a standard tableau, in which the dual basic variables, the corresponding primal nonbasic variables, and the corresponding vectors and matrices are partitioned according to the dual basic variables which stay in the basis given by the vector v^1 and the dual basic variables given by the vector v^2, which are going to be replaced by their corresponding primal nonbasic variables in x^2. The new tableau is obtained by block-pivoting on the matrix $-Q_{22}$; the resulting tableau is tableau II. It may easily be established by observing the mat-

Table 3

Transformation of a standard tableau (1)

Tableau	Basic variables	Values basic variables	x^1	x^2	u
I	F	q_0	$q^{1'}$	$q^{2'}$	$-r'$
	v^1	q^1	$-Q_{11}$	$-Q_{12}$	$-P_1'$
	v^2	q^2	$-Q_{21}$	$\underline{-Q_{22}}$	$-P_2'$
	y	r	P_1	P_2	$-R$
			x^1	v^2	u
II	F	$q_0 + q^{2'}Q_{22}^{-1}q^2$	$q^{1'} - q^{2'}Q_{22}^{-1}Q_{21}$	$q^{2'}Q_{22}^{-1}$	$-r' - q^{2'}Q_{22}^{-1}P_2'$
	v^1	$q^1 - Q_{12}Q_{22}^{-1}q^2$	$-Q_{11} + Q_{12}Q_{22}^{-1}Q_{21}$	$-Q_{12}Q_{22}^{-1}$	$-P_1' + Q_{12}Q_{22}^{-1}P_2'$
	x^2	$-Q_{22}^{-1}q^2$	$Q_{22}^{-1}Q_{21}$	$-Q_{22}^{-1}$	$Q_{22}^{-1}P_2'$
	y	$r + P_2 Q_{22}^{-1}q^2$	$P_1 - P_2 Q_{22}^{-1}Q_{21}$	$P_2 Q_{22}^{-1}$	$-R - P_2 Q_{22}^{-1}P_2'$

rices within dashed lines that the symmetry and skew-symmetry properties of a standard tableau are still valid for this tableau. Furthermore we
may write

$$- Q_{11} + Q_{12} Q_{22}^{-1} Q_{21} = [I - Q_{12} Q_{22}^{-1}] \begin{bmatrix} -Q_{11} & -Q_{12} \\ -Q_{21} & -Q_{22} \end{bmatrix} \begin{bmatrix} I \\ -Q_{22}^{-1} Q_{21} \end{bmatrix}, \quad (14)$$

so that this matrix is negative semi-definite. For the other symmetric matrix we write

$$\begin{bmatrix} -Q_{22}^{-1} & Q_{22}^{-1} P_2' \\ P_2 Q_{22}^{-1} & -R - P_2 Q_{22}^{-1} P_2' \end{bmatrix} = \begin{bmatrix} 0 & 0 \\ 0 & -R \end{bmatrix} + \begin{bmatrix} I \\ P_2 \end{bmatrix} [-Q_{22}^{-1}] [I \quad P_2']; \quad (15)$$

this is the sum of two negative semi-definite matrices, so that this matrix
is negative semi-definite. Hence tableau II has all properties of a standard
tableau.

A transformation of the type (2) is performed in table 4. Tableau I is
partitioned according to the primal basic variables, which are replaced by
their corresponding dual variables; these primal variables are given by
the vector y' and their corresponding dual variables by u'. A transformation with $-R_{11}$ as block-pivot gives tableau II. It is easily observed that
symmetry and skew-symmetry exists as indicated by the dashed lines.
That the symmetric matrices are negative semi-definite is proved by

$$\begin{bmatrix} -Q - P_1' R_{11}^{-1} P_1 & -P_1' R_{11}^{-1} \\ -R_{11}^{-1} P_1 & -R_{11}^{-1} \end{bmatrix} = \begin{bmatrix} -Q & 0 \\ 0 & 0 \end{bmatrix} + \begin{bmatrix} P_1' \\ I \end{bmatrix} [-R_{11}^{-1}] [P_1 \quad I], \quad (16)$$

and

$$- R_{22} + R_{21} R_{11}^{-1} R_{12} = [-R_{21} R_{11}^{-1} \quad I] \begin{bmatrix} -R_{11} & -R_{12} \\ -R_{21} & -R_{22} \end{bmatrix} \begin{bmatrix} -R_{11}^{-1} R_{12} \\ I \end{bmatrix}. \quad (17)$$

The third type of transformation occurs if a primal basic variable is
replaced by another primal variable, after which the corresponding dual
basic variable of the last primal variable is replaced by the corresponding
dual nonbasic variable of the first primal variable; instead of just one
variable of each kind, there may be a number of them. Such a transformation can be interpreted as two transformations, one of the first type in
which a new primal variable replaces its corresponding dual variable and
one of the second type in which a dual nonbasic variable replaces its
corresponding primal basic variable. However, this can only be done if
the relevant (block-)pivots are nonzero (non-singular). In the case of
linear programming, the matrix C is zero, and so will the matrices in any
other standard tableau for this problem be. Similar cases may occur in

Table 4

Transformation of a standard tableau (2)

Tableau	Basic variables	Values basic variables	x	u^1	u^2
I	F	q_0	q	$-r^{1\prime}$	$-r^{2\prime}$
	v	q	$-Q$	$-P_1'$	$-P_2'$
	y^1	r^1	P_1	$-R_{11}$	$-R_{12}$
	y^2	r^2	P_2	$-R_{21}$	$-R_{22}$
			x	y^1	u^2
II	F	$q_0 - r^{1\prime}R_{11}^{-1}r^1$	$q - r^{1\prime}R_{11}^{-1}P_1$	$-r^{1\prime}R_{11}^{-1}$	$-r^{2\prime} + r^{1\prime}R_{11}^{-1}R_{12}$
	v	$q - P_1'R_{11}^{-1}r^1$	$-Q - P_1'R_{11}^{-1}P_1$	$-P_1'R_{11}^{-1}$	$-P_2' + P_1'R_{11}^{-1}R_{12}$
	u^1	$-R_{11}^{-1}r^1$	$-R_{11}^{-1}P_1$	R_{11}^{-1}	$R_{11}^{-1}R_{12}$
	y^2	$r^2 - R_{21}R_{11}^{-1}r^1$	$P_2 - R_{21}R_{11}^{-1}P_1$	$-R_{21}R_{11}^{-1}$	$-R_{22} + R_{21}R_{11}^{-1}R_{12}$

quadratic programming. Tableau I of table 5 illustrates this situation. In the new tableau we should have the primal-variables in x^2 as basic variables instead of those of y^1 and the dual variables in u^1 instead of those in v^2. It is not possible to interchange first the variables in x^2 and v^2 or the variables in u^1 and y^1, since the matrices which should serve as block-pivots are both zero. The desired tableau may be obtained by block-pivoting first on the matrix P_{12} and then on its skew-symmetric counterpart $-P'_{12}$. But before doing so, we should note that because $-Q_{22} = 0$, also $-Q_{12}$ and $-Q_{21}$ are zero, since $-Q$ is a negative definite matrix. Since $-R_{11} = 0$, we conclude for the same reason that $-R_{12} = -R_{21} = 0$. This, of course, simplifies the two block-pivoting operations to a great extent. Tableau III gives the final result of these transformations, from which we may immediately conclude that it has the properties of a standard tableau.

We shall now prove that if a standard solution is found which has nonnegative primal and dual basic variables, this is an optimal solution. According to the representation of a general standard tableau, the following equations are valid

$$
\begin{aligned}
q_0 &= q'x - r'u + F, \\
q &= -Qx - P'u + v, \\
r &= Px - Ru + y.
\end{aligned}
\tag{18}
$$

Hence the objective function F may be expressed as follows after introduction of the nonlinear terms:

$$
F = q_0 - q'x + r'u - v'x - u'y.
\tag{19}
$$

A substitution for v and y according to the last two sets of equations of (18) results in

$$
\begin{aligned}
F &= q_0 - q'x + r'u - x'(q + Qx + P'u) - u'(r - Px + Ru) \\
&= q_0 - 2q'x - x'Qx - u'Ru.
\end{aligned}
\tag{20}
$$

From this it follows that if the elements of q are nonnegative, no positive value of any x- or u-variable can increase the objective function. Since also the values of the primal basic variables given by r are nonnegative, the present basic solution is an optimal one in the wide sense.

Conversely, it can be proved by construction that if a finite optimal solution to the problem exists, a standard tableau with nonnegative basic variables always can be found; in the next sections it will be shown that

Table 5

Transformation of a standard tableau (3)

Tableau	Basic variables	Values basic variables	x^1	x^2	u^1	u^2
I	F	q_0	$q^{1\prime}$	$q^{2\prime}$	$-r^{1\prime}$	$-r^{2\prime}$
	v^1	q^1	$-Q_{11}$	0	$-P'_{11}$	$-P'_{21}$
	v^2	q^2	0	0	$-P'_{12}$	$-P'_{22}$
	y^1	r^1	P_{11}	P_{12}	0	0
	y^2	r^2	P_{21}	P_{22}	0	$-R_{22}$
			x^1	y^1	u^1	u^2
II	F	$q_0 - q^{2\prime}P_{12}^{-1}r^1$	$q^{1\prime} - q^{2\prime}P_{12}^{-1}P_{11}$	$-q^{2\prime}P_{12}^{-1}$	$-r^{1\prime}$	$-r^{2\prime}$
	v^1	q^1	$-Q_{11}$	0	$-P'_{11}$	$-P'_{21}$
	v^2	q^2	0	0	$-P'_{12}$	$-P'_{22}$
	x^2	$P_{12}^{-1}r^1$	$P_{12}^{-1}P_{11}$	P_{12}^{-1}	0	0
	y^2	$r^2 - P_{22}P_{12}^{-1}r^1$	$P_{21} - P_{22}P_{12}^{-1}P_{11}$	$-P_{22}P_{12}^{-1}$	0	$-R_{22}$
			x^1	y^1	v^2	u^2
III	F	$q_0 - 2q^{2\prime}P_{12}^{-1}r^1$	$q^{1\prime} - q^{2\prime}P_{12}^{-1}P_{11}$	$-q^{2\prime}P_{12}^{-1}$	$-r^{1\prime}P_{12}^{-1}$	$-r^{2\prime} + r^{1\prime}P_{12}^{-1}P'_2$
	v^1	$q^1 - P'_{11}P'^{-1}_{12}q^2$	$-Q_{11}$	0	$-P'_{11}P'^{-1}_{12}$	$-P'_{21} + P'_{11}P'^{-1}_{12}P'_2$
	u^1	$-P'^{-1}_{12}q^2$	0	0	$-P'^{-1}_{12}$	$P'^{-1}_{12}P'_{22}$
	x^2	$P_{12}^{-1}r^1$	$P_{12}^{-1}P_{11}$	P_{12}^{-1}	0	0
	y^2	$r^2 - P_{22}P_{12}^{-1}r^1$	$P_{21} - P_{22}P_{12}^{-1}P_{11}$	$-P_{22}P_{12}^{-1}$	0	$-R_{22}$

the simplex and the dual method for quadratic programming always lead to a standard tableau with nonnegative basic variables if a finite optimal solution exists. Hence a necessary and sufficient condition for an optimal solution is the existence of a standard tableau with nonnegative basic variables.

A slightly more general formulation of the conditions for an optimal solution is as follows. Necessary and sufficient conditions for an optimal solution are

$$Ax + y = b, \tag{21}$$

$$v = -p + Cx + A'u, \tag{22}$$

$$v'x + u'y = 0, \tag{23}$$

$$x, \ y, \ u, \ v \geq 0. \tag{24}$$

These are the so-called *Kuhn-Tucker conditions* for convex nonlinear programming, applied to convex quadratic programming. (21) and (22) are simply the equations of the set-up tableaux, minus the equation for F. Any solution of a quadratic simplex tableau will satisfy (21)–(22). Eq. (23) will be satisfied by a solution of any standard tableau. It is obvious that a nonnegative solution of a standard tableau satisfies (21)–(24).

9.2. The first symmetric variant of the simplex method for quadratic programming

An obvious way to develop a quadratic programming method is to try to generalize linear programming methods. A number of methods for quadratic programming have been presented as generalizations of the simplex for linear programming, but some are closer to it than others. Here we shall consider three variants of a method which is called the simplex method for quadratic programming; this name is given because it seems to be the most obvious and natural generalization of the simplex method for linear programming. The three variants are equivalent in the sense that they have the same sequence of solutions, but they have different pivots. The first variant to be considered is the first symmetric variant; later the asymmetric and the second symmetric variant will be treated.

Let us consider the following example. Maximize

$$f = 2x_1 + 2x_2 - 2x_1^2 + 2x_1x_2 - 2x_2^2 \tag{25}$$

subject to

$$x_1 + x_2 \leq 1,$$
$$-x_1 + 6x_2 \leq 2,$$
$$x_1, x_2 \geq 0. \tag{26}$$

The equation system for this problem is after the introduction of slack variables and dual variables as follows.

$$0 = -2x_1 - 2x_2 - u_1 - 2u_2 + F,$$
$$-2 = -4x_1 + 2x_2 - u_1 + u_2 + v_1,$$
$$-2 = 2x_1 - 4x_2 - u_1 - 6u_2 + v_2,$$
$$1 = x_1 + x_2 + y_1,$$
$$2 = - x_1 + 6x_2 + y_2. \tag{27}$$

This equation system can be found in tableau-form in tableau 0 of table 6. Note that the first equation of (27) is only valid if $-v'x - u'y = 0$; this is so for the basic solution of tableau 0.

Let us consider an increase of x_1 from 0 to some positive value θ. The values of the basic variables then become

$$0 + 2\theta = F,$$
$$-2 + 4\theta = v_1,$$
$$-2 - 2\theta = v_2,$$
$$1 - \theta = y_1,$$
$$2 + \theta = y_2. \tag{28}$$

Table 6

An application of the first symmetric variant of the simplex method for quadratic programming

Tabl.	Bas. var.	Values basic variables			Ratio	x_1	x_2	u_1	u_2
		c-term	θ-term	$\theta = \frac{1}{2}$					
	F_c	0	2	1		-2	-2	-1	-2
	F_θ	2	-4	0	$\frac{1}{2}$	4	-2	1	-1
0	v_1	-2	4	0	$\frac{1}{2}$	-4	2	-1	1
	v_2	-2	-2	-3		2	-4	-1	-6
	y_1	1	-1	$\frac{1}{2}$	1	1	1	0	0
	y_2	2	1	$2\frac{1}{2}$		-1	6	0	0

Table 6   (*continued*)

Tabl.	Bas. var.	Values basic variables — c-term	θ-term	$u=\frac{1}{3}$		Ratio	v_1	x_2	u_1	u_2
	F_c	1	3	2			$-\frac{1}{2}$	-3	$-\frac{1}{2}$	$-2\frac{1}{2}$
	F_θ	3	-3	2		1	$-\frac{1}{2}$	3	$1\frac{1}{2}$	$5\frac{1}{2}$
1	x_1	$\frac{1}{2}$	$\frac{1}{2}$	$\frac{2}{3}$			$-\frac{1}{4}$	$-\frac{1}{2}$	$\frac{1}{4}$	$-\frac{1}{4}$
	v_2	-3	3	-2		1	$\frac{1}{2}$	-3	$-1\frac{1}{2}$	$-5\frac{1}{2}$
	y_1	$\frac{1}{2}$	$-1\frac{1}{2}$	0		$\frac{1}{3}$	$\frac{1}{4}$	$1\frac{1}{2}$	$-\frac{1}{4}$	$\frac{1}{4}$
	y_2	$2\frac{1}{2}$	$-5\frac{1}{2}$	$\frac{2}{3}$		$\frac{5}{11}$	$-\frac{1}{4}$	$5\frac{1}{2}$	$\frac{1}{4}$	$-\frac{1}{4}$
				$\theta=\frac{1}{3}$	$\theta=\frac{3}{7}$		v_1	x_2	y_1	u_2
	F_c	0	6	2	$2\frac{4}{7}$		-1	-6	-2	-3
	F_θ	6	-12	2	$\frac{6}{7}$	$\frac{1}{2}$	1	12	6	7
2	x_1	1	-1	$\frac{2}{3}$	$\frac{4}{7}$	1	0	1	1	0
	v_2	-6	12	-2	$-\frac{6}{7}$	$\frac{1}{2}$	-1	-12	-6	-7
	u_1	-2	6	0	$\frac{4}{7}$		-1	-6	-4	-1
	y_2	3	-7	$\frac{2}{3}$	0	$\frac{3}{7}$	0	7	1	0
							v_1	y_2	y_1	u_2
	F_c	$2\frac{4}{7}$					-1	$\frac{6}{7}$	$-1\frac{1}{7}$	-3
3a	x_1	$\frac{4}{7}$					0	$-\frac{1}{7}$	$\frac{6}{7}$	0
	v_2	$-\frac{6}{7}$					-1	$1\frac{5}{7}$	$-4\frac{2}{7}$	-7
	u_1	$\frac{4}{7}$					-1	$\frac{6}{7}$	$-3\frac{1}{7}$	-1
	x_2	$\frac{3}{7}$					0	$\frac{1}{7}$	$\frac{1}{7}$	0
							v_1	v_2	y_1	u_2
	F	3					$-\frac{1}{2}$	$-\frac{1}{2}$	1	$\frac{1}{2}$
3	x_1	$\frac{1}{2}$					$-\frac{1}{12}$	$\frac{1}{12}$	$\frac{1}{2}$	$-\frac{7}{12}$
	x_2	$\frac{1}{2}$					$\frac{1}{12}$	$-\frac{1}{12}$	$\frac{1}{2}$	$\frac{7}{12}$
	u_1	1					$-\frac{1}{2}$	$-\frac{1}{2}$	-1	$2\frac{1}{2}$
	y_2	$-\frac{1}{2}$					$-\frac{7}{12}$	$\frac{7}{12}$	$-2\frac{1}{2}$	$-4\frac{1}{12}$
							v_1	v_2	y_1	y_2
	F	$2\frac{46}{49}$					$-\frac{4}{7}$	$-\frac{3}{7}$	$\frac{34}{49}$	$\frac{6}{49}$
4	x_1	$\frac{4}{7}$					0	0	$\frac{6}{7}$	$-\frac{1}{7}$
	x_2	$\frac{3}{7}$					0	0	$\frac{1}{7}$	$\frac{1}{7}$
	u_1	$\frac{34}{49}$					$-\frac{6}{7}$	$-\frac{1}{7}$	$-2\frac{26}{49}$	$\frac{30}{49}$
	u_2	$\frac{6}{49}$					$\frac{1}{7}$	$-\frac{1}{7}$	$\frac{30}{49}$	$-\frac{12}{49}$

In the tableau the same result is obtained by copying in a "θ-term" column of the values of basic variables the column of x_1 with a minus-sign added, see tableau 0; the other column is now called "constant-term" column.

However, for $\theta = x_1 \neq 0$, the first equation of (28) does not give the correct value of F, since the terms $-v'x - u'y$ are no longer zero. Hence we should add to the left side of the first equation of (28)

$$-v_1 x_1 = -(-2 + 4\theta)\theta; \tag{29}$$

the right-hand side is obtained by substitution according to the second equation of (28). F is therefore given by the following expression:

$$0 + 4\theta - 4\theta^2 = F. \tag{30}$$

The same expression may be obtained in the tableau by copying in a new row, indicated by F_θ, the row of v_1 with a minus-sign; this row should be thought of as being multiplied by θ and added to the proper row of the objective function, which row is now indicated by F_c. We find therefore for the elements in the rows of F_c and F_θ and in the columns of c-term and θ-term in tableau 0:

	c-term	θ-term
F_c	0	2
F_θ	2	-4

From this we find for the value of the objective function again eq. (30).

The value of θ for a maximum of the objective function is found from

$$dF/d\theta = 4 - 8\theta = 0$$

or $\theta = \frac{1}{2}$. The same value is found by equating the terms in row F_θ to zero:

$$2 - 4\theta = 0 \qquad \text{or} \qquad \theta = \frac{1}{2}.$$

Since row F_θ is equal to minus the v_1-row, the same value can be found from the v_1-row.

As in linear programming, increasing values of a nonbasic variable may render the solution infeasible. Hence an upper bound for θ may be found from the rows of the primal basic variables. In tableau 0 we find that if θ exceeds 1, y_1 becomes negative. However, this is larger than the value of θ giving a maximum of the objective function which is $\frac{1}{2}$. We therefore introduce x_1 into the basis, replacing v_1 as a basic variable by pivoting on the element -4. The result is tableau 1, which has indeed a solution $x_1 = \frac{1}{2}$ and an increased value of F which corresponds with (30) for $\theta = \frac{1}{2}$. The θ-term column is not transformed, neither is the F_θ-row.

Note that in tableau 1 there are three primal basic variables. In quadratic programming the number of primal basic variables is not always equal to m, the number of constraints, as in linear programming, but may vary from m to $m + n$, where n is the number of x-variables.

Figure 1 illustrates the situation. A is the solution of tableau 0. An increase of x_1 means a movement along the x_1 axis. In B the increase in the value of the objective function for increasing x_1 is zero and turns into a decrease; note that B is not an extreme point of the feasible region.

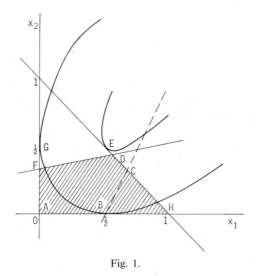

Fig. 1.

We could just as well have chosen x_2 as the new basic variable. In that case we would have found for F

$$0 + 4\theta - 4\theta^2 = F,$$

but y_2 would have become zero for $\theta = \frac{1}{3}$, so that in the figure the solution moves from A to F. In general we could have chosen as the new basic variable any nonbasic primal variable with a negative coefficient in the F-row, or, what amounts to the same, because of the symmetry properties, any nonbasic primal variable having a corresponding dual basic variable with a negative value. As a general rule we shall introduce the primal nonbasic variable with the most negative element in the F-row. If there are no such negative elements in the F-row, the optimal solution has been found. If a new basic variable has been found, but no leaving basic variable can be found in the manner indicated, the problem has no finite solution. Note that there is a close correspondence with the simplex method for linear programming.

It is interesting to know what happens if a dual nonbasic variable with a negative element in the F-row is increased. Let us assume that u_2 is increased to a level θ. The values of basic variables are in this case

$$0 + 2\theta = F,$$
$$-2 - \theta = v_1,$$
$$-2 + 6\theta = v_2,$$
$$1 \qquad = y_1,$$
$$2 \qquad = y_2. \tag{31}$$

Adding the nonzero nonlinear term, we find for F:

$$F = 0 + 2\theta - u_2 y_2 = 2\theta - \theta \times 2 = 0.$$

It is easily shown that in the expression for F the linear terms in dual nonbasic variables vanish: if there is a quadratic term, it is negative, see (20).

We shall now proceed with tableau 1, introducing into the basis the primal nonbasic variable with the most negative coefficient in the F-row; this is x_2. The column of x_2 is copied with a minus-sign in the θ-term column and the v_2-row is copied with a minus-sign in the F_θ-row. Hence F is expressed as follows in terms of $\theta = x_2$:

$$F = 1 + 6\theta - 3\theta^2.$$

F obtains its maximum value for $\theta = 1$; this value is also found from the row of v_2. The values of θ for which primal basic variables become zero are $\frac{1}{3}$ for y_1 and $\frac{5}{11}$ for y_2. The minimum value of θ is $\frac{1}{3}$, which is found in the row of y_1. This corresponds with point C in figure 1.

Now we would be inclined to introduce x_2 into the basis, replacing y_1. By doing so, a nonstandard tableau would be obtained because both v_2 and x_2 would be basic. Since nonstandard tableaux do not have the properties of symmetry, skew-symmetry and negative semi-definiteness of certain matrices it would be difficult to proceed. However, it is possible to retain standard tableaux by treating the new basic variable x_2

as a parameter θ. For $x_2 = \theta = \frac{1}{3}$, the value of y_1 is zero. If for $\theta = \frac{1}{3}$, y_1 leaves the basis and u_1 enters it, a new standard tableau is found which for $\theta = \frac{1}{3}$ has the same values of basic variables as the previous one for $\theta = \frac{1}{3}$. This can be done by pivoting on the element in the row of y_1 and the column of u_1, $-\frac{1}{4}$. This element is negative or zero because it is a principal element of a negative semi-definite matrix. A case in which such a principal element is zero occurs in the next iteration. Tableau 2 is generated; in this tableau θ still stands for x_2.

The basic variable y_1 blocked the increase of x_2, but now it has been replaced by its corresponding dual variable u_1. This means that we are now moving along the constraint $x_1 + x_2 = 1$. For $x_2 = 0$, point H is found and for $x_2 = \frac{1}{3}$, C is found. The expression for F is now

$$F = 0 + 12\theta - 12\theta^2.$$

F is maximal for $\theta = \frac{1}{2}$; this corresponds with point E. From the rows of x_1 and y_2 we find, that x_1 becomes zero for $\theta = 1$ and y_2 becomes zero for $\theta = \frac{3}{7}$. The minimum is $\frac{3}{7}$, which is found for y_2; this is in accordance with the fact that E lies outside the feasible region. Now y_2 should leave the basis, but it cannot be replaced by its corresponding dual variable u_2 since the element which should serve as a pivot is zero.

Then we may have the following double nonstandard iteration. x_2 enters the basis, replacing y_2, with the element 7 as a pivot; after this iteration all primal basic variables are still nonnegative. This iteration will have had no effect of the column of u_2, since its element in the pivot row is zero. The result is tableau 3a, which is a nonstandard tableau. Next u_2 enters the basis replacing v_2, so that -7 is the pivot; the result is tableau 4, which is a standard tableau. Note that the values of the primal basic variables have not changed in the last iteration, because the elements in their rows in the column of u_2 are zero. This is no coincidence, since it was found that the element in the row of y_2 and the column of u_2 was zero, this is a principal element in the negative semi-definite matrix $-R$, so that the elements in the same column also must be zero. Hence the values of primal basic variables do not change during the last iteration.

Instead of having the two iterations indicated, we may first have an iteration in which x_2 replaces its corresponding variable v_2 and then an iteration in which u_2 replaces its corresponding variable y_2; in this case only principal pivots are used. This is only possible if the first pivot to be used is nonzero. In this case this pivot is -12, so that it can be used. The result is tableau 3, which is a standard tableau. In the following iteration

u_2 replaces y_2, with the element $-4\frac{1}{12}$ as a pivot; this results in tableau 4. Since this is a standard tableau with nonnegative primal and dual variables, it must contain the optimal solution. In case the elements in the row of x_2 and in the column of v_2 in tableau 2 would have been zero, no principal pivot would have been possible, but, as explained before when a transformation of a standard tableau of type 3 was discussed, the required standard tableau can be written down immediately after the first transformation.

The symmetry and skew-symmetry properties of standard tableaux imply that it is not necessary to write down all elements of a tableau; only the elements on and below or on and above the main diagonal of the simplex tableaux are required. The θ-term column and F_θ-row can of course be deleted. Table 7 gives the successive tableaux for the example in its most concise representation.

For a general statement of the method we shall use the general representation of a standard tableau, as given in table 2. Unlike in linear programming, a distinction must be made between the variable which is increased to a level θ and the variable which actually enters the basis. This is illustrated by tableau 1 of table 6, where x_2 is increased to $\frac{1}{3}$, but in which u_1 turned out to be the new basic variable. In what follows we shall call the variable which is increased in a certain tableau the *incoming variable* of that tableau. For some tableaux the incoming variable is given because it is the same as for the previous tableau; for others it should first be determined.

The rules for the method may be stated as follows. It is assumed that the initial tableau is standard and contains a vector r with nonnegative elements.

INCOMING VARIABLE:

If a tableau has no incoming variable select as such the primal non-basic variable connected with

$$\text{Min}_i (q_i | q_i < 0) = q_k. \tag{32}$$

If all q_i are nonnegative the optimal solution has been found.

TRANSFORMATIONS:

Determine

$$\text{Min}_i (q_k / - q_{kk}, \ r_i / p_{ik} | - q_{kk} \neq 0, \ p_{ik} > 0). \tag{33}$$

If $-q_{kk} = 0$ and $p_{ik} \leq 0$ for all i, the problem has no finite optimal solution. If the minimum is $q_k / - q_{kk}$, pivot on $-q_{kk}$. If the minimum is $r_\ell / p_{\ell k}$ and

Table 7

Triangular quadratic simplex tableaux

Tableau	Basic var.	Ratio	Values basic variables					
0	F		0					
	v_1	$\frac{1}{2}$	-2	-4				
	v_2		-2	2	-4			
	y_1	1	1	1	1	0		
	y_2		2	-1	6	0	0	
1	F		1					
	x_1		$\frac{1}{2}$	$-\frac{1}{4}$				
	v_2	1	-3	$\frac{1}{2}$	-3			
	y_1	$\frac{1}{3}$	$\frac{1}{2}$	$\frac{1}{4}$	$1\frac{1}{2}$	$-\frac{1}{4}$		
	y_2	$\frac{5}{11}$	$2\frac{1}{2}$	$-\frac{1}{4}$	$5\frac{1}{2}$	$\frac{1}{4}$	$-\frac{1}{4}$	
2	F		0					
	x_1	1	1	0				
	v_2	$\frac{1}{2}$	-6	-1	-12			
	u_1	$\frac{3}{7}$	-2	-1	-6	-4		
	y_2		3	0	7	1	0	
3	F		3					
	x_1		$\frac{1}{2}$	$-\frac{1}{12}$				
	x_2		$\frac{1}{2}$	$\frac{1}{12}$	$-\frac{1}{12}$			
	u_1		1	$-\frac{1}{2}$	$-\frac{1}{2}$	-1		
	y_2		$-\frac{1}{2}$	$-\frac{7}{12}$	$\frac{7}{12}$	$-2\frac{1}{2}$	$-4\frac{1}{12}$	
4	F		$2\frac{46}{49}$					
	x_1		$\frac{4}{7}$	0				
	x_2		$\frac{3}{7}$	0	0			
	u_1		$\frac{34}{49}$	$-\frac{6}{7}$	$-\frac{1}{7}$	$-2\frac{26}{49}$		
	u_2		$\frac{6}{49}$	$\frac{1}{7}$	$-\frac{1}{7}$	$\frac{30}{49}$	$-\frac{12}{49}$	

$-r_{\ell\ell} < 0$, *pivot on* $-r_{\ell\ell}$ *and retain the same incoming variable. If* $-r_{\ell\ell} = 0$, *and* $-q_{kk} < 0$, *pivot first on* $-q_{kk}$ *and then on the element in the place of* $-r_{\ell\ell}$. *If* $-q_{kk} = 0$, *pivot first on* $p_{\ell k}$ *and then on the element* $-p_{\ell k}$.

A formal proof for the convergence to the optimal solution in a finite number of steps can be given as follows. First we shall prove that the value of the objective function increases in each iteration. It should be observed that the value of the objective function for a certain tableau without a predetermined incoming variable is the same as the value of the objective function of the previous tableau evaluated at the upper bound for the incoming variable; the same is obviously true for tableaux with a predetermined incoming variable. Hence we only have to show that an increase of the incoming variable to its upper bound increases the objective function. From the previous discussion this is obvious, but it may formally be stated as follows. In terms of a general standard tableau we have

$$F = q^0 - 2q_k\theta - q_{kk}\theta^2, \tag{34}$$

which may be written as

$$F = q^0 + q_k^2/q_{kk} - q_{kk}(q_k/-q_{kk} - \theta)^2. \tag{35}$$

This is an increasing function of θ for

$$0 < \theta \leq q_k/-q_{kk} \tag{36}$$

if $-q_{kk} \neq 0$; otherwise it follows immediately from (34). It is obvious that $q_k/-q_{kk}$ is the upper bound found in the row of the corresponding dual variable. If degeneracy occurs, θ may not increase in two successive iterations, but in this case perturbation techniques may be used.

Since the objective function increases in any iteration, the same solution cannot reoccur. Because there is only a finite number of combinations of the $m + n$ primal and dual basic variables among the $2(m + n)$ variables, the method must terminate in a finite number of iterations.

Since the simplex method for quadratic programming requires a basic solution of the set-up tableau having nonnegative values of primal basic variables, it should be indicated how such a solution should be found if one is not immediately available; this may happen if the constraints are of the form

$$Ax = b;$$

$$x \geq 0. \tag{37}$$

In this case the vector of artificial variables z is introduced and the sum of the z-variables is minimized:

$$\text{Min } f = e'z \tag{38}$$

subject to

$$Ax + z = b, \tag{39}$$

$$x, z \geq 0. \tag{40}$$

This problem can be solved by the simplex method for linear programming. In the quadratic simplex tableau the same solution is generated by using the same pivots both in the matrix A and in its transpose A'. The introduction and elimination of the artificial variables may also take place in the quadratic simplex tableau. We shall not go into details.

9.3. The asymmetric variant

The asymmetric variant of the simplex method for quadratic programming was first discovered by Dantzig[6], who called it a variant of the Wolfe–Markowitz procedures. One and a half year later it was discovered independently[1] by the author[67], who with Whinston in [71] developed a complete proof for the convergence of the method in a finite number of steps. It turns out that it is much more convenient to prove convergence for the symmetric variant and then prove the equivalence of both variants than to give a direct proof of convergence for the asymmetric variant; in fact, the proof for the convergence of the asymmetric variant implicitly relied on the symmetric variant.

The asymmetric method does not have the parametric variation of the incoming variable; instead this variable enters the basis as soon as it is assigned a nonzero value. As a consequence (and a drawback) nonstandard tableaux occur. Let us consider the same example as in the previous section, see table 8. The first iteration would have been entirely the same for both variants, but in tableau 1 it is found that y_1 becomes zero first for increasing values of x_2. In the symmetric variant x_2 is considered as a parameter θ and y_2 is replaced by its corresponding variable, if possible. In the asymmetric variant x_2 enters the basis replacing y_1.

The result is tableau 2, which is a nonstandard tableau since both x_2 and v_2 occur together as basic variables; x_2 and v_2 are called the *basic pair* of corresponding variables. If there is such a basic pair, there must also be a *nonbasic pair* of corresponding variables; in this tableau these are y_1 and u_1. In a nonstandard tableau the asymmetric variant introduces into the basis *the dual variable of the nonbasic pair*; this variable is always the corresponding dual variable of the primal variable which just left the basis. In tableau 2 it is u_1.

The leaving basic variable is determined by taking the minimum of the nonnegative ratio's in rows of primal basic variables and in the row of the

---

[1] Before the report[67] was written, Graves and Wolfe advised the author to read Dantzig's work, which was then available as a paper, which the author did after he wrote his paper.

Table 8

An application of the asymmetric variant of the simplex method for quadratic programming

Tabl.	Basic var.	Values bas. var.	Ratio	x_1	x_2	u_1	u_2
	F	0		-2	-2	-1	-2
	v_1	-2	$\frac{1}{2}$	-4	2	-1	1
0	v_2	-2		2	-4	-1	-6
	y_1	1	1	1	1	0	0
	y_2	2		-1	6	0	0

				v_1	x_2	u_1	u_2
	F	1		$-\frac{1}{2}$	-3	$-\frac{1}{2}$	$-2\frac{1}{2}$
	x_1	$\frac{1}{2}$		$-\frac{1}{4}$	$-\frac{1}{2}$	$\frac{1}{4}$	$-\frac{1}{4}$
1	v_2	-3	1	$\frac{1}{2}$	-3	$-1\frac{1}{2}$	$-5\frac{1}{2}$
	y_1	$\frac{1}{2}$	$\frac{1}{3}$	$\frac{1}{4}$	$1\frac{1}{2}$	$-\frac{1}{4}$	$\frac{1}{4}$
	y_2	$2\frac{1}{2}$	$\frac{5}{11}$	$-\frac{1}{4}$	$5\frac{1}{2}$	$\frac{1}{4}$	$-\frac{1}{4}$

				v_1	y_1	u_1	u_2
	F	2		0	2	-1	-2
	x_1	$\frac{2}{3}$	4	$-\frac{1}{6}$	$\frac{1}{3}$	$\frac{1}{6}$	$-\frac{1}{6}$
2	v_2	-2	1	1	2	-2	-5
	x_2	$\frac{1}{3}$		$\frac{1}{6}$	$\frac{2}{3}$	$-\frac{1}{6}$	$\frac{1}{6}$
	y_2	$\frac{2}{3}$	$\frac{4}{7}$	$-1\frac{1}{6}$	$-3\frac{2}{3}$	$1\frac{1}{6}$	$-1\frac{1}{6}$

				v_1	y_1	y_2	u_2
	F	$2\frac{4}{7}$		-1	$-1\frac{1}{7}$	$\frac{6}{7}$	-3
	x_1	$\frac{4}{7}$		0	$\frac{6}{7}$	$-\frac{1}{7}$	0
3	v_2	$-\frac{6}{7}$	$\frac{6}{49}$	-1	$-4\frac{2}{7}$	$1\frac{5}{7}$	-7
	x_2	$\frac{3}{7}$		0	$\frac{1}{7}$	$\frac{1}{7}$	0
	u_1	$\frac{4}{7}$		-1	$-3\frac{1}{7}$	$\frac{6}{7}$	-1

				v_1	y_1	y_2	v_2
	F	$2\frac{46}{49}$		$-\frac{4}{7}$	$\frac{34}{49}$	$\frac{6}{49}$	$-\frac{3}{7}$
	x_1	$\frac{4}{7}$		0	$\frac{6}{7}$	$-\frac{1}{7}$	0
4	u_2	$\frac{6}{49}$		$\frac{1}{7}$	$\frac{30}{49}$	$-\frac{12}{49}$	$-\frac{1}{7}$
	x_2	$\frac{3}{7}$		0	$\frac{1}{7}$	$\frac{1}{7}$	0
	u_1	$\frac{34}{49}$		$-\frac{6}{7}$	$-2\frac{26}{49}$	$\frac{30}{49}$	$-\frac{1}{7}$

dual variable of the basic pair. In tableau 2 the following ratio's are considered:

$$(\tfrac{2}{3}/\tfrac{1}{6}, \; -2/-2, \; \tfrac{2}{3}/1\tfrac{1}{6}) = (4, \, 1, \, \tfrac{4}{7}).$$

The minimum is found in the row of y_2, so that this variable leaves the basis.

The result is tableau 3, which is nonstandard with x_2 and v_2 as its basic pair and y_2 and u_2 as its nonbasic pair. Again the dual variable of the nonbasic pair is introduced into the basis. Taking ratio's in rows of primal variables and in the row of the dual variable of the basic pair, we find that there is only one relevant ratio, that in the row of v_2, so that this variable leaves the basis. The result is tableau 4, which is standard and, because it has nonnegative basic variables, contains the optimal solution.

Note that in each nonstandard tableau there is only one basic pair and one nonbasic pair; further that in a sequence of nonstandard tableaux the basic pair always remains the same, but that the nonbasic pair changes in each iteration.

The equivalence between the symmetric and the asymmetric variant may be shown as follows. Let us consider a tableau in the symmetric method which does not have an incoming variable inherited from the previous tableau; for instance, this may be the initial tableau. This tableau is the same for the asymmetric variant. The incoming basic variable in the symmetric variant will be the same as the new basic variable in the asymmetric variant. If the minimum ratio is connected with the dual variable of the incoming variable, both variants have the same iterations resulting in the same tableaux.

Now suppose that the minimum ratio occurs in the row of a primal basic variable, say y_b; the element in this row and in the column of the corresponding dual variable u_b may be denoted by $-r_{bb}$. In case $-r_{bb} = 0$, the double nonstandard iteration of the symmetric variant and the two successive iterations of the asymmetric version are equivalent; both variants have the same resulting tableaux.

This can be shown as follows. Let x_k be the new basic variable, y_b the leaving basic variable and y_i a typical other basic variable; v_k, u_b and u_i are the corresponding dual variables. The relevant part of the tableau is

		x_k	u_b	u_i
v_k	q_k	$-q_{kk}$	$-p_{bk}$	$-p_{ik}$
y_b	r_b	p_{bk}	$r_{bb} = 0$	$r_{bi} = 0$
y_i	r_i	p_{ik}	$r_{ib} = 0$	$-r_{ii}$

$$(41)$$

Since $r_{bb} = 0$ and $-R$ is negative semi-definite, r_{bi} and r_{ib} are zero for all i. In the first iteration of both the symmetric and the asymmetric variants p_{bk} is used as a pivot. This results in the following tableau:

		y_b	u_b	u_i
v_k	$q_k - q_{kk} p_{bk}^{-1} r_b$	$q_{kk} p_{bk}^{-1}$	$-p_{bk}$	$-p_{ik}$
x_k	$p_{bk}^{-1} r_b$	p_{bk}^{-1}	0	0
y_i	$r_i - p_{ik} p_{bk}^{-1} r_b$	$-p_{ik} p_{bk}^{-1}$	0	$-r_{ii}$

(42)

The symmetric method automatically takes $-p_{bk}$ as a pivot. In the asymmetric variant u_b is the new basic variable, the leaving basic variable is determined by taking ratio's in the rows of primal basic variables and in the row of v_k. Since all elements in the column of u_b and in the rows of primal basic variables are zero, v_k must be the leaving basic variable.

In case $-r_{bb} \neq 0$, the symmetric variant pivots on this element, introducing u_b into the basis replacing y_b. The relevant parts of the resulting tableau are then as follows:

		x_k	y_b	u_i
v_k	q_k	$-q_{kk}$	$-q_{kb}$	$-p_{ik}$
u_b	q_b	$-q_{bk}$	$-q_{bb}$	$-p_{ib}$
y_i	r_i	p_{ik}	p_{ib}	$-r_{ii}$

(43)

x_k is the incoming variable, v_k its dual variable; y_b was the leaving basic variable and u_b its dual variable; y_i is a typical primal basic variable and u_i its dual variable. The value of x_k for this tableau is $\theta = -q_b q_{bk}^{-1}$, so that the values of basic variables for this value of θ are

$$v_k = q_k - q_b q_{bk}^{-1} q_{kk},$$
$$u_b = q_b - q_b q_{bk}^{-1} q_{bk} = 0,$$
$$y_i = r_i + q_b q_{bk}^{-1} p_{ik}.$$

(44)

The tableau which would have been generated by the asymmetric variant can be found from tableau of (43) by pivoting on the element $-q_{bk}$. The relevant parts of the resulting tableau are

		u_b	y_b
v_k	$q_k - q_{kk} q_{bk}^{-1} q_b$	$-q_{kk} q_{bk}^{-1}$	$-q_{kb} + q_{kk} q_{bk}^{-1} q_{bb}$
x_k	$-q_{bk}^{-1} q_b$	$-q_{bk}^{-1}$	$q_{bk}^{-1} q_{bb}$
y_i	$r_i + p_{ik} q_{bk}^{-1} q_b$	$p_{ik} q_{bk}^{-1}$	$p_{ib} - p_{ik} q_{bk}^{-1} q_{bb}$

(45)

The values of basic variables are the same in (44) and (45). Now $q_{bk}^{-1} q_{bb} > 0$, since it was the pivot in the iteration of the asymmetric variant in which x_k replaced y_b; hence $q_{bk} > 0$. In the asymmetric variant u_b is introduced into the basis, which in the symmetric variant entered the basis in the previous iteration. The leaving basic variable is the same in both cases, which can be shown as follows. In the symmetric variant, the leaving basic variable is connected with the minimum of the ratios

$$q_k / - q_{kk}, \qquad r_i / p_{ik} \tag{46}$$

for all i with $p_{ik} > 0$. In the asymmetric variant it is connected with the minimum of the ratio's

$$\frac{q_k - q_{kk} q_{bk}^{-1} q_b}{- q_{kk} q_{bk}^{-1}}, \qquad \frac{r_i + p_{ik} q_{bk}^{-1} q_b}{p_{ik} q_{bk}^{-1}} \tag{47}$$

for all i with $p_{ik} q_{bk}^{-1} > 0$. These ratio's may be written, since $q_{bk} > 0$, as

$$q_b + q_{bk}(q_k / - q_{kk}), \qquad q_b + q_{bk}(r_i / p_{ik}), \tag{48}$$

for all i with $p_{ik} > 0$. Comparing (48) with (46), it is obvious that the same row must be selected. If the minimum ratio is connected with v_k, the same standard tableau is found in both methods; otherwise the same primal basic variable leaves the basis in both methods. The same analysis is valid for consecutive tableaux of the symmetric and the asymmetric variant. The conclusion is that both variants are equivalent.

It is interesting to give a geometrical interpretation of the asymmetric variant and compare this with an interpretation of the symmetric variant. The values of the primal variables in a standard tableau can be interpreted as constituting an optimal solution to the quadratic programming problem with the added constraints that the nonbasic primal variables must have the value zero; this is referred to as a restricted optimal solution. In case the value of a dual variable corresponding to a primal nonbasic variable is negative, then by increasing that variable the objective function is increased. The algorithm then aims for a new restricted optimal solution to the same problem except that the constraint on the selected nonbasic variable is removed. However, it is possible that as the new variable is increased, the solution vector hits one of the constraints. This constraint is added by introducing into the basis its dual variable; and the algorithm moves in the direction of the new restricted optimal solution. Either the solution vector moves to the new restricted optimal solution or a new constraint must be added and the process repeated.

Eventually a case must arise where no further constraints need be added and the algorithm achieves a restricted optimal solution. In this case we have again a standard tableau.

The symmetric version alters the sequence of pivots. Again, we may start from a standard tableau. The algorithm proceeds by finding the first constraint to be encountered if the nonbasic variable, which is called the incoming variable, were to be increased from zero. If no constraint would be encountered while increasing the incoming variable to its optimal value then the variable is actually introduced and we have a restricted optimal solution. In case a constraint would be encountered, this constraint is first imposed by removing the appropriate variable from the basis; a new restricted optimal solution is then obtained. We continue the hypothetical increase of the incoming variable until either its optimal value is achieved with respect to the current set of constraints or a new constraint is encountered. We finally reach a case where the incoming variable may be hypothetically increased to its optimal value. When this occurs it is actually introduced and a restricted optimal solution results.

The main difference between the two versions may be summarized as follows. Starting from a standard tableau the asymmetric version increases the nonbasic variable and successively imposes the required constraints. After each constraint is imposed, the algorithm moves in the direction of a restricted optimal solution relative to the constraints imposed. The symmetric version first imposes all required constraints and with the imposition of each constraint a restricted optimal solution is found, only after this, the incoming variable is introduced.

9.4. The second symmetric variant

It is also possible to use another symmetric variant of the simplex method for quadratic programming. In order to distinguish between the two variants, the variant described earlier will be called the first symmetric variant and the alternative method which will presently be described the second symmetric variant of the simplex method for quadratic programming. Just as the first symmetric variant, the second symmetric variant has in each iteration the same solution as the asymmetric variant but the pivots are different.

The tableaux of the three variants are identical for standard tableaux in the asymmetric variant. The variants may start to generate different tableaux when a basic primal variable leaves the basis. If in a standard tableau the new basic variable, say x_k, drives a basic primal variable, say y_b, to zero first, the second symmetric variant follows the asymmetric variant by replacing y_b by x_k in the basis, but it has immediately afterwards another iteration in which the corresponding dual variable of y_b, u_b, replaces the corresponding dual variable of x_k, which is v_k; hence we have a double nonstandard iteration. This second iteration may have made basic primal variables other than x_k negative. The relevant part of the resulting standard tableau is as follows:

		x_b	v_k	u_i
u_b	q_b	$-q_{bb}$	$-p_{bk}$	$-p_{ik}$
x_k	r_k	p_{bk}	$-r_{kk}$	$-r_{ki}$
y_i	r_i	p_{ik}	$-r_{ik}$	$-r_{ii}$

$$(49)$$

From this tableau we may generate the previous tableau by pivoting on the element $-q_{bk}$; note that this is the current tableau of the asymmetric variant. The relevant parts of this tableau are:

		y_b	u_b	u_i
v_k	$-p_{bk}^{-1}q_b$	$p_{bk}^{-1}q_{bb}$	$-p_{bk}^{-1}$	$p_{bk}^{-1}q_{ik}$
x_k	$r_k - r_{kk}p_{bk}^{-1}q_b$	$p_{bk} + r_{kk}p_{bk}^{-1}q_{bb}$	$-r_{kk}p_{bk}^{-1}$	$-r_{ki} + r_{kk}p_{bk}^{-1}p_{ik}$
y_i	$r_i - r_{ik}p_{bk}^{-1}q_b$	$p_{ik} + r_{ik}p_{bk}^{-1}q_{bb}$	$-r_{ik}p_{bk}^{-1}$	$-r_{ii} + r_{ik}p_{bk}^{-1}p_{ik}$

$$(50)$$

If in this tableau y_b replaces x_k, the original standard tableau is obtained. The first pivot of the double standard iteration was therefore

$$q_{bk} + r_{kk}p_{bk}^{-1}q_{bb},$$

which is positive. From this we deduce that $p_{bk} > 0$. Now the solution of (50) may be obtained in (49) if v_k is given the negative value $\theta = -p_{bk}^{-1}q_b$; the values are then:

$$u_b = q_b + \theta p_{bk} = q_b - p_{bk}^{-1}q_b p_{bk} = 0,$$
$$x_k = r_k + \theta r_{kk} = r_k - p_{bk}^{-1}q_b r_{kk},$$
$$y_i = r_i + \theta r_{ik} = r_i - p_{bk}^{-1}q_b r_{ik}. \qquad (51)$$

This solution is a feasible one since the solution of the asymmetric variant is feasible. Since the value of the nonbasic variable v_k is taken to be nonzero in (49) in order to have a feasible solution, we call (49) a *blocked* tableau; v_k is called the *outgoing variable* of the blocked tableau. If the value θ of v_k can be increased to zero without making the solution infeasible, the tableau is unblocked. The second symmetric variant increases v_k to zero, pivoting out any primal basic variables which blocks such an increase. Hence the leaving basic variable in a blocked tableau is determined by the minimum of the ratios

$$r_i/-r_{ik} \tag{52}$$

for $r_i < 0$ and $-r_{ik} > 0$. Let the minimum be negative and connected with the primal basic variable y_b. We now pivot on the element $-r_{bb}$, which is negative because $-r_{bk} > 0$ and the matrix $-R$ negative semi-definite. If all r_i are nonnegative, the tableau is unblocked, otherwise (52) is reapplied.

We shall now show that in the second symmetric variant and in the asymmetric variant the same primal basic variable, if any, leaves the basis. In the asymmetric variant the leaving basic variable is determined by the minimum of the ratios (see (50))

$$\frac{-p_{bk}^{-1}q_b}{-p_{bk}^{-1}}, \qquad \frac{r_i - r_{ik}p_{bk}^{-1}q_b}{-r_{ik}p_{bk}^{-1}}, \tag{53}$$

for all i with $-r_{ik}p_{bk}^{-1} > 0$; this reduces to

$$q_b, \qquad p_{bk}(r_i/-r_{ik}) + q_b, \tag{54}$$

for all i with $-r_{ik} > 0$. From this, it follows the minimum ratio will be q_b, unless there are some negative r_i. If q_b is the minimum ratio, u_b replaces v_k in the basis of the tableau of (50) and the tableau of (49) is obtained. If there are negative r_i, the same choice of the leaving basic variable is made as in (52). The same arguments can be used to prove the equivalence of the two variants if there are a number of consecutive tableaux with the same outgoing variable.

If the second symmetric variant is applied to the example used before, the first iteration would be the same as in table 8. In tableau 1, however, we would pivot on the element in the row of y_1 and the column of x_2 which is $1\frac{1}{2}$, and in the next iteration which is the second half of a double standard iteration we would pivot on the element in the row of v_2 and in the column of u_1, which is -2. Alternatively, diagonal pivots could have been used by first replacing v_2 by x_2 and then y_1 by u_1. The resulting tableau would be as table 9. The outgoing variable is v_2. Taking ratios according to (52) we find only one ratio, $-\frac{6}{7}$. Hence y_2 leaves the basis and u_2 enters it. The next tableau contains the optimal solution, it is the same as tableau 4 of table 8.

Table 9

B.v.	V.b.v.			v_1	y_1	v_2	u_2
F	3	$-\frac{6}{7}$		$-\frac{1}{2}$	1	$-\frac{1}{2}$	$\frac{1}{2}$
x_1	$\frac{1}{2}$			$-\frac{1}{12}$	$\frac{1}{2}$	$\frac{1}{12}$	$-\frac{7}{12}$
u_1	1			$-\frac{1}{2}$	-1	$-\frac{1}{2}$	$2\frac{1}{2}$
x_2	$\frac{1}{2}$			$\frac{1}{12}$	$\frac{1}{2}$	$-\frac{1}{12}$	$\frac{7}{12}$
y_2	$-\frac{1}{2}$	$-\frac{6}{7}$		$-\frac{7}{12}$	$-2\frac{1}{2}$	$\frac{7}{12}$	$-4\frac{1}{12}$

The second symmetric variant has the same advantages over the asymmetric method as the first symmetric method, namely symmetric tableaux. Compared with the first symmetric method the second symmetric method is somewhat less easier to explain. Computationally there does not seem to be much difference.

9.5. Duality in quadratic programming

In linear programming it was shown that any method had its equivalent dual method, unless the method was symmetric, in which case the equivalent dual method was identical to the method. In quadratic programming the same is true. However, before treating dual methods, it is necessary to deal with the subject of quadratic duality.

We consider the typical convex quadratic maximization problem: Maximize f

$$f = p'x - \tfrac{1}{2}x'Cx,$$
$$Ax \le b,$$
$$x \ge 0. \tag{55}$$

The dual problem is: Minimize g

$$g = b'u + \tfrac{1}{2}x'Cx,$$
$$A'u \ge p - Cx,$$
$$u \ge 0. \tag{56}$$

This can be proved as follows. Every problem with C positive semidefinite can be transformed in such a manner that C can be partitioned as follows:

$$C = \begin{bmatrix} C_{11} & 0 \\ 0 & 0 \end{bmatrix}$$

where C_{11} is a positive definite matrix. The Kuhn–Tucker conditions for problem (56) can be found by first writing this problem as a typical maximization problem: Maximize $-g$

$$-g = -b'u - \tfrac{1}{2}x'Cx,$$
$$-Cx - A'u \leq -p,$$
$$u \geq 0. \tag{57}$$

This problem may be written as: Maximize $-g$

$$-g = \begin{bmatrix} -b \\ 0 \end{bmatrix}' \begin{bmatrix} u \\ x \end{bmatrix} - \tfrac{1}{2} \begin{bmatrix} u \\ x \end{bmatrix}' \begin{bmatrix} 0 & 0 \\ 0 & C \end{bmatrix} \begin{bmatrix} u \\ x \end{bmatrix}$$

$$[-A' \quad -C] \begin{bmatrix} u \\ x \end{bmatrix} \leq -p,$$

$$u \geq 0. \tag{58}$$

Note that the x-variables are unrestricted. The "dual equations", which for the general problem (55) are

$$-p = -Cx - A'u + v, \tag{59}$$

are in this case

$$\begin{bmatrix} b \\ 0 \end{bmatrix} = -\begin{bmatrix} 0 & 0 \\ 0 & C \end{bmatrix} \begin{bmatrix} u \\ x \end{bmatrix} - \begin{bmatrix} A \\ C \end{bmatrix} w^1 + \begin{bmatrix} w^2 \\ w^3 \end{bmatrix}; \tag{60}$$

the vectors w^1 and $[w^2 \quad w^3]'$ play the role of the vectors u and v in (59). Since the x-variables are nonrestricted, the variables in w^3 should be equal to zero in the optimal solution.

The second set of equations in (60) then may be written as follows

$$0 = Cx - Cw^1, \tag{61}$$

or, after partitioning C,

$$\begin{bmatrix} 0 \\ 0 \end{bmatrix} = \begin{bmatrix} C_{11} & 0 \\ 0 & 0 \end{bmatrix} \begin{bmatrix} x^1 \\ x^2 \end{bmatrix} - \begin{bmatrix} C_{11} & 0 \\ 0 & 0 \end{bmatrix} \begin{bmatrix} w^{11} \\ w^{12} \end{bmatrix}, \tag{62}$$

where w^1 is partitioned into the subvectors w^{11} and w^{12}. From the first set of equations of (62) we conclude, since C_{11} was assumed to be positive definite and therefore nonsingular

$$w^{11} = x^1. \tag{63}$$

The variables in x^2 and y do in fact not occur in problem (58); let us denote w^{12} as x^2 and let us denote w^2 as y. Hence the first set of equations in (60) may be written as

$$b = Ax + y. \tag{64}$$

These are the dual constraints of problem (58) corresponding with the constraints (59) for problem (55). The complete Kuhn-Tucker conditions for the dual problem (56) are therefore

$$b = Ax + y,$$
$$-p = -Cx - A'u + v,$$
$$u'y + v'x = 0,$$
$$u, v, x, y \geq 0. \tag{65}$$

These are the same conditions as those for the primal problem (55). Hence both problems must have the same optimal solution.

Let us substitute in g for Cx according to the second set of equations of (65) and in the result let us substitute for Ax according to the first set of equations of (65):

$$\begin{aligned} g &= b'u + \tfrac{1}{2}x'(p + v - A'u) \\ &= b'u + \tfrac{1}{2}p'x + \tfrac{1}{2}v'x - \tfrac{1}{2}u'(b - y) \\ &= \tfrac{1}{2}p'x + \tfrac{1}{2}b'u + \tfrac{1}{2}u'y + \tfrac{1}{2}v'x. \end{aligned} \tag{66}$$

For an optimal solution the last two terms are equal to zero, so that

$$\hat{g} = \tfrac{1}{2}p'\hat{x} + \tfrac{1}{2}b'\hat{u}. \tag{67}$$

The solution $\hat{x}$, $\hat{u}$ of the dual problem is also an optimal solution of the primal problem, since the conditions for both problems are (65). In general we have for a solution to the primal problem

$$f = \tfrac{1}{2}p'x + \tfrac{1}{2}b'u - \tfrac{1}{2}u'y - \tfrac{1}{2}v'x. \tag{68}$$

For the solution $\hat{x}$, $\hat{u}$ we have therefore

$$f = \tfrac{1}{2}p'\hat{x} + \tfrac{1}{2}b'\hat{u}, \tag{69}$$

$$\hat{f} = \hat{g}, \tag{70}$$

and from this we have

$$\text{Max } f = \text{Min } g. \tag{71}$$

Hence problem (56) has the properties of a dual problem of the general problem (55).

From (70) we find

$$\hat{f} = p'\hat{x} - \tfrac{1}{2}\hat{x}'C\hat{x} = b'\hat{u} + \tfrac{1}{2}\hat{x}'C\hat{x} = \hat{g} \tag{72}$$

or

$$(p - C\hat{x})'\hat{x} = b'\hat{u}. \tag{73}$$

Hence at the optimum there is an equality between the revenue of the

optimal solution evaluated at the marginal revenues for the various products and the value of resources evaluated at their shadow prices. The
interpretation of the dual problem is here much less obvious than in
linear programming.

The set-up tableau and the general standard tableau for the dual problem are entirely the same as for the primal problem since the conditions
(65) are the same. The sufficiency of the conditions (65) may be proved
using a general standard tableau in which the basic variables denoted by
v and y are nonnegative. This standard tableau implies the equations

$$2g^0 = q'x - r'u + G,$$
$$q = -Qx - P'u + v,$$
$$r = Px - Ru + y. \tag{74}$$

The first equation can be written after addition of the nonlinear terms
according to (66)

$$2g = G = 2g^0 - q'x + r'u + u'y + v'x. \tag{75}$$

After substitution for v and y according to the last two sets of equations
of (74) we obtain

$$G = 2r'u + u'Ru + x'Qx. \tag{76}$$

Since $r \geq 0$ and Q and R are positive semi-definite any solution with
$u > 0$ and $x \neq 0$ increases G, so that the basic solution $v = q$ and $y = r$ is
indeed optimal.

9.6. Dual methods for quadratic programming

Just as the simplex method for linear programming can be generalised
into the simplex method for quadratic programming, so can the dual
method for linear programming be generalised into the dual method for
quadratic programming. Just as in the simplex method for quadratic
programming three variants may be distinguished for the dual method for
quadratic programming.

The symmetric variant of the dual method for quadratic programming
will be explained by means of the following example. Maximize f

$$f = -2x_1 - 2x_2 - 2x_1^2 + 2x_1x_2 - 2x_2^2,$$
$$x_1 + x_2 \geq 1,$$
$$-x_1 + 6x_2 \geq 2,$$
$$x_1, x_2 \geq 0. \tag{77}$$

Figure 2 gives a geometrical representation of the problem. E is the unrestricted maximum; contour lines for the objective function are elipses centered on E. Since the objective function is symmetric in x_1 and x_2, it is obvious from the figure that D is the optimal solution.

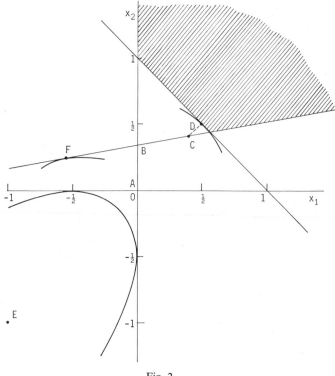

Fig. 2.

The set-up tableau is the same as it would be in the corresponding simplex method; it is given in tableau 0 of table 10. The G_θ-row and θ-term column should for a moment be disregarded. Instead of twice the objective function of the primal problem, which was denoted by F, we now consider twice the objective function of the dual problem, which is denoted by G; the coefficients in the first row are the same, however.

In linear programming the dual method can be derived in two ways. It may be done directly by working with the tableau of primal variables; primal variables with negative values are replaced by other variables in such a way that the optimality in the narrow sense is preserved. Alternatively, the dual problem is solved by means of the simplex method and the dual method then consists of making the iterations in the tableaux with the primal variables corresponding with those of the simplex method in the tableaux with the dual variables. In quadratic programming

Table 10

An application of the symmetric variant of the dual method for quadratic programming

Tabl.	Bas. var.	Values bas. var.		Ratio				
		c-term	θ-term		x_1	x_2	u_1	u_2
	G	0	-2		2	2	1	2
	G_θ	-2	0		1	-6	0	0
0	v_1	2	1		-4	2	1	-1
	v_2	2	-6	$\frac{1}{3}$	2	$\underline{-4}$	1	6
	y_1	-1	0		-1	-1	0	0
	y_2	-2	0		1	-6	0	0
					x_1	v_2	u_1	u_2
	G	1	-5		3	$\frac{1}{2}$	$1\frac{1}{2}$	5
	G_θ	-5	9		-2	$-1\frac{1}{2}$	$-1\frac{1}{2}$	-9
1	v_1	3	-2	$1\frac{1}{2}$	-3	$\frac{1}{2}$	$1\frac{1}{2}$	2
	x_2	$-\frac{1}{2}$	$1\frac{1}{2}$		$-\frac{1}{2}$	$-\frac{1}{4}$	$-\frac{1}{4}$	$-1\frac{1}{2}$
	y_1	$-1\frac{1}{2}$	$1\frac{1}{2}$		$-1\frac{1}{2}$	$-\frac{1}{4}$	$-\frac{1}{4}$	$-1\frac{1}{2}$
	y_2	-5	9	$\frac{5}{9}$	-2	$-1\frac{1}{2}$	$-1\frac{1}{2}$	$\underline{-9}$
					x_1	v_2	u_1	y_2
	G	$-1\frac{7}{9}$	$-\frac{2}{3}$		$1\frac{8}{9}$	$-\frac{1}{3}$	$\frac{2}{3}$	$\frac{5}{9}$
	G_θ	$-\frac{2}{3}$	0		$-1\frac{1}{6}$	0	0	$-\frac{1}{6}$
2	v_1	$1\frac{8}{9}$	$-1\frac{1}{6}$	$1\frac{13}{21}$	$-3\frac{4}{9}$	$\frac{1}{6}$	$1\frac{1}{6}$	$\frac{2}{9}$
	x_2	$\frac{1}{3}$	0		$-\frac{1}{6}$	0	0	$-\frac{1}{6}$
	y_1	$-\frac{2}{3}$	0		$-1\frac{1}{6}$	0	0	$-\frac{1}{6}$
	u_2	$\frac{5}{9}$	$\frac{1}{6}$	$3\frac{1}{3}$	$\frac{2}{9}$	$\frac{1}{6}$	$\frac{1}{6}$	$-\frac{1}{9}$
					v_1	v_2	u_1	y_2
	G	$-\frac{23}{31}$	$-1\frac{19}{62}$		$\frac{17}{31}$	$-\frac{15}{62}$	$1\frac{19}{62}$	$\frac{21}{31}$
	G_θ	$-1\frac{19}{62}$	$\frac{49}{124}$		$-\frac{21}{62}$	$-\frac{7}{124}$	$-\frac{49}{124}$	$-\frac{15}{62}$
3	x_1	$-\frac{17}{31}$	$\frac{21}{62}$		$-\frac{9}{31}$	$-\frac{3}{62}$	$-\frac{21}{62}$	$-\frac{2}{31}$
	x_2	$\frac{15}{62}$	$\frac{7}{124}$		$-\frac{3}{62}$	$-\frac{1}{124}$	$\frac{7}{124}$	$-\frac{11}{62}$
	y_1	$-1\frac{19}{62}$	$\frac{49}{124}$	$3\frac{15}{49}$	$-\frac{21}{62}$	$-\frac{7}{124}$	$\frac{49}{124}$	$-\frac{15}{62}$
	u_2	$\frac{21}{31}$	$-\frac{15}{62}$	$2\frac{4}{5}$	$\frac{2}{31}$	$\frac{11}{62}$	$\frac{15}{62}$	$-\frac{3}{31}$

Table 10 (continued)

Tabl.	Bas. var.	Values bas. var. c-term	Values bas. var. θ-term	Ratio	v_1	v_2	u_1	u_2
	G	4	-3		1	1	3	7
	G_θ	-3	1		$-\frac{1}{2}$	$-\frac{1}{2}$	-1	$-2\frac{1}{2}$
4	x_1	-1	$\frac{1}{6}$		$-\frac{1}{3}$	$-\frac{1}{6}$	$-\frac{1}{2}$	$-\frac{2}{3}$
	x_2	-1	$\frac{1}{6}$		$-\frac{1}{6}$	$-\frac{1}{3}$	$-\frac{1}{2}$	$-1\frac{5}{6}$
	y_1	-3	1	3	$-\frac{1}{2}$	$-\frac{1}{2}$	-1	$-2\frac{1}{2}$
	y_2	-7	$2\frac{1}{2}$		$-\frac{2}{3}$	$-1\frac{5}{6}$	$-2\frac{1}{2}$	$-10\frac{1}{3}$

Tabl.	Bas. var.	Values bas. var. c-term	Values bas. var. θ-term	Ratio	v_1	v_2	y_1	u_2
	G	-5			$-\frac{1}{2}$	$-\frac{1}{2}$	3	$-\frac{1}{2}$
5	x_1	$\frac{1}{2}$			$-\frac{1}{12}$	$\frac{1}{12}$	$-\frac{1}{2}$	$\frac{7}{12}$
	x_2	$\frac{1}{2}$			$\frac{1}{12}$	$-\frac{1}{12}$	$-\frac{1}{2}$	$-\frac{7}{12}$
	u_1	3			$\frac{1}{2}$	$\frac{1}{2}$	-1	$2\frac{1}{2}$
	y_2	$\frac{1}{2}$			$\frac{7}{12}$	$-\frac{7}{12}$	$-2\frac{1}{2}$	$-4\frac{1}{12}$

the tableaux are a combination of primal and dual tableaux, so that both derivations may be illustrated by the same tableau. Here the interpretation of the dual method as an application of the simplex method to the dual problem is used.

The solution of tableau 0 is a feasible one for the dual problem since the values of the dual basic variables v_1 and v_2 are nonnegative. It is not optimal because the values of the primal variables are negative. Let us consider an increase in the dual nonbasic variable u_2. The effect on the objective function follows from the G-row of tableau 0:

$$G = 0 - 2u_2,$$

but we should add to this the nonlinear terms of G, which are in general $v'x + u'y$ and in this case $u_2 y_2$. For y_2 we may write according to its row in tableau 0:

$$y_2 = -2 + 0u_2,$$

so that G may be expressed in terms of the nonbasic variable u_2 as

$$G = 0 - 4u_2. \tag{78}$$

The same results may be obtained in a symbolic way by copying the u_2-column with a minus-sign in a θ-term column and copying the y_2-row as it is in an additional row G_θ. In this case the coefficient of $\theta^2 = u_2^2$ is zero, so that no upper bound for $\theta = u_2$ is found from the objective function. Since the dual variables should remain nonnegative, other upper bounds for θ may be derived. In tableau 0 the upper bound $\theta = \frac{1}{3}$ is derived from v_2. Since this is the only ratio, v_2 should leave the basis. The element in the row of v_2 and the column of x_2 is -4, so that x_2 will replace v_2. Tableau 1 results, which has a feasible solution in terms of the dual variables for $\theta = \frac{1}{3}$, since the value of v_2 and x_2 was zero for $\theta = \frac{1}{3}$.

It is interesting to find out what would have happened if a primal variable, say x_1, would have been increased from zero to a positive value θ. In this case we would have found for G:

$$G = -2x_1 + v_1 x_1 = -2\theta_1 + (2 + 4\theta)\theta = 4\theta^2. \tag{79}$$

It is obvious that the linear terms vanish and only a positive quadratic term remains. Hence we should select as the new basic variable a dual variable with a positive coefficient in the θ-row; in order to make our choice unique, the dual nonbasic variable with the largest positive coefficient in the G-row should be selected. Because of the skew-symmetry of a standard tableau this may also be stated as the selection of a corresponding dual variable of the primal basic variable with the most negative value.

In tableau 1, θ is increased from $\frac{1}{3}$. The expression for G is now

$$G = 1 - 10\theta + 9\theta^2; \tag{80}$$

G is at its minimum for

$$dG/df = -10 + 18\theta = 0,$$

or $\theta = \frac{5}{9}$. The same result can be found from taking the ratio of constant-term and θ-term in the row of y_2. Other upper bounds for θ are found from rows of dual basic variables. In this case this is the ratio $1\frac{1}{2}$ in the row of v_1. The minimum ratio is $\frac{5}{9}$ in the row of y_1. Since the element in the row of y_1 and the column of u_1 is nonzero, it can be used as a pivot.

In tableau 2, u_1 is selected as the new basic variable since its coefficient in the G-row is the only positive one. Ratio's are taken in the rows of y_1, v_1 and u_2; it turns out that the smallest ratio occurs in the row of v_1. The element in this row and in the column of its corresponding primal variable x_1 is nonzero, so that it may be used as a pivot. Tableau 3 results, which still has u_1 as the incoming variable. Ratio's are taken in the rows of y_1 and u_2; the smallest is found in the row of u_2. Again a principal pivot is possible. In tableau 4 u_1 is still the incoming variable. There is only one ratio, namely that in the row of y_1, since all dual variables are nonbasic. Hence y_1 leaves the basis and because the element in this row and in the column of u_1 is nonzero, it is replaced by u_1. Tableau 5 contains the optimal solution.

The solution path can be followed in an incomplete manner in figure 2. Point A corresponds with the solution of tableau 0. The increase of u_2 in tableau 0 does not change primal basic variables, so that the solution remains in A. In tableau 1, x_2 changes for increasing u_2; we arrive in B. In tableau 2, the increase of u_1^1 does not affect x_2, but in tableau 3 it affects both x_1 and x_2, so that the solution moves to C. In tableau 4 the solution moves from C to D.

The rules for the symmetric variant of the dual method for quadratic programming are not difficult to state. They are the same as the rules of the symmetric variant of the simplex method for quadratic programming if the words "primal" and "dual" are interchanged. As in the corresponding simplex method, it is also possible to develop an asymmetric and a second symmetric variant of the dual method for quadratic programming. As before, it may be proved that all three variants are equivalent.

CHAPTER 10

PARAMETRIC QUADRATIC PROGRAMMING

10.1. The symmetric variant of parametric quadratic programming

Let us consider quadratic programming problems in which the constant terms of the constraints, or the coefficients of the linear terms of the objective function, or both, are dependent on a parameter λ. The optimal solution of these problems is required for a given range of values of λ. The problem may be formulated as follows. Maximize f

$$f = (p + p_\lambda \lambda)'x - \tfrac{1}{2}x'Cx,$$
$$Ax \le b + b_\lambda \lambda,$$
$$x \ge 0, \qquad\qquad (1)$$

for $\lambda_l \le \lambda \le \lambda_u$, where λ_l and λ_u are given upper and lower bounds. Frequently we have $\lambda_l = 0$ and $\lambda_u \to \infty$. As before, it is required that the matrix C is positive semi-definite. Usually one of the vectors b_λ and p_λ will be zero.

The set-up tableau for this problem is given in tableau I of table 1. Since the constant terms of the constraints and the coefficients of the linear terms of the objective function depend linearly on λ, the column "Values of basic variables" consists of a part independent of λ and a part which should be multiplied by λ. Similarly the row of the objective function consists of two parts, one part given in a row indicated by F_c, which gives coefficients independent of λ, and one part in a row F_λ, consisting of coefficients which should be multiplied by λ. The value of F is a quadratic function of λ, the coefficients of which are given by the 2×2 matrix, which is zero in the set-up tableau.

The parametric method is started by determining the optimal solution for a fixed value of λ, for instance $\lambda = 0$ or $\lambda = \lambda_l$. For this purpose the simplex method, the dual method or any other method for quadratic programming may be used. The result is a standard tableau as given in tableau II of table 1. The vector v contains all dual basic variables and the vector y all primal basic variables; x and u are the vectors of primal

297

Table 1

Set-up and general standard tableau for a parametric quadratic programming problem

| Tableau | Basic variables | Values basic variables | | x | u |
		c-term	λ-term		
I	F_c	0	0	$-p'$	$-b'$
	F_λ	0	0	$-p'_\lambda$	$-b'_\lambda$
	v	$-p$	$-p_\lambda$	$-C$	$-A'$
	y	b	b_λ	A	0
II	F_c	S_{cc}	$S_{c\lambda}$	q'	$-r'$
	F_λ	$S_{\lambda c}$	$S_{\lambda\lambda}$	q'_λ	$-r'_\lambda$
	v	q	q_λ	$-Q$	$-P'$
	y	r	r_λ	P	$-R$

and dual nonbasic variables. Since the tableau is optimal for $\lambda = \lambda_l$, we have

$$q_i + q_{\lambda i}\lambda_l \geq 0,$$
$$r_i + r_{\lambda i}\lambda_l \geq 0,$$

for all i.

Now the parametric procedure is started. We determine the upper bound of λ for which the present solution is optimal. If λ is increasing, basic variables will decrease for negative elements $q_{\lambda i}$ and $r_{\lambda i}$. Hence the upper bound of λ is found from

$$\bar{\lambda} = \operatorname*{Min}_i (q_i/- q_{\lambda i}|q_{\lambda i} < 0, r_i/- r_{\lambda i}|r_{\lambda i} < 0). \tag{2}$$

The present solution is optimal for $\lambda_l \leq \lambda \leq \bar{\lambda}$.

Let us assume that $\bar{\lambda}$ is associated with v_b, so that

$$q_b + q_{\lambda b}\bar{\lambda} = 0. \tag{3}$$

Hence v_b is the first basic variable to become negative for increasing λ. Since for $\lambda = \bar{\lambda}$, the value of v_b as well as the element in the F-row and in the column of x_b is zero, x_b may be introduced into the basis. Let us consider the case that $-q_{bb} \neq 0$. The elements in the rows of F for $\lambda = \bar{\lambda}$ and of v_b and in the columns of values of basic variables for $\lambda = \bar{\lambda}$ and of x_b are in this case:

B.v.	V.b.v.	x_b
F	$s_c + 2s_{c\lambda}\bar\lambda + s_{\lambda\lambda}\bar\lambda^2$	0
v_b	0	$-q_{bb}$

$$(4)$$

It is obvious that if $-q_{bb}$ is used as a pivot, the resulting tableau will have the same values of basic variables, including the objective function, for $\lambda = \bar\lambda$. The values of x_b will be in this tableau

$$r_b{}^* + r_{\lambda b}^*\lambda = -q_{bb}^{-1}q_b - q_{bb}^{-1}q_{\lambda b}\lambda \qquad (5)$$

which is zero for $\lambda = \bar\lambda$ but positive for $\lambda > \bar\lambda$, since $-q_{bb}^{-1}q_{\lambda b} > 0$. Hence $\bar\lambda$ is a lower bound for the new tableau. We may proceed by finding an upper bound for this tableau.

If $-q_{bb} = 0$, the situation is somewhat more complicated. Table 2 gives a representation of such a tableau. Since $-q_{bb} = 0$, all elements in the same row and the same column of the matrix-Q are zero. In a third column of values of basic variables, the values are given for $\lambda = \bar\lambda$; the same is true for a third row indicated by $\bar F$. If x_b is increased the value of F is for $\lambda = \bar\lambda$:

$$F = \bar s_{\lambda\lambda} + 0x_b - v_b x_b = \bar s_{\lambda\lambda} + 0x_b - (0 + 0x_b)x_b = \bar s_{\lambda\lambda}.$$

Hence the value of the objective function does not change. Neither do the values of dual basic variables, since the elements in their rows in the columns of x_b are zero. The values of primal basic variables change,

Table 2

A double nonstandard iteration in parametric quadratic programming

| Bas. var. | Values basic var. | | | x_b | x_j | u_r | u_i |
	c-term	λ-term	$\lambda = \bar\lambda$				
F_c	s_c	$s_{c\lambda}$	$\bar s_c$	q_b	q_j	$-r_r$	$-r_i$
F_λ	$s_{\lambda c}$	$s_{\lambda\lambda}$	$\bar s_\lambda$	$q_{\lambda b}$	$q_{\lambda j}$	$-r_{\lambda r}$	$-r_{\lambda i}$
$\bar F$	$\bar s_c$	$\bar s_\lambda$	$\bar s_{\lambda\lambda}$	0	$\bar q_j$	$-\bar r_r$	$-r_i$
v_b	q_b	$q_{\lambda b}$	0	0	0	$-p_{rb}$	$-p_{ib}$
v_j	q_j	$q_{\lambda j}$	$\bar q_j$	0	$-q_{ji}$	$-p_{rj}$	$-p_{ij}$
y_r	r_r	$r_{\lambda r}$	$\bar r_r$	p_{rb}	p_{ri}	$-r_{rr}$	$-r_{ri}$
y_i	r_i	$r_{\lambda i}$	$\bar r_i$	p_{ib}	p_{ij}	$-r_{ir}$	$-r_{ii}$

however; since they should stay nonnegative while increasing x_b, ratio's should be determined as follows:

$$\text{Min}_{i} \ (\bar{r}_i/p_{ib}|p_{ib} > 0), \tag{6}$$

where $\bar{r}_i$ is the value of y_i for $\lambda = \bar{\lambda}$.

Let the minimum be found in the row of y_r. Then p_{rb} is used as a pivot for a transformation in which x_b replaces y_r as a basic variable. Only the primal basic variables have changed, but they remain nonnegative for $\lambda = \bar{\lambda}$; the resulting tableau is a nonstandard one. Then another iteration is performed with $-p_{rb}$ as a pivot, so that u_r replaces v_b as a basic variable; this results in a standard tableau. In the last iteration none of the values of basic variables have changed for $\lambda = \bar{\lambda}$, since the value of v_b is then zero. Now we have arrived at a new solution which is still optimal for $\lambda = \bar{\lambda}$, but which has $\bar{\lambda}$ as its upper bound for λ, since we have for u_r:

$$u_r = -p_{rb}^{-1} q_b - p_{rb}^{-1} q_{\lambda b} \lambda. \tag{7}$$

Since p_{rb} was a pivot in the first iteration, it is positive, so that $-p_{rb}^{-1} q_{\lambda b} > 0$. If $-r_{rr} \neq 0$, principal pivots and symmetric tableaux can be maintained by pivoting first on $-r_{rr}$ and then on the element in the row of v_b and the column of x_b. If $-r_{rr} = 0$, one iteration is sufficient to compute all desired elements of the next standard tableau.

Note that the case in which the upper bound of λ is found in a row of a dual variable and in which the principal element is zero is very similar to linear programming with a parametric objective function. In the latter case the ratio's are not taken in the values of basic variables, but in the rows of F_c and F_λ, but because of the symmetry properties, the same ratio's will be found.

If $\bar{\lambda}$ in (2) is connected with a primal basic variable, say y_b, the procedure is similar to the one described for a dual basic variable. If the principal element $-r_{bb} \neq 0$, it should be used as a pivot. If $-r_{bb} = 0$, the corresponding dual variable u_b should be introduced into the basis, replacing the variable connected with

$$\text{Min}_{j} \ (\bar{q}_j/-p_{bj}|-p_{bj} > 0), \tag{8}$$

where $\bar{q}_j$ is the value of v_j evaluated at $\lambda = \bar{\lambda}$.

Let this variable be v_r, so that $-p_{br}$ is the pivot. Another iteration in which x_r enters the basis, replacing y_b results in a standard tableau which is optimal for a new range of λ, with $\bar{\lambda}$ as its lower bound. The case

$- r_{bb} = 0$ is similar to linear programming with parametric constant terms. In the linear programming case we have compared ratio's as in the dual method for linear programming, which amounts to taking ratio's of elements in the F-row for $\lambda = \bar{\lambda}$ and in the y_b-row. The same ratio's are found as in (8).

As an example of application of the described method, the following problem will be solved. Maximize f for $0 \leq \lambda \leq \infty$

$$f = 2x_1 + 2x_2 - 2x_1^2 + 2x_1 x_2 - 2x_2^2,$$

$$x_1 + x_2 \leq \lambda,$$

$$- x_1 + 6x_2 \leq 2. \tag{9}$$

Note that this is just a parametric version of a problem treated before. An application of parametric programming to this problem may be viewed as a solution method for the same problem with $\lambda = 1$; this is Houthakker's capacity method which adds a "capacity" constraint $e'x \leq \lambda$ to the problem and solves it for $\lambda \to \infty$; if such a constraint is present in the problem, it is not necessary to add it.

Tableau 0 of table 3 gives the set-up tableau for the problem. First an optimal solution for $\lambda = 0$ should be generated. This can be done in two iterations, by first introducing the primal nonbasic variable with the most negative dual variable into the basis, for which x_1 may be taken, replacing y_1; then in another iteration u_1 enters the basis replacing v_1. Since the element of tableau 0 in the row of v_1 and the column of x_1 is nonzero, this may be used as a principal pivot in a first iteration; in the second iteration u_1 replaces y_1. The result is in both cases tableau 2.

Now the parametric variation can start. In tableau 2 the smallest ratio is found in the row of v_2. Since this row has a nonzero principal element, it should be used as a pivot. In tableau 3, the smallest ratio is found in the row of y_2. This row has a nonzero principal element, which is used as a pivot. The result is tableau 4, which has as its single upper bound $\lambda = 1\frac{17}{62}$, which is found in the row of u_1. This row has a principal element, which may be used as a pivot. Since y_1 is entering the basis again, the λ-term column consists of a unit vector, so that an increase of λ does not affect the solution any more.

Figure 1 gives a geometrical interpretation of the successive solutions. Point A, the origin, is optimal for $\lambda = 0$, see tableau 2. For $0 \leq \lambda \leq \frac{4}{5}$, the solution moves from A to B, see tableau 3. For $\frac{4}{5} \leq \lambda \leq 1\frac{17}{62}$, the solution moves from B to D, see tableau 4. Point C is the solution for $\lambda = 1$,

Table 3

An example of parametric quadratic programming

Tabl.	Basic variables	Values basic variables		Ratio				
		c-term	λ-term					
					x_1	x_2	u_1	u_2
0	F_c	0	0		-2	-2	0	-2
	F_λ	0	0		0	0	-1	0
	v_1	-2	0		$\underline{-4}$	2	-1	1
	v_2	-2	0		$\underline{2}$	-4	-1	-6
	y_1	0	1		1	1	0	0
	y_2	2	0		-1	6	0	0
					v_1	x_2	u_1	u_2
1	F_c	1	0		$-\frac{1}{2}$	-3	$\frac{1}{2}$	$-2\frac{1}{2}$
	F_λ	0	0		0	0	-1	0
	x_1	$\frac{1}{2}$	0		$-\frac{1}{4}$	$-\frac{1}{2}$	$\frac{1}{4}$	$-\frac{1}{4}$
	v_2	-3	0		$\frac{1}{2}$	-3	$-1\frac{1}{4}$	$-5\frac{1}{4}$
	y_1	$-\frac{1}{2}$	1		$\frac{1}{4}$	$1\frac{1}{2}$	$-\frac{1}{4}$	$\frac{1}{4}$
	y_2	$2\frac{1}{2}$	0		$-\frac{1}{4}$	$5\frac{1}{2}$	$-\frac{1}{4}$	$-\frac{1}{4}$
					v_1	x_2	y_1	u_2
2	F_c	0	2		0	0	2	-2
	F_λ	2	-4		-1	-6	-4	-1
	x_1	0	1		0	1	1	0
	v_2	0	-6	0	-1	$\underline{-12}$	-6	-7
	u_1	2	-4	$\frac{1}{2}$	-1	$\overline{-6}$	-4	-1
	y_2	2	1		0	7	1	0
					v_1	v_2	y_1	u_2
3	F_c	0	2		0	0	2	-2
	F_λ	2	-1		$-\frac{1}{2}$	$-\frac{1}{2}$	-1	$2\frac{1}{2}$
	x_1	0	$\frac{1}{2}$		$-\frac{1}{12}$	$\frac{1}{12}$	$\frac{1}{2}$	$-\frac{7}{12}$
	x_2	0	$\frac{1}{2}$		$\frac{1}{12}$	$-\frac{1}{12}$	$\frac{1}{2}$	$\frac{7}{12}$
	u_1	2	-1	2	$-\frac{1}{2}$	$-\frac{1}{2}$	-1	$2\frac{1}{2}$
	y_2	2	$-2\frac{1}{2}$	$\frac{4}{5}$	$-\frac{7}{12}$	$\frac{7}{12}$	$-2\frac{1}{2}$	$\underline{-4\frac{1}{12}}$
					v_1	v_2	y_1	y_2
4	F_c	$-\frac{48}{49}$	$3\frac{11}{49}$		$\frac{2}{7}$	$-\frac{2}{7}$	$3\frac{11}{49}$	$-\frac{24}{49}$
	F_λ	$3\frac{11}{49}$	$-2\frac{26}{49}$		$-\frac{6}{7}$	$-\frac{1}{7}$	$-2\frac{26}{49}$	$\frac{30}{49}$
	x_1	$-\frac{2}{7}$	$\frac{6}{7}$		0	0	$\frac{6}{7}$	$-\frac{1}{7}$
	x_2	$\frac{2}{7}$	$\frac{1}{7}$		0	0	$\frac{1}{7}$	$\frac{1}{7}$
	u_1	$3\frac{11}{49}$	$-2\frac{26}{49}$	$1\frac{17}{62}$	$-\frac{6}{7}$	$-\frac{1}{7}$	$\underline{-2\frac{26}{49}}$	$\frac{30}{49}$
	u_2	$-\frac{24}{49}$	$\frac{30}{49}$		$\frac{1}{7}$	$-\frac{1}{7}$	$\frac{30}{49}$	$-\frac{12}{49}$

Table 3   (*continued*)

Tabl.	Basic variables	Values basic variables		Ratio	v_1	v_2	u_1	y_2
		c-term	λ-term					
5	F_c	$3\frac{4}{31}$	0		$-\frac{25}{31}$	$-\frac{29}{62}$	$1\frac{17}{62}$	$\frac{9}{31}$
	F_λ	0	0		0	0	-1	0
	x_1	$\frac{25}{31}$	0		$-\frac{9}{31}$	$-\frac{3}{62}$	$\frac{21}{62}$	$\frac{2}{31}$
	x_2	$\frac{29}{62}$	0		$-\frac{3}{62}$	$-\frac{1}{124}$	$\frac{7}{124}$	$\frac{11}{62}$
	y_1	$-1\frac{17}{62}$	1		$\frac{21}{62}$	$\frac{7}{124}$	$-\frac{49}{124}$	$-\frac{15}{62}$
	u_2	$\frac{9}{31}$	0		$-\frac{2}{31}$	$-\frac{11}{62}$	$\frac{15}{62}$	$-\frac{3}{31}$

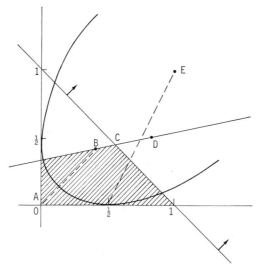

Fig. 1.

which was found previously. Point D remains the optimal solution for $\lambda > 1\frac{17}{62}$. Figure 2 gives the relation between F and λ; it is observed that F is an increasing function of λ until $\lambda = 1\frac{17}{62}$. Between two critical values of λ, F is a quadratic function of λ. Figure 3 gives $dF/d\lambda$ as a function of λ; this is, as in linear programming, a nonincreasing function of λ. In linear programming, this function is discontinuous, but in quadratic programming it may be a piecewise linear continuous function as it is in this case.

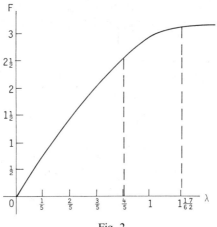

Fig. 2.

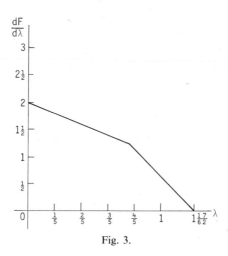

Fig. 3.

10.2. The asymmetric and second symmetric variants of parametric quadratic programming

Let us again consider the general parametric quadratic programming problem: Maximize f

$$f = (p + p_\lambda \lambda)' x - \tfrac{1}{2} x' C x$$

$$Ax \leq b + b_\lambda \lambda,$$

$$x \geq 0, \tag{10}$$

for $0 \leq \lambda \leq \infty$. Now λ may be interpreted as an additional x-variable, say the first one. The problem (10) may therefore be written as: Maximize f

$$f = [0 \quad p'] \begin{bmatrix} \lambda \\ x \end{bmatrix} - \tfrac{1}{2} [\lambda \quad x'] \begin{bmatrix} 0 & -p'_\lambda \\ -p_\lambda & -C \end{bmatrix} \begin{bmatrix} \lambda \\ x \end{bmatrix},$$

$$[-b_\lambda \quad A] \begin{bmatrix} \lambda \\ x \end{bmatrix} \leq b,$$

$$\lambda, x \geq 0. \tag{11}$$

A set-up tableau for this problem is constructed in the usual manner. Then parametric methods may be developed as follows. First the problem is solved by an application of any quadratic programming method, but λ is kept as a nonbasic variable and its corresponding dual variable as a basic variable. An optimal solution of the problem is obtained for $\lambda = 0$. Now the parametric variation is started by increasing λ, keeping values of primal and dual basic variables nonnegative. The resulting changes of basis may be performed in three ways.

Firstly, the first symmetric variant of the simplex method for quadratic programming may be followed by increasing λ as if it were a parameter; any primal or dual variables which block the increase of λ are pivoted out using principal pivots. Secondly, the asymmetric variant may be employed by introducing λ into the basis, where it replaces the primal or dual variable which first becomes negative; thereafter the corresponding dual or primal variable of the variable which just left the basis is introduced into the basis, and so on, until a standard tableau or an infinite solution is found. Thirdly, the second symmetric variant may be followed, which has the same initial iteration as the asymmetric variant but which right after this performs another iteration, which brings the tableau into standard form again. The resulting tableau may be infeasible, but if so, it is feasible for some negative value of the dual variable of λ. This negative value is decreased by pivoting out primal or dual basic variables until a solution is obtained which is nonnegative for a zero value of the dual variable of λ.

An example is used to explain these ideas in a more detailed manner. The same problem is considered as in the previous section. The set-up tableau for this problem is found as indicated by (11), see tableau 0 of table 4. The optimal solution for $\lambda = 0$ is found by the symmetric variant of the simplex method in two iterations, see tableau 2.

First, the method corresponding with the first symmetric variant of the

Table 4

An alternative formulation of the first symmetric and the asymmetric variant of parametric quadratic programming

Tableau	Bas. var.	V.b.v.	Ratio	λ	x_1	x_2	u_1	u_2
0	F	0		0	-2	-2	0	-2
	v_λ	0		0	0	0	1	0
	v_1	-2		0	-4	2	-1	-1
	v_2	-2		0	2	-4	1	-6
	y_1	0		-1	1	1	0	0
	y_2	2		0	-1	6	0	0
				λ	v_1	x_2	u_1	u_2
1	F	1		0	$-\frac{1}{2}$	-3	$\frac{1}{2}$	$-2\frac{1}{2}$
	v_λ	0		0	0	0	1	0
	x_1	$\frac{1}{2}$		0	$-\frac{1}{4}$	$-\frac{1}{2}$	$\frac{1}{4}$	$-\frac{1}{4}$
	v_2	-3		0	$\frac{1}{2}$	-3	$-1\frac{1}{2}$	$-5\frac{1}{2}$
	y_1	$-\frac{1}{2}$		-1	$\frac{1}{4}$	$1\frac{1}{2}$	$-\frac{1}{4}$	$\frac{1}{4}$
	y_2	$2\frac{1}{2}$		0	$-\frac{1}{4}$	$5\frac{1}{2}$	$\frac{1}{4}$	$-\frac{1}{4}$
				λ	v_1	x_2	y_1	u_2
2	F	0		-2	0	0	2	-2
	v_λ	-2	$\frac{1}{2}$	-4	1	6	4	1
	x_1	0		-1	0	1	1	0
	v_2	0	0	6	-1	-12	-6	-7
	u_1	2	$\frac{1}{2}$	4	-1	-6	-4	-1
	y_2	2		-1	0	7	1	0
				λ	v_1	x_2	y_1	u_2
3a, c	F	0		-2	0	0	2	-2
	v_λ	-2		-1	$\frac{1}{2}$	$\frac{1}{2}$	1	$-2\frac{1}{2}$
	x_1	0		$-\frac{1}{2}$	$-\frac{1}{12}$	$\frac{1}{12}$	$\frac{1}{2}$	$-\frac{7}{12}$
	x_2	0		$-\frac{1}{2}$	$\frac{1}{12}$	$-\frac{1}{12}$	$\frac{1}{2}$	$\frac{7}{12}$
	u_1	2	2	1	$-\frac{1}{2}$	$-\frac{1}{2}$	-1	$2\frac{1}{2}$
	y_2	2	$\frac{4}{5}$	$2\frac{1}{2}$	$-\frac{7}{12}$	$\frac{7}{12}$	$-2\frac{1}{2}$	$-4\frac{1}{12}$
				λ	v_1	v_2	y_1	y^2
4a	F	$-\frac{48}{49}$		$-3\frac{11}{49}$	$\frac{2}{7}$	$-\frac{2}{7}$	$3\frac{11}{49}$	$-\frac{24}{49}$
	v_λ	$-3\frac{11}{49}$		$-2\frac{26}{49}$	$\frac{6}{7}$	$\frac{1}{7}$	$2\frac{26}{49}$	$-\frac{30}{49}$
	x_1	$-\frac{2}{7}$		$-\frac{6}{7}$	0	0	$\frac{6}{7}$	$-\frac{1}{7}$
	x_2	$\frac{2}{7}$		$-\frac{1}{7}$	0	0	$\frac{1}{7}$	$\frac{1}{7}$
	u_1	$3\frac{11}{49}$	$1\frac{17}{62}$	$2\frac{26}{49}$	$-\frac{6}{7}$	$-\frac{1}{7}$	$-2\frac{26}{49}$	$\frac{30}{49}$
	u_2	$-\frac{24}{49}$		$-\frac{30}{49}$	$\frac{1}{7}$	$-\frac{1}{7}$	$\frac{30}{49}$	$-\frac{12}{49}$

Table 4   (*continued*)

Tableau	Bas. var.	V.b.v.	Ratio	λ	v_1	v_2	u_1	y_2
5a	F	$3\frac{4}{31}$		0	$-\frac{25}{31}$	$-\frac{29}{62}$	$1\frac{17}{62}$	$\frac{9}{31}$
	v_λ	0		0	0	0	1	0
	x_1	$\frac{9}{31}$		0	$-\frac{3}{31}$	$-\frac{3}{62}$	$\frac{21}{62}$	$\frac{2}{31}$
	x_2	$\frac{29}{62}$		0	$-\frac{3}{62}$	$-\frac{1}{124}$	$\frac{7}{124}$	$\frac{11}{62}$
	y_1	$-\frac{17}{62}$		-1	$\frac{21}{62}$	$\frac{7}{124}$	$-\frac{49}{124}$	$-\frac{15}{62}$
	u_2	$\frac{9}{31}$		0	$-\frac{2}{31}$	$-\frac{11}{62}$	$\frac{15}{62}$	$-\frac{3}{31}$

				v_2	v_1	x_2	y_1	u_2
3b	F	0		$\frac{1}{3}$	$-\frac{1}{3}$	-4	0	$-4\frac{1}{3}$
	v_λ	-2	1	$\frac{2}{3}$	$\frac{1}{3}$	-2	0	$-3\frac{2}{3}$
	x_1	0		$\frac{1}{6}$	$-\frac{1}{6}$	-1	0	$-1\frac{1}{6}$
	λ	0		$\frac{1}{6}$	$-\frac{1}{6}$	-2	-1	$-1\frac{1}{6}$
	u_1	2	1	$-\frac{2}{3}$	$-\frac{1}{3}$	2	0	$3\frac{2}{3}$
	y_2	2	$\frac{2}{5}$	$\frac{1}{6}$	$-\frac{1}{6}$	5	0	$-1\frac{1}{6}$

				v_2	v_1	y_2	y_1	u_2
4b	F	$1\frac{3}{5}$		$\frac{7}{15}$	$-\frac{7}{15}$	$\frac{4}{5}$	0	$-5\frac{4}{15}$
	v_λ	$-1\frac{1}{5}$		$\frac{11}{15}$	$\frac{4}{15}$	$\frac{2}{5}$	0	$-4\frac{2}{15}$
	x_1	$\frac{2}{5}$		$\frac{1}{5}$	$-\frac{1}{5}$	$\frac{1}{5}$	0	$-1\frac{2}{5}$
	λ	$\frac{4}{5}$		$\frac{7}{30}$	$-\frac{7}{30}$	$\frac{2}{5}$	-1	$-1\frac{19}{30}$
	u_1	$1\frac{1}{5}$	$\frac{9}{31}$	$-\frac{11}{15}$	$-\frac{4}{15}$	$-\frac{2}{5}$	0	$4\frac{2}{15}$
	x_2	$\frac{2}{5}$		$\frac{1}{30}$	$-\frac{1}{30}$	$\frac{1}{5}$	0	$-\frac{7}{30}$

				v_2	v_1	y_2	y_1	u_1
5b	F	$3\frac{4}{31}$		$-\frac{29}{62}$	$-\frac{25}{31}$	$\frac{9}{31}$	0	$1\frac{17}{62}$
	v_λ	0		0	0	0	0	1
	x_1	$\frac{25}{31}$		$-\frac{3}{62}$	$-\frac{9}{31}$	$\frac{2}{31}$	0	$\frac{21}{62}$
	λ	$1\frac{17}{62}$		$-\frac{7}{124}$	$-\frac{21}{62}$	$\frac{15}{62}$	-1	$\frac{49}{124}$
	u_2	$\frac{9}{31}$		$-\frac{11}{62}$	$-\frac{2}{31}$	$-\frac{3}{31}$	0	$\frac{15}{62}$
	x_2	$\frac{29}{62}$		$-\frac{1}{124}$	$-\frac{3}{62}$	$\frac{11}{62}$	0	$\frac{7}{124}$

simplex method is considered; this method turns out to be an alternative formulation of the method described in the previous section. In tableau 2, λ is considered the incoming variable. The minimum ratio is found for v_2, so that this variable should leave the basis. Since the corresponding diagonal element is -12 and hence nonzero, it may be used as a pivot; this results in tableau 3a. In this tableau λ still is the incoming variable; the minimum ratio is found in the row of y_2, so that this variable should leave the basis. Since there is a nonzero diagonal element in this row, u_2

enters the basis. Tableau 4a results, which, if λ is considered as the incoming variable, has only one ratio, namely that in the row of u_1, so that this variable leaves the basis y_1 replaces u_1 and tableau 5a results. In this tableau no leaving basic variable can be found since λ can be increased indefinitely. This is therefore the optimal solution for $\lambda \geq 1\frac{17}{62}$. Note that all tableaux of table 4 are the same as the corresponding ones of table 3, apart from the λ-column and the v_λ row, which are equal to minus the λ-column and F_λ-row of the tableaux of table 3. Table 4 can therefore be considered an alternative formulation of table 3.

In the asymmetric method the changes of basic variables occur in a way similar to that of the asymmetric variant of the simplex method. This means that, proceeding from tableau 2 of table 4, λ enters the basis, and replaces v_2. The result is tableau 3b. Now the corresponding variable of the variable which just left the basis is introduced; this is x_2. Taking ratio's we find that the leaving basic variable is y_2. This results in tableau 4b, where u_2 is the new basic variable. u_1 is found as the leaving basic variable. The resulting tableau is tableau 5b, which has y_1 as its new basic variable. No leaving basic variable can be found, which means that y_1 can be increased indefinitely. The solution of the last tableau is optimal for $\lambda \geq 1\frac{17}{62}$; the solution of 3b and 4b is optimal for $\lambda = 0$ and $\lambda = \frac{4}{5}$. Solutions for intermediate values of λ are found by linear interpolation between the adjacent solutions.

As for the symmetric and the asymmetric variants of the simplex method, it can be proved that the symmetric and asymmetric methods for parametric quadratic programming lead to the same sequence of solutions and can therefore be considered to be equivalent.

It is interesting to observe that the asymmetric method is almost the same as the long form of Wolfe's quadratic programming method, see section 11.3 and [80]. The difference is that, after an optimal solution for $\lambda = 0$ has been reached, Wolfe adds an objective function, which maximizes λ and only allows variables into the basis which do not have a corresponding variable which is basic. Since there is only one pair of corresponding nonbasic variables, the corresponding variable of the variable which just left the basic must enter the basis.

In the second symmetric variant, the first iteration is the same as that in the asymmetric method, but immediately after that another iteration follows, which brings the tableau back into standard form. This means that the underlined element -2 of tableau 3b should be used as a pivot. The same result is obtained if the underlined element -1 of tableau 3a is

Table 5

The second symmetric method for parametric quadratic programming

Tabl.	B.v.	V.b.v.	Ratio	v_λ	v_1	v_2	y_1	u_2
4c	F	4		2	-1	-1	0	3
	λ	2		-1	$-\frac{1}{2}$	$-\frac{1}{2}$	-1	$2\frac{1}{2}$
	x_1	1		$-\frac{1}{2}$	$-\frac{1}{3}$	$-\frac{1}{6}$	0	$\frac{2}{3}$
	x_2	1		$-\frac{1}{2}$	$-\frac{1}{6}$	$-\frac{1}{3}$	0	$1\frac{5}{6}$
	u_1	0		1	0	0	0	0
	y_2	-3	$-1\frac{1}{5}$	$2\frac{1}{2}$	$\frac{2}{3}$	$1\frac{5}{6}$	0	$-10\frac{1}{3}$

				v_λ	v_1	v_2	y_1	y_2
5c	F	$3\frac{4}{31}$		$-1\frac{17}{62}$	$-\frac{25}{31}$	$-\frac{29}{62}$	0	$\frac{9}{31}$
	λ	$1\frac{17}{62}$		$-\frac{49}{124}$	$-\frac{21}{62}$	$-\frac{7}{124}$	-1	$\frac{15}{62}$
	x_1	$\frac{25}{31}$		$-\frac{21}{62}$	$-\frac{9}{31}$	$-\frac{3}{62}$	0	$\frac{2}{31}$
	x_2	$\frac{29}{62}$		$-\frac{7}{124}$	$-\frac{3}{62}$	$-\frac{1}{124}$	0	$\frac{11}{62}$
	u_1	0		1	0	0	0	0
	u_2	$\frac{9}{31}$		$-\frac{15}{62}$	$-\frac{2}{31}$	$-\frac{11}{62}$	0	$-\frac{3}{31}$

used as a pivot. Since in the latter case only standard tableaux are used, tableau 4c of table 5 is generated in this manner. The basic variables of this tableau are nonnegative for $v_\lambda = -1\frac{1}{5}$. This variable is parametrically increased to zero. u_2 enters the basis, replacing y_2. In tableau 5c a solution with nonnegative basic variables is found. In this solution, y_1 may be increased indefinitely, thus increasing λ indefinitely without changing the other basic variables. This means that this solution is optimal for $\lambda \geq 1\frac{17}{62}$. Optimal solutions for lower values of λ can be found by giving v_λ negative values in the previous tableaux.

CHAPTER 11

PARAMETRIC METHODS FOR QUADRATIC PROGRAMMING

11.1. Introduction

Parametric methods may be used to solve quadratic programming problems. The basic concept is to modify a problem by the inclusion of a certain parameter. An initial solution to the problem is available for one value of the parameter, while the optimal solution of the original problem is given for another value of the parameter. By tracing the optimal solution from the first value to the second one, the optimal solution to the original problem is found.

This may be illustrated by Houthakker's capacity method[57], in which the constraint

$$x_1 + x_2 + \dots + x_n \leq \lambda \tag{1}$$

is added to the problem. If the constraints are of the form

$$Ax \leq b$$

with $b \geq 0$, an initial solution is easily available; then λ is varied upwards until an optimal solution is found for $\lambda \to \infty$, which should be the solution of the original problem.

The parameter λ may be introduced into the problem in a number of ways. In the capacity method it is introduced as the constant term of an additional constraint, but in the primal-dual method it is introduced as a coefficient of artificial variables. Though in principle parametric methods are general in the sense that they can be used to solve any type of convex quadratic programming problem, most are suitable only for a certain type of problem since for other problems first an initial solution of the appropriate type should be obtained.

However, some methods are general in the sense that no initial solution of a certain type is required. These are Lemke's method and Cottle's index method. This does not mean that these methods are preferable to other methods. After all, in linear programming, Dantzig's self-dual parametric method for linear programming, of which Lemke's method is the generalization for quadratic programming, is not more attractive than

310

the simplex method for linear programming in two phases. The simplex method for quadratic programming, coupled with a method for finding an initial feasible solution to the constraints, is a perfectly general method which may be more attractive.

The content of this chapter is arranged as follows. Section 11.2 deals with the generalisation of the primal-dual, gradient and capacity methods for linear programming to quadratic programming problems. Section 11.3 deals with parametric and dual equivalence in quadratic programming and shows that Wolfe's long form is equivalent to a variant of the capacity method. Section 11.4 treats the generalisation of the self-dual parametric method for linear programming; equivalent methods have been proposed by Lemke and Graves. Cottle's method, which is described in section 11.5, combines a parametric approach with a different criterion of convergence.

11.2. The primal-dual, gradient and capacity methods for quadratic programming

All of the methods mentioned in chapter 6 may be generalized for quadratic programming. As an example, let us take the primal-dual method. Consider the problem: Minimize f

$$f = p'x + \tfrac{1}{2}x'Cx,$$
$$Ax \geq b,$$
$$x \geq 0. \tag{2}$$

As in linear programming, the extended problem is formulated:

$$f = p'x + \lambda e'z + \tfrac{1}{2}x'Cx, \tag{3a}$$
$$Ax - y + z = b, \tag{3b}$$
$$x, y, z \geq 0. \tag{3c}$$

The z-variables can be eliminated from (3a) by adding to it $\lambda e'$ times (3b). This results in

$$f = \lambda e'b + (p' - \lambda e'A)x + \lambda e'y + \tfrac{1}{2}x'Cx. \tag{4}$$

The equations for the set-up tableau are now

$$2\lambda e'b = -(p' - \lambda e'A)x - \lambda e'y - b'u + F, \tag{5a}$$
$$-p + \lambda A'e = Cx - A'u + v, \tag{5b}$$

$$- \lambda e = u + w, \tag{5c}$$

$$b = Ax - y + z. \tag{5d}$$

Table 1 gives the set-up tableau. If the elements of p and b are nonnegative, the solution is optimal for $\lambda = 0$ (note that this is a minimization problem, so that dual variables should have nonpositive values).

The optimal solution to the problem is found by varying λ parametrically from 0 to ∞. For $\lambda \to \infty$, the z-variables should be nonbasic or there is no feasible solution.

Table 1

Set-up tableau for the primal-dual method for quadratic programming

Basic variables	Values basic variables		Nonbasic variables		
	c-term	λ-term	x	y	u
F_c	0	$e'b$	$-p'$	0	$-b'$
F_λ	$e'b$	0	$e'A$	$-e'$	0
v	$-p$	$A'e$	C	0	$-A'$
w	0	$-e$	0	0	I
z	b	0	A	$-I$	0

There are three variants for parametric quadratic programming, namely the first symmetric variant, the asymmetric variant and the second symmetric variant. Furthermore, as will be indicated in the next section, there are three methods equivalent with the primal-dual method. Hence there are a total of twelve methods which are related to the primal-dual method.

Rather than go into details, we shall argue that, since the primal-dual method adds a number of rows and variables to the problem it increases the size of the problem considerably and is therefore less suitable. In linear programming, the primal-dual method only added variables but not equations, and additional variables are not so difficult to handle. Additional equations or rows as in the quadratic programming case, increase memory capacity substantially and methods which generate a number of additional rows are therefore less suitable.

The constraints (5c) are in fact upper bound constraints and can therefore be handled implicitly. The columns of the y-variables have a very simple structure and may therefore be treated in a similar manner. This, however, complicates the methods considerably and makes them less attractive.

In quadratic programming, the gradient vector of the objective function f

$$f = p'x - \tfrac{1}{2}x'Cx \tag{6}$$

is

$$\partial f / \partial x = p - Cx. \tag{7}$$

Hence the gradient is no longer constant as in linear programming. As in linear programming, we might constrain the solution by a parametric constraint involving the gradient

$$x \leq \lambda(p - Cx), \tag{8}$$

but this is a nonlinear parametric constraint which is difficult to handle.

Instead of this, the gradient at the initial solution $x = 0$ may be taken. This results in the constraint $x \leq \lambda p$, which can be handled easily. But since additional constraints are created, the method has the same disadvantages as the primal-dual method. This is, of course, true also for all methods equivalent with this gradient method. As in the case of the primal-dual method for quadratic programming, a total of 12 methods can be distinguished. Hence none of the 24 methods discussed in this section is very useful.

As indicated before, Houthakker's capacity method adds the constraint $\Sigma x_i \leq \lambda$ to the problem, then finds an optimal solution for $\lambda = 0$, after which λ is increased until $\lambda \to \infty$. An example has been given in the previous chapter, which included the constraint $x_1 + x_2 \leq \lambda$. The capacity method has the advantage that it adds only one row and one column to the problem. The extension of the capacity method for linear programming to that of quadratic programming is straightforward. Parametrically and dually equivalent methods can be found by making use of parametric and dual equivalence which is treated in the next section.

11.3. Parametric equivalence in quadratic programming

A key concept is the notion of parametric equivalence in quadratic programming which is similar to the same concept in linear programming; this is the subject of this section.

Let us consider the following problem which will be called problem I. Maximize f for $0 \leq \lambda \leq \infty$,

$$f_1 = (p - \lambda q)'x - \tfrac{1}{2}x'Cx,$$
$$Ax \leq b,$$
$$x \geq 0. \tag{9}$$

Consider also the following problem, to be called problem II. Maximize f for $0 \leq \mu \leq \infty$,

$$f_{\text{II}} = p'x - \tfrac{1}{2}x'Cx,$$

$$Ax \leq b,$$

$$q'x \leq \mu,$$

$$x \geq 0. \tag{10}$$

The equations for the set-up tableau of problem I are

$$0 = -(p - \lambda q)'x - b'u + F_1, \tag{11a}$$

$$-p + \lambda q = -Cx - A'u + v, \tag{11b}$$

$$b = Ax + y. \tag{11c}$$

For problem II these equations are

$$0 = -p'x - b'u - \mu u_\mu + F_{\text{II}}, \tag{12a}$$

$$-p = -Cx - A'u - qu_\mu + v, \tag{12b}$$

$$b = Ax + y, \tag{12c}$$

$$\mu = q'x + y_\mu. \tag{12d}$$

According to the Kuhn–Tucker conditions each problem has an optimal solution if the equations are satisfied and the variables have nonnegative values and meet the complementary slackness conditions.

Assume that problem I has an optimal solution for a given value of λ indicated by $\bar{\lambda}$; the corresponding values of the variables are indicated by $\bar{x}$, $\bar{y}$, $\bar{u}$ and $\bar{v}$. This solution will satisfy eqs. (11b) and (11c), be nonnegative, and have the complementary slackness property. Consider now the same solution as a solution of problem II, with $u_\mu = \bar{\lambda}$ and $\mu = q'\bar{x}$; the latter equation implies $y_\mu = 0$. Since this solution satisfies (12b), (12c) and (12d), is nonnegative and has the complementary slackness property (also $u_\mu y_\mu = 0$), it must be an optimal solution of problem II. The solution for $\lambda = 0$ corresponds with a solution of problem II with $q'\bar{x} \leq \mu$.

Note that in quadratic programming all variables are in general a function of λ; this is different in linear programming, where only the dual variables are a function of λ, while the primal variables are not. For increasing values of λ, the basic variables of the solution of problem I will change at a certain critical value of λ. This new basic solution of problem I will then also be a solution of problem II for a new value of μ, which will be lower. It can be shown that $q'x$ is a nonincreasing function of λ. Hence to the solutions of problem I for increasing values of λ, are also solutions of problem II for decreasing values of μ.

Alternatively, problem II may be taken as point of departure. Consider an optimal solution for a given value of μ indicated by $\bar{\mu}$. If y_μ is basic in this solution, then it must also be optimal for problem I for $\lambda = 0$. If y_μ is nonbasic, the solution must also be optimal for problem I for $\lambda = u_\mu$.

Because both problems have the same solutions for increasing values of λ and decreasing values of μ, they are said to be *parametrically equivalent*.

Note that the values of F are not the same for both problems; the following relation exists between the value of F for problem I, F_I and the value of F for problem II, F_{II} for the same solution:

$$F_{II} = F_I + \mu \bar{u}_\mu + \lambda q' \bar{x} = F_I + 2\mu\lambda. \tag{13}$$

This implies for the values of the objective function f_I and F_{II}:

$$f_{II} = f_I + \mu\lambda. \tag{14}$$

In economic terms, this is explained by the fact that in problem I the decisionmaker has to pay a price λ for the resource which is available in an unlimited quantity; in problem II the quantity limitation in the resource results in the same solution, but the decisionmaker does not have to pay for it.

Consider now the following pair of problems. Problem III: Maximize f for $0 \leq \lambda \leq \infty$,

$$f_{III} = p'x - \tfrac{1}{2}x'Cx,$$
$$Ax \leq b + \lambda a,$$
$$x \geq 0, \tag{15}$$

and problem IV: Maximize f for $0 \leq \mu \leq \infty$,

$$f_{IV} = p'x - \tfrac{1}{2}x'Cx - \mu x_\mu,$$
$$Ax - ax_\mu \leq b,$$
$$x \geq 0. \tag{16}$$

In problem III, the resources available are increased in proportions given by the vector a, while in problem IV a variable x_μ indicates the possibility of buying a "package" of resources with quantities indicated by the vector a at a price μ.

The equations for the set-up tableau of problem III are

$$0 = -p'x - (b + \lambda a)'u + F_{\text{III}}, \tag{17a}$$

$$-p = -Cx - A'u + v, \tag{17b}$$

$$b + \lambda a = Ax + y, \tag{17c}$$

and for problem IV,

$$0 = -p'x + \mu x_\mu - b'u + F_{\text{IV}}, \tag{18a}$$

$$-p = -Cx - A'u + v, \tag{18b}$$

$$\mu = a'u + v_\mu, \tag{18c}$$

$$b = Ax - ax_\mu + y. \tag{18d}$$

Consider an optimal solution of problem III for $\lambda = \bar{\lambda} \neq 0$. This solution must also be optimal for problem IV for $x_\mu = \bar{\lambda}$ and $\mu = a'\bar{u}$. For $\lambda = 0$, the optimal solution of problem III is also optimal for problem IV for $x_\mu = 0$ and $\mu \geq a'\bar{u}$. Conversely, an optimal solution of problem IV for $\mu = \bar{\mu}$ is optimal for problem III for $\lambda = \bar{x}_\mu$; if $\bar{v}_\mu$ is basic, then $\bar{x}_\mu$ is nonbasic and the solution of problem IV is optimal for problem III for $\lambda = 0$.

Also in this case, both problems are said to be parametrically equivalent. In this case the following relation exists between the values of F for the same solution:

$$F_{\text{IV}} = F_{\text{III}} - \mu x_\mu - \lambda a'u = F_{\text{III}} - 2\mu\lambda, \tag{19}$$

which implies for the values of the objective function

$$f_{\text{IV}} = f_{\text{III}} - \mu\lambda. \tag{20}$$

This can be made clear intuitively by pointing out that in problem III the decisionmaker does not have to pay for the increased amounts of resources, whereas in problem IV he has to pay.

The above relationships indicate how for the primal-dual, the gradient and the capacity method for quadratic programming a parametrically equivalent method may be designed. We shall not go into details here.

Two methods are said to be dually equivalent if the roles of primal and dual variables are interchanged; again we shall not go into details.

As an application of parametric equivalence of methods, a method proposed by Wolfe will be linked to the capacity method.

Consider the following quadratic programming problem: Maximize f

$$f = p'x - \tfrac{1}{2}x'Cx,$$
$$Ax \le b,$$
$$x \ge 0. \tag{21}$$

The capacity method would add to this problem the parametric constraint

$$e'x \le \lambda \tag{22}$$

for positive elements of p. Instead of giving all relevant x-variables a coefficient of 1 in the capacity constraint, we may consider giving it the same coefficients as in the objective function. If all coefficients of p are positive, the additional constraint would be

$$p'x \le \lambda. \tag{23}$$

This method was discussed in linear programming, but it turned out to be equivalent with the simplex method. For quadratic programming this is not true; the resulting method is different from the simplex method for quadratic programming. The method may be called the gradient variant of the capacity method for quadratic programming.

Let us now consider the parametric equivalent of this method. The problem is then: Maximize f

$$f = (p - \lambda p)'x - \tfrac{1}{2}x'Cx,$$
$$Ax \le b,$$
$$x \ge 0. \tag{24}$$

with λ varying from ∞ to 0. The objective function may be written as

$$f = (1 - \lambda)p'x - \tfrac{1}{2}x'Cx = \lambda^*p'x - \tfrac{1}{2}x'Cx,$$

with λ^* varying from $-\infty$ to 1.

This method was proposed by Wolfe[80], who called it the long form of the simplex method for quadratic programming.[1] There are two slight differences: instead of taking as a starting-point $\lambda^* = -\infty$, which is the most convenient one for typical maximization problems, he takes $\lambda^* = 0$; furthermore, he uses the asymmetric variant of parametric quadratic programming.

---

[1] In his article Wolfe links two different methods together. The first he calls the short form; it can only be applied if C is positive definite or p is zero; this method is not described here. The second method he calls the long form, which can be considered a separate and completely general method for quadratic programming.

11.4. The self-dual parametric method for quadratic programming

For linear programming, Dantzig[2] has proposed a self-dual parametric method. In this method both primal infeasibilities and dual infeasibilities in the set-up tableau are compensated by a positive term in a parameter λ. This parameter is then decreased parametrically until an optimal solution is found for $\lambda = 0$, which must be the optimal solution of the original problem.

This method can easily be generalized for quadratic programming. In quadratic programming, both primal and dual variables appear explicitly, so that the method is even more straightforward than in linear programming. The reason for devoting a section of this book to this method is that both Lemke[107] and Graves[54] have proposed methods which are equivalent to the quadratic generalization of the self-dual parametric method for linear programming.

First an example will be given of the method in which the first variant of parametric quadratic programming is used; after that, Lemke's method and Graves' method are discussed.

As an example, let us consider the problem: Maximize f,

$$f = 2x_1 + 2x_2 - 2x_1^2 + 2x_1x_2 - 2x_2^2$$
$$x_1 + x_2 \geq 1,$$
$$-x_1 + 6x_2 \geq 3,$$
$$x_1, x_2 \geq 0. \tag{25}$$

Tableau 0 of table 2 gives the set-up tableau. All variables, primal as well as dual ones, are negative, so that parametric terms are added to each of the variables. In fact, we are now considering the extended problem: Maximize f^*

$$f^* = (2 - \lambda)x_1 + (2 - \lambda)x_2 - 2x_1^2 + 2x_1x_2 - 2x_2^2$$
$$x_1 + x_2 \geq 1 - \lambda,$$
$$-x_1 + 6x_2 \geq 3 - \lambda,$$
$$x_1, x_2 \geq 0. \tag{26}$$

For $\lambda \geq 3$, the solution of the set-up tableau is feasible. λ is now varied downwards, until an optimal solution for $\lambda = 0$ is obtained.

The critical value of λ is found in the row of y_2. Since this row has a zero principal pivot, u_2 is considered the new basic variable and the

---
[2] See [6], p. 245.

Table 2

Application of the self-dual parametric method

Tabl.	B.v.	Values basic variables						
		c-term	λ-term	λ = 3				
					x_1	x_2	u_1	u_2
	F	0	0	0	-2	-2	1	3
	v_1	-2	1	1	-4	2	1	-1
0	v_2	-2	1	1	2	-4	1	6
	y_1	-1	1	2	-1	-1	0	0
	y_2	-3	1	0	1	-6	0	0
					x_1	v_2	u_1	u_2
	F	1	$-\frac{1}{2}$		-3	$-\frac{1}{2}$	$\frac{1}{2}$	0
	v_1	-3	$1\frac{1}{2}$		-3	$\frac{1}{2}$	$1\frac{1}{2}$	2
1	x_2	$\frac{1}{2}$	$-\frac{1}{4}$		$-\frac{1}{2}$	$-\frac{1}{4}$	$-\frac{1}{4}$	$-1\frac{1}{2}$
	y_1	$-\frac{1}{2}$	$\frac{3}{4}$		$-1\frac{1}{2}$	$-\frac{1}{4}$	$-\frac{1}{4}$	$-1\frac{1}{2}$
	y_2	0	$-\frac{1}{2}$		-2	$-1\frac{1}{2}$	$-1\frac{1}{2}$	$\underline{-9}$
					x_1	v_2	u_1	y_2
	F	1	$-\frac{1}{2}$		-3	$-\frac{1}{2}$	$\frac{1}{2}$	0
	v_1	-3	$1\frac{7}{18}$		$-3\frac{4}{9}$	$\frac{1}{6}$	$1\frac{1}{6}$	$\frac{2}{9}$
2	x_2	$\frac{1}{2}$	$-\frac{1}{6}$		$-\frac{1}{6}$	0	0	$-\frac{1}{6}$
	y_1	$-\frac{1}{2}$	$\frac{5}{6}$		$-1\frac{1}{6}$	0	0	$-\frac{1}{6}$
	u_2	0	$\frac{1}{18}$		$\frac{2}{9}$	$\frac{1}{6}$	$\frac{1}{6}$	$-\frac{1}{9}$
					v_1	v_2	u_1	y_2
	F	$3\frac{19}{31}$	$-1\frac{22}{31}$		$-\frac{27}{31}$	$-\frac{20}{31}$	$-\frac{16}{31}$	$-\frac{6}{31}$
	x_1	$\frac{27}{31}$	$-\frac{25}{62}$		$-\frac{9}{31}$	$\frac{3}{62}$	$-\frac{21}{62}$	$-\frac{2}{31}$
3	x_2	$\frac{20}{31}$	$-\frac{29}{124}$		$-\frac{3}{62}$	$-\frac{1}{124}$	$-\frac{7}{124}$	$-\frac{11}{62}$
	y_1	$\frac{16}{31}$	$\frac{45}{124}$		$-\frac{21}{62}$	$\frac{7}{124}$	$\frac{49}{124}$	$\frac{15}{62}$
	u_2	$-\frac{6}{31}$	$\frac{9}{62}$		$\frac{2}{31}$	$\frac{11}{62}$	$\frac{15}{62}$	$\underline{-\frac{3}{31}}$
					v_1	v_2	u_1	u_2
	F	4	-2		-1	-1	-1	-2
	x_1	1	$-\frac{1}{2}$		$-\frac{1}{3}$	$-\frac{1}{6}$	$-\frac{1}{2}$	$-\frac{2}{3}$
4	x_2	1	$-\frac{1}{2}$		$-\frac{1}{6}$	$-\frac{1}{3}$	$-\frac{1}{2}$	$-1\frac{5}{6}$
	y_1	1	0		$-\frac{1}{2}$	$-\frac{1}{2}$	-1	$-2\frac{1}{2}$
	y_2	2	$-1\frac{1}{2}$		$-\frac{2}{3}$	$-1\frac{5}{6}$	$-2\frac{1}{2}$	$-10\frac{1}{3}$

values of basic variables are fixed at $\lambda = 3$. The smallest ratio is then found for v_2, so that a double nonstandard iteration should follow with u_2 replacing v_2 and x_2 replacing y_2. Instead we may, using principal pivots, first let x_2 replace v_2 and then u_2 replace y_2.

This results in tableau 2. The critical value of λ is now found in the row of v_1. Since the principal element in this row is nonzero, x_1 replaces v_1. In tableau 3, the critical value of λ is found in the row of u_2, and since the principal element is found in this row is nonzero, y_2 replaces u_2. The solution of tableau 4 is optimal for $\lambda = 0$, that this must be the optimal solution of the original problem.

The method proposed by Graves[54] is, apart from minor differences, exactly the same as the first symmetric variant of the self-dual parametric method, though he does not recognize this. Instead, he describes his method as one in which "a curious function is constructed" which is "shown to decrease strictly with a proper choice of trial solutions". The method as proposed in his article is slightly more general because the use of a lexicographic positive matrix B is proposed instead of a nonnegative vector for the coefficients of λ. Graves also gives an interesting adaption of the method for nonsymmetric complementarity problems, which will be treated in chapter 16.

Lemke's method[107] uses the asymmetric variant of parametric programming.[3] λ is now considered a variable and put together with the other nonbasic variables. The set-up tableau is then as indicated in tableau 0 of table 3. It is identical to tableau 0 of table 2, except that the λ-term column is put at the other side of the equation sign implied in the tableau and that λ is now a variable instead of a parameter. The solution of this tableau is optimal for $\lambda \geq 3$; for $\lambda = 3$, y_2 becomes 0, so that λ enters the basis, replacing y_2.

This results in tableau 1. The corresponding variable of the variable which just left the basis is now the new basic variable. The leaving basic variable is determined as in the simplex method for linear programming, with the dual variables and λ being considered on the same basis as primal variables.

After five iterations, λ leaves the basis so that obviously an optimal solution for $\lambda = 0$ is obtained (see tableau 5).

Comparing the asymmetric variant with the symmetric variant, we find differences in two respects. The rules for the asymmetric variant are more simple than those for the symmetric variant. On the other hand, the asymmetric method does not make any use of existing symmetry, which the symmetric variant does; use of the symmetry reduces the required

---

[3] Lemke devised his method for solving the linear complementarily problem; here its application to convex quadratic programming is discussed.

Table 3
Application of the asymmetric variant of the self-dual method

Tableau	B.v.	V.b.v.	λ	x_1	x_2	u_1	u_2
0	F	0	0	-2	-2	1	3
	v_1	-2	-1	-4	2	1	-1
	v_2	-2	-1	2	-4	1	6
	y_1	-1	-1	-1	-1	0	0
	y_2	-3	$\underline{-1}$	1	-6	0	0

Tableau	B.v.	V.b.v.	y_2	x_1	x_2	u_1	u_2
1	F	0	0	-2	-2	1	3
	v_1	1	-1	-5	8	1	-1
	v_2	1	-1	1	2	1	6
	y_1	2	-1	-2	5	0	$\underline{0}$
	λ	3	-1	-1	6	0	0

Tableau	B.v.	V.b.v.	y_2	x_1	x_2	u_1	v_2
2	F	$-\frac{1}{2}$	$\frac{1}{2}$	$-2\frac{1}{2}$	-3	$\frac{1}{2}$	$-\frac{1}{2}$
	v_1	$1\frac{1}{6}$	$-1\frac{1}{6}$	$-4\frac{5}{6}$	$8\frac{1}{3}$	$1\frac{1}{6}$	$\frac{1}{6}$
	u_2	$\frac{1}{6}$	$-\frac{1}{6}$	$\frac{1}{6}$	$\frac{1}{3}$	$\frac{1}{6}$	$\frac{1}{6}$
	y_1	2	-1	-2	5	0	0
	λ	3	-1	-1	6	0	0

Tableau	B.v.	V.b.v.	y_2	x_1	v_1	u_1	v_2
3	F	$-\frac{2}{25}$	$\frac{2}{25}$	$-4\frac{6}{25}$	$\frac{9}{25}$	$\frac{23}{25}$	$-\frac{11}{25}$
	x_2	$\frac{7}{50}$	$-\frac{7}{50}$	$-\frac{29}{50}$	$\frac{3}{25}$	$\frac{7}{50}$	$\frac{1}{50}$
	u_2	$\frac{9}{25}$	$-\frac{9}{25}$	$\frac{9}{25}$	$-\frac{1}{25}$	$\frac{3}{25}$	$\frac{4}{25}$
	y_1	$1\frac{3}{10}$	$-\frac{3}{10}$	$\frac{9}{10}$	$-\frac{3}{5}$	$-\frac{7}{10}$	$-\frac{1}{10}$
	λ	$2\frac{4}{25}$	$-\frac{4}{25}$	$2\frac{12}{25}$	$-\frac{18}{25}$	$-\frac{21}{25}$	$-\frac{3}{25}$

Tableau	B.v.	V.b.v.	y_2	u_2	v_1	u_1	v_2
4	F	$1\frac{1}{3}$	$-1\frac{1}{3}$	$11\frac{7}{9}$	$-\frac{1}{9}$	$2\frac{1}{3}$	$1\frac{4}{9}$
	x_2	$\frac{1}{3}$	$-\frac{1}{3}$	$1\frac{11}{18}$	$\frac{1}{18}$	$\frac{1}{3}$	$\frac{5}{18}$
	x_1	$\frac{1}{3}$	$-\frac{1}{3}$	$2\frac{7}{9}$	$-\frac{1}{9}$	$\frac{1}{3}$	$\frac{4}{9}$
	y_1	1	0	$-2\frac{1}{2}$	$-\frac{1}{2}$	-1	$-\frac{1}{2}$
	λ	$\frac{4}{3}$	$\frac{2}{3}$	$-6\frac{8}{9}$	$-\frac{4}{9}$	$-1\frac{1}{3}$	$-1\frac{2}{9}$

Tableau	B.v.	V.b.v.	λ	u_2	v_1	u_1	v_2
5	F	4	2	-2	-1	-1	-1
	x_2	1	$\frac{1}{2}$	$-1\frac{5}{6}$	$-\frac{1}{6}$	$-\frac{1}{2}$	$-\frac{1}{3}$
	x_1	1	$\frac{1}{2}$	$-\frac{2}{3}$	$-\frac{1}{3}$	$-\frac{1}{2}$	$-\frac{1}{6}$
	y_1	1	0	$-2\frac{1}{2}$	$-\frac{1}{2}$	-1	$-\frac{1}{2}$
	y_2	2	$1\frac{1}{2}$	$-10\frac{1}{3}$	$-\frac{2}{3}$	$-2\frac{1}{2}$	$-1\frac{5}{6}$

computations by nearly 50%. Since the difference in the complication of the rules is slight, it may be concluded that the symmetric variant is preferable for symmetric problems.

11.5. Cottle's index method

The quadratic programming methods considered so far are extensions of the simplex, dual and parametric methods for linear programming. In the simplex methods for linear and quadratic programming, the initial solution possesses primal feasibility. This solution is changed in order to obtain dual feasibility while maintaining primal feasibility. As the objective function increases monotonically in maximization problems, cycling is impossible and the optimal solution is obtained in a finite number of iterations. In the corresponding dual methods, the same is true if primal and dual feasibility are interchanged and the objective function of the dual problem is used.

In parametric methods for linear and quadratic programming, the original problem is changed by terms depending on a variable parameter, of which a certain value gives both primal and dual feasibility. Then the parameter is varied while maintaining primal and dual feasibility until it has reached a value for which solutions of the modified problem and the original problem coincide. Convergence is in this case based on the monotonic increase or decrease of the parameter.

Apart from somewhat more intricate rules for pivot selection, there is not much difference between methods for linear programming and methods for quadratic programming. But the tableaux are different. In linear programming only the primal variables appear explicitly as variables in the primal problem, while in quadratic programming, both primal and dual variables appear explicitly as variables.

There is a difference between the simplex and dual methods for quadratic programming on one hand and parametric methods for quadratic programming on the other hand. The simplex and the dual method deal with primal and dual variables separately, while the parametric methods treat them on the same basis.

The question then arises whether it is not possible to use for problems of which the initial solution is neither primally nor dually feasible, a mixture of steps of the simplex method and the dual method. Such a symmetric method has been proposed for linear programming by

Talacko.[4] The difficulty with such a method is that the behavior of the objective function is no longer monotonic; in maximization problems, the objective function would increase with each iteration of the simplex method while it would decrease with each iteration of the dual method. Hence the objective function can no longer be used to prove convergence for such a method.

If such a method is developed, convergence should be proved in a different manner. Cottle [44] uses in his method, which he calls the principal pivoting method, the number of primal and dual infeasibilities of a basic solution, which he calls the *index* of that solution. As the usage of this index is the characteristic which distinguishes Cottle's method from others, and not the principal pivots (which are not exactly principal anyway), we shall call this method the *index method for quadratic programming*. The two main characteristics of Cottle's methods are firstly a monotonic decrease of the index in each cycle (to be explained) of the method and secondly the fact that no distinction is made between primal and dual infeasibilities. It should be noted that Cottle has developed his method for a more general type of problem than the quadratic programming problem, problems which will be treated in chapters 16–18; however, other methods for quadratic programming can equally well be applied to these problems, as will be shown.

In the remainder of this section, we shall describe the symmetric variant of the method; the method as proposed by Cottle is actually the asymmetric variant.

The principle of the method is to eliminate an infeasibility (either primal or dual) while maintaining all existing primal and dual feasibility. After this has been done, the number of infeasibilities, which is called the index, has decreased by at least one. Then another infeasibility is eliminated in the same manner, and so on, until no infeasibilities remain, in which case the optimal solution must have been obtained.

The way in which each infeasibility is eliminated is quite simple. Let a basic variable have a value $- b_i$ and let this be the infeasibility which is to be eliminated.

Then add to this value the variable parameter λ. For $\lambda = b_i$, the infeasibility is eliminated, while existing feasibility is maintained. Then λ is varied downwards in such a way that existing feasibility is maintained, until $\lambda = 0$. After this, another infeasibility is attacked in the same man-

---

[4] See WOLFE and CUTLER [30].

ner. Hence the method boils down to successive applications of parametric quadratic programming.

There is one possible difficulty. Solutions to the problem in which one infeasibility is eliminated and existing feasibility is maintained do not always exist, even though the original problem has a finite optimal solution. This difficulty is remedied by imposing lower bounds on the infeasibilities.

To illustrate the method fully, we shall use three small examples. The first one is as follows. Maximize f:

$$f = 6x_1 - x_2 - 2x_1^2 + 2x_1x_2 - 2x_2^2,$$
$$x_1 + x_2 \leq 2,$$
$$x_1 \geq 0, \ x_2 \geq 0. \tag{27}$$

Tableau 0 of table 4 gives the set-up tableau. There is one infeasibility, namely $v_1 = -6$. Hence we add λ to this value, which results in a λ-term column of values of basic variables having a unit in the row of v_1 and zeros elsewhere. Since the λ-term column will always be equal to the v_1-column, the v_1-column could be used, but for the sake of clarity this column has been retained.

For $\lambda = 6$, the infeasibility is eliminated. Then λ is decreased parametrically. Following the rules for parametric quadratic programming, the lower critical value of λ is found to be 6 in the row of v_1. Since there is a nonzero principal element in this row, this is used for the pivot of a transformation, which results in tableau 1. There the lower critical value of λ is found to be 4 in the row of v_1. This results in selecting the principal element -3 as a pivot. In tableau 2, the lower critical value of λ is found to be 1 in the row of y, and the principal element of this row is used as a pivot. The resulting tableau 3 has a lower critical value of λ of -5, which is lower than 0, so that the solution for $\lambda = 0$ is found. Since there was only one infeasibility, which now has been eliminated, this must be the optimal solution.

In the first example, there was always a nonzero principal element in the row of the critical value of λ, so that these elements could be used as a pivot. In linear programming problems, which are special cases of quadratic programming problems, all principal elements are zero. The following problem will now be considered. Maximize f:

$$f = -18x_1 - 10x_2 + 15x_3,$$
$$-2x_1 - x_2 + x_3 \leq 0,$$

Table 4

First example of the index method

Tableau	Basic var.	c-term	λ-term	x_1	x_2	u
0	F	0	0	-6	1	-2
	v_1	-6	1	$\underline{-4}$	2	-1
	v_2	1	0	$\underline{2}$	-4	-1
	y	2	0	1	1	0
				v_1	x_2	u
1	F	9	$-1\frac{1}{2}$	$-1\frac{1}{4}$	-2	$-\frac{1}{2}$
	x_1	$1\frac{1}{2}$	$-\frac{1}{4}$	$-\frac{1}{4}$	$-\frac{1}{2}$	$\frac{1}{4}$
	v_2	-2	$\frac{1}{2}$	$\frac{1}{2}$	$\underline{-3}$	$-1\frac{1}{2}$
	y	$\frac{1}{2}$	$\frac{1}{4}$	$\frac{1}{4}$	$1\frac{1}{2}$	$-\frac{1}{4}$
				v_1	v_2	u
2	F	$10\frac{1}{3}$	$-1\frac{5}{6}$	$-1\frac{5}{6}$	$-\frac{2}{3}$	$\frac{1}{2}$
	x_1	$1\frac{5}{6}$	$-\frac{1}{3}$	$-\frac{1}{3}$	$-\frac{1}{6}$	$\frac{1}{2}$
	x_2	$\frac{2}{3}$	$-\frac{1}{6}$	$-\frac{1}{6}$	$-\frac{1}{3}$	$\frac{1}{2}$
	y	$-\frac{1}{2}$	$\frac{1}{2}$	$\frac{1}{2}$	$\frac{1}{2}$	$\underline{-1}$
				v_1	v_2	y
3	F	$10\frac{1}{12}$	$-1\frac{7}{12}$	$-1\frac{7}{12}$	$-\frac{5}{12}$	$\frac{1}{2}$
	x_1	$1\frac{7}{12}$	$-\frac{1}{12}$	$-\frac{1}{12}$	$\frac{1}{12}$	$\frac{1}{2}$
	x_2	$\frac{5}{12}$	$\frac{1}{12}$	$\frac{1}{12}$	$-\frac{1}{12}$	$\frac{1}{2}$
	u	$\frac{1}{2}$	$-\frac{1}{2}$	$-\frac{1}{2}$	$-\frac{1}{2}$	-1

$$- \quad x_2 + \quad x_3 \leq -5,$$
$$x_1 + \quad x_2 + \quad x_3 \leq 10,$$
$$x_1 \geq 0,\ x_2 \geq 0,\ x_3 \geq 0. \tag{28}$$

The set-up tableau is given in tableau 0 of table 5. There are two infeasibilities, $v_3 = -15$ and $y_2 = -5$. Let us first eliminate $v_3 = -15$. As before, λ is added to the value of v_3, but for reasons of space the λ-column is not shown in the tableaux. For $\lambda = 15$, the infeasibility has disappeared. λ is varied downwards and the lower critical value of λ is 15 and is found in the row of v_3. Since there is no principal element in the row of v_3, x_3 should enter the basis. The leaving basic variable is

Table 5
Example with block pivots

Tableau 0

Tableau	Basic var.	V.b.v.	x_1	x_2	x_3	u_1	u_2	u_3
0	F	0	18	10	−15	0	5	−10
	v_1	18	0	0	0	2	0	−1
	v_2	10	0	0	0	1	1	−1
	v_3	−15	0	0	0	−1	−1	−1
	y_1	0	−2	−1	1	0	0	0
	y_2	−5	0	−1	1	0	0	0
	y_3	10	1	1	1	0	0	0

Tableau 1

Tableau	Basic var.	V.b.v.	x_1	x_2	y_1	v_3	u_2	u_3
1	F	0	−12	−5	15	0	5	−10
	v_1	−12	0	0	0	2	−2	−3
	v_2	−5	0	0	0	1	0	−2
	u_1	15	0	0	0	−1	1	1
	x_3	0	−2	−1	1	0	0	0
	y_2	−5	2	0	−1	0	0	0
	y_3	10	3	−2	−1	0	0	0

Tableau 2

Tableau	Basic var.	V.b.v.	y_3	x_2	y_1	v_3	u_2	v_1
2	F	80	4	3	11	$-6\frac{2}{3}$	$11\frac{2}{3}$	$-3\frac{1}{3}$
	u_3	4	0	0	0	$-\frac{2}{3}$	$\frac{2}{3}$	$-\frac{1}{3}$
	v_2	3	0	0	0	$-\frac{1}{3}$	$1\frac{1}{3}$	$-\frac{2}{3}$
	u_1	11	0	0	0	$-\frac{1}{3}$	$\frac{1}{3}$	$\frac{1}{3}$
	x_3	$6\frac{2}{3}$	$\frac{2}{3}$	$\frac{1}{3}$	$\frac{1}{3}$	0	0	0
	y_2	$-11\frac{2}{3}$	$-\frac{2}{3}$	$-1\frac{1}{3}$	$-\frac{1}{3}$	0	0	0
	x_1	$3\frac{1}{3}$	$\frac{1}{3}$	$\frac{2}{3}$	$-\frac{1}{3}$	0	0	0

Tableau 3

Tableau	Basic var.	V.b.v.	y_3	y_2	y_1	v_3	v_2	v_1
3	F	$27\frac{1}{2}$	$2\frac{1}{2}$	$2\frac{1}{4}$	$10\frac{1}{4}$	$-3\frac{3}{4}$	$-8\frac{3}{4}$	$2\frac{1}{2}$
	u_3	$2\frac{1}{2}$	0	0	0	$-\frac{1}{2}$	$-\frac{1}{4}$	0
	u_2	$2\frac{1}{4}$	0	0	0	$-\frac{1}{4}$	$\frac{3}{4}$	$-\frac{1}{2}$
	u_1	$10\frac{1}{4}$	0	0	0	$-\frac{1}{4}$	$-\frac{1}{4}$	$\frac{1}{2}$
	x_3	$3\frac{3}{4}$	$\frac{1}{2}$	$\frac{1}{4}$	$\frac{1}{4}$	0	0	0
	x_2	$8\frac{3}{4}$	$\frac{1}{2}$	$-\frac{3}{4}$	$\frac{1}{4}$	0	0	0
	x_1	$-2\frac{1}{2}$	0	$\frac{1}{2}$	$-\frac{1}{2}$	0	0	0

Tableau 4

Tableau	Basic var.	V.b.v.	y_3	y_2	x_1	v_3	v_2	u_1
4	F	−75	0	0	0	$-2\frac{1}{2}$	$-7\frac{1}{2}$	−5
	u_3	$2\frac{1}{2}$	0	0	0	$-\frac{1}{2}$	$-\frac{1}{2}$	0
	u_2	$12\frac{1}{2}$	0	0	0	$-\frac{1}{2}$	$\frac{1}{2}$	1
	v_1	$2\frac{1}{2}$	0	0	0	$-\frac{1}{2}$	$-\frac{1}{2}$	2
	x_3	$2\frac{1}{2}$	$\frac{1}{2}$	$\frac{1}{2}$	$\frac{1}{2}$	0	0	0
	x_2	$7\frac{1}{2}$	$\frac{1}{2}$	$-\frac{1}{2}$	$\frac{1}{2}$	0	0	0
	y_1	5	0	−1	−2	0	0	0

determined by taking ratios for rows of feasible basic variables at $\lambda = 15$. Hence we have to compare:

$$y_1: (0 + 0 \times 15)/1 = 0,$$
$$y_3: (10 + 0 \times 15)/1 = 10.$$

y_1 leaves the basis and is replaced by x_3 so that the element 1 is used as a pivot. This is the first half of a double nonstandard iteration; for the second half the skew-symmetric element -1 is used. The result is tableau 1. The values of basic variables of this tableau are, if it is taken into account that the λ-term column is identical to the v_3-column:

$$v_1 = -12 + 2\lambda,$$
$$v_2 = -5 + \lambda,$$
$$u_1 = 15 - \lambda,$$
$$x_3 = 0 + 0\lambda,$$
$$y_2 = -5 + 0\lambda,$$
$$y_3 = 10 + 0\lambda.$$

The lower critical value of λ is now found to be 6 in the row of v_1. Hence x_1 should enter the basis. The leaving basic variable is determined by putting $\lambda = 6$ and ignoring infeasible basic variables; y_3 is found as the leaving basic variable. A double nonstandard iteration follows, which results in tableau 2.

The values of basic variables are now:

$$u_3 = 4 - \tfrac{2}{3}\lambda,$$
$$v_2 = 3 - \tfrac{1}{3}\lambda,$$
$$u_1 = 11 - \tfrac{1}{3}\lambda,$$
$$x_3 = 6\tfrac{2}{3} + 0\lambda,$$
$$y_2 = -11\tfrac{2}{3} + 0\lambda,$$
$$x_1 = 3\tfrac{1}{3} + 0\lambda.$$

No lower critical value of λ is found, so that λ may be put equal to 0. One cycle in which the infeasibility $v_3 = -15$ has been eliminated is completed. The remaining infeasibility $y_2 = -11\tfrac{2}{3}$ is eliminated in the same fashion.

Let us now consider a case in which the elimination of an infeasibility, if other infeasibilities are ignored, is impossible. As an example we shall use the following linear programming problem. Maximize f:

$$f = 2x_1 + x_2,$$
$$-\tfrac{1}{4}x_1 + x_2 \leq 2,$$
$$x_1 - x_2 \leq -1,$$
$$x_1 \geq 0,\ x_2 \geq 0. \tag{29}$$

Tableau 0 of table 6 gives the set-up tableau. To keep the tableaux simple, the F-row has been deleted. There are three infeasibilities of which $v_1 = -2$ will be eliminated first. λ is added to this variable, the lower critical value 2 is found in the row of v_1; there is no principal element in this row. Hence x_1 should enter the basis. The values of basic variables are

$$v_1 = -2 + 2 = 0,$$

Table 6

Index method for case in which lower bounds are imposed

| Tableau | B.v. | Values basic variables | | | | | |
		c-term	λ-term	x_1	x_2	u_1	u_2
0	v_1	-2	1	0	0	$\tfrac{1}{4}$	-1
	v_2	-1	0	0	0	-1	1
	y_1	2	0	$-\tfrac{1}{4}$	1	0	0
	y_2^*	$(-1)0$	0	1	-1	0	0

				y_2^*	x_2	u_1	v_1
1	u_2	2	0	0	0	$-\tfrac{1}{4}$	-1
	v_2^*	$(-3)0$	0	0	0	$-\tfrac{3}{4}$	1
	y_1	2	$-\tfrac{1}{4}$	$\tfrac{1}{4}$	$\tfrac{3}{4}$	0	0
	x_1	0	-1	1	-1	0	0

		c-term	λ-term	c-term	λ-term	$y_2^{(*)}$	x_1	u_1	v_2^*
2	u_2	2	0	2	-1	0	0	-1	1
	v_1	0	0	0	-1	0	0	$-\tfrac{3}{4}$	1
	y_1	2	-1	1	0	1	$\tfrac{3}{4}$	0	0
	x_2	0	1	1	0	-1	-1	0	0

		c-term	λ-term			y_2	y_1	v_1	v_2
3	u_2	2	$\tfrac{1}{3}$	3		0	0	$-1\tfrac{1}{3}$	$-\tfrac{1}{3}$
	u_1	0	$1\tfrac{1}{3}$	4		0	0	$-1\tfrac{1}{3}$	$-1\tfrac{1}{3}$
	x_1	$1\tfrac{1}{3}$	0	$1\tfrac{1}{3}$		$1\tfrac{1}{3}$	$1\tfrac{1}{3}$	0	0
	x_2	$2\tfrac{1}{3}$	0	$2\tfrac{1}{3}$		$\tfrac{1}{3}$	$1\tfrac{1}{3}$	0	0

$$v_2 = -1 + 0 \times 2 = -1,$$
$$y_1 = 2 + 0 \times 2 = 2,$$
$$y_2 = -1 + 0 \times 2 = -1.$$

The only positive element in the column of x_1 is in the row of y_2, but this row should be ignored since y_2 is infeasible. No leaving basic variable can be found. The reason for this is obvious if we look at (29): x_1 can be increased indefinitely if the second constraint is ignored.

In this case we introduce a lower bound on y_2: $y_2 \geq -1$. Note that this lower bound is redundant since the nonnegativity constraint $y_2 \geq 0$ implies it. Let us measure y_2 as a deviation from -1, which is done by substitution of $y_2^* = y_2 - (-1)$ into the tableau. y_2 is then replaced by y_2^* which now has a value of 0, and is therefore feasible, so that ratios are taken also in this row. This leads to a double nonstandard iteration with the underlined elements 1 and -1 as pivots. The result is tableau 1, of which the stars and brackets should be ignored for the moment. This tableau maintains existing feasibility for $\lambda = 0$, so that the first cycle is finished.

Now there are two infeasibilities, $v_2 = -3$ and $y_2^* = 0$, which implies $y_2 = -1$. Let us eliminate the second infeasibility first. This is done by increasing y_2^* from 0 to 1, which is done parametrically. The values of basic variables as a function of $y_2^* = \lambda$ are:

$$u_2 = \quad 2 + 0\lambda,$$
$$v_2 = -3 + 0\lambda,$$
$$y_1 = \quad 2 - \tfrac{1}{4}\lambda,$$
$$x_1 = \quad 0 - \lambda.$$

Increasing λ parametrically, we find that the critical value of λ is found in the x_1-row at $\lambda = 0$. Hence v_1 should enter the basis. The column of v_1 has one positive element in the v_2-row, but this row should be ignored since $v_2 = -3$ for $\lambda = 0$ and hence infeasible. No pivot can be found.

As before, a redundant lower bound is imposed on an infeasible basic variable. We define $v_2^* = v_2 - (-3)$ and substitute for this in the tableau. Now $v_2^* = 0$ is feasible so that a pivot is found. After a double nonstandard iteration tableau 2 is found.

The upper critical value of λ is now 2, which is found for y_1. This exceeds 1, so that y_2 may replace y_2^* by substitution of $y_2^* = y_2 + 1$ in the equations underlying the tableau. This amounts to adding 1 times the λ-term column to the c-term column, which results in a new c-term column; the star of y_2^* is now deleted.

The only infeasibility remaining is now the value of $v_2^* = 0$, which corresponds to $v_2 = -3$. Hence v_2^* should be increased parametrically to 3. Constructing a new λ-term column, which is equal to minus the v_2^*-column, we find as the upper critical value of λ, 0 in the row of v_1. Since there is no principal element in this row, x_1 is introduced, the leaving basic variable is y_1 and a double nonstandard iteration follows.

In tableau 3, there is no upper critical value of λ, so that the solution must be optimal for $\lambda = 3$. The solution for $\lambda = 3$ then gives the values of basic variables after v_2 has replaced v_2^*.

CHAPTER 12

SOME OTHER QUADRATIC PROGRAMMING METHODS

12.1. Beale's quadratic programming method

In this chapter some other quadratic programming methods are treated, namely Beale's method [32, 33] and the Theil–Van de Panne method [78]. Both methods have in common that they are based on interesting and elegant ideas that are worth considering.

In this chapter it will be shown that Beale's method is closely related to the simplex method for quadratic programming. In a number of cases its successive solutions are the same as those of the simplex method, but the method is somewhat less efficient computationally. In other cases Beale's method may have intermediate solutions which do not appear in the simplex method, while the possibility exists that the solution path may become different altogether.

The Theil–Van de Panne or combinatorial method does not use simplex tableaux and is therefore not very efficient; even if simplex tableaux are used it is not efficient, but the ideas on which it is based are so general that a treatment in the framework of quadratic programming methods is interesting. It turns out that the Theil–Van de Panne method is related to the dual method for quadratic programming.

In this section Beale's method will be given in its original formulation. In the following section this method is reformulated in the framework of quadratic simplex tableaux. An application of Beale's method and the simplex method to a relatively simple example then give indications as to the differences of both methods which are discussed more formally afterwards.

The problem to be considered is as before: Maximize f,

$$f = p'x - \tfrac{1}{2}x'Cx$$
$$Ax + y = b,$$
$$x, y \geq 0, \tag{1}$$

with C positive semi-definite.

Beale's method also requires an initial feasible solution. An initial feasible solution is immediately available if $b \geq 0$; the y-variables are in this case basic, the x-variables nonbasic. The constraints then give an expression of the basic variables in terms of the nonbasic ones:

$$y = b - Ax. \tag{2}$$

The objective function is expressed as a quadratic form in terms of the nonbasic variables:

$$f = f_0 + p'x - \tfrac{1}{2}x'Cx, \tag{3}$$

where $f_0 = 0$ in the initial solution.

Beale's method in its original formulation is based on eqs. (2) and (3), which can be put in tableau form. For each solution both the basic variables and the objective function are expressed in terms of the nonbasic variables, as in (2) and (3). In each iteration a new basic variable and a leaving variable are chosen, after which the constraints and the objective function are transformed accordingly. This is done as follows. If, for example, x_1 is the new basic variable and y_1 the leaving basic variable, then the first equation of (2), which is

$$y_1 = b_1 - a_{11}x_1 - a_{12}x_2 - \ldots - a_{1n}x_n, \tag{4}$$

is written as

$$x_1 = (b_1/a_{11}) - (1/a_{11})y_1 - (a_{12}/a_{11})x_2 - \ldots - (a_{1n}/a_{11})x_n; \tag{5}$$

this expression is used to substitute for x_1 in (2) and (3). This results in another expression of basic variables and objective function in terms of the nonbasic variables.

The substitution of an expression like (4) in the constraints is equivalent to a simplex transformation with a pivot element in ordinary linear programming. For the objective function the transformation is more complicated, but it is clear that another quadratic form in the nonbasic variables will be obtained. In the next section it will be shown that the transformation of constraints and of objective function is performed by two successive simplex transformations.

New basic variables are selected as follows. The vector of partial derivatives of the objective function with respect to the nonbasic variables is in the initial situation, in which the x-variables are nonbasic,

$$\partial f/\partial x = p - Cx; \tag{6}$$

the term Cx cancels because the x-variables are nonbasic and therefore zero.

If an element of p is positive, an increase in the value of the corresponding x-variable will increase the value of the objective function. Hence, the nonbasic variable having the largest positive coefficient in the linear term of the objective function is chosen as the new basic variable. Let the new basic variable be x_j. We have then for the corresponding equation in (6),

$$\partial f / \partial x_j = p_j - c_{1j}x_1 - \ldots - c_{jj}x_j - \ldots - c_{nj}x_n. \tag{7}$$

It is clear that $\partial f / \partial x_j$ stays positive until

$$x_j = p_j / c_{jj}. \tag{8}$$

On the other hand, the present basic variables should remain nonnegative, so that, as in linear programming, x_j can at most be increased until one of the present basic variables becomes zero, which is the value of x_j corresponding with

$$\min_i (b_i / a_{ij} | a_{ij} > 0). \tag{9}$$

Comparing the values of (8) and (9) we choose the smaller one.

If the smaller value occurs in (9), the corresponding basic variable leaves the basis and an iteration is performed as described above. If (8) yields the smaller value, eq. (7) is used as follows. Let us write $\partial f / \partial x_j = z_1$; z_1 can then be considered a basic variable in (7), which is going to become nonbasic instead of x_j. This means that we have added (7) to the constraints of the problem. In this case (7) is transformed into

$$x_j = (p_j / c_{jj}) - (c_{1j} / c_{jj})x_1 - \ldots - (1 / c_{jj})z_1 - \ldots - (c_{nj} / c_{jj})x_n. \tag{10}$$

This equation is now used to substitute for x_j in the constraints and the objective function as indicated before.

As a nonbasic variable we now have z_1, which is called a *free variable*, since it is not constrained by nonnegativity constraints. This means that, if such a variable occurs in the objective function with a nonzero linear part, then it may be made basic at a profit, since if it has a negative partial derivative, it can be decreased, thus increasing the objective function, whereas if it has a positive partial derivative, it may be increased, again increasing the objective function. A solution containing nonbasic free variables with nonzero partial derivatives is denoted by Beale as a *nonstandard solution*. It is required in Beale's method that in any nonstandard solution a free variable enters the basis in preference to any constrained variable.

The optimal solution is found when the partial derivatives of the objective function with respect to the nonbasic variables are nonpositive if it concerns a constrained variable, and zero if it concerns a free variable.

12.2. A reformulation

The original formulation of Beale's method has the advantage that it is straightforward; the transformations for the iterations are indicated as substitutions. However, for theoretical and computational reasons it is useful to give the transformations as those of a tableau which are performed by pivoting on a tableau-element, as in the simplex method for linear programming and in the simplex method for quadratic programming. Here a reformulation of Beale's method will be given in which the transformations are of this type. The reformulated method will amount to an application of the simplex method for quadratic programming to a modified quadratic programming problem. In the following we must therefore carefully distinguish between three in many respects similar methods: Beale's method in its original formulation, its reformulation, and the simplex method for quadratic programming. In the following it is convenient to consider always the asymmetric variant of the last method.

This reformulation will now be described. First a set-up tableau must be given, which is entirely the same as in the simplex method for quadratic programming; see table 1. Comparing this tableau with Beale's formulation a close correspondence can be noted. The constraints in the rows of the y-variables in the simplex method and those in Beale's method are the same. The v-rows and the F-row in the simplex method correspond with Beale's representation of the objective function, except that in Beale's representation the columns of the nonbasic dual variables, in the initial tableau the u-variables, are missing. The basic dual variables in the simplex tableaux and the linear part of the quadratic form in Beale's formulation [the vector p in (6)] are the same, apart from sign.

Hence the rule that the nonbasic variable having the largest positive coefficient in the linear part of the objective function should enter the basis corresponds with the rule that the nonbasic variable having the largest nonnegative dual variable enters the basis. The rule for the leaving basic variable is in the reformulation similar to that in the simplex method, but not quite the same. The same positive ratios are considered, namely the ratio in the row of the dual variable corresponding to the new basic variable and the ratios in the rows of primal basic variables. If the minimum ratio is connected with a primal basic variable, then that variable leaves the basis, as in the simplex method.

If, however, the ratio is connected with the dual variable, a new constraint is added to the problem, as it was in Beale's own formulation; see (7). This new constraint is added by copying in a new row the elements of the row of the dual variable concerned, except for the elements in the columns of nonbasic dual variables, which are replaced by zeros. The basic variable of the new row is denoted by z_1; it is considered to be a primal variable.

In quadratic simplex tableaux the constraints do not only occur in the rows of basic primal variables, but also, with a minus-sign, in the column of nonbasic dual variables (the u-variables in table 1). Hence a nonbasic dual variable w_1 is added, which is the corresponding dual variable of z_1. The elements of this column in the rows of primal basic variables are zero.

After the new row and column are added, z_1 is taken as the leaving basic variable and the tableau is transformed accordingly. In both cases, when an ordinary primal basic variable leaves the basic and when a free variable leaves the basis, a nonstandard tableau in the terminology of the simplex method is generated. In both cases, the rule of the simplex method is now applied, which says that the dual variable of the nonbasic pair should enter the basis. Thus, if y_1 left the basis, u_1 should enter it and if z_1 left the basis, w_1 should enter it. The leaving basic variable is also determined according to the rules of the simplex method. Hence positive ratios should be compared in the row of the dual variable of the basic pair and in the rows of the primal basic variable. Now Beale's method is constructed in such a way that the elements in the rows of primal vari-

Table 1

Set-up tableau for a quadratic programming problem

Basic variables	Values basic variables	x	u	F	v	y
F	0	$-p'$	$-b'$	1	0	0
v	$-p$	$-C$	$-A'$	0	I	0
y	b	A	0	0	0	I

ables and in the columns of nonbasic dual variables are always zero. Hence the dual variable of the basic pair must leave the basis.

Other iterations of the reformulated Beale method are similar. New constraints are added whenever the ratio in the row of the dual variable of the new basic variable proves to be the smallest, so that a number of z- and w-variables may occur in the tableaux. Iterations are always performed in pairs. In the first iteration of a pair, a primal variable enters the basis and another primal variable, which is either a constrained variable or the free variable of an added constraint, leaves the basis. In the second iteration the corresponding dual variable of the variable which left the basis, enters it, and the corresponding dual variable of the variable that entered the basis leaves it. The first iteration of a pair is therefore called the *primal iteration*, the second the *dual iteration*. Iterations of the same type occur when the simplex method for quadratic programming is applied to a linear programming problem. One pair of iterations in the reformulated version corresponds to one iteration in the original version. This does not mean that the reformulated version requires more work than the original formulation; if proper use is made of the properties of symmetry and skew-symmetry of standard tableaux, both versions require exactly the same computations.

A nonstandard solution in Beale's sense occurs when there are one or more nonzero w-variables. In the case the corresponding z-variables or free variables should be introduced into the basis *in preference to* the other constrained primal variables. If a w-variable is positive, then the corresponding z-variable should be introduced in a negative direction, while in the case it is negative, it is introduced in a positive direction. After a pair of iterations in which a z-variable entered the basis and a w-variable left it, the row of this z-variable and the column of this w-variable can be deleted.

Beale's method can be seen as an extreme-point method; each solution is found to be an extreme point of the constraints; if a solution is not an extreme point of the original constraints, additional constraints are generated in such a way that the solution is an extreme point of original and additional constraints. Hence it can be said that Beale's method is in this respect very close to the simplex method for linear programming, in which each solution is also an extreme point of the constraints.

The reformulated Beale method will be applied to an example. This application has two purposes, first, to explain the reformulation of Beale's method and second, to serve as a concrete example for a comparison of Beale's method and the simplex method for quadratic programming. The problem is as follows. Maximize f

$$f = 3x_1 + 2x_2 + x_3 - 1\tfrac{1}{2}x_1^2 - x_2^2 - \tfrac{1}{6}x_3^2,$$

$$x_1 + 2x_2 + x_3 \leq 4,$$

$$x_1, x_2, x_3 \geq 0. \tag{11}$$

As indicated, the set-up tableau is the same as that for the simplex method (see tableau 0 of table 2); the z_1-row and w_1-column are added later. Since v_1 is the largest negative dual variable, x_1 enters the basis. Comparing the ratios $\frac{-3}{-3}$ and $\frac{4}{1}$ we find that v_1 becomes zero first when x_1 is increased. The following constraint is therefore added to the problem,

$$-3 = -3x_1 - 0x_2 - 0x_3 + z_1, \tag{12}$$

which is done by copying the row of v_1 in a new row, with z_1 instead of v_1 as a basic variable and having a zero instead of -1 in the column of u; further a column with a nonbasic variable w_1 is added, which has the same elements as the z_1-row, apart from sign. The underlined element -3 in the z_1-row and the x_1-column is the pivot of the primal iteration, which gives tableau 1. In the corresponding dual iteration w_1 enters the basis and v_1 leaves it, so that the underlined element 3 is the pivot of the transformation. Tableau 2 is then again in standard form.

The solution of tableau 2 is a standard one in Beale's sense, since w_1 is zero. v_2 is then found to be the largest negative dual variable; hence x_2 must enter the basis. Comparing the ratios $\frac{-2}{-2}$ and $\frac{3}{2}$, we find that v_2 becomes zero first. We therefore add the constraint

$$-2 = 0x_1 - 2x_2 - 0x_3 + z_2 \tag{13}$$

in the form of an additional row and an additional column. z_2 leaves the basis and, after pivoting, tableau 3 results. In the corresponding dual iteration w_2 enters the basis and z_2 leaves it, which results in tableau 4. Again the tableau is in standard form; further the solution is a standard one in Beale's sense since w_1 and w_2 are both zero.

We then find that v_3 is the only negative dual variable, so that x_3 must enter the basis. Comparing the ratios $-1/-\tfrac{1}{3}$ and $\frac{-1}{-1}$, we find that y must leave the basis, so that in this case no constraint is added. The primal iteration gives tableau 5 and the corresponding dual one tableau 6.

This tableau is in standard form, but it has a nonstandard solution in Beale's sense, since w_1 and w_2 are both nonzero. Since w_1 is positive, z_1 is

Table 2

An application of the reformulated Beale method to the illustrative problem

Tableau	Basic var.	Values bas. var.	Nonbasic variables				
			x_1	x_2	x_3	u	w_1
0	F	0	-3	-2	-1	-4	3
	v_1	-3	-3	0	0	-1	3
	v_2	-2	0	-2	0	-2	0
	v_3	-1	0	0	$-\frac{1}{3}$	-1	0
	y	4	1	2	1	0	0
	z_1	-3	$\underline{-3}$	0	0	0	0

			z_1	x_2	x_3	u	w_1
1	F	3	-1	-2	-1	-4	3
	v_1	0	-1	0	0	-1	3
	v_2	-2	0	-2	0	-2	0
	v_3	-1	0	0	$-\frac{1}{3}$	-1	0
	y	3	$\frac{1}{3}$	2	1	0	0
	x_1	1	$-\frac{1}{3}$	0	0	0	0

			z_1	x_2	x_3	u	v_1	w_2
2	F	3	0	-2	-1	-3	-1	2
	w_1	0	$-\frac{1}{3}$	0	0	$-\frac{1}{3}$	$\frac{1}{3}$	0
	v_2	-2	0	-2	0	-2	0	2
	v_3	-1	0	0	$-\frac{1}{3}$	-1	0	0
	y	3	$\frac{1}{3}$	2	1	0	0	0
	x_1	1	$-\frac{1}{3}$	0	0	0	0	0
	z_2	-2	0	$\underline{-2}$	0	0	0	0

			z_1	z_2	x_3	u	v_1	w_2
3	F	5	0	-1	-1	-3	-1	2
	w_1	0	$-\frac{1}{3}$	0	0	$-\frac{1}{3}$	$\frac{1}{3}$	0
	v_2	0	0	-1	0	-2	0	2
	v_3	-1	0	0	$-\frac{1}{3}$	-1	0	0
	y	1	$\frac{1}{3}$	1	1	0	0	0
	x_1	1	$-\frac{1}{3}$	0	0	0	0	0
	x_2	1	0	$-\frac{1}{2}$	0	0	0	0

			z_1	z_2	x_3	u	v_1	v_2
4	F	5	0	0	-1	-1	-1	-1
	w_1	0	$-\frac{1}{3}$	0	0	$-\frac{1}{3}$	$\frac{1}{3}$	0
	w_2	0	0	$-\frac{1}{2}$	0	-1	0	$\frac{1}{2}$
	v_3	-1	0	0	$-\frac{1}{3}$	-1	0	0
	y	1	$\frac{1}{3}$	1	1	0	0	0
	x_1	1	$-\frac{1}{3}$	0	0	0	0	0
	x_2	1	0	$-\frac{1}{2}$	0	0	0	0

Table 2 (continued)

			z_1	z_2	y	u	v_1	v_2
	F	6	$\frac{1}{3}$	1	1	-1	-1	-1
	w_1	0	$-\frac{1}{3}$	0	0	$-\frac{1}{3}$	$\frac{1}{3}$	0
	w_2	0	0	$-\frac{1}{2}$	0	-1	0	$\frac{1}{2}$
5	v_3	$-\frac{2}{3}$	$\frac{1}{9}$	$\frac{1}{3}$	$\frac{1}{3}$	$\underline{-1}$	0	0
	x_3	1	$\frac{1}{3}$	1	1	0	0	0
	x_1	1	$-\frac{1}{3}$	0	0	0	0	0
	x_2	1	0	$-\frac{1}{2}$	0	0	0	0

			z_1	z_2	y	v_3	v_1	v_2	w_3
	F	$6\frac{2}{3}$	$\frac{2}{9}$	$\frac{2}{3}$	$\frac{2}{3}$	-1	-1	-1	$-\frac{2}{9}$
	w_1	$\frac{2}{9}$	$-\frac{10}{27}$	$-\frac{1}{9}$	$-\frac{1}{9}$	$-\frac{1}{3}$	$\frac{1}{3}$	0	$\frac{10}{27}$
	w_2	$\frac{2}{3}$	$-\frac{1}{9}$	$-\frac{5}{6}$	$-\frac{1}{3}$	-1	0	$\frac{1}{2}$	$\frac{1}{9}$
6	u	$\frac{2}{3}$	$-\frac{1}{9}$	$-\frac{1}{3}$	$-\frac{1}{3}$	-1	0	0	$\frac{1}{9}$
	x_3	1	$\frac{1}{3}$	1	1	0	0	0	0
	x_1	1	$-\frac{1}{3}$	0	0	0	0	0	0
	x_2	1	0	$-\frac{1}{2}$	0	0	0	0	0
	z_3	$\frac{2}{9}$	$\underline{-\frac{10}{27}}$	$-\frac{1}{9}$	$-\frac{1}{9}$	0	0	0	0

			z_3	z_2	y	v_3	v_1	v_2	w_3
	F	$6\frac{4}{5}$	$\frac{3}{5}$	$\frac{3}{5}$	$\frac{3}{5}$	-1	-1	-1	$\frac{2}{9}$
	w_1	0	-1	0	0	$-\frac{1}{3}$	$\frac{1}{3}$	0	$\frac{10}{27}$
	w_2	$\frac{3}{5}$	$-\frac{3}{10}$	$-\frac{4}{5}$	$-\frac{3}{10}$	-1	0	$\frac{1}{2}$	$\frac{1}{9}$
7	u	$\frac{3}{5}$	$-\frac{3}{10}$	$-\frac{3}{10}$	$-\frac{3}{10}$	-1	0	0	$\frac{1}{9}$
	x_3	$1\frac{1}{5}$	$\frac{9}{10}$	$\frac{9}{10}$	$\frac{9}{10}$	0	0	0	0
	x_1	$\frac{4}{5}$	$-\frac{9}{10}$	$\frac{1}{10}$	$\frac{1}{10}$	0	0	0	0
	x_2	1	0	$-\frac{1}{2}$	0	0	0	0	0

			z_3	z_2	y	v_3	v_1	v_2	w_4
	F	$6\frac{4}{5}$	0	$\frac{3}{5}$	$\frac{3}{5}$	$-1\frac{1}{5}$	$-\frac{4}{5}$	-1	$-\frac{3}{5}$
	w_3	0	$-2\frac{7}{10}$	0	0	$-\frac{9}{10}$	$\frac{9}{10}$	0	0
	w_2	$\frac{3}{5}$	0	$-\frac{4}{5}$	$-\frac{3}{10}$	$-\frac{9}{10}$	$-\frac{1}{10}$	$\frac{1}{2}$	$\frac{4}{5}$
8	u	$\frac{3}{5}$	0	$-\frac{3}{10}$	$-\frac{3}{10}$	$-\frac{9}{10}$	$-\frac{1}{10}$	0	$\frac{3}{10}$
	x_3	$1\frac{1}{5}$	$\frac{9}{10}$	$\frac{9}{10}$	$\frac{9}{10}$	0	0	0	0
	x_1	$\frac{4}{5}$	$-\frac{9}{10}$	$\frac{1}{10}$	$\frac{1}{10}$	0	0	0	0
	x_2	1	0	$-\frac{1}{2}$	0	0	0	0	0
	z_4	$\frac{3}{5}$	0	$\underline{-\frac{4}{5}}$	$-\frac{3}{10}$	0	0	0	0

			z_3	z_4	y	v_3	v_1	v_2	w_4
	F	$7\frac{1}{4}$	0	$\frac{3}{4}$	$\frac{3}{8}$	$-1\frac{1}{5}$	$-\frac{4}{5}$	-1	$-\frac{3}{5}$
	w_3	0	$-2\frac{7}{10}$	0	0	$-\frac{9}{10}$	$\frac{9}{10}$	0	0
	w_2	0	0	-1	0	$-\frac{9}{10}$	$-\frac{1}{10}$	$\frac{1}{2}$	$\frac{4}{5}$
9	u	$\frac{3}{8}$	0	$-\frac{3}{8}$	$-\frac{3}{16}$	$-\frac{9}{10}$	$-\frac{1}{10}$	0	$\frac{3}{10}$
	x_3	$1\frac{7}{8}$	$\frac{9}{10}$	$\frac{3}{8}$	$\frac{9}{16}$	0	0	0	0
	x_1	$\frac{7}{8}$	$-\frac{9}{10}$	$\frac{1}{8}$	$\frac{1}{16}$	0	0	0	0
	x_2	$\frac{5}{8}$	0	$-\frac{5}{8}$	$\frac{3}{16}$	0	0	0	0

Table 2   (*continued*)

Tableau	Basic var.	Values bas. var.	Nonbasic variables					
			z_3	z_4	y	v_3	v_1	v_2
	F	$7\frac{1}{4}$	0	0	$\frac{3}{8}$	$-1\frac{7}{8}$	$-\frac{7}{8}$	$-\frac{5}{8}$
	w_3	0	$-2\frac{7}{10}$	0	0	$-\frac{9}{10}$	$\frac{9}{10}$	0
	w_4	0	0	$-\frac{5}{4}$	0	$-1\frac{1}{8}$	$-\frac{1}{8}$	$\frac{5}{8}$
10	u	$\frac{3}{8}$	$\frac{3}{8}$	0	0	$-\frac{3}{16}$	$-\frac{9}{16}$	$-\frac{3}{16}$
	x_3	$1\frac{7}{8}$	$\frac{9}{10}$	$1\frac{1}{8}$	$\frac{9}{16}$	0	0	0
	x_1	$\frac{7}{8}$	$-\frac{9}{10}$	$\frac{1}{8}$	$\frac{1}{16}$	0	0	0
	x_2	$\frac{5}{8}$	0	$-\frac{5}{8}$	$\frac{3}{16}$	0	0	0

entered in the basis in a negative direction. Comparing the ratios $-\frac{2}{9}/-\frac{10}{27}$ and $-1/-\frac{1}{3}$, we find that w_1 becomes zero first. Since w_1 is a dual variable, the following constraint is added:

$$\tfrac{2}{9} = -\tfrac{10}{27}z_1 - \tfrac{1}{9}z_2 - \tfrac{1}{9}y + z_3, \tag{14}$$

which results in an additional row and column in tableau 6. x_3 leaves the basis and the next iteration gives tableau 7, which is transformed in the corresponding dual iteration in tableau 8. Note that the row of z_1 and the column of w_1 are deleted. Since in tableau 8 w_2 is nonzero, z_2 is introduced into the basis. Comparing the ratios $-\frac{3}{5}/-\frac{4}{5}$ and $-1/-\frac{1}{2}$, we find that w_2 becomes zero first, so that another constraint,

$$\tfrac{3}{5} = 0z_3 - \tfrac{4}{5}z_2 - \tfrac{3}{10}y + z_4 \tag{15}$$

has to be added. z_2 enters the basis and z_4 leaves it and after the corresponding dual iteration tableau 10 is obtained. The row of z_2 and column of w_2 are deleted. The solution of tableau 10 has w-variables with zero values and nonnegative dual variables, so that the optimal solution has been obtained.

12.3. Comparison with the simplex method

In order to compare Beale's method with the simplex method for quadratic programming, the same example is solved by the asymmetric version of the last method in table 3. The set-up tableau is then as given in tableau 0. The solution of the set-up tableau is obviously a feasible one, so that the simplex method can be applied straightforwardly. Since v_1 is the largest negative dual variable, x_1 is introduced into the basis. Comparing the ratios $\frac{-3}{-3}$ and $\frac{4}{1}$, we find that v_1 must leave the basis, so that -3 is the pivot of the transformation which results in tableau 1. This is a standard

Table 3

An application of the simplex method for quadratic programming to the example

Tableau	Basic variables	Values basic variables	Nonbasic variables			
			x_1	x_2	x_3	u
0	F	0	-3	-2	-1	-4
	v_1	-3	-3	0	0	-1
	v_2	-2	0	-2	0	-2
	v_3	-1	0	0	$-\frac{1}{3}$	-1
	y	4	1	2	1	0
			v_1	x_2	x_3	u
1	F	3	-1	-2	-1	-3
	x_1	1	$-\frac{1}{3}$	0	0	$\frac{1}{3}$
	v_2	-2	0	-2	0	-2
	v_3	-1	0	0	$-\frac{1}{3}$	-1
	y	3	$\frac{1}{3}$	2	1	$-\frac{1}{3}$
			v_1	v_2	x_3	u
2	F	5	-1	-1	-1	-1
	x_1	1	$-\frac{1}{3}$	0	0	$\frac{1}{3}$
	x_2	1	0	$-\frac{1}{2}$	0	1
	v_3	-1	0	0	$-\frac{1}{3}$	-1
	y	1	$\frac{1}{3}$	1	1	$-2\frac{1}{3}$
			v_1	v_2	y	u
3	F	6	$-\frac{2}{3}$	0	1	$-3\frac{1}{3}$
	x_1	1	$-\frac{1}{3}$	0	0	$\frac{1}{3}$
	x_2	1	0	$-\frac{1}{2}$	0	1
	v_3	$-\frac{2}{3}$	$\frac{1}{9}$	$\frac{1}{3}$	$\frac{1}{3}$	$-1\frac{7}{9}$
	x_3	1	$\frac{1}{3}$	1	1	$-2\frac{1}{3}$
			v_1	v_2	y	v_3
4	F	$7\frac{1}{4}$	$-\frac{7}{8}$	$-\frac{5}{8}$	$\frac{3}{8}$	$-1\frac{7}{8}$
	x_1	$\frac{7}{8}$	$-\frac{5}{16}$	$\frac{1}{16}$	$\frac{1}{16}$	$\frac{3}{16}$
	x_2	$\frac{5}{8}$	$\frac{1}{16}$	$-\frac{5}{16}$	$\frac{3}{16}$	$\frac{9}{16}$
	u	$\frac{3}{8}$	$-\frac{1}{16}$	$-\frac{3}{16}$	$-\frac{3}{16}$	$-\frac{9}{16}$
	x_3	$1\frac{7}{8}$	$\frac{3}{16}$	$\frac{9}{16}$	$\frac{9}{16}$	$-1\frac{5}{16}$

tableau, so that, since v_2 is the largest negative dual variable, x_2 must be introduced into the basis. Comparing the ratios $\frac{-2}{-2}$ and $\frac{3}{2}$, we find that v_2 leaves the basis. After transformation with -2 as a pivot, tableau 2 is found. x_3 must now enter the basis, but when comparing the ratios $-1/-\frac{1}{3}$

and $\frac{1}{1}$, we find that y must leave the basis. Tableau 3 is a nonstandard tableau, so that u must enter the basis. When comparing the ratios $1/\frac{1}{3}$, $\frac{1}{1}$ and $-\frac{2}{3}/-1\frac{7}{9}$, we find that v_3 must leave the basis. Tableau 4 is in standard form and optimal.

Let us first compare the application of both methods to the same problem as given in tables 2 and 3. Both methods start with the same set-up tableau and in both methods x_1 enters the basis. Further it is found in both methods that v_1 becomes zero first when x_1 is increased. In the simplex method v_1 leaves the basis and a new standard tableau is obtained. In Beale's method the new basic variable is prevented from increasing beyond the point where the partial derivative becomes negative by the addition of a new constraint with z_1 as its slack variable and w_1 as its dual variable. x_1 then replaces z_1 in the primal iteration and in the corresponding dual iteration w_1 replaces v_1. The solutions of tableau 1 of table 3 and tableau 2 of table 2 are the same apart from the occurrence of w_1, which has a zero value.

The second iteration of the simplex method and the third and the fourth iteration of Beale's method are of a similar kind. In both methods x_2 enters the basis. In the simplex method v_2 leaves the basis and tableau 2 of table 3 is again in standard form. In Beale's method another constraint is added with slack variable z_2 and dual variable w_2. z_2 leaves the basis in the primal iteration; in the dual iteration w_2 enters and v_2 leaves the basis.

Both methods then introduce x_3 into the basis and it is found in both methods that y becomes zero first, so that it leaves the basis in both cases. In the asymmetric variant of the simplex method for quadratic programming a nonstandard tableau is obtained. In Beale's method the iteration is followed by its corresponding dual iteration in which u enters the basis and v_3 leaves it. The solutions of tableau 3 of table 3 and tableau 6 of table 2 are the same for the primal variables, but not for the dual variables. Tableau 3 of table 3 is a nonstandard tableau, so in the next iteration u must be introduced into the basis. When this is done, v_3 leaves the basis and the optimal solution is found. In Beale's method, however, we find that the solution of tableau 6 of table 2, though it is contained in a standard tableau, is a nonstandard one in Beale's sense, since w_1 and w_2 are both nonzero. This means that both the added constraints can be deleted with advantage. The following four iterations are used to discard the previously added constraints and generate new ones which have w-variables with zero values; this means that these constraints do not restrict the solution from obtaining the maximum value of the objective function. The final solution, that of tableau 10 of table 2 is, of course, the same as that of tableau 4 of table 3.

In order to describe the various possible cases that may occur in the simplex method in a concise way, the following definitions are used. An iteration in which a primal variable enters the basis and its corresponding dual variable leaves it is called a *standard iteration*; if another basic variables leaves the basis instead, the iteration is called a *nonstandard iteration*. If a dual variable enters the basis and another dual variable leaves it, the iteration is called a *dual standard iteration*; if a primal basic variable leaves the basis instead, the iteration is called a *dual nonstandard iteration*. In the simplex method the following cycles of iterations may occur, where a cycle is counted from a standard tableau to the next standard tableau. We may have a standard iteration, which carries a standard tableau into another standard tableau; see, for example, the first two iterations in the example. We may have a nonstandard iteration and a dual standard iteration; see, for example, the third and the fourth iterations in the example. Finally we may have a nonstandard iteration, a series of dual nonstandard iterations and a dual standard iteration. The Beale method, viewed in the context of the simplex method, consists of pairs of nonstandard and dual standard iterations.

We shall now explore all possibilites of successive iterations in both methods. We assume that both methods start with the same basic feasible solution to the constraints; hence, if there are m constraints, the initial solution contains m basic variables. The first iteration in the simplex method may be a standard or a nonstandard one, depending on whether the dual variable of the new basic variable or some primal variable leaves the basis. If a primal variable leaves the basis, the next iteration will be a dual standard one, since all elements in the columns of nonbasic dual variables and the rows of basic primal variables will be zero, as for example in the set-up tableau. After this a number of similar iterations may occur, which all consist of pairs of nonstandard and dual standard iterations. In iterations like these, Beale's method will generate exactly the same solutions and tableaux. If C is a zero matrix, so that we have a linear programming problem, the iterations will all be of this type and both methods are equivalent to a simultaneous application of the simplex method for linear programming to the problem and the dual method for linear programming to the dual of the problem.

Next we consider the situation in which, proceeding from a basic feasible solution to the constraints, the simplex method performs a standard iteration. This happens in the first iteration of table 3. In this case Beale's method will add a constraint and perform a primal and a dual iteration. The resulting tableaux are the same, apart from the elements connected with the z- and w-variables in Beale's method. The w-variable will have a value zero, because the dual variable of the new basic variable became zero in the primal iteration. The solutions of corresponding tableaux in the simplex method and in Beale's method must be the same, since they are both solutions of the following problem: Maximize the objective function subject to the equality constraints (excluding possible added constraints), and the condition that the nonbasic variables are zero. We may call such a solution the *provisional maximum solution* for the tableau. The other common tableau-elements of corresponding tableaux must be the same for similar reasons.

Let us now consider the case of a number of successive standard iterations in the simplex method. Beale's method will for any standard iteration in the simplex method add a constraint and perform a primal and a dual iteration. All w-variables will have a value zero because the matrix of elements in the column of the z-variables and the rows of the w-variables is diagonal; this in its turn results from the addition of the constraints. The solutions of corresponding tableaux in the simplex method and in Beale's method are again the same and for the same reasons as indicated before. The number of successive standard iterations cannot exceed the rank of C. This means that in Beale's method the number of added constraints cannot exceed the rank of C.

We shall now consider the remaining cases. In these, the simplex method, starting from a standard tableau, performs a nonstandard iteration, possibly one or more dual nonstandard iterations and finally a dual standard iteration, which brings the tableau back into standard form. We shall assume that the simplex method and Beale's method both start with the same solution, apart from the w-variables in Beale's method which must have a zero value. Note that we have already treated the special case when we start out with a basic feasible solution to the constraints; in this case the methods proved to be completely equivalent. Here we treat the remaining cases, where we have a solution having more than m primal basic variables, so that in the simplex method a number of standard iterations must already have occurred. In Beale's method we must then have added so many constraints as to make the number of primal basic variables equal to the total number of constraints. In the example iterations of this type occur in the last two iterations of the simplex method and the last six iterations in Beale's method.

In order to find out what happens during these iterations, we must take a closer look at what both methods set out to do. When in the simplex method a new basic variable enters the basis, a provisional maximum is sought with the new basic variable plus the old basic primal variables in the basis. This maximum is found when the dual variable of the new basic variable leaves the basis. However, this may result in some of the basic variables obtaining negative values. This is prevented in the simplex method by stipulating that the first primal basic variable that becomes zero leaves the basis. After this, a new provisional maximum is sought, which is obtained by entering the dual variable of the nonbasic pair into the basis. If the dual variable of the basic pair leaves the basis, the provisional maximum is obtained; otherwise another provisional maximum is sought, which has one primal basic variable less, and so on.

When in Beale's method, starting from a standard solution in Beale's sense, a new basic variable enters the basis, the same provisional maximum is sought as in the simplex method. Also here it turns out that a basic variable, the same as in the simplex method, threatens to become negative; this variable therefore leaves the basis. Also Beale's method then modifies its aims: a new provisional maximum is sought in which the variable that threatened to become negative is nonbasic. After the first primal and dual iteration it is found that this provisional maximum is evidently not obtained, because it is found that the constraints that were added in previous iterations can be dropped with advantage. This is then done in the next iterations and other constraints are added in such a way that the provisional maximum is obtained; this maximum would be the same as in the simplex method. However, it might be that, when the constraints are dropped one by one, some basic variables become negative. Hence these variables become nonbasic and new provisional maxima are sought.

Comparing the simplex method and Beale's method, we see that the iterations of both methods have the same aim: obtaining a particular provisional maximum. In the simplex method this is achieved by the dual standard iteration. In Beale's method previously added constraints are dropped one by one and new constraints are added in such a way that the provisional maximum is an extreme point of added and original constraints. Now it may be that in the process of deleting and adding constraints, basic variables threaten to become negative, which are not negative in the provisional maximum. Hence a different provisional maximum may be found in Beale's method; this maximum would have a lower value than that obtained in the simplex method. Now it may be that the primal variable(s) concerned would enter the basis in the next iteration(s), in which case the same solution would be obtained as in the simplex method. However, this is not necessarily so. Hence it must be concluded that the simplex method and Beale's method have solution-paths that are possibly different.

One may wonder why in a nonstandard solution in Beale's sense it is necessary to replace all previously added constraints by new ones. It turns out that the added constraints are linear combinations of the equations

$$- p + Au - v = Cx,\qquad(16)$$

$$b = Ax + y,\qquad(17)$$

for specific values of u and v; actually only the equations of (16) with corresponding basic x-variables play a role. If a nonstandard solution appears in Beale's method, the values of u-variables change, and because of this the added constraints must all be replaced. Hence all complications of Beale's method appear to be due to not carrying along in the tableau the columns of nonbasic dual variables.

12.4. A combinatorial method

The quadratic programming method developed by Theil and Van de Panne [78] can be considered an extension of classical quadratic maximization, that is maximization of a quadratic form subject to linear equalities. This method can be called a combinatorial method for reasons which will become clear in the following. The combinatorial method requires that C is positive definite.

The quadratic programming problem can be formulated as the maximization of a quadratic form with

$$f = p'x - \tfrac{1}{2}x'Cx\qquad(18)$$

with C positive definite subject to a set S of linear equality constraints

$$Ax + y = b,\qquad(19)$$

and a set T of inequality constraints

$$x \geq 0, \quad y \geq 0. \tag{20}$$

The set S may be empty.

The combinatorial method is based on the idea that if the constraints in T which are binding (that is, satisfied in equality form) for the optimal solution are known, this solution may be obtained by maximization subject to the constraints in S and the binding ones in T in equality form. The method tries to find this set of constraints which is binding for the optimal solution.

As a starting-point, the quadratic form maximized subject to the equality constraints in S. It is supposed that this maximum exists; a sufficient condition for this is that the matrix of the quadratic form is negative definite. Let the resulting solution be $x^{(0)}$. If $x^{(0)}$ satisfies all constraints in T, it is the optimal solution. If not, let the set of violated constraints of T be T_0. It can be proved that the optimal solution satisfies at least one of the constraints as an equality. This means that T_0 contains at least one of the binding constraints of the optimal solution. If this constraint or these constraints were known, they could be added to the set S and another maximum of the quadratic form subject to the constraints in S could be found which perhaps would satisfy the remaining inequality constraints. But since it is not known which inequality constraints are binding, each of the constraints of T_0 is added seperately in equality form to S and the maximum of the quadratic form subject to the augmented S-sets is found. This means that as many solutions are generated as the number of inequalities in T_0. If none of these solutions satisfies the inequality constraints, then each solution is treated as $x^{(0)}$ was treated. This means that for each solution all constraints violated by that solution are, each one separately, added in equality form to the set S, so that these sets now contain two inequality constraints in equation form. The objective function is maximized subject to these new S-sets. If none of the resulting solutions satisfy all inequality constraints, the violated constraints are added separately to form new sets S, and so on.

If a solution is found which does not violate the inequality constraints, it is not necessarily the optimal solution; a necessary condition for the optimal solution is that if each one of the inequality constraints imposed on this solution is deleted, the resulting solutions should violate the deleted inequality constraint.

As an illustration, let us consider a problem with five inequality constraints, numbered 1, 2, 3, 4, 5, and let $x^{(i,j,...)}$ indicate the solution of a maximization of the quadratic form subject to the equality constraints and the inequality constraints i, j, ... in equality form. First $x^{(0)}$ is computed. Let $x^{(0)}$ violate constraints 1, 3, and 4. Then $x^{(1)}$, $x^{(3)}$ and $x^{(4)}$ are computed. Let $x^{(1)}$ violate constraints 3 and 5, x^3 violate constraints 1 and 4 and x^4 violate constraints 1 and 2. Then we should compute $x^{(1,3)}$ and $x^{(1,5)}$; $x^{(1,3)}$ and $x^{(3,4)}$; $x^{(1,4)}$ and $x^{(2,4)}$. $x^{(1,3)}$ should, of course, be only computed once. If $x^{(2,4)}$ violates no constraint, it is the optimal solution if $x^{(4)}$ violates constraint 2 and $x^{(2)}$ violates constraint 4. Figure 1 gives an indication of the sequence of these computations.

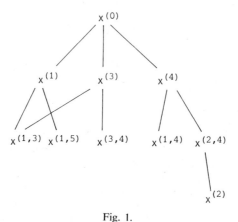

Fig. 1.

It was shown in chapter 8 that all computations for a maximization of a quadratic form, subject to linear equalities if there are any, may conveniently be performed in simplex tableaux using pivot transformations. This can be used with advantage in the combinatorial method. As an example for this method, the problem that served as an example for the simplex method for quadratic programming is used: Maximize

$$f = 2x_1 + 2x_2 - 2x_1^2 + 2x_1x_2 - 2x_2^2 \qquad (21)$$

subject to

$$x_1 + x_2 \le 1,$$
$$-x_1 + 6x_2 \le 2,$$
$$x_1, x_2 \ge 0. \qquad (22)$$

Figure 2 gives a geometric representation of this problem. By using this figure, it can easily be indicated how the method works. First the

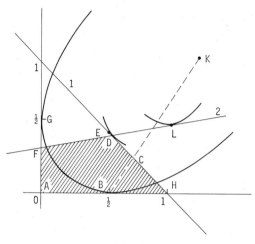

Fig. 2.

unconstrained maximum $x^{(0)}$ is determined which is found in K. Since $x^{(0)}$ violates both constraints 1 and 2, $x^{(1)}$ and $x^{(2)}$ are determined; these are found in E and L, respectively. $x^{(1)}$ violates 2, so that $x^{(1,2)}$ should be determined; $x^{(2)}$ violates 1 so that also from this the determination of $x^{(1,2)}$ follows. $x^{(1,2)}$ is found in D. In order to ascertain that this is the optimal solution, it should be checked that $x^{(2)}$ violates constraint 1 and $x^{(1)}$ violates constraint 2, which is indeed the case.

Table 4 presents the computations for the method in simplex tableau format. Tableau 0 is the set-up tableau. The unconstrained maximum x^0 is found by introducing into the basis x_1 and x_2, which replace v_1 and v_2. Tableau 2 contains $x^{(0)}$.

Since y_1 and y_2 are negative, solutions should be generated in which each of these variables is replaced seperately by their corresponding dual variables u_1 and u_2; in this way constraints 1 and 2 are imposed in equality form. This results in tableaux 3 and 4. In tableau 3, y_2 is negative, so that it should be replaced by u_2 and in tableau 4, y_1 is negative, so that it should be replaced by u_1. In both cases tableau 5 results, which contains $x^{(1,2)}$, a solution which is nonnegative in the primal variables. The constraints imposed in equality form are indicated by the basic dual variables in the solution, u_1 and u_2. If constraint 1 is deleted, which is done by replacing u_1 by y_1, tableau 4 would result, in which y_1 is negative; and if constraint 2 is deleted by replacing u_2 by y_2, tableau 3 would result, with y_2 negative. Hence the optimal solution is found.

Table 4

Simplex tableaux for an application of the combinatorial method

Tableau	Basic var.	Values bas. var.	x_1	x_2	u_1	u_2
	F	0	-2	-2	-1	-2
	v_1	-2	$\underline{-4}$	2	-1	1
0	v_2	-2	$\underline{2}$	-4	-1	-6
	y_1	1	1	1	0	0
	y_2	2	-1	6	0	0
			v_1	x_2	u_1	u_2
	F	1	$-\frac{1}{2}$	-3	$-\frac{1}{2}$	$-2\frac{1}{2}$
	x_1	$\frac{1}{2}$	$-\frac{1}{4}$	$-\frac{1}{2}$	$\frac{1}{4}$	$-\frac{1}{4}$
1	v_2	-3	$\frac{1}{2}$	$\underline{-3}$	$-1\frac{1}{2}$	$-5\frac{1}{2}$
	y_1	$\frac{1}{2}$	$\frac{1}{4}$	$1\frac{1}{2}$	$-\frac{1}{4}$	$\frac{1}{4}$
	y_2	$2\frac{1}{2}$	$-\frac{1}{4}$	$5\frac{1}{2}$	$\frac{1}{4}$	$-\frac{1}{4}$
			v_1	v_2	u_1	u_2
	F	4	-1	-1	1	3
	x_1	1	$-\frac{1}{3}$	$-\frac{1}{6}$	$\frac{1}{2}$	$\frac{2}{3}$
2	x_2	1	$-\frac{1}{6}$	$-\frac{1}{3}$	$\frac{1}{2}$	$1\frac{5}{6}$
	y_1	-1	$\frac{1}{2}$	$\frac{1}{2}$	-1	$-2\frac{1}{2}$
	y_2	-3	$\frac{2}{3}$	$1\frac{5}{6}$	$-2\frac{1}{2}$	$\underline{-10\frac{1}{3}}$
			v_1	v_2	y_1	u_2
	F	3	$-\frac{1}{2}$	$-\frac{1}{2}$	1	$\frac{1}{2}$
	x_1	$\frac{1}{2}$	$-\frac{1}{12}$	$\frac{1}{12}$	$\frac{1}{2}$	$-\frac{7}{12}$
3	x_2	$\frac{1}{2}$	$\frac{1}{12}$	$-\frac{1}{12}$	$\frac{1}{2}$	$\frac{7}{12}$
	u_1	1	$-\frac{1}{2}$	$-\frac{1}{2}$	-1	$2\frac{1}{2}$
	y_2	$-\frac{1}{2}$	$-\frac{7}{12}$	$\frac{7}{12}$	$-2\frac{1}{2}$	$\underline{-4\frac{1}{12}}$
			v_1	v_2	u_1	y_2
	F	$3\frac{4}{31}$	$-\frac{25}{31}$	$-\frac{29}{62}$	$\frac{17}{62}$	$\frac{9}{31}$
	x_1	$\frac{25}{31}$	$-\frac{9}{31}$	$-\frac{3}{62}$	$\frac{21}{62}$	$\frac{3}{31}$
4	x_2	$\frac{29}{62}$	$-\frac{3}{62}$	$-\frac{1}{124}$	$\frac{7}{124}$	$\frac{11}{62}$
	y_1	$-\frac{17}{62}$	$\frac{21}{62}$	$\frac{7}{124}$	$\underline{\frac{49}{124}}$	$-\frac{15}{62}$
	u_2	$\frac{9}{31}$	$-\frac{2}{31}$	$-\frac{11}{62}$	$\frac{15}{62}$	$-\frac{3}{31}$
			v_1	v_2	y_1	y_2
	F	$2\frac{46}{49}$	$-\frac{4}{7}$	$-\frac{3}{7}$	$\frac{34}{49}$	$\frac{6}{49}$
	x_1	$\frac{4}{7}$	0	0	$\frac{6}{7}$	$-\frac{1}{7}$
5	x_2	$\frac{3}{7}$	0	0	$\frac{1}{7}$	$\frac{1}{7}$
	u_1	$\frac{34}{49}$	$-\frac{6}{7}$	$-\frac{1}{7}$	$-2\frac{26}{49}$	$\frac{30}{49}$
	u_2	$\frac{6}{49}$	$\frac{1}{7}$	$-\frac{1}{7}$	$\frac{30}{49}$	$-\frac{12}{49}$

It must be observed that the optimality check on a solution, which is nonnegative in the primal variables, is just a roundabout way to check the signs of the basic dual variables. Since in the transformation corresponding with the deletion of an imposed constraint a negative pivot (principal element of a standard tableau) is used, the sign of the new basic (primal) variable is negative if that of the leaving basic (dual) variable was positive.

The combinatorial method can be compared with the dual method for quadratic programming because both methods use a starting solution, which has nonnegative dual variables. The unconstrained maximum (if it exists, which it guaranteed if C is positive definite) always provides such a solution for the simple reason that it does not contain basic dual variables. The combinatorial method then uses principal pivots for all negative primal variables and does the same for all resulting tableaux until the optimal solution is obtained. In the dual method each tableau leads to only one subsequent tableau; furthermore it may also occur that a dual variable is replaced by a primal variable in case of a blocked tableau, which does not occur in the combinatorial method.

For most problems, the combinatorial method will be computationally inefficient compared with other methods for quadratic programming. Furthermore, there is the condition that the maximum $x^{(0)}$ must exist, which is the case of no equality constraints amounts to the condition that the matrix of the quadratic form is negative definite. Under some special circumstances, for instance cases in which the vectors $x^{(0)}, x^{(i)}, x^{(j)} \ldots$ are given, the method may be useful. The most interesting aspect of the method seems to be that the monotonic increase or decrease of the objective function or a parameter is not used, which is used in almost all other methods. It is obvious that the price for not using this property is rather high.

CHAPTER 13

MULTIPARAMETRIC LINEAR AND QUADRATIC PROGRAMMING

13.1. Linear problems with multiparametric constant terms

So far, only parametric problems with just one parameter were considered. In this chapter, methods are developed for linear and quadratic problems in a number of parameters $\lambda_1, \lambda_2, ..., \lambda_l$, which are elements of a vector λ. These parameters may occur in the constant terms of the constraints or in the coefficients of the linear part of the objective function.

The linear problem with multiparametric constant terms may be formulated as follows: Maximize

$$f = p'x$$

subject to

$$Ax \leq b + B\lambda,$$

$$x \geq 0,$$

where λ is a vector of l parameters $\lambda_1, ..., \lambda_l$ and B is an $m \times l$ matrix. Usually, optimal solutions are required for certain ranges of the λ's, for instance for $0 \leq \lambda \leq \bar{\lambda}$, where $\bar{\lambda}$ is a vector of given upper bounds.

To clarify the nature of the problem, assume that λ contains two parameters λ_1, and λ_2. The problem then requires an optimal solution for each pair of values of λ_1 and λ_2 within the indicated range, that is for each point of a rectangle in the λ_1, λ_2-plane.

Fortunately, simplifications are possible. The equation system for the problem may be written as

$$f_0 + q'\lambda = -p'x + f, \tag{1}$$

$$b + B\lambda = Ax + y, \tag{2}$$

where $f_0 = 0$, $q = 0$. Now consider any basic solution to the problem. The equation system for which this matrix is basic will have the same form as (1)–(2) with x standing for the vector of nonbasic variables (containing a mixture of the x- and y-variables of the original problem) and y standing for the vector of basic variables (containing the remaining x- and y-variables of the original problem); all coefficients of the original problem will have been changed by transformation.

If the basic solution is optimal in the narrow sense, then $-p \geq 0$. If it is optimal in the wide sense, then in addition its values of basic variables should be nonnegative:

$$b + B\lambda \geq 0. \tag{3}$$

This system of inequalities indicates a region in the λ-space for which this basic solution is optimal. The multiparametric problem can be solved by generating for basic solutions with $-p \geq 0$, the different regions of the form (3). These regions together form that part of the λ-space for which optimal solutions are required. The basic solutions and the corresponding regions can be generated in a systematic manner. We shall follow the main aspects of a method developed by Gal and Nedoma[14].

To explain the method, the following example is used: Maximize for all $0 \leq \lambda_1 = \infty, \ 0 \leq \lambda_2 \leq \infty$,

$$f = 2x_1 + x_2$$

subject to

$$x_1 + x_2 \leq \lambda_1,$$
$$2x_1 - x_2 \leq 1 - \lambda_2,$$
$$4x_1 + x_2 \leq 4,$$
$$x_1 \geq 0, \quad x_2 \geq 0.$$

This example has only two parameters, so that the different regions for which a particular basic solution is optimal can conveniently be indicated in a two-dimensional space.

Tableau 0 of table 1 gives the set-up tableau. Let us first generate an optimal solution for $\lambda_1 = \lambda_2 = 0$, which is done by solving the problem while ignoring the λ-term columns. In one step of the simplex method the optimal solution is obtained, see tableau 1.

The basic solution in terms of the λ's is now:

$$x_1 = 0 + \lambda_1 + 0\lambda_2,$$
$$y_2 = 1 - 2\lambda_1 - \lambda_2,$$
$$y_3 = 4 - 4\lambda_1 + 0\lambda_2.$$

The region in the λ-space for which this solution is feasible and therefore

Table 1

Generation of solutions for multiparametric problem

Tableau	Bas. var.	Val. bas. var.			x_1	x_2
		$c-t$	λ_1-t	λ_2-t		
0	f	0	0	0	-2	-1
	y_1	0	1	0	1	1
	y_2	1	0	-1	2	-1
	y_3	4	0	0	4	1
					y_1	x_2
1	f	0	2	0	2	1
	x_1	0	1	0	1	1
	y_2	1	-2	-1	-2	-3
	y_3	4	-4	0	-4	-3
					y_1	y_2
2	f	$\frac{1}{3}$	$1\frac{1}{3}$	$-\frac{1}{3}$	$1\frac{1}{3}$	$\frac{1}{3}$
	x_1	$\frac{1}{3}$	$\frac{1}{3}$	$-\frac{1}{3}$	$\frac{1}{3}$	$\frac{1}{3}$
	x_2	$-\frac{1}{3}$	$\frac{2}{3}$	$\frac{1}{3}$	$\frac{2}{3}$	$-\frac{1}{3}$
	y_3	3	-2	1	-2	-1
					y_1	y_3
3	f	$1\frac{1}{3}$	$\frac{2}{3}$	0	$\frac{2}{3}$	$\frac{1}{3}$
	x_1	$1\frac{1}{3}$	$-\frac{1}{3}$	0	$-\frac{1}{3}$	$\frac{1}{3}$
	x_2	$-1\frac{1}{3}$	$1\frac{1}{3}$	0	$1\frac{1}{3}$	$-\frac{1}{3}$
	y_2	-3	2	-1	2	-1
					x_1	y_3
4	f	4	0	0	2	1
	y_1	-4	1	0	-3	-1
	x_2	4	0	0	4	1
	y_2	5	0	-1	6	1

optimal in the wide sense is now given by $x_1 \geq 0$, $y_2 \geq 0$, $y_3 \geq 0$ or

$$-\lambda_1 \leq 0, \tag{4}$$

$$2\lambda_1 + \lambda_2 \leq 1, \tag{5}$$

$$4\lambda_1 \leq 4. \tag{6}$$

Furthermore, we only require solutions for $\lambda_1 \geq 0$, $\lambda_2 \geq 0$. The result is region I of figure 1. This region is bounded by (5) and $\lambda_1 \geq 0$, $\lambda_2 \geq 0$. The constraints (4) and (6) are redundant.

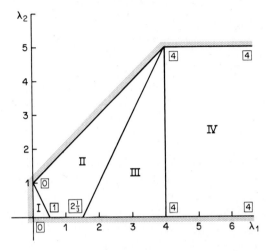

Fig. 1.

Now optimal solutions for regions adjacent to region I are generated. Since solutions for negative values of λ_1, and λ_2 are not required, only solutions for which constraint (5) is not satisfied are considered, that is, solutions for which

$$2\lambda_1 + \lambda_2 > 1. \tag{7}$$

In other words, solutions are considered for points on the other side of the boundary $2\lambda_1 + \lambda_2 = 1$.

Such solutions are generated by selecting a negative pivot in the row of y_2 in tableau 1. If there is no negative element in this row in the columns of the nonbasic variables, then no solution exists for (7), so that for points in the region on the other side of the constraint there is no solution.

In this case there are two negative elements. Since we are interested in optimal solutions only, the rule for the dual method should be followed when selecting the pivot. After comparison of

$$\frac{2}{-(-2)}, \frac{1}{-(-3)}$$

-3 is selected. Transformation results in tableau 2.

The basic solution of this tableau is optimal for

$$\tfrac{1}{3} + \tfrac{1}{3}\lambda_1 - \tfrac{1}{3}\lambda_2 \geq 0,$$
$$-\tfrac{1}{3} + \tfrac{2}{3}\lambda_1 + \tfrac{1}{3}\lambda_2 \geq 0,$$
$$3 - 2\lambda_1 + \lambda_2 \geq 0,$$

or

$$-\tfrac{1}{3}\lambda_1 + \tfrac{1}{3}\lambda_2 \leq \tfrac{1}{3}, \tag{8}$$

$$-\tfrac{2}{3}\lambda_1 - \tfrac{1}{3}\lambda_2 \leq -\tfrac{1}{3}, \tag{9}$$

$$2\lambda_1 - \lambda_2 \leq 3. \tag{10}$$

Together with $\lambda_2 \geq 0$, these constraints delineate region II of figure 1. Note that in this case there are no redundant constraints.

Now regions adjacent to region II should be found. First the region on the other side of (8) is considered. Since in the row of x_2 in tableau 2, there are no negative elements in the columns of nonbasic variables, the region adjacent to region II on this side is infeasible. The region on the other side of (9) is region I. We are not interested in points with $\lambda_2 < 0$, so that the region below region II is not considered. Coming to constraint (10), we find that there are two negative elements in the row of y_3 in tableau 2. After comparison of

$$\frac{1\tfrac{1}{3}}{-(-2)}, \frac{\tfrac{1}{3}}{-(-1)}$$

we decide that -1 is the pivot and generate tableau 3.

The basic solution of this tableau is optimal for

$$\tfrac{1}{3}\lambda_1 \leq 1\tfrac{1}{3}, \tag{11}$$

$$-1\tfrac{1}{3}\lambda_1 \leq -1\tfrac{1}{3}, \tag{12}$$

$$-2\lambda_1 + \lambda_2 \leq -3. \tag{13}$$

Together with $\lambda_2 \geq 0$, then constraints determine region III. The constraint (12) is redundant.

Adjacent regions are generated by considering the various nonredundant constraints. (11) leads to inspection of the row of x_1 in tableau 3. The single negative element can be used as a pivot which results in tableau 4. (13) leads back to tableau 2.

The region for which the solution of tableau 4 is optimal is described by

$$- \lambda_1 \le - 4, \tag{14}$$

$$0 \le 4, \tag{15}$$

$$\lambda_2 \le 5, \tag{16}$$

and $\lambda_2 \ge 0$. This is region IV of figure 1. Constraint (14) leads back to region III, (15) is redundant, and the region $\lambda_2 > 5$ is infeasible. No new regions can be generated, so that the problem is solved. Figure 2 gives the contour lines for the objective function for optimal solutions. Note that the lines are parallel and straight for the same region, but that they are piecewise linear over all regions taken together.

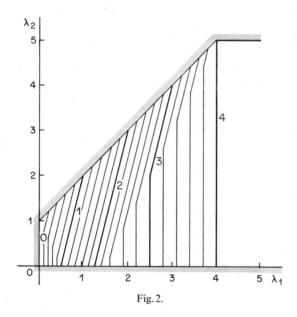

Fig. 2.

Now a general description of the method will be given. As a first step, an optimal solution must be generated for at least one point within that part of the λ-space for which solutions are required; in most cases $\lambda = 0$ can be used. Then an equation system of the form (1)–(2) will be obtained with a region in the λ-space described by (3) or

$$- B\lambda \le b, \tag{17}$$

and possible nonegativity constraints and upper bounds of the λ's. Let us assume that the problem has such constraints.

Now adjacent regions in the λ-space should be found for each non-redundant constraint. This means that each constraint of (17) should be tested for redundancy, which is done as follows. The system (17) together with the nonnegativity and upper-bound constraints in the λ's can be written as

$$- b_{11}\lambda_1 - ... - b_{1l}\lambda_l + z_1 = b_1,$$

$$- b_{m1}\lambda_1 - ... - b_{ml}\lambda_l + z_m = b_m,$$

$$0 \leq \lambda_1 \leq \bar{\lambda}_1, ..., 0 \leq \lambda_l \leq \bar{\lambda}_l, z_1 \geq 0, ..., z_m \geq 0.$$

First, a feasible solution is generated. Then, in order to test the ith constraint for redundancy, z_i is made basic if it is not yet basic. z_i is now considered the variable of an objective function which is to be minimized. If a minimum $\hat{z}_i \geq 0$ is found, the ith constraint is redundant, otherwise it is not. The minimum is, of course, characterized by having nonpositive elements in the z_i-row.

For each nonredundant constraint we find out whether there is an adjacent feasible region in the λ-space by searching for negative elements in its row in the columns of nonbasic variables in (2). If there are no such elements, there is no adjacent feasible region, if there is one, then it leads to another basic solution with adjacent feasible region. If there are more negative elements, the pivot is selected as in the dual method and another basic solution with adjacent feasible region results.

Each region of the form (3) generated in this way is treated in the same manner. If no more basic solutions with corresponding regions can be found, the problem is solved as required.

From the way the regions are generated, it is obvious that for a point within a region the corresponding basic solution is optimal. That a basic solution cannot be optimal for a point outside its corresponding region follows from the fact that outside the region at least one basic variable is negative since at least one constraint of (3) must be violated outside the region.

The order in which the various basic solutions are generated can be chosen to suit computational convenience. Gal and Nedoma[14] propose to take among all basic solutions to be generated the one which is closest to the one generated before in terms of pivot steps.

It is obvious that the number of basic solutions and corresponding regions will grow increasingly with the number of parameters l and the number of constraints of the problem, m. To keep the number of regions down to a reasonable number, the number of parameters should perhaps not exceed 4 or 5. It should be realized that the usefulness of the vector decreases as the number of regions increases, since it must then be increasingly difficult to find out to which region any given vector λ belongs.

13.2. A node method for multiparametric programming

The solution of a multiparametric problem takes the form of a number of basic solutions of the type (1)–(2) which are optimal for corresponding regions of the type (3). For any given vector λ it is in general not immediately clear to which region it belongs, except in the case of two parameters, where a two-dimensional figure can be used. The only way of finding out to which region, if any, a given vector λ belongs is to substitute it in all regions of the form (3). This points out the fact that as the number of regions increases the accessibility of solutions for any given vector λ decreases.

In a number of cases, the parametric problem is used to explore the consequences of having different values of a number of parameters. The regions and the corresponding solutions are then not very informative without substitution of specific values of the λ's. If we want to generate optimal solutions for some specific values of the λ's, then the extreme points of the regions in the λ-space are very informative. We may, therefore, consider the generation of the optimal solutions for these points form (3) for the different basic solutions.

If this would be done, the solution for each extreme point would be generated as many times as there are regions for which it is an extreme point, which is rather inefficient. In the following, a direct method for the generation of these points is developed.

For any extreme point, it is easy to generate the various regions of which it is an extreme point, as will be shown later. This will be useful if it is decided that the neighbourhood of a certain extreme point should be explored in more detail.

One of the possible applications is in post-optimality analysis, where all extreme-point feasible solutions are required having at least a certain f-value. The motivation of this analysis is that solutions which are close to the optimum are also interesting, because it may be difficult to formulate an objective function which exactly reflects the preferences of the decision maker with respect to the variables; analysis of different near-optimal solutions gives the decisionmaker an opportunity to correct for this. The decisionmaker is therefore given all extreme points of the feasible region, which have at least a minimum value of f, so that he may select one or a linear combination. The drawback is the large number of solution generated for even small problems.

However, it is not likely that the decisionmaker is uncertain about the specification of the objective function with respect to all variables, but only with respect to a few of these. This means that solutions are required which are optimal with respect to all variables, except these few variables. If these few variables are interpreted as parameters, the method of this section will generate only extreme points which are optimal with respect to the other variables. This will reduce the number of extreme points to be generated to a large extent, thus making near-optimality analysis more attractive.

The equation system for a general basic solution of the multiparametric problem can be written as follows:

$$f_0 + q'\lambda = p'x + f, \tag{18}$$

$$b + B\lambda = Ax + y, \tag{19}$$

where x is the vector of nonbasic variables and y that of basic variables. If $p \geq 0$, the solution is optimal. The region in the λ-space for which the solution is optimal is given by

$$- B\lambda \leq b,$$

$$0 \leq \lambda \leq \bar{\lambda}. \tag{20}$$

After introduction of slack variables, this can be written out as follows:

$$- b_{11}\lambda_1 - \ldots - b_{1l}\lambda_l + z_1 = b_1,$$

$$- b_{m1}\lambda_1 - \ldots - b_{ml}\lambda_l + z_m = b_m,$$

$$0 \leq \lambda_1 \leq \bar{\lambda}_1, \ldots, 0 \leq \lambda_l \leq \bar{\lambda}_l,$$

$$z_1 \geq 0, \ldots, z_m \geq 0. \tag{21}$$

The various extreme-point solutions of (21) can be generated by making the appropriate λ's basic, replacing certain z-variables and putting λ's at their upper bounds.

Let us now consider the λ's as ordinary variables, which means that the λ-terms are moved to the right in (18) and (19). The system then becomes

$$f_0 = - q'\lambda + p'x + f,$$

$$b = - B\lambda + Ax + y. \tag{22}$$

Then the same extreme points can be generated as before by making the same λ's basic replacing now y-variables instead of z-variables. Instead of generating extreme points of (20) for different optimal solutions of (18)–(19), we may generate these points directly from (22).

But not all extreme points of (22) are optimal with respect to the x- and y-variables and only these are required. Hence each extreme point of (22) should be tested for optimality. If this requires the generation and testing of each extreme point of (22), this method would be very ineffi- cient, but fortunately all optimal extreme points form a connected graph, so that only extreme points which are neighbours of optimal extreme points have to be generated and tested.

The relationship between the region of (21) for a particular basic solu- tion of (19) and the feasible region of (22) is that the former is a projection of the latter on the λ-space. Any extreme point of (21) is adjacent to at least one other extreme point for the same basic solution. Furthermore, from the way the regions for different basic solutions were generated, it follows that each of these regions is adjacent to at least one other region if there is more than one region, and that they have at least one extreme point in common. Hence each extreme point of (21) is linked, directly or indirectly, to all other extreme points. The same must be true for the corresponding extreme points of (22), so that each optimal extreme point must be linked, directly or indirectly to all other extreme points.

The optimality test for extreme points of (22) is now described. If no λ's are basic in the extreme point, then the point is optimal if the equation system which has this extreme point as basic solution has $p \geq 0$; otherwise it is not optimal.

If one or more λ's are basic, then the optimality of the point is proved if it is possible to express the point as an extreme point of (20) for a basic solution of (22) with $p \geq 0$. Consider the system of which the ex- treme point is a basic solution. This system may, after partitioning and rearrangement, be represented as follows:

$$b_0 = q'\lambda^1 + p'x + f, \tag{23}$$

$$b^1 = B_1\lambda^1 + A_1x + \lambda^2, \tag{24}$$

$$b^2 = B_2\lambda^1 + A_2x + y. \tag{25}$$

This system should be transformed into a system of the form (22) with the λ's in λ^2 nonbasic, the same variables as in (25) basic, and all coefficients of nonbasic variables in the f-equation nonnegative, if the point is optimal. To find out whether this is possible we consider the system

$$0 = p'x + f,$$
$$0 = A_1x + \lambda^2, \tag{26}$$

and find out whether an optimal solution exists with the λ's in λ^2 nonbasic.

The dual of this problem is: Find a basic feasible solution to

$$- A_1'u + v = p,$$
$$v \geq 0, \tag{27}$$

in which the u-variables are basic. If $p \geq 0$, the u-variables can be made basic while retaining feasibility by making them basic in a positive or negative direction; the case that a column of $-A_1'$ contains zeros only cannot happen, because then the rows of A_1 are dependent, which is impossible because in the original problem the λ's were nonbasic; hence if $p \geq 0$, the point is optimal.

If some elements of p are negative, artificial variables are introduced into the corresponding rows and the first-phase simplex method is used. If a feasible solution is obtained, the extreme point is optimal; if not, it is not. It is not necessary to make all u-variables basic after a feasible solution has been obtained for the same reason as indicated in the previous paragraph.

The same computations can be performed in the primal framework (26). If $p \geq 0$, the point is optimal, if it contains some negative elements, the system can be written as

$$0 = p^{1\prime}x^1 + p^{2\prime}x^2 + f,$$
$$0 = A_{11}x^1 + A_{12}x^2 + \lambda^2, \tag{28}$$

with $p^1 < 0$ and $p^2 \geq 0$. Then artificial constraints $x^1 \leq e$ are added, where e is a vector of units. After addition of slack variables, this constraint becomes

$$e = x^1 + z, \quad z \geq 0.$$

If the variables in x^1 are made basic, the system becomes

$$-p^{1\prime}e = -p^{1\prime}z + p^{2\prime}x^2 + f,$$
$$-A_{11}e = -A_{11}z + A_{12}x^2 + \lambda^2,$$
$$e = z \qquad\qquad + x^1,$$
$$z \geq 0, \, x^1 \geq 0, \, x^2 \geq 0. \tag{29}$$

The dual method is now used to minimize f. If in the final solution the z-variables are all basic, the point is optimal; otherwise it is not. Again it is not necessary to eliminate all λ's once the z-variables are all basic.

If λ^2 contains only one element, the following optimality test suffices: If

$$\underset{j}{\text{Max}} \left\{ \frac{p_j}{-a_{rj}} \,\middle|\, a_{rj} > 0 \right\} \leq \left\{ \underset{j}{\text{Min}} \, \frac{p_j}{-a_{rj}} \,\middle|\, a_{rj} < 0 \right\}, \tag{30}$$

then the point concerned is optimal, otherwise it is not. a_{rj} indicates the jth element in the rth row of A_1, in which the λ is basic.

13.3. Numerical example

The same problem as in section 1 will be used for a numerical example. Tableau 0 of table 2 gives the set-up tableau. After one iteration the optimal solution for $\lambda_1 = \lambda_2 = 0$ is found, see tableau 1. This is the first extreme-point optimal solution. Introduction of y_1 or x_2 evidently leads to nonoptimal solutions, so that only the introduction of λ_1 and λ_2 is considered. This leads to the solutions of tableaux 2 and 3. The solution of tableau 3 is optimal, while that of tableau 2 can be shown to be optimal by means of (30).

When the neighbours of tableau 2 are considered, it is found that there is no unique leaving basic variable when y_2 or λ_2 are introduced, but when lexicographic ordering is used, it is found that λ_1 is the leaving basic variable. The introduction of y_2 then leads to tableau 1, that of λ_2 to tableau 3, that of y_1 to no finite solution and that of x_2 to tableau 4. The solution of tableau 4 is shown to be optimal by means of (30).

We consider then the neighbours of tableau 3. The introduction of λ_1 and y_2 leads to tableaux 2 and 1, that of y_1 and x_2 to the new tableaux 5 and 6, the first of which is nonoptimal and the second of which is optimal. Turning to tableau 4, we find that it has as neighbours new tableaux 7 and 8, which are both optimal, and tableau 2. Introducing λ_1 in tableau 6, we find again tableau 8. Introduction of y_2 leads to a nonoptimal solution, that of y_1 to tableau 5, which was nonoptimal. Introduction of x_1 leads to tableau 2. In the same manner it can be checked that tableaux 7 and 8 do not lead to new optimal solutions.

Hence, we have generated all optimal extreme-point solutions. The solution $\lambda_1 = 0$, $\lambda_2 = 1$ has been generated twice, which is due to the degeneracy in the system. There are no extreme-point solutions for $\lambda_1 \rightarrow \infty$; if some finite bound $\lambda_1 = \bar{\lambda}_1$ had been given, two more solutions

Table 2

Tableaux for node generation

Tab.	B.v.	V.b.v.	λ_1	λ_2	x_1	x_2
0	f	0	0	0	-2	-1
	y_1	0	-1	0	1	1
	y_2	1	0	1	2	-1
	y_3	4	0	0	4	1

			λ_1	λ_2	y_1	x_2
1	f	0	-2	0	2	1
	x_1	0	-1	0	1	1
	y_2	1	2	1	-2	-3
	y_3	4	4	0	-4	-3

			y_2	λ_2	y_1	x_2
2	f	1	1	1	0	-2
	x_1	$\frac{1}{2}$	$\frac{1}{2}$	$\frac{1}{2}$	0	$-\frac{1}{2}$
	λ_1	$\frac{1}{2}$	$\frac{1}{2}$	$\frac{1}{2}$	-1	$-1\frac{1}{2}$
	y_3	2	-2	-2	0	3

			λ_1	y_2	y_1	x_2
3	f	0	-2	0	2	1
	x_1	0	-1	0	1	1
	λ_2	1	2	1	-2	-3
	y_3	4	4	0	-4	-3

Tab.	B.v.	V.b.v.	y_2	λ_2	y_1	y_3
4	f	$2\frac{1}{3}$	$-\frac{1}{3}$	$-\frac{1}{3}$	0	$\frac{2}{3}$
	x_1	$\frac{5}{6}$	$\frac{1}{6}$	$\frac{1}{6}$	0	$\frac{1}{6}$
	λ_1	$1\frac{1}{2}$	$-\frac{1}{2}$	$-\frac{1}{2}$	-1	$\frac{1}{2}$
	x_2	$\frac{2}{3}$	$-\frac{2}{3}$	$-\frac{2}{3}$	0	$\frac{1}{3}$

			λ_1	y_2	x_1	x_2
5	f	0	0	0	-2	-1
	y_1	0	-1	0	1	1
	λ_2	1	0	1	2	-1
	y_3	4	0	0	4	1

			λ_1	y_2	y_1	x_1
6	f	0	-1	0	1	-1
	x_2	0	-1	0	1	1
	λ_2	1	-1	1	1	3
	y_3	4	1	0	-1	3

			x_1	λ_2	y_1	y_3
7	f	4	2	0	0	1
	y_2	5	6	1	0	1
	λ_1	4	3	0	-1	1
	x_2	4	4	0	0	1

			y_2	x_1	y_1	y_3
8	f	4	0	2	0	1
	λ_2	5	1	6	0	1
	λ_1	4	0	3	-1	1
	x_2	4	0	4	0	1

would have been found, namely $f = 4$, $\lambda_1 = \bar{\lambda}_1$, $\lambda_2 = 0$, and $f = 4$, $\lambda_1 = \bar{\lambda}_1$, $\lambda_2 = 5$.

All these extreme points can be found in figure 1, together with their f-values. Note that the different regions are not generated explicitly by the method. However, it is easy to generate for any extreme point the regions of which it is an extreme point.

This can be done in the following manner. As point of departure, the tableau in which the node is generated is used. Then the same pivot transformation as in the optimality checking procedure, as indicated in

section 2 are performed in that tableau, so that an optimal solution results with nonbasic λ's. The basic solution in terms of the λ's provides
one region in the λ-space.

Adjacent feasible regions of which the node was an extreme point are
found by generating only those neighbouring regions which can be found
by operating on rows which give an equality for the λ-values of the node.
An example will clarify this.

Let us generate the feasible regions which have the point $\lambda_1 = 4$, $\lambda_2 = 5$
as an extreme point, so that tableau 8 of table 2 is the point of departure,

Table 3

Generation of feasible regions at a node

Tabl.	B.v.	V.b.v.	y_2	x_2	y_1	y_3
0	f	4	0	2	0	1
	λ_2	5	1	6	0	1
	λ_1	4	$\underline{0}$	3	-1	1
	x_2	4	0	4	0	1

			λ_2	x_1	y_1	y_3
1	f	4	0	2	0	1
	y_2	5	1	6	0	1
	λ_1	4	0	3	$\underline{-1}$	1
	x_2	4	0	4	$\underline{0}$	1

			λ_2	x_1	λ_1	y_3
2	f	4	0	2	0	1
	y_2	5	1	6	0	1
	y_1	-4	0	$\underline{-3}$	-1	-1
	x_2	4	0	$\underline{4}$	0	1

			λ_2	y_1	λ_1	y_3
3	f	$1\frac{1}{3}$	0	$\frac{2}{3}$	$-\frac{2}{3}$	$\frac{1}{3}$
	y_2	-3	1	2	-2	-1
	x_1	$1\frac{1}{3}$	0	$-\frac{1}{3}$	$\frac{1}{3}$	$\frac{1}{3}$
	x_2	$-1\frac{1}{3}$	0	$1\frac{1}{3}$	$-1\frac{1}{3}$	$-\frac{1}{3}$

			λ_2	y_1	λ_1	y_2
4	f	$\frac{1}{3}$	$\frac{1}{3}$	$1\frac{1}{3}$	$-1\frac{1}{3}$	$\frac{1}{3}$
	y_3	3	-1	-2	2	-1
	x_1	$\frac{1}{3}$	$\frac{1}{3}$	$\frac{1}{3}$	$-\frac{1}{3}$	$\frac{1}{3}$
	x_2	$-\frac{1}{3}$	$-\frac{1}{3}$	$\frac{2}{3}$	$-\frac{2}{3}$	$-\frac{1}{3}$

see tableau 0 of table 3. To find a corresponding optimal tableau with λ_1 and λ_2 as nonbasic variables is rather easy in this case.

In tableau 2, an optimal solution with nonbasic λ's is obtained. The region in the λ-space connected with this solution is given by the inequalities which are derived from the constant term and the λ-columns:

$$0\lambda + \lambda_2 \leq 5,$$
$$-\lambda_1 + 0\lambda_2 \leq -4,$$
$$0\lambda_1 + 0\lambda_2 \leq 4.$$

This region is easily identified as region IV of figure 1.

Now adjacent regions with the same extreme point should be generated. We note that both the first and the second inequality are binding for $\lambda_1 = 4$, $\lambda_2 = 5$. To generate adjacent regions we try to change basic variables in the corresponding rows, while retaining the optimality. Since the row of y_2 in tableau 2 does not contain negative elements in the columns of x- and y-variables we conclude that there is no feasible region in the half-space $0\lambda_1 + \lambda_2 > 5$, which is also obvious from figure 1.

In the row of y_1 there are two negative elements, so that, to retain optimality we compare $\frac{2}{3}$ and $\frac{1}{1}$ and conclude that x_1 should replace y_1. Tableau 3 results, which has as feasible region

$$-2\lambda_1 + \lambda_2 \leq -3,$$
$$\tfrac{1}{3}\lambda_1 + 0\lambda_2 \leq 1\tfrac{1}{3},$$
$$-1\tfrac{1}{3}\lambda_1 + 0\lambda_2 \leq -1\tfrac{1}{3}.$$

This is region III of figure 1.

Both the first and the second constraints are binding for $\lambda_1 = 4$, $\lambda_2 = 5$. First we try to replace the basic variable of the first constraint, y_2, which is done by pivoting on the element -1. The result is tableau 4, which has as feasible region

$$2\lambda_1 - \lambda_2 \leq 3,$$
$$-\tfrac{1}{3}\lambda_1 + \tfrac{1}{3}\lambda_2 \leq \tfrac{1}{3},$$
$$-\tfrac{2}{3}\lambda_1 - \tfrac{1}{3}\lambda_2 \leq -\tfrac{1}{3}.$$

This is region II of figure 1.

Proceeding from the second constraint of region III, we only find back region IV. In region II, again the first two constraints are binding for $\lambda_1 = 4$, $\lambda_2 = 5$. The first constraint leads back to region III, while the second constraint leads to no feasible region. Our search is now completed.

13.4. A multiparametric objective function

We consider a linear problem with a multiparametric objective function: Maximize

$$f = (p + P\lambda)'x$$

subject to

$$Ax \leq b,$$
$$x \geq 0,$$

where λ is a vector of l parameters and P is an $n \times l$ matrix. Optimal solutions are required for all values of the vector λ or for the values $0 \leq \lambda \leq \bar{\lambda}$, where $\bar{\lambda}$ is a given vector of upper bounds.

The method for these problems is quite similar to the method developed for multiparametric constant term. First an optimal solution for $\lambda = 0$ or for any given value of λ is generated. Let the equation system for this solution be written in a general solution as:

$$f^0 + q'\lambda = (p + P\lambda)'x + f,$$
$$b = Ax + y. \tag{31}$$

This solution is optimal for $p + P\lambda \geq 0$ or

$$-P\lambda \leq p. \tag{32}$$

(32) describes a region in the λ-space for which the basic solution of (31) is optimal. To find optimal solutions for adjacent regions, first all non-redundant constraints of (32) should be determined, which is done as indicated in section 13.1.

For each non-redundant constraint, the corresponding nonbasic variable should enter the basic. The leaving basic variable is determined as in the simplex method. If no leaving basic variable can be found, the adjacent region has no finite optimal solution. After transformation another solution of the form (31) is obtained, which is optimal for a region in the λ-space of the form (32) and which is adjacent to the previous region. This process continues until no new basic solutions of the form (31) can be found.

It is interesting to note that for all points in the same region the optimal solution is exactly the same in terms of the x- and y-variables; the value of f varies with the λ's. As the method is relatively straightforward, no numerical example is given. A node method for extreme points in the λ-space can be developed similar to the node method of the previous sections.

13.5. Multiparametric quadratic programming

The multiparametric quadratic programming problem can be formulated as follows: Maximize

$$f = (p + P\lambda)'x - \tfrac{1}{2}x'Cx$$

subject to

$$Ax \le b + B\lambda,$$

$$x \ge 0,$$

for all λ or $0 \le \lambda \le \bar{\lambda}$; λ is a vector of l parameters; C is positive semi-definite and symmetric.

The set-up tableau for the problem is shown in table 4.

Table 4

B.v.	V.b.v.			
	$c - t$	$\lambda - t$	x	u
f_c	0	0	$-p'$	$-b'$
f_λ	0	0	$-P'$	$-B'$
v	$-p$	$-P$	$-C$	$-A'$
y	b	B	A	0

Note that the values of basic variables and the coefficients of nonbasic variables in the objective function are linear expressions in λ, while the value of f is a quadratic form in the λ's.

First, an optimal solution for some value of λ, say $\lambda = 0$, is found. This results in table 5.

Table 5

B.v.	V.b.v.			
	$c - t$	$\lambda - t$	x	u
f_c	s_0	s'	$-q'$	$-r'$
f_λ	s	S	$-D$	$-E$
v	q	D	Q	$-P'$
y	r	E	P	R

In table 5 Q and R are negative semi-definite and symmetric and S is symmetric. Note that the table has the usual symmetry and skew-symmetry properties. v and y are vectors of dual and primal basic variables, and x and u are the vectors of corresponding primal and dual nonbasic variables.

The basic solution $v = q + D\lambda$, $y = r + E\lambda$ is optimal in the wide sense for

$$- D\lambda \le q,$$
$$- E\lambda \le r. \tag{33}$$

This is a region in the λ-space.

To find optimal solutions for adjacent regions, first all nonredundant constraints of (33) are determined in the usual fashion. For each of these constraints, the adjacent region or regions of the type (33) must be determined.

If the diagonal element in the row of the constraint is negative, this element is used as a pivot of a transformation, which results in another tableau for which an adjacent region is optimal. In this case there is only one adjacent region, since any point on the boundary of the original region will also belong to the newly generated region.

If the diagonal element is zero, the situation is more complicated. Consider the case in which a constraint is connected with a dual variable, say v_k of a general tableau. If $q_{kk} = 0$, then, since Q is negative semi-definite, $q_{ik} = q_{kj} = 0$ for all i and j. Now x_k should enter the basis, replacing a primal variable; the dual variable corresponding to the latter should enter the basis replacing v_k, where the pivot of the latter transformation should be negative in order to unblock v_k. Hence, we are looking for the pivots p_{ik} and $-p_{ik} > 0$ of a double pivot transformation, such that the points on the boundary

$$q_k + d_{k1}\lambda_1 + ... + d_{kl}\lambda_l = 0 \tag{34}$$

are feasible in the original region and in the region(s) of (the) new basic solution(s) to be generated.

We should therefore consider the system consisting of (34) and

$$r_i + e_{i1}\lambda_1 + ... + e_{il}\lambda_l - p_{ik}x_k \ge 0 \tag{35}$$

for all i with $p_{ik} > 0$, and the nonnegativity constraint on x_k and possible bounds in the λ's. After introduction of slack variables in (35), these inequalities may be tested for redundancy. For each nonredundant inequality of (35), a double pivot p_{ik} and $- p_{ik}$ is chosen. Hence, there are as many adjacent regions as there are nonredundant constraints of the type (35).

If the nonredundant constraint of (33) is connected with a primal variable, the same method applies *mutatis mutandis.*

As an example of application of multiparametric quadratic programming let us use the slightly modified example for the simplex method for quadratic programming. Maximize

$$f = 2x_1 + 2x_2 - 2x_1^2 + 2x_1 x_2 - 2x_2^2$$

subject to

$$x_1 + x_2 \le 1 + \lambda_1,$$
$$-x_1 + 6x_2 \le 2 + \lambda_2,$$
$$x_1 \ge 0,\ x_2 \ge 0,\ 0 \le \lambda_1 \le \infty,\ 0 \le \lambda_2 \le \infty.$$

Since the columns of λ_1 and λ_2 are identical to those of the slack variables y_1 and y_2, they do not have to be written down separately. Tableau 1 of table 6 gives the optimal tableau for $\lambda_1 = \lambda_2 = 0$.

The region in the λ-space for which this basic solution is optimal is determined by

$$-\tfrac{6}{7}\lambda_1 + \tfrac{1}{7}\lambda_2 \le \tfrac{4}{7},$$
$$-\tfrac{1}{7}\lambda_1 - \tfrac{1}{7}\lambda_2 \le \tfrac{3}{7},$$
$$2\tfrac{26}{49}\lambda_1 - \tfrac{30}{49}\lambda_2 \le \tfrac{34}{49},$$
$$-\tfrac{30}{49}\lambda_1 + \tfrac{12}{49}\lambda_2 \le \tfrac{6}{49}.$$

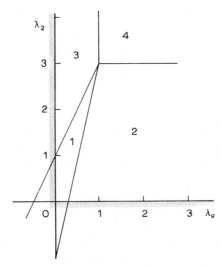

Fig. 3.

Table 6

An example of multiparametric quadratic programming

Tableau	B.v.	V.b.v.	v_1	v_2	y_1	y_2
	F	$2\frac{46}{49}$	$-\frac{4}{7}$	$-\frac{3}{7}$	$\frac{34}{49}$	$\frac{6}{49}$
	x_1	$\frac{4}{7}$	0	0	$\frac{6}{7}$	$-\frac{1}{7}$
1	x_2	$\frac{3}{7}$	0	0	$\frac{1}{7}$	$\frac{1}{7}$
	u_1	$\frac{34}{49}$	$-\frac{6}{7}$	$-\frac{1}{7}$	$-2\frac{26}{49}$	$\frac{30}{49}$
	u_2	$\frac{6}{49}$	$\frac{1}{7}$	$-\frac{1}{7}$	$\frac{30}{49}$	$-\frac{12}{49}$

			v_1	v_2	u_1	y_2
	F	$3\frac{4}{31}$	$-\frac{25}{31}$	$-\frac{29}{62}$	$\frac{17}{62}$	$\frac{9}{31}$
	x_1	$\frac{25}{31}$	$-\frac{9}{31}$	$-\frac{3}{62}$	$\frac{21}{62}$	$\frac{2}{31}$
2	x_2	$\frac{29}{62}$	$-\frac{3}{62}$	$-\frac{1}{124}$	$\frac{7}{124}$	$\frac{11}{62}$
	y_1	$-\frac{17}{62}$	$\frac{21}{62}$	$\frac{7}{124}$	$-\frac{49}{124}$	$\frac{15}{62}$
	u_2	$\frac{9}{31}$	$-\frac{2}{31}$	$-\frac{11}{62}$	$\frac{15}{62}$	$-\frac{3}{31}$

			v_1	v_2	y_1	u_2
	F	3	$-\frac{1}{2}$	$-\frac{1}{2}$	1	$\frac{1}{2}$
	x_1	$\frac{1}{2}$	$-\frac{1}{12}$	$\frac{1}{12}$	$\frac{1}{2}$	$-\frac{7}{12}$
3	x_2	$\frac{1}{2}$	$\frac{1}{12}$	$-\frac{1}{12}$	$\frac{1}{2}$	$\frac{7}{12}$
	u_1	1	$-\frac{1}{2}$	$-\frac{1}{2}$	-1	$2\frac{1}{2}$
	y_2	$-\frac{1}{2}$	$-\frac{7}{12}$	$\frac{7}{12}$	$-2\frac{1}{2}$	$-4\frac{1}{12}$

			v_1	v_2	u_1	u_2
	F	4	-1	-1	1	3
	x_1	1	$-\frac{1}{3}$	$-\frac{1}{6}$	$\frac{1}{2}$	$\frac{2}{3}$
4	x_2	1	$-\frac{1}{6}$	$-\frac{1}{3}$	$\frac{1}{2}$	$1\frac{5}{6}$
	y_1	-1	$\frac{1}{2}$	$\frac{1}{2}$	-1	$-2\frac{1}{2}$
	y_2	-3	$\frac{2}{3}$	$1\frac{5}{6}$	$-2\frac{1}{2}$	$-10\frac{1}{3}$

The first constraints are redundant, the last two, together with $\lambda_1 \geq 0$, $\lambda_2 \geq 0$, give region 1 in figure 3.

The rows of u_1 and u_2 in tableau 1 correspond to the nonredundant constraints. Since both have negative diagonal elements, these can be used to generate new solutions for adjacent regions. The resulting optimal tableaux are tableaux 2 and 3.

For tableau 2 we find

$$-\tfrac{2}{31}\lambda_2 \leq \tfrac{25}{31},$$

$$-\tfrac{11}{62}\lambda_2 \leq \tfrac{29}{62},$$

$$-\lambda_1 + \tfrac{15}{62}\lambda_2 \leq -\tfrac{17}{62},$$

$$\tfrac{3}{31}\lambda_2 \leq \tfrac{9}{31}.$$

Again the first two constraints are redundant. The latter two together with $\lambda_2 \geq 0$ determine region 2 of figure 3.

For tableau 3 we find

$$-\tfrac{1}{2}\lambda_1 \leq \tfrac{1}{2},$$

$$-\tfrac{1}{2}\lambda_1 \leq \tfrac{1}{2},$$

$$\lambda_1 \leq 1,$$

$$2\tfrac{1}{2}\lambda_1 - \lambda_2 \leq -\tfrac{1}{2}.$$

The first two constraints are redundant, the latter two, together with $\lambda_1 \geq 0$ lead to region 3.

Tableau 4 can be generated either from tableau 2 or tableau 3. The region is determined by

$$0\lambda_1 + 0\lambda_2 \leq 1,$$

$$0\lambda_1 + 0\lambda_2 \leq 1,$$

$$-\lambda_1 \qquad \leq -1,$$

$$-\lambda_2 \leq -3.$$

This is region 4. No new regions can be found, so that the problem is solved.

The value of the objective function for the various regions can be easily found. We have

$$F_1 = 2\tfrac{46}{49} + 1\tfrac{19}{49}\lambda_1 + \tfrac{12}{49}\lambda_2 - 2\tfrac{26}{49}\lambda_1^2 + 1\tfrac{11}{49}\lambda_1\lambda_2 - \tfrac{12}{49}\lambda_2^2,$$

$$F_2 = 3\tfrac{4}{31} + \tfrac{18}{31}\lambda_2 - \tfrac{3}{31}\lambda_2^2,$$

$$F_3 = 3 + 2\lambda_1 - \lambda_1^2,$$

$$F_4 = 4.$$

CHAPTER 14

CONCAVE QUADRATIC PROGRAMMING

14.1. Convex and concave functions

Let $f(x)$ be a function of a vector x. Consider two points in the x-space, x^1 and x^2, and points x^* on the line joining these two points:

$$x^* = \theta x^1 + (1 - \theta)x^2 \qquad \text{for } 0 \le \theta \le 1.$$

The function $f(x)$ is said to be *convex* if

$$f(x^*) \le \theta f(x^1) + (1 - \theta)f(x^2). \tag{1}$$

In other words, a function is convex if linear interpolation between two points never underestimates the function value at that point.

Figure 1 gives an illustration for the case that vector x contains one variable. The straight line between A and B gives the linear interpolation $\bar{f}(x)$ and is never below the curve of $f(x)$ if $f(x)$ is convex.

The expression *convex function* should not be confused with the expression *convex region*. A region is called convex if for any two points within that region, any point on the line joining these two points is also in the region. In symbols: a region R is convex if for all $x^1 \in R$ and $x^2 \in R$, we have $x^* \in R$, where

$$x^* = \theta x^1 + (1 - \theta)x^2 \qquad \text{for } 0 \le \theta \le 1.$$

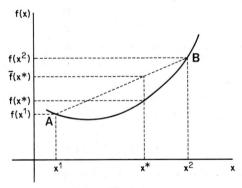

Fig. 1.

373

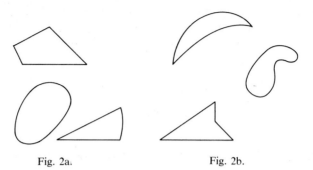

Fig. 2a.   Fig. 2b.

Figure 2a gives some examples of convex regions, and figure 2b of nonconvex regions. Obviously the region of points satisfying linear inequality and equality constraints is convex.

If $f(x)$ is a convex function, it has a unique minimum over a convex region or the same minimum value is found for a convex subset of the convex region. For suppose there are two local minima x^1 and x^2 in the convex region with values of $f(x^1)$ and $f(x^2)$. For any point x^* between x^1 and x^2 we have

$$f(x^*) \le \theta f(x^1) + (1 - \theta)f(x^2).$$

Let $f(x^2) < f(x^1)$. We then find

$$\theta f(x^1) + (1 - \theta)f(x^2) < \theta f(x^1) + (1 - \theta)f(x^1) = f(x^1), 0 \le \theta < 1,$$

so that $f(x^*) < f(x^1)$; hence x^1 cannot have been a local minimum. In case $f(x^1) = f(x^2)$, we have

$$f(x^*) \le f(x^1), f(x^*) \le f(x^2).$$

Since x^1 and x^2 are assumed to be local minima, we conclude

$$f(x^*) = f(x^1) = f(x^2).$$

The above property is of great importance, since if we have found a local minimum of a convex function over a convex region, we can be sure that this is the global minimum.

Let us now consider the *maximum* of a convex function over a convex region. From (1) we see immediately that for any point x^* between x^1 and x^2, we have

$$f(x^*) \le \text{Max} [f(x^1), f(x^2)]. \qquad (2)$$

From this we may conclude that if there is a unique maximum it must be at an extreme point of the feasible region. Furthermore, a number of local maxima may exist, which are all extreme points, or if they are not, then they can be written as a nonnegative linear combination of local maxima with the same value which are extreme points.

This important property allows us to restrict the search for maxima of convex functions to extreme points of the feasible region.

A function is said to be *concave* if

$$f(x^*) \le \theta f(x^1) + (1 - \theta)f(x^2) \tag{3}$$

for

$$x^* = \theta(x^1) + (1 - \theta)x^2 \qquad \text{for } 0 \le \theta \le 1.$$

Obviously, $f(x)$ is concave if $-f(x)$ is convex. For a concave function linear interpolation between two points never overestimates the value of the function. Figure 3 gives a representation of a concave function.

If $f(x)$ is a concave function, it has a unique maximum over a convex region or the same maximum value is found for a convex subset of the convex region. Furthermore, the minimum of a concave function over a convex region occurs at an extreme point of that region, or otherwise points which are not extreme points can be written as linear combinations of extreme point minima.

A linear function $f = p'x$ is both convex and concave, as can be verified from (1) and (3). The maximum or the minimum of a linear function has the combined properties of concave and convex functions, namely it has a unique optimum at an extreme point or the same optimum value for a convex subset of the region such that any interior point can be written as a nonnegative linear combination of extreme points which are optima.

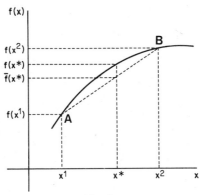

Fig. 3.

Let us now consider a quadratic function

$$f(x) = p'x + \tfrac{1}{2}x'Cx. \tag{4}$$

For this function we find for any two points x^1 and x^2:

$$f\{\theta x^1 + (1-\theta)x^2\} = p'\{\theta x^1 + (1-\theta)x^2\}$$
$$+ \tfrac{1}{2}\{\theta x^1 + (1-\theta)x^2\}'C\{\theta x^1 + (1-\theta)x^2\}$$
$$= \theta(p'x^1 + \tfrac{1}{2}x^{1\prime}Cx^1) + (1-\theta)(p'x^2 + \tfrac{1}{2}x^{2\prime}Cx^2)$$
$$- \tfrac{1}{2}\theta(1-\theta)(x^1 - x^2)'C(x^1 - x^2).$$

From this it follows that if C is positive semi-definite, then (1) applies, so that the functions is convex. If C is negative semi-definite, then (3) applies and the function is concave.

The problem: Minimize

$$f = p'x + \tfrac{1}{2}x'Cx \tag{5}$$

subject to

$$ax \leq b,$$
$$x \geq 0, \tag{6}$$

is called a *convex quadratic programming problem* if C is positive semi-definite, so that f is convex. Since this problem can be written in this form: Maximize

$$f = -p'x - \tfrac{1}{2}x'Cx \tag{7}$$

subject to

$$Ax \leq b,$$
$$x \geq 0, \tag{8}$$

with C positive semi-definite, the list problem is also a convex quadratic programming problem. Problems of the above kind have been treated in previous chapters.

Now consider problems of the form (5)–(6), in which C is negative semi-definite, so that f is concave. Such problems are called *concave quadratic programming problems* and have not been treated before. The equivalent maximization problem is: Maximize

$$f = p'x + \tfrac{1}{2}x'Cx$$

subject to

$$Ax \leq b,$$
$$x \geq 0,$$

where C is positive semi-definite. Of these problems we know that the optimum will occur at an extreme point, and that a number of local optima may exist. These problems are more difficult to solve than convex problems.

Problems in which no assumption is made about C are called *general quadratic programming problems*. Here the local optima are not necessarily unique nor do they necessarily occur at extreme points, which makes the solution of these problems even more difficult.

14.2. The simplex method for concave problems

The remainder of this chapter deals with methods for concave quadratic programming problems. This section discusses the use of the simplex method for quadratic programming to find local optima. The remaining sections explain Ritter's method applied to concave problems and Tui's method for the maximization of a convex nonlinear objective function subject to linear constraints.[1] As will be explained in the next chapter, Ritter's method can also be used for general quadratic problems, but then one part of the method is considerably more complicated. Tui's method will only be explained for the case of a convex quadratic objective function.

Let us consider the problem: Maximize

$$f = p'x + \tfrac{1}{2}x'Cx$$

subject to

$$Ax \le b,$$
$$x \ge 0,$$

where C is positive semi-definite. The Kuhn-Tucker conditions are the same as before and are necessary but not sufficient for a local optimum. As before, the following equation system is used:

$$0 = -p'x - b'u + F, \tag{9}$$
$$-p = Cx - A'u + v, \tag{10}$$
$$b = Ax + y. \tag{11}$$

---

[1] Recently, a number of new methods for concave and for general quadratic programming have been developed by BALAS[84], BURDET[86], BALAS and BURDET[85], GONCALVES[92] and KONNO[93], which could not be included in this book.

If b has some negative elements, the basic solution of (9)–(11) is not feasible; a basic feasible solution to the constraints may then first be found in the usual manner, after which the problem can be reformulated in a form with $b \geq 0$.

If $b \geq 0$, but some elements of p are positive, the simplex method for quadratic programming can be used to generate a local optimum. The rules for the method are the same as in convex quadratic programming, except that no ratios have to be taken in the row of the dual variable of the incoming variable, since the diagonal element in that row in nonnegative, because C is positive semi-definite. This means that, as in linear programming, the transformations are always of the double-pivot type.

The general form of the equation system is, apart from the 0, the same as that of (9)–(11). It may be written as

$$q_0 = q'x - r'u + F, \tag{12}$$

$$q = Qx - P'u + v, \tag{13}$$

$$r = Px \qquad + y, \tag{14}$$

where x and y are the vectors of nonbasic and basic primal variables, and v and u are the vectors of corresponding basic and nonbasic dual variables. The number of basic primal variables is always the same as the number of constraints of the original problem. Since basic solutions are feasible in the simplex method, $r \geq 0$. If some elements of q are negative, the nonbasic primal variable corresponding to it is introduced into the basis. The objective function as a function of the x-variables may be written as

$$F = 2f = q_0 - 2q'x + x'Qx, \tag{15}$$

so that $\partial F / \partial x_k = - q_k + q_{kk} x_k$ for $x_i = 0$ for $i \neq k$. $q_{kk} \geq 0$, because C is positive semi-definite and q_{kk} can be written as a ratio of two principal determinants of C. Hence F is a nondecreasing function of x_k, so that there is no reason for taking a ratio in the row of q_k. The leaving basic variable is therefore always a primal variable. The matrix R which appears in convex quadratic programming is here 0 and stays 0. The transformations are then always of the double-pivot type.

Application of the simplex method now leads to strictly increasing values of F in nondegenerate cases, until all elements of q are nonnegative.

If $q > 0$, then according to (15) a local optimum is obtained. If $q \geq 0$, but some elements of q are zero, the solution is not a local optimum if the corresponding elements of Q are nonzero, but it is if the problem is suitably perturbed.

The simplex method finds easily a local optimum, but in concave programming problems usually many local optima exist. To find the global optimum, the simplest method is to enumerate all extreme points of the feasible region and to evaluate them. As the number of extreme points is very large in most cases, this method is usually either very expensive or cannot be used at all. Nevertheless, to provide insight into the nature of the problem, a small example will be given.

We consider the problem: Maximize

$$f = - 1\tfrac{1}{2}x_1 - 1\tfrac{1}{2}x_2 + \tfrac{1}{2}x_1^2 + \tfrac{1}{2}x_1x_2 + \tfrac{1}{2}x_2^2$$

subject to

$$-\tfrac{1}{2}x_1 + x_2 \leq 2,$$
$$x_2 \leq 3,$$
$$2x_1 - x_2 \leq 7,$$
$$4x_1 + x_2 \leq 17,$$
$$x_1 \geq 0_1, \; x_2 \geq 0.$$

A three-dimensional representation of the problem is given in figure 4; f is measured from its minimum, $-1\tfrac{1}{2}$. See also figure 7.

Tableau 0 of table 1 gives the set-up tableau; the columns of dual nonbasic variables have been deleted as they contain the matrices $-A'$ and 0, which are known. The basic solution of tableau 0 is a local optimum, since the dual variables are positive. From the figure it is obvious that all extreme points can be generated without repetition in one path; in the tableaux this finds its expression in the fact that there are only two nonbasic primal variables.

If x_1 is introduced into the basis in tableau 0, y_3 leaves the basis; after a double pivot, tableau 1 results, which is not a local maximum. Introduction of x_2 with a double pivot leads to tableau 2, which again is not a local maximum. y_3 then replaces y_2, and u_2 replaces u_3, which results in tableau 3; this is a local maximum. Tableaux 4 and 5 are generated in the same manner. Introduction of y_1 in tableau 5 leads back to tableau 0. All extreme points have now been generated.

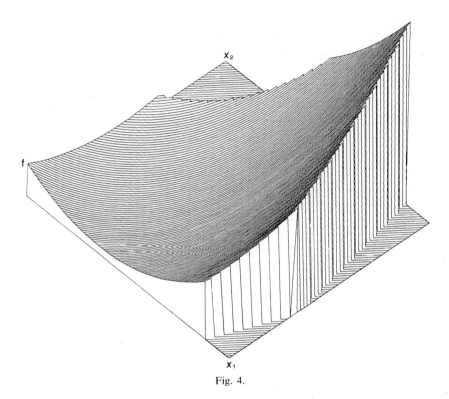

Fig. 4.

The solutions of tableaux 0, 3, and 5 are local maxima with f-values of 0, $6\frac{1}{8}$, and -1; the global maximum is given by the basic solution of tableau 3, and is $x_1 = 3\frac{1}{2}$, $x_2 = 3$, $f = 6\frac{1}{8}$. Note that the local maximum of tableau 5 with a value of -1 is, in fact, very close to the absolute minimum of $-1\frac{1}{2}$.

In any practical set-up, it is probably better to use any method for the generation of extreme points, while finding the value of f by direct substitution. The methods to be used would be the reverse simplex method with any objective function or the methods of Balinski[1], or Manas and Nedoma[22].

For larger concave problems, the method of Ritter and the method of Tui should be used, which are explained in the following sections, while for concave problems in which C has a low rank the parametric method for general quadratic programming explained in the next chapter could be attractive.

Table 1

Enumeration of extreme points

Tabl.	B.v.	V.b.v.	x_1	x_2	Tabl.	B.v.	V.b.v.	y_2	y_4
	F	0	$1\frac{1}{2}$	$1\frac{1}{2}$		F	$12\frac{1}{4}$	$2\frac{3}{8}$	$\frac{7}{8}$
	v_1	$1\frac{1}{2}$	1	$\frac{1}{2}$		u_2	$2\frac{3}{8}$	$\frac{13}{16}$	$\frac{1}{16}$
	v_2	$1\frac{1}{2}$	$\frac{1}{2}$	1		u_4	$\frac{7}{8}$	$\frac{1}{16}$	$\frac{1}{16}$
0	y_1	2	$-\frac{1}{2}$	1	3	y_1	$\frac{3}{4}$	$-1\frac{1}{8}$	$\frac{1}{8}$
	y_2	3	0	1		y_3	3	$1\frac{1}{2}$	$-\frac{1}{2}$
	y_3	7	2	-1		x_1	$3\frac{1}{2}$	$-\frac{1}{4}$	$\frac{1}{4}$
	y_4	17	4	1		x_2	3	1	0
			y_3	x_2				y_2	y_1
	F	$1\frac{3}{4}$	1	$-1\frac{1}{4}$		F	4	$6\frac{1}{2}$	-4
	u_3	1	$\frac{1}{4}$	$-\frac{1}{2}$		u_2	$6\frac{1}{2}$	7	-5
	v_2	$-1\frac{1}{4}$	$-\frac{1}{2}$	$1\frac{3}{4}$		u_1	-4	-5	4
1	y_1	$3\frac{3}{4}$	$\frac{1}{4}$	$\frac{3}{4}$	4	y_4	6	9	8
	y_2	3	0	1		y_3	6	-3	4
	x_1	$3\frac{1}{2}$	$\frac{1}{2}$	$-\frac{1}{2}$		x_1	2	2	-2
	y_4	3	-2	3		x_2	3	1	0
			y_3	y_4				x_1	y_1
	F	6	$-\frac{1}{2}$	1		F	-2	$\frac{1}{4}$	$\frac{1}{2}$
	u_3	$-\frac{1}{2}$	$\frac{13}{36}$	$-\frac{2}{9}$		v_1	$\frac{1}{4}$	$1\frac{3}{4}$	-1
	u_4	1	$-\frac{2}{9}$	$\frac{7}{36}$		u_1	$\frac{1}{2}$	-1	1
2	y_1	3	$\frac{3}{4}$	$-\frac{1}{4}$	5	y_4	15	$4\frac{1}{2}$	-1
	y_2	2	$\frac{2}{3}$	$-\frac{1}{3}$		y_3	9	$1\frac{1}{2}$	1
	x_1	4	$\frac{1}{6}$	$\frac{1}{6}$		y_2	1	$\frac{1}{2}$	-1
	x_2	1	$-\frac{2}{3}$	$\frac{1}{3}$		x_2	2	$-\frac{1}{2}$	1

14.3. Ritter's method for concave quadratic programming

As enumeration of extreme points is very inefficient, methods should be found which limit the parts of the feasible region which have not been searched. An obvious idea in this respect is the use of cutting planes by means of which parts of the feasible region which have been searched are separated from the parts which have not yet been searched. Both Ritter's method and Tui's method are of this kind. Ritter's method is devised for general quadratic programming, but in the case of concave problems important simplifications can be made so that this case is treated first.

The presentation of Ritter's method given in this chapter and the next one differs substantially from the one given by Ritter himself, see [98, 99], without, however, changing the essence of the method. The presentation of Ritter's method by Cottle and Mylander[90] goes partly in the same direction.

The method starts by finding a local optimum to the problem. Ritter employs for this purpose a kind of capacity method. As the simplex method for quadratic programming is simpler and more straightforward, we shall assume that a local optimum is found in this manner. In concave problems this results in the basic solution of an equation system of the form:

$$q_0 = q'x - r'u + F, \tag{16}$$

$$q = Qx - P'u + v, \tag{17}$$

$$r = Px \qquad + y, \tag{18}$$

where x and y are the vectors of nonbasic and basic primal variables of this solution, and u and v are the vectors of nonbasic and basic dual variables. Since the solution is optimal, we have $q \geq 0$ and $r \geq 0$, Q is positive semi-definite; from this it follows that for a local maximum, if $q_i = 0$, then $q_{ii} = 0$ and also $q_{ij} = 0$ for each j.

The objective function can be written as:

$$F = q_0 - 2q'x + x'Qx. \tag{19}$$

Since $q \geq 0$, small increases in the x-variables will not increase F, but decrease it for any x_i with $q_i > 0$, but if x-variables are considered with nonzero terms in $x'Qx$, then larger increases will eventually lead to higher values of F.

Instead of the maximization of (19) subject to (18), we may formulate an *auxiliary problem*, where (19) is maximized subject to the constraint

$$q'x \leq \lambda. \tag{20}$$

This constraint is of the capacity type; the reason why q is chosen instead of e will become clear below. For this auxiliary problem, a global maximum is required for all values of λ. The solution of this problem will now first be discussed.

The Kuhn-Tucker equations for the auxiliary problem may be written as

$$q_0 = q'x - \lambda u + F, \tag{21}$$

$$q = Qx - qu + v, \tag{22}$$

$$\lambda = q'x \qquad + y. \tag{23}$$

The basic solution of these equations, $v = q$, $y = \lambda$, $F = 0$ gives a local maximum. In any other local maximum, y must be nonbasic, which is obvious from the fact that any positive semi-definite quadratic form cannot have an interior maximum. Hence, for other local maxima we have $q'x = \lambda$. After subtraction of this equation from (21) and combination of the terms in q (these manipulations are made possible by the choice of q in the capacity constraint), the system (21)–(23) becomes

$$q_0 - 2\lambda = -\lambda(u + 1) + F, \tag{24}$$

$$0 = Qx - q(u + 1) + v, \tag{25}$$

$$\lambda = q'x + y. \tag{26}$$

Let us divide (25) and (26) by λ; this results in

$$0 = Qx^* - qu^* + v^*, \tag{27}$$

$$1 = q'x^* + y^*, \tag{28}$$

where $x^* = x/\lambda$, $y^* = y/\lambda$, $v^* = v/\lambda$, $u^* = (u + 1)/\lambda$. According to (24), F may be expressed for any complementary solution as

$$F = q_0 - 2\lambda + \lambda^2 u^*. \tag{29}$$

The global maximum of F is therefore found for a local maximum with the highest value of u^*.

Since the auxiliary problem is concave, its maxima must be extreme points. Since there is only one constraint, there are only as many extreme points as there are x-variables. These extreme-point solutions can be found from the system (27)–(28) by introduction of each of the x^*-variables instead of y^*, and introducing u^* replacing the v^*-variable corresponding to the x^*-variable. All nonnegative solutions resulting from this should be compared for the value of u^*, and the one with the highest value of u^* should be chosen as the global maximum. Let this highest value by $\hat{u}^*$.

If $u = \lambda\hat{u}^* - 1 < 0$ or $\lambda < 1/\hat{u}^*$, the solution is not a local maximum, because it is not a nonnegative solution of (21)–(23). This means that for these values of λ the local maximum $x = 0$, $y = \lambda$ is the global maximum.

From (29) we conclude that the local maximum $x = 0$, $y = \lambda$ is dominated by the global solution of the auxiliary problem if $q_0 - 2\lambda + \lambda^2\hat{u}^* > q_0$ or $\hat{u}^* > 2/\lambda$ or $\lambda > 2/\hat{u}^*$. Let us define

$$\lambda_1 = 2/\hat{u}^*.$$

Then for $\lambda \leq \lambda_1$, the auxiliary problem gives no higher value than $F = q_0$. If therefore the constraint

$$q'x \geq \lambda_1 \tag{30}$$

is added to the original problem, no local optima with $F > q_0$ are eliminated.

But we may improve upon this by cutting an even larger piece from the feasible region. Let the solution of (27)–(28) corresponding with $\hat{u}*$ be $\hat{x}*$. The global maximum of (19) subject to (20) for $\lambda > 2/\hat{u}* = \lambda_1$ is then $\lambda\hat{x}*$. In the original problem of maximizing (19) subject to (18), the solution $\lambda\hat{x}*$ may be inserted and λ varied upwards from 0 until the solution leaves the feasible region; this value of λ is defined as λ_2.

λ_2 can be found as follows. Let us proceed from the equations

$$r = Px + y,$$
$$\lambda\hat{x}* = x + z,$$

where z is a vector of artificial variables. After the vector x has been made basic instead of z, the system becomes

$$r - P\hat{x}*\lambda = -Pz + y,$$
$$\hat{x}*\lambda = z + x.$$

Let us write $s = -P\hat{x}*$; we have then for λ_2:

$$\lambda_2 = \text{Min}\left(\frac{r_i}{-s_i}\middle| s_i < 0\right). \tag{31}$$

If $\lambda_2 < \lambda_1$, no higher value for F than q_0 can be found within $q'x \leq \lambda_2$ in the feasible region of the original problem, so that λ_1 should be used. On the other hand, if $\lambda_2 > \lambda_1$, then the value of $F = q_0 - 2\lambda_2 + \lambda_2^2\hat{u}*$ exceeds q_0 and the solution $x = \lambda_2\hat{x}*$ is the global maximum of (19) subject to (18) and $q'x \leq \lambda_2$. In this case we add to the original constraints the constraint

$$q'x \leq \lambda_2. \tag{32}$$

Hence, if $\lambda_2 > \lambda_1$, a larger part of the feasible region is eliminated, or, in other words, the cut is deeper; moreover, a higher value of F is found. Figures 5a and 5b illustrate the two possibilities.

To summarize, we determine

$$\bar{\lambda} = \text{Max}(\lambda_1, \lambda_2)$$

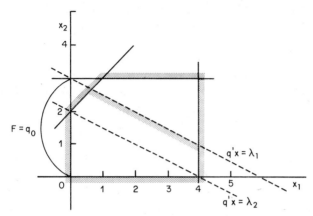

Fig. 5a.

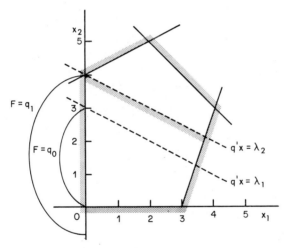

Fig. 5b.

and carry on with solving the problem of maximizing (19) subject to (18) and the additional constraint

$$q'x \geq \bar{\lambda}. \tag{33}$$

This problem is solved in the same manner as before by first determining a local maximum, and then formulating and finding the global maximum of the auxiliary problem; from this, a new cutting constraint of the type (33) is determined, and so on. There is a difference in the determination of λ_1. Instead of selecting λ_1, such that the maximum value of F in the auxiliary problem is equal to the value of F for $x = 0$, q_0, it may be made equal to the best value of F found so far; by doing this we do not eliminate solutions with a higher value of F.

Let this best value of F found so far be $\hat{F}$; evidently $\hat{F} \geq q_0$. We then have

$$q_0 - 2\lambda_1 + \lambda_1^2 u^* = \hat{F}$$

or

$$\lambda_1 = \{1 + \sqrt{1 + \hat{u}^*(F - q_0)}\}/\hat{u}^*.$$

The method terminates if, after the addition of another constraint, the feasible region turns out to be empty. The global maximum is then the solution with the highest value of F found so far.

Ritter[99] proved that his method would terminate in a finite number of steps, but Zwart[101] constructed a counter-example, which is given in the next section, and pointed out an error in the proof. The difficulty is that the cutting planes may cut off even smaller pieces of the feasible region without necessarily converging to the global maximum.

14.4. Numerical example and counter-example

For an illustration of Ritter's method, the same example is used as for the enumeration of extreme points method. Tableau 0 of table 1 gives the set-up tableau which gives a local maximum. The auxiliary problem is now: Maximize

$$f = -1\tfrac{1}{2}x_1 - 1\tfrac{1}{2}x_2 + \tfrac{1}{2}x_1^2 + \tfrac{1}{2}x_1x_2 + \tfrac{1}{2}x_2^2$$

subject to

$$1\tfrac{1}{2}x_1 + 1\tfrac{1}{2}x_2 \leq \lambda,$$

$$x_1 \geq 0, \ x_2 \geq 0.$$

For this problem a global maximum is required. The problem is modified as described in section 14.3; the set-up tableau is then as given in tableau 0 of table 2. The two extreme points are found without difficulty, see tableaux 1 and 2.

The global maximum of the auxiliary problem is not unique, since F has the same values in tableux 1 and 2; this means that either one can be chosen. Let us select the solution of tableau 1. We find $\lambda_1 = 2/\hat{u}^* = 1\tfrac{1}{8}$.

Table 2

Solutions of first auxiliary problem

	B.v.	V.b.v.	x_1	x_2	u
	F	0	0	0	-1
0	v_1	0	1	$\frac{1}{2}$	$-1\frac{1}{2}$
	v_2	0	$\frac{1}{2}$	1	$-1\frac{1}{2}$
	y	1	$1\frac{1}{2}$	$1\frac{1}{2}$	0
			y	x_2	v_1
	F	$\frac{4}{9}$	$\frac{4}{9}$	$\frac{1}{3}$	$-\frac{2}{3}$
1	u	$\frac{4}{9}$	$\frac{4}{9}$	$\frac{1}{3}$	$-\frac{2}{3}$
	v_2	$\frac{1}{3}$	$\frac{1}{3}$	1	-1
	x_1	$\frac{2}{3}$	$\frac{2}{3}$	1	0
			x_1	y	v_2
	F	$\frac{4}{9}$	$\frac{1}{3}$	$\frac{4}{9}$	$-\frac{2}{3}$
2	v_1	$\frac{1}{3}$	1	$\frac{1}{3}$	-1
	u	$\frac{4}{9}$	$\frac{1}{3}$	$\frac{4}{9}$	$-\frac{2}{3}$
	x_2	$\frac{2}{3}$	1	$\frac{2}{3}$	0

For λ_2 we find:

$$P = \begin{bmatrix} -\frac{1}{2} & 1 \\ 0 & 1 \\ 2 & -1 \\ 4 & 1 \end{bmatrix}, \qquad \hat{x}^* = \begin{bmatrix} \frac{2}{3} \\ 0 \end{bmatrix}, \qquad s = -P\hat{x}^* = \begin{bmatrix} \frac{1}{3} \\ 0 \\ -1\frac{1}{3} \\ -2\frac{2}{3} \end{bmatrix},$$

$$\lambda_2 = \text{Min} \left[\frac{7}{1\frac{1}{3}}, \frac{17}{2\frac{2}{3}} \right] = 5\frac{1}{4}.$$

Hence

$$\bar{\lambda} = \text{Max} \,(\lambda_1, \lambda_2) = 5\frac{1}{4}.$$

The cutting constraint

$$1\tfrac{1}{2}x_1 + 1\tfrac{1}{2}x_2 \geq 5\tfrac{1}{4}$$

now passes through the point $(3\frac{1}{2}, 0)$. The best value of F is

$$F = q_0 - 2\lambda + \lambda^2 \hat{u}^* = 0 - 2 \times 5\tfrac{1}{4} + (5\tfrac{1}{4})^2 \times \tfrac{4}{9} = 1\tfrac{3}{4},$$

which was also found in tableau 1 of table 1.

The cutting constraint is added to the main problem as

$$-5\tfrac{1}{4} = -1\tfrac{1}{2}x_1 - 1\tfrac{1}{2}x_2 + y_5.$$

To save space, use will be made of the tableaux of table 1. The next constraint is added as an additional row and column to tableau 0. The extreme point $(3\frac{1}{2}, 0)$ is generated in tableau 1, for which the y_5-row should be

$$0 = \tfrac{3}{4}y_3 - \tfrac{3}{4}x_2 + y_5.$$

The solution of this tableau is feasible but not optimal, so that the simplex method for quadratic programming is applied. The first iteration results in tableau 2, which should have as y_5-row

$$2\tfrac{1}{4} = -\tfrac{3}{4}y_3 - \tfrac{3}{4}y_4 + y_5.$$

Since $u_3 = -\frac{1}{2}$, the solution is not yet optimal. The next iteration results in tableau 3, which has as y_5-row

$$4\tfrac{1}{2} = 1\tfrac{1}{8}y_2 + \tfrac{3}{8}y_4 + y_5.$$

This is a local maximum.

The set-up tableau for the auxiliary problem is tableau 0 of table 3. The two extreme-point solutions are given in tableaux 1 and 2. The maximum value of $\hat{u}^*$ is given by tableau 1 and is $\frac{52}{361}$.

Table 3

Solution of second auxiliary problem

Tableau	B.v.	V.b.v.	y_2	y_4	u
0	F	0	0	0	-1
	u_2	0	$\frac{13}{16}$	$\frac{1}{16}$	$-2\frac{3}{8}$
	u_4	0	$\frac{1}{16}$	$\frac{1}{16}$	$-\frac{7}{8}$
	y	1	$2\frac{3}{8}$	$\frac{7}{8}$	0
			y	y_4	u_2
1	F	$\frac{52}{361}$	$\frac{52}{361}$	$\frac{36}{361}$	$-\frac{8}{19}$
	u	$\frac{52}{361}$	$\frac{52}{361}$	$\frac{36}{361}$	$-\frac{8}{19}$
	u_4	$\frac{36}{361}$	$\frac{36}{361}$	$3\frac{126}{361}$	$-\frac{7}{19}$
	y_2	$\frac{8}{19}$	$\frac{8}{19}$	$\frac{7}{19}$	0
			y_2	y	u_4
2	F	$\frac{4}{49}$	$\frac{6}{49}$	$\frac{4}{49}$	$-1\frac{1}{7}$
	u_2	$\frac{6}{49}$	$\frac{103}{196}$	$\frac{6}{49}$	$-2\frac{5}{7}$
	u	$\frac{4}{49}$	$\frac{6}{49}$	$\frac{4}{49}$	$-1\frac{1}{7}$
	y_4	$1\frac{1}{7}$	$2\frac{5}{7}$	$1\frac{1}{7}$	0

The best value of F found so far is $12\frac{1}{4}$, which belongs to the current local maximum. Hence, $\lambda_1 = 2/\hat{u}^* = 13\frac{23}{26}$.

To find λ_2, we have

$$P = \begin{bmatrix} -1\frac{1}{8} & \frac{1}{8} \\ 1\frac{1}{2} & -\frac{1}{2} \\ -\frac{1}{4} & \frac{1}{4} \\ 1 & 0 \\ 1\frac{1}{8} & \frac{3}{8} \end{bmatrix}, \qquad \hat{x}^* = \begin{bmatrix} \frac{8}{19} \\ 0 \end{bmatrix} \qquad s = -P\hat{x}^* = \begin{bmatrix} \frac{9}{19} \\ -\frac{12}{19} \\ \frac{2}{19} \\ -\frac{8}{19} \\ -\frac{9}{19} \end{bmatrix}$$

$$\lambda_2 = \text{Min} \left(\frac{3}{\frac{12}{19}}, \frac{3}{\frac{8}{19}}, \frac{4\frac{1}{2}}{\frac{9}{19}} \right) = (4\frac{3}{4}, 7\frac{1}{8}, 9\frac{1}{2}) = 4\frac{3}{4}.$$

Hence,

$$\bar{\lambda} = \text{Max} (\lambda_1, \lambda_2) = 13\frac{23}{26}.$$

The additional constraint is therefore

$$2\frac{3}{8}y_2 + \frac{7}{8}y_4 \geq 13\frac{23}{26}.$$

Then a new local maximum should be found which satisfies also this constraint. The new constraint can be written in the form

$$-13\frac{23}{26} = -2\frac{3}{8}y_2 - \frac{7}{8}y_4 + y_6.$$

The other constraints were (see tableau 3)

$$\begin{aligned}
\tfrac{3}{4} &= -1\tfrac{1}{8}y_2 + \tfrac{1}{8}y_4 + y_1, \\
3 &= 1\tfrac{1}{2}y_2 - \tfrac{1}{2}y_4 + y_3, \\
3\tfrac{1}{2} &= -\tfrac{1}{4}y_2 + \tfrac{1}{4}y_4 + x_1, \\
3 &= y_2 \quad\quad + x_2, \\
4\tfrac{1}{2} &= 1\tfrac{1}{8}y_2 + \tfrac{3}{8}y_4 + y_5.
\end{aligned}$$

If y_6 is maximized subject to the other constraints, it is found that no feasible solution exists. This means that the best solution found so far, $F = 12\frac{1}{4}$, $x_1 = 3\frac{1}{2}$, $x_2 = 3$ is the global maximum.

Ritter's method does not necessarily find the global maximum is a finite number of steps, nor does it necessarily converge towards it. This has been demonstrated by Zwart[101] in an ingeniously designed counter-example, of which a simplified version is given here.

The simplified problem is: Maximize

$$f = 2x_2 + 2x_1^2 + x_1x_2$$

subject to

$$x_1 + 2x_2 \leq 1,$$
$$x_1 \geq 0, x_2 \geq 0.$$

The objective function is indefinite, but the main difference of the method for concave and for general problems resides in the solution of the auxiliary problem, which in the general case is more complicated; for this example the auxiliary problem is easily solved.

Tableau 0 of table 4 gives the set-up tableau for $k = 0$, and tableau 1 gives the local optimum resulting from one double pivot iteration of the simplex method. Figure 6 gives a graphical representation of the prob-

Table 4

Generation of local maximum for counter-example

Tableau	B.v.	V.b.v.	x_1	x_2	u
0	F	0	0	-2	-2^{-k}
	v_1	0	4	1	-2^{-k}
	v_2	-2	1	0	$\dfrac{-2}{}$
	y	2^{-k}	2^{-k}	2	0
			x_1	y	v_2
1	F	2^{-k+1}	2^{-k-1}	1	-2^{-k-1}
	v_1	2^{-k-1}	$4-2^{-k}$	-2^{-1}	-2^{-k-1}
	u	1	-2^{-1}	0	-2^{-1}
	x_2	2^{-k-1}	2^{-k-1}	2^{-1}	0

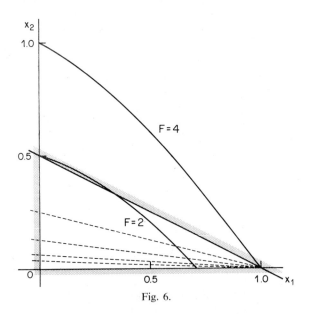

Fig. 6.

lem. The point $(0,\frac{1}{2})$ with $F = 2$ is a local maximum while the point $(1,0)$ with $F = 4$ is evidently the global maximum.

The set-up tableau for the auxiliary problem is then tableau 0 of table 5 for $k = 0$; the F-row and the values of basic variables column have been deleted since they are trivial. Since x_1 is the only variable with positive terms in the objective function, and since there is only one constraint with positive coefficients, x_1 must be the only basic variable in the global maximum of the auxiliary problem. This global maximum is generated in tableau 1, where the values of basic variables column is the same as the y-column and the F-row the same as the u-row.

Table 5

Solution of auxiliary problem

Tableau	B.v.	x_1	y	u_a
0	v_1	$4 - 2^{-k}$	-2^{-1}	-2^{-k-1}
	u	-2^{-1}	0	-1
	y_a	2^{-k-1}	1	0
		y_a	y	v_1
1	u_a	$2^{2k+4} - 2^{k+2}$	$2^{2k+4} - 3.2^k$	-2^{k+1}
	u	$2^{2k+4} - 3.2^k$	$2^{2k+4} - 2^{k+1}$	-2^{k+1}
	x_1	2^{k+1}	2^{k+1}	0

According to this solution, x_1 should be increased as much as possible, while y should be kept as zero; we should therefore move along the constraint in the direction of the point $(1,0)$, until both of the following has happened: (1) $F = 2$, (2) the end of the feasible region is reached.

In this case, $F = 2$ happens first, so that we continue to the point $(1, 0)$, where the boundary of the feasible region is found. This means $\lambda_2 > \lambda_1$.

For λ_2 we find:

$$- P\hat{x}^* = - 2^{-k-1}.2^{k+1} = - 1, \qquad \text{Min}\left(\frac{2^{-k-1}}{1}\right) = 2^{-k-1},$$

so that the cutting constraint is

$$2^{-k-1}x_1 + y \geq 2^{-k-1},$$

or

$$-2^{-k-1} = - 2^{-k-1}x_1 - y + y_k.$$

This constraint should be added to tableau 1 of table 4. If this constraint is transformed back to tableau 0, it becomes

$$2^{-k-1} = 2^{-k-1}x_1 + 2x_2 + y_k,$$

which for $k = 0$ becomes

$$\tfrac{1}{2} = \tfrac{1}{2}x_1 + 2x_2 + y_0.$$

This constraint in figure 6 is indicated by the upper dashed line. This new constraint makes the original constraint redundant, so that this last constraint may be deleted.

For the modified problem, a new local maximum must be found. Tableau 0 of table 4 gives the set-up tableau for $k = 1$, and tableau 1 gives the local maximum. The auxiliary problem is then formulated and solved as in table 5 for $k = 1$. From the figure it is obvious that again $\lambda_2 > 1$, and $\lambda_2 = 2^{-k-1}$ for $k = 1$, as computed before. The new cutting constraint is added, making the existing constraint redundant. The new problem is then solved as indicated before, with $k = 2$, and so on.

It is obvious that the cutting constraints cut off ever smaller pieces of the feasible region. The local maximum converges to the point $(0, 0)$, while the global maximum is $(1, 0)$.

This example demonstrates that Ritter's method does not work in all cases. This does not mean that the method is useless. The method can be modified to avoid convergence to points which are not global maxima by stipulating that, if the local maxima found in successive steps converge, then a different local maximum should be found.

14.5. Tui's method for concave quadratic programming

Tui's method[100] has been developed for concave nonlinear problems with linear constraints, but here only the case of concave quadratic programming will be treated; the generalization to concave nonlinear programming is straightforward. First the simple cutting plane version of the method will be treated, after which what may be called the composite version follows.

Let us consider the same example as for Ritter's method. First a local maximum should be found which Tui does by comparing adjacent extreme points. The simplex method for quadratic programming can be used to the same effect, except that a local maximum may have an adjacent extreme point with a higher value.

Tableau 0 of table 6 gives the required local maximum[2] for the example (the y_s-row should be ignored) and figure 7 gives a graphical representation of the problem.

Consider now an increase of x_1. F initially decreases, but then starts to increase again until at $x_1 = 3$, $F = 0$ again, see figure 7. This can also be

Table 6

Application of Tui's method

Tableau	B.v.	V.b.v.	x_1	x_2	Tableau	B.v.	V.b.v.	y_2	y_3
	F	0	$1\frac{1}{2}$	$1\frac{1}{2}$		F	$12\frac{1}{4}$	$2\frac{3}{8}$	$\frac{7}{8}$
	v_1	$1\frac{1}{2}$	1	$-\frac{1}{2}$		u_2	$2\frac{3}{8}$	$\frac{13}{16}$	$\frac{1}{16}$
	v_2	$1\frac{1}{2}$	$\frac{1}{2}$	1		u_3	$\frac{7}{8}$	$\frac{1}{16}$	$\frac{1}{16}$
0	y_1	2	$-\frac{1}{2}$	1	1	y_1	$\frac{3}{4}$	$-1\frac{1}{8}$	$\frac{1}{8}$
	y_2	3	0	1		y_4	3	$1\frac{1}{2}$	$-\frac{1}{2}$
	y_3	17	4	1		x_1	$3\frac{1}{2}$	$-\frac{1}{4}$	$\frac{1}{4}$
	y_4	7	2	-1		x_2	3	1	0
	y_5	-1	$-\frac{1}{3}$	$-\frac{1}{3}$		y_5	$1\frac{1}{6}$	$\frac{1}{4}$	$\frac{1}{12}$

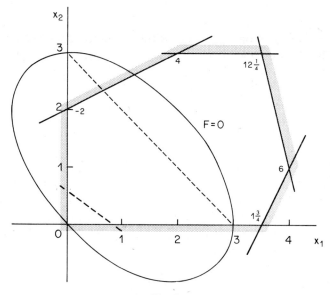

Fig. 7.

---

[2] The point $(3\frac{1}{2}, 0)$ has a higher f-value, but in order to be able to use the same example as in the Ritter method, this is overlooked.

concluded from tableau 0:

$$F = 0 - 3x_1 - 3x_2 + x_1^2 + x_1x_2 + x_2^2;$$

for $F = 0$, $x_2 = 0$, we find

$$-3x_1 + x_1^2 = 0,$$

so that $x_1 = 0$ or 3. If x_2 is increased from the point $(0, 0)$ we find in quite the same way that F decreases and then becomes equal to 0 again for $x_2 = 3$. We may now put a constraint through the points $(3, 0)$ and $(0, 3)$:

$$\tfrac{1}{3}x_1 + \tfrac{1}{3}x_2 \leq 1.$$

The region bounded by this constraint and $x_1 \geq 0$, $x_2 \geq 0$ has at all its extreme points $F = 0$. Since F is concave, no exterior points of this region can have a higher value. Hence, we may exclude the region from consideration by imposing the cutting constraint

$$\tfrac{1}{3}x_1 + \tfrac{1}{3}x_2 \geq 1.$$

If in the remaining feasible region no higher value of F is found, then $(0, 0)$ is a global maximum, otherwise it is not.

Let us proceed with the example in this manner. The cutting constraint is adjoined in tableau 0 of table 6. A feasible solution to the modified problem is obtained by selecting the underlined element $-\tfrac{1}{3}$ as the pivot of a double pivot transformation. Then after three iterations of the simplex method for quadratic programming, the local maximum $F = 12\tfrac{1}{4}$, $x_1 = 3\tfrac{1}{2}$, $x_2 = 3$ is found, see tableau 1.

Now another cutting plan should be constructed. The highest value of F, obtained so far, indicated by $\hat{F}$, is $12\tfrac{1}{4}$. First y_2 is increased until $F = 12\tfrac{1}{4}$, which is found from

$$12\tfrac{1}{4} - 4\tfrac{3}{4}y_2 + \tfrac{13}{16}y_2^2 = 12\tfrac{1}{4}$$

or $y_2 = 5\tfrac{11}{13}$. For y_3 we find in the same manner $y_3 = 28$. The cutting constraint is therefore

$$\tfrac{13}{76}y_2 + \tfrac{1}{28}y_3 \geq 1.$$

This is added to the constraints of table 1. It then turns out that there is no feasible solution, which is also clear from figure 7, where the constraint is drawn in. Hence the global maximum is $F = 12\tfrac{1}{4}$, $x_1 = 3\tfrac{1}{2}$, $x_2 = 3$.

To describe the method more formally, let the tableau of the current local maximum be

B.v.	V.b.v.	x	u
F	q_0	q'	$-r'$
v	q	Q	$-P'$
y	r	P	0

The points $\hat{x}_i$ through which the cutting plane goes are then determined for each nonbasic variable as follows. If $q_{ii} > 0$, then $\hat{x}_i$ is the largest root of the equation

$$q_0 - 2q_i x_i + q_{ii} x_i^2 = \hat{F},$$

where $\hat{F}$ is the best value of F found so far. Hence

$$\hat{x}_i = \frac{q_i}{q_{ii}} + \sqrt{\left(\frac{q_i}{q_{ii}}\right)^2 + \frac{\hat{F} - q_0}{q_{ii}}}. \tag{34}$$

The cutting constraint then becomes

$$\sum_{i | q_{ii} > 0} (1/\hat{x}_i) \geq 1. \tag{35}$$

In order to avoid accumulation of constraints, it is useful to test each of the constraints for redundancy at regular intervals. After the first constraint has been added in the example, the constraint $x_1 \geq 0$ is redundant, which means that the x_1-row in the various tableaux can be deleted.

Some improvement can be made by introducing a modification inspired by the Ritter method. To determine the initial cutting plane in the example, x_1 was increased until $F = 0$, which happens at $x_1 = 3$. However, x_1 could have been increased until $x_1 = 3\frac{1}{2}$, where $F = 1\frac{3}{4}$. Then x_2 is also increased until $F = 1\frac{3}{4}$, which happens at $x_2 = 3\frac{1}{2}$. The best solution found so far is then $F = 1\frac{3}{4}$, $x_1 = 3\frac{1}{2}$, $x_2 = 0$, and the cutting constraint is

$$\tfrac{2}{7}x_1 + \tfrac{2}{7}x_2 \geq 1.$$

Formally, this modification can be described as follows. The $\hat{x}_i$ are determined as before, but we also determine for each i

$$\bar{x}_i = \left(\underset{j}{\text{Min}} \, \frac{r_j}{p_{ji}} \,\middle|\, p_{ji} > 0\right).$$

We then determine

$$\bar{F} = \underset{i}{\text{Max}} \, (-2q_i + q_{ii} \,|\, \bar{x}_i > \hat{x}_i)$$

and

$$\hat{\hat{F}} = \text{Max} \, (\hat{F}, \bar{F}).$$

If $\hat{\hat{F}} \neq \hat{F}$, (34) is used again with $\hat{\hat{F}}$ instead of $\hat{F}$, and the resulting $\hat{x}_i$ are used for the cutting plane.

If this modification is made to Tui's method, the cuts of this method are deeper than those of Ritter's method or are the same. Figure 8 gives an illustration of this for the case

$$q = \begin{bmatrix} 1 \\ 1 \end{bmatrix}.$$

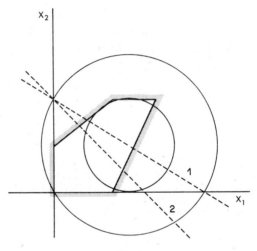

Fig. 8.

The reason for this is that in Tui's method the coefficients of the constraint are determined in such a way that the cut-off region is as large as possible, whereas in Ritter's method these coefficients are fixed.

14.6. The composite version of Tui's method

But Tui's method as proposed in his article [100], which here is called the composite method, is more complicated than a simple cutting plane method. The method will be explained mainly geometrically since the computations are rather involved. The same example will be used as before, see figure 9.

The starting-point is again the local maximum $(0, 0)$ and the cutting plane is constructed as in the simple version. Let the cutting plane be $d'x = 1$; for the problem at hand it is $\frac{1}{3}x_1 + \frac{1}{3}x_2 = 1$. Now an auxiliary prob-

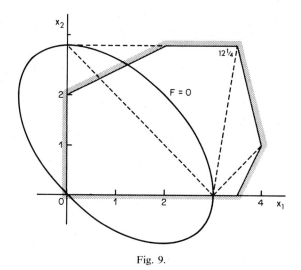

Fig. 9.

lem is formulated in which $d'x$ is maximized over the feasible region. In geometric terms this means that we try to find the point of the feasible region on the other side of the cutting plane with a maximal distance to that plane. In figure 9, this is the point $(3\frac{1}{2}, 3)$. The value of F in this point is evaluated, which is in the example $12\frac{1}{4}$.

If the objective function is convex, the maximum over a polygon is always found in an extreme point. Since the values of F are known for each of the extreme points of the polygon $(0, 0)$, $(3, 0)$, $(3\frac{1}{2}, 3)$, $(0, 3)$, the part of the feasible region within this polygon is explored.

The remainder of the feasible region is explored by formulating new auxiliary problems. In the example there are two such problems. In the first, the cutting plane through $(3, 0)$ and $(0, 3)$ is replaced by a plane through $(3\frac{1}{2}, 3)$ and $(3, 0)$, and the distance to this plane is maximized. The solution of this problem is point $(4, 1)$ with $F = 6$. In the second of these new problems, the cutting plane is replaced by a plane through $(3\frac{1}{2}, 3)$ and $(0, 3)$ and again the distance to the plane is maximized. In this case there is no feasible region outside the plane, which means that at this side no feasible region is unexplored.

Returning to the auxiliary problem with solution $(4, 1)$, we now have explored also the triangle $(3, 0)$, $(3\frac{1}{2}, 3)$, $(1, 4)$ and must find out whether on these sides there are parts of the feasible region left unexplored. The plane through $(3, 0)$ and $(3\frac{1}{2}, 3)$ is replaced by two planes, the first one going through $(3, 0)$ and $(4, 1)$ and the second going through $(3\frac{1}{2}, 3)$ and $(4, 1)$. Then two new auxiliary problems are formulated in which the distances to these planes are maximized. The first auxiliary problem discovers the new point $(3\frac{1}{2}, 0)$ with $F = 1\frac{3}{4}$ and the second discovers no new point.

From the figure we see that we have explored the entire feasible region, but formally we should first replace the plane through $(3, 0)$ and $(4, 1)$ by a plane through $(3, 0)$ and $(3\frac{1}{2}, 0)$ and one through $(3\frac{1}{2}, 0)$ and $(4, 1)$. In the resulting auxiliary problems no new points are found, so that we have explored the entire feasible region. The solution is the best point found so far, $(3\frac{1}{2}, 3)$ with $F = 12\frac{1}{4}$.

More precisely, new auxiliary problems are created as follows. If the solution of an auxiliary problem results in a new point (a new point is found if the objective function of the auxiliary problem is positive), then a new auxiliary problem is found by replacing one of the points through which the old plane went by the new point. This means that as many new problems are created as there are variables in the problem; in the example this is two.

Now that the principles of the method are treated, we can go further and explain the method exactly as Tui has proposed it. If in a second or later step a higher value of F is found, all points and hence the planes going through these points may be adjusted to take this higher value into account.

In the example we found for the solution of the first auxiliary problem the point $(3\frac{1}{2}, 3)$ with $F = 12\frac{1}{4}$. Now the points of the initial cutting plane were based on $F = 0$. The point $(3, 0)$ may now be replaced by $(3k, 0k)$, where k is determined such that $F(3k, 0k) = 12\frac{1}{4}$; the solution of this is the point $(1\frac{1}{2} + \frac{1}{2}\sqrt{29}, 0)$. Similarly, the point $(0, 3)$ is replaced by $(0, 1\frac{1}{2} + \frac{1}{2}\sqrt{29})$.

This results in a polygon $(0, 0)$, $(1\frac{1}{2} + \frac{1}{2}\sqrt{29}, 0)$, $(3\frac{1}{2}, 3)$, $(0, 1\frac{1}{2} + \frac{1}{2}\sqrt{29})$, which has as the highest F-value $12\frac{1}{4}$, see figure 10. The two resulting auxiliary problems have no new point, so that the optimal solution has been found, as is obvious from figure 10.

To summarize the last modification: after a new set of auxiliary problems has been solved, each point is multiplied by a multiple such that a point results with an F-value equal to the highest value found so far.

Tui's method is not necessarily finite. Zwart[101] has constructed an example in which the optimal solution is not found by the method; in this

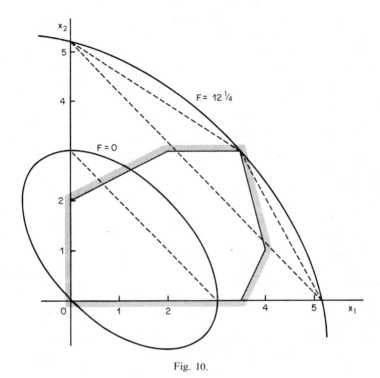

Fig. 10.

example the same auxiliary problems are found again and again. As for the Ritter method this does not necessarily render the method useless.

CHAPTER 15

GENERAL QUADRATIC PROGRAMMING

15.1. The simplex method for finding a local maximum

This chapter deals with methods for the general quadratic programming problem; in this case no assumptions are made about the matrix of the quadratic form. Since methods for convex and concave problems, in which the objective functions were negative and positive semi-definite, have been treated before, this means that now methods are given for the indefinite case.

This section deals with the application of the simplex method for quadratic programming to find a local maximum. The results of this section are used constructively in sections 15.2 and 15.3, where conditions are given for local optima. These conditions are used in the section 15.4 for finding a global maximum of the auxiliary problem in Ritter's method. The remainder of this chapter deals with a new parametric method for general quadratic programming.

We consider again the problem: Maximize

$$f = p'x + \tfrac{1}{2}x'Cx$$

subject to

$$Ax \le b,$$
$$x \ge 0.$$

C is a general symmetric matrix. To this problem the simplex method for quadratic programming can be applied. The equations of the set-up tableau are:

$$0 = -p'x - b'u + F, \tag{1}$$
$$-p = Cx - A'u + v, \tag{2}$$
$$b = Ax + y. \tag{3}$$

As before, we have

$$1F = 2f = 2p'x + x'Cx = p'x + x'(p + Cx)$$
$$= p'x + x'(A'u - v) = p'x - v'x + u'Ax$$
$$= p'x - v'x + u'(b - y) = p'x + b'u - v'x - u'y. \tag{4}$$

If $b \geq 0$, the basic solution of (1)–(3) is feasible; if $b \not\geq 0$, first a feasible solution to $Ax \leq b$, $x \geq 0$ should be found, after which the problem can be reformulated in such a way that $b \geq 0$.

Then the x-variable corresponding to a negative element of q is introduced into the basis; let this be x_k. Then ratios are taken in the rows of primal basic variables and in the row of v_k, the latter only if the corresponding diagonal element is negative, and so on.

To describe a general iteration, the equations of a general tableau of the method are used:

$$q_0 = q'x - r'u + F, \tag{5}$$

$$q = Qx - P'u + v, \tag{6}$$

$$r = Px + Ru + y, \tag{7}$$

where x and y are the vectors of nonbasic and basic primal variables, and v and u are the vectors of the corresponding basic and nonbasic dual variables. For the successive basic solutions we have $r \geq 0$ and R negative semi-definite. Note that this is true for the initial tableau (1)–(3).

F as given by (5) is a transformation of the linear part of (4); for a complete expression of F in terms of the nonbasic variables the terms $-v'x - u'y$ should be added, see (4); the change in naming of variables is irrelevant. We then obtain

$$\begin{aligned} F &= q_0 - q'x + r'u - v'x - u'y \\ &= q_0 - q'x + r'u - x'(q - Qx + P'u) - u'y \\ &= q_0 - 2q'x + x'Qx + u'(r - y - Px) \\ &= q_0 - 2q'x + x'Qx + u'Ru. \end{aligned} \tag{8}$$

From this it follows that if $q > 0$, the current basic solution is a local maximum since R is negative semi-definite. If some elements of q are 0, but the corresponding submatrix of Q is negative semi-definite, the solution is also a local maximum. The cases when $q \geq 0$ and $q_i = 0$ for some i can always be resolved by suitable perturbation techniques.

If some element of q, say q_k, is negative, F is increased by increasing x_k, since according to (8) F is the following function of x_k:

$$F(x_k) = -2q_k x_k + q_{kk} x_k^2. \tag{9}$$

If $q_{kk} \geq 0$, F increases with increasing x_k for all $x_k \geq 0$, but if $q_{kk} < 0$, the maximum of F is reached for

$$\frac{df(x_k)}{dx_k} = -2q_k + 2q_{kk} x_k = 0 \text{ or } x_k = \frac{q_k}{q_{kk}}.$$

However, the increase of x_k may have been stopped by one of the primal variables dropping to 0. Hence, the maximum value of x_k is determined by

$$\text{Min}_i \left(\frac{q_k}{q_{kk}} \middle| q_{kk} < 0, \frac{r_i}{p_{ik}} \middle| p_{ik} > 0 \right).$$

If q_k/q_{kk} is the minimum, x_k replaces v_k, the tableau is transformed and the method continues by selecting another negative element of q_k. If the minimum is connected with a primal variable y_r, then if $r_{rr} < 0$, r_{rr} is used as a pivot and the method continues by increasing x_k. If $r_{rr} = 0$, a double pivot transformation follows, after which another negative element of q_k is selected, if there is one. The rules are therefore entirely the same as in the simplex method for convex quadratic programming.

The proof for R being negative semi-definite is also entirely the same, see section 9.1. Q is, of course, not necessarily negative semi-definite in this case.

From this it may be concluded that the simplex method for quadratic programming finds a local maximum of a general quadratic programming problem in a finite number of steps in nondegenerate cases. The cases of degeneracy, suitable perturbation techniques may be employed.

15.2. Conditions for local optima

According to (8), F can, for any basic solution, be expressed in terms of the nonbasic variables as

$$F = q_0 - 2q'x + x'Qx + u'Ru.$$

From this it can be concluded that $r > 0$, $q > 0$, and R negative semi-definite are sufficient conditions for a strict local maximum in nondegenerate cases; a local maximum is strict if all points in its neighbourhood have a lower value of F. These conditions are also necessary, which can be shown as follows.

Let, in the notation of the original problem, $\hat{x}$ be a strict local maximum. The objective function may then be written as:

$$F = 2p'x + x'Cx = q_0 + 2(p + C\hat{x})'x^* + x^{*\prime}Cx^*,$$

where $x^* = x - \hat{x}$ and $q_0 = 2p'\hat{x} + \hat{x}'C\hat{x}$. The constraints can be written as

$$Ax^* \le b - A\hat{x},$$
$$x^* \ge \quad -\hat{x}.$$

The equations of the set-up tableau for the problem in this form are

$$q_0 = -(p + C\hat{x})'x^* - (b - A\hat{x})'u + F,$$
$$-(p + C\hat{x}) = Cx^* - A'u + v,$$
$$b - A\hat{x} = Ax^* + y.$$

The simplex method for quadratic programming can be applied to this problem, bearing in mind that instead of $x \geq 0$, we have $x^* \geq -\hat{x}$. Then first all variables x_j^* with $\hat{x}_j > 0$ are introduced into the basis; if $v_j < 0$, x_j^* is introduced in a positive direction and if $v_j > 0$, it is introduced in a negative direction, while if $v_j = 0$, it can be introduced in either direction. The leaving basic variable should always have a zero value, because if it is not, then if $v_j \neq 0$, the solution is not a local maximum and if $v_j = 0$, the local maximum is not strict. As soon as a variable x_j^* with $\hat{x}_j > 0$ has become basic, we substitute $x_j^* = x_j - \hat{x}_j$.

After all x_j^* with $x_j > 0$ have become basic in this manner, the simplex method for quadratic programming is applied in the usual fashion by introducing primal variables with negative dual basic variables. Again the leaving basic variable should always have a zero value. In the final result we should have a basic solution with $r > 0$, $q > 0$, and R negative semi-definite where r constraints $x = \hat{x}$ and $y = b - A\hat{x}$, or we have to conclude that $\hat{x}$ is not a strict local maximum.

Local maxima which are not strict can easily be recognized as linear combinations of different basic solutions with $q \geq 0$, $r \geq 0$, and R negative semi-definite with Q_{11} negative semi-definite, where Q_{11} is the submatrix of Q corresponding with elements of q which are zero.

In the same way it can be shown that necessary and sufficient conditions for a local minimum are that the minimum is a basic solution or a linear combination of basic solutions with $r \geq 0$, $q \leq 0$, and R and Q_{11} positive semi-definite, where Q_{11} is the submatrix of Q corresponding with $q_i = 0$.

15.3. Optimal solutions and latent roots

In concave quadratic programming and in linear programming optima are extreme points, while in convex quadratic programming they may be interior points, which means that all primal variables are basic. In this section, a much more general result is derived, namely that the maximum number of primal basic variables in an optimal solution is $m + n(C)$, where $n(C)$ is the number of negative latent roots of C. If $n(C) = n$, then C is negative definite and an interior solution is possible. If $n(C) > 0$, an optimal solution is not necessarily an extreme point.

First it will be shown that under the transformations of the simplex method for quadratic programming F can always be expressed as a quadratic form in the nonbasic variables with the same number of positive and negative latent roots.

Let us consider a general complementary solution of the Kuhn-Tucker equations augmented by the linearized F-equation:

$$q_0 = q'x - r'u + F,$$
$$q = Qx - P'u + v,$$
$$r = Px + Ru + y, \tag{10}$$

where x and u are the vectors of nonbasic primal and dual variables, and y and v those of basic primal and dual variables.

The objective function can be expressed as follows:

$$F = q_0 - q'x + r'u - v'x - u'y = q_0 - 2q'x + x'Qx + u'Ru.$$

Consider the matrix of this quadratic form:

$$\begin{bmatrix} Q & 0 \\ 0 & R \end{bmatrix}. \tag{11}$$

The set-up tableau with $Q = C$ and $R = 0$ is a particular case. It will be shown that any principal pivot transformation of (10) leads to a new complementary equation system of the same form with a corresponding matrix of the form (11), which has the same number of positive and negative roots.

Consider first a transformation of (10) in which the last diagonal element of Q is the pivot. Q is partitioned accordingly and the coefficients of (10) are then transformed as follows:

Q_{11}	Q_{12}	$-P_1'$
Q_{21}	q_{22}	$-P_2'$
P_1	P_2	R
$Q_{11} - Q_{12}q_{22}^{-1}Q_{21}$	$-Q_{12}q_{22}^{-1}$	$-P_1' + Q_{12}q_{22}^{-1}P_2'$
$q_{22}^{-1}Q_{21}$	q_{22}^{-1}	$-q_{22}^{-1}P_2'$
$P_1 - P_2q_{22}^{-1}Q_{21}$	$-P_2q_{22}^{-1}$	$R + P_2q_{22}^{-1}P_2'$

The matrix corresponding to (11) is now:

$$\begin{bmatrix} Q^* & 0 \\ 0 & R^* \end{bmatrix} = \begin{bmatrix} Q_{11} - Q_{12}q_{22}^{-1}Q_{21} & 0 & 0 \\ 0 & q_{22}^{-1} & -q_{22}^{-1}P_2' \\ 0 & -P_2q_{22}^{-1} & R + P_2q_{22}^{-1}P_2' \end{bmatrix}. \quad (12)$$

This is a congruent transformation of the original matrix (11):

$$\begin{bmatrix} I & -Q_{12}q_{22}^{-1} & 0 \\ 0 & q_{22}^{-1} & 0 \\ 0 & -P_2q_{22}^{-1} & I \end{bmatrix} \begin{bmatrix} Q_{11} & Q_{12} & 0 \\ Q_{21} & q_{22} & 0 \\ 0 & 0 & R \end{bmatrix} \begin{bmatrix} I & 0 & 0 \\ -q_{22}^{-1}Q_{21} & q_{22}^{-1} & -q_{22}^{-1}P_2' \\ 0 & 0 & I \end{bmatrix}$$

Hence, the number of positive roots and of negative roots must be the same in (11) and (12).

If a diagonal element of R is used as a pivot, the same proof applies by interchange of R and Q.

Consider now double pivot transformations in which there is a block pivot of the type

$$\begin{bmatrix} q_{ii} & -p_{ij} \\ p_{ij} & r_{jj} \end{bmatrix}$$

If q_{ii} and r_{jj} are not both zero, the transformation can be represented as the successive application of single diagonal transformations.

In the remaining cases, the transformation of the coefficients of (10) by a double pivot operation can be represented as follows:

Q_{11}	Q_{12}	$-P_{11}'$	$-P_{21}'$
Q_{21}	0	$-p_{12}$	$-P_{22}'$
P_{11}	p_{12}	0	0
P_{21}	P_{22}	0	R_{22}

| $\begin{array}{c} Q_{11} - Q_{12}p_{12}^{-1}P_{11} - P_{11}'p_{12}^{-1}Q_{21} \\ -p_{12}^{-1}Q_{21} \\ p_{12}^{-1}P_{11} \\ P_{21} - P_{22}p_{12}^{-1}P_{11} \end{array}$ | $\begin{array}{c} -Q_{12}p_{12}^{-1} \\ 0 \\ p_{12}^{-1} \\ -P_{22}p_{12}^{-1} \end{array}$ | $\begin{array}{c} -P_{11}'p_{12}^{-1} \\ -p_{12}^{-1} \\ 0 \\ 0 \end{array}$ | $\begin{array}{c} -P_{21} + P_{11}'p_{12}^{-1}P_{22}' \\ p_{12}^{-1}P_{22}' \\ 0 \\ R_{22} \end{array}$ |

Since R is negative semi-definite and $r_{11} = 0$, we have $r_{1j} = r_{i1} = 0$ for all i and j.

The matrix (11) after transformation can be written as

$$
\begin{bmatrix} Q^* & 0 \\ 0 & R^* \end{bmatrix} = \begin{bmatrix} Q_{11} - Q_{12}p_{12}^{-1}P_{11} - P'_{11}p_{12}^{-1}Q_{21} & -Q_{12}p_{12}^{-1} & 0 & 0 \\ -p_{12}^{-1}Q_{21} & 0 & 0 & 0 \\ 0 & 0 & 0 & 0 \\ 0 & 0 & 0 & R_{22} \end{bmatrix}.
$$

This may be written as a congruent transformation of the previous matrix (11):

$$
\begin{bmatrix} I & -P'_{11}p_{12}^{-1} & 0 & 0 \\ 0 & -p_{12}^{-1} & 0 & 0 \\ 0 & 0 & 1 & 0 \\ 0 & 0 & 0 & I \end{bmatrix} \begin{bmatrix} Q_{11} & Q_{12} & 0 & 0 \\ Q_{21} & 0 & 0 & 0 \\ 0 & 0 & 0 & 0 \\ 0 & 0 & 0 & R \end{bmatrix} \begin{bmatrix} I & 0 & 0 & 0 \\ -p_{12}^{-1}P_{11} & -p_{12}^{-1} & 0 & 0 \\ 0 & 0 & 1 & 0 \\ 0 & 0 & 0 & I \end{bmatrix}
$$

Hence the number of positive and negative latent roots has not changed.

Consider now a successive application of these transformations. Initially, the m slack variables are basic, $Q = C$, and $R = 0$. Each time a negative diagonal element of Q is chosen as a pivot, the number of primal basic varibles increases by one, the number of negative roots of Q decreases by one, and the number of negative roots of R increases by one. The reverse is true for a pivot in a diagonal element of R. Double pivot operations do not affect the number of primal basic variables. Hence, the maximum number of primal basic variables is equal to $m + n(C)$.

15.4. The solution of Ritter's auxiliary problem

The equations of Ritter's auxiliary problem are

$$
\begin{aligned}
0 &= 0'x - u + F, \\
0 &= Cx - qu + v, \\
1 &= q'x + y.
\end{aligned} \tag{13}
$$

A global maximum should be found with y nonbasic. Since C is a general symmetric matrix, a number of local maxima may exist, which should all be compared.

Ritter proposes the following search method to find all local maxima. First, assume that x_1 is a basic variable in the maximum. x_1 may then be introduced into the basis, replacing y, after which u is introduced replacing v_1. The resulting system is of the form:

$$q_0 = q'x - rv_1 + F,$$
$$q = Qx^* - pv_1 + v^*,$$
$$r = p'x^* \qquad + x_1, \tag{14}$$

where

$$x^* = \begin{bmatrix} x_2 \\ \cdot \\ \cdot \\ \cdot \\ x_r \end{bmatrix}, \qquad v^* = \begin{bmatrix} v_2 \\ \cdot \\ \cdot \\ v_r \end{bmatrix}.$$

Consider now a local maximum for the system (14) in which x_1 is basic. In such a maximum, a number of variables of x^* will be basic, having replaced the corresponding variables of v^*. Let these variables be arranged in a vector x^{1*} and let us rearrange and partition the vectors and matrices of (14) accordingly. The system is then in tableau-form:

B.v.	V.b.v.	x^{1*}	x^{2*}	v_1
F	q_0	$q^{1\prime}$	$q^{2\prime}$	$-r$
v^{1*}	q^1	Q_{11}	Q_{12}	$-p^1$
v^{2*}	q^2	Q_{21}	Q_{22}	$-p^2$
x_1	r	$p^{1\prime}$	$p^{2\prime}$	0

The local maximum is now the basic solution of the equation system

B.v.	V.b.v.	v^{1*}	x^{2*}	v_1
F	$q_0 - q^{1\prime}Q_{11}^{-1}q^1$	$-q^{1\prime}Q_{11}^{-1}$	$q^{2\prime} - q^{1\prime}Q_{11}^{-1}Q_{12}$	$-r + q^{1\prime}Q_{11}^{-1}p^1$
x^{1*}	$Q_{11}^{-1}q^1$	Q_{11}^{-1}	$Q_{11}^{-1}Q_{12}$	$-Q_{11}^{-1}p^1$
v^{2*}	$q^2 - Q_{21}Q_{11}^{-1}q^1$	$-Q_{21}Q_{11}^{-1}$	$Q_{22} - Q_{21}Q_{11}^{-1}Q_{12}$	$-p^2 + Q_{21}Q_{11}^{-1}p^1$
x^1	$r - p^{1\prime}Q_{11}^{-1}q^1$	$-p^{1\prime}Q_{11}^{-1}$	$p^{2\prime} - p^{1\prime}Q_{11}^{-1}Q_{12}$	$p^{1\prime}Q_{11}^{-1}p^1$

The local maximum is characterized by

$$x^{1*} = Q_{11}^{-1}q^1 \geq 0, \; v^{2*} = q^2 - Q_{21}Q_{11}^{-1}q^1 \geq 0, \; x_1 = r - p^{1\prime}Q_{11}^{-1}q^1 \geq 0,$$

and the matrix

$$\begin{bmatrix} Q_{11}^{-1} & -Q_{11}^{-1}p^1 \\ -p^{1\prime}Q_{11}^{-1} & p^{1\prime}Q_{11}^{-1}p^1 \end{bmatrix} = \begin{bmatrix} 1 \\ -p^1 \end{bmatrix}' Q_{11}^{-1} \begin{bmatrix} 1 \\ -p^1 \end{bmatrix},$$

which corresponds to the matrix R in a general tableau, is negative semi-definite. Hence, Q_{11}^{-1} is negative definite, so that Q_{11} is negative definite. From this we conclude that $q^{1\prime}Q_{11}^{-1}q^1 < 0$ for $q^1 \neq 0$. Since $Q_{11}^{-1}q^1 = x^{1*} \geq 0$, q^1 must contain at least one negative element. We must therefore consider in (14) all basic solutions corresponding with a negative definite submatrix Q_{11} and q^1 with at least one negative element.

First, the idea of the Theil–van de Panne method for convex quadratic programming (see chapter 12), as applied to this situation will be explained, after which the original Ritter proposal is treated.

Proceeding from (14), for each $q_i < 0$, with $q_{ii} < 0$, x_i^* replaces v_i^*. The resulting equation systems again have the form (14), except that there are now two primal basic variables and $R \neq 0$. Then the same procedure is applied to these equation systems, and so on, until no $q_i < 0$ with $q_{ii} < 0$ is left. All local maxima which include x_1 as a basic variable must now have been generated. x_1 is deleted from (13) and x_2 takes its place, and so on.

Ritter does essentially the same, though in a different sequence. The point of departure is again (14). One $q_i < 0$, with $q_{ii} < 0$ is selected and the corresponding x_i^* replaces v_i^*. In the resulting system this is repeated, and in the next tableau this is repeated again, until no more $q_i < 0$ with $q_{ii} < 0$ are found. Then the x_i^*, which was last introduced, is replaced again by v_i^*. In the resulting system some $q_i < 0$ with $q_{ii} < 0$ may again occur, which are treated as before. When these occur no more, the last but are x_i^* is replaced by v_i^*, and so on. This continues until all x_i^* which have ever become basic have been made nonbasic again. Then x_1 is deleted from (13), x_2 takes its place, and so on.

After this, all possible solutions must be gathered and compared, so that the global maximum can be chosen. Note that Ritter's proposal leads to a sequential computation, whereas the idea of the Theil–van de Panne method branches out.

Ritter's auxiliary problem can also be solved by the parametric method for general quadratic programming which is explained in the remainder of this chapter.

15.5. Outline of the parametric method for general quadratic programming

In the remaining sections a new method for general quadratic programming is presented. It is called the parametric method because it uses multiparametric quadratic programming as a subroutine.

The method is based in the following idea. Let the problem be formulated as follows: Maximize

$$f = p'x + \tfrac{1}{2}x'Cx$$

subject to

$$Ax \le b,$$

$$x \ge 0.$$

Assume that the matrix C can be partitioned, possibly after rearrangement, in the following manner:

$$C = \begin{bmatrix} C_{11} & C_{12} \\ C_{21} & C_{22} \end{bmatrix},$$

with C_{22} negative semi-definite. If C cannot be partitioned in this manner, then, as is shown later, a diagonalization procedure may be used to the same effect. The vectors p and x and the matrix A may be partitioned correspondingly:

$$p = \begin{bmatrix} p^1 \\ p^2 \end{bmatrix}, \qquad x = \begin{bmatrix} x^1 \\ x^2 \end{bmatrix}, \qquad A = [A_1 \quad A_2].$$

The problem may now be formulated as follows: Maximize

$$f = 2p^{1\prime}x^1 + x^{1\prime}C_{11}x^1 + 2(p^1 + C_{12}x^1)'x^2 + x^{2\prime}C_{22}x^2$$

subject to

$$A_2x^2 \le b - A_1x^1,$$

$$x^1 \ge 0, \; x^2 \ge 0.$$

The variables in x^1 may now be interpreted as variable parameters. The problem is then a multiparametric convex quadratic programming problem. We may now determine regions in the space of the x^1-variables, for which basic solutions of the above problems are optimal. Finding the best point in each of these regions gives rise to another general quadratic programming problem, which is smaller in size than the original problem. These subproblems and possible subproblems arising from them are solved in the same fashion.

It is obvious that the method will be fastest if the number of variables in x^1 is small. It will be shown that the minimum number of variables in x^1 is equal to the number of positive roots of C. If the matrix C of the original problem or any of the subproblems is positive semi-definite, any of the existing methods for concave quadratic programming can be used.

Let us now consider the method in some more detail. In a number of cases it is advantageous first to generate a local optimal solution. Let the equation system for such a solution be represented as follows:

$$q_0 = q'x - r'u + F,$$
$$q = Qx - P'u + r,$$
$$r = Px + Ru + y, \tag{16}$$

where x is the vector of nonbasic primal variables, and so on. Since the solution is a local optimum, we have $q \geq 0$ and $r \geq 0$. In order to avoid complications of degeneracy, we assume $q > 0$ and $r > 0$.

Consider now a partition of Q

$$Q = \begin{bmatrix} Q_{11} & Q_{12} \\ Q_{21} & Q_{22} \end{bmatrix},$$

such that Q_{22} is negative semi-definite or a diagonalization which has the same effect. The vectors x, q, and v, and the matrix P may be partitioned accordingly:

$$x = \begin{bmatrix} x^1 \\ x^2 \end{bmatrix}, \qquad q = \begin{bmatrix} q^1 \\ q^2 \end{bmatrix}, \qquad v = \begin{bmatrix} v^1 \\ v^2 \end{bmatrix}, \qquad P = [P_1 \quad P_2].$$

The variables in x^1 are now considered as variable parameters. Eqs. (16) then become

$$q_0 - 2q^{1\prime}x^1 + x^{1\prime}Q_{11}x^1 = (q^2 - Q_{21}x^1)'x^2 - (r - P_1x^1)'u + f,$$
$$q^2 - Q_{21}x^1 = Q_{22}x^2 - P_2'u + v,$$
$$r - P_1x^1 = P_2x^2 + Ru + y. \tag{17}$$

After a change in notation, this may be written as

$$q_0 + 2s't + t'St = (q + Dt)'x - (r + Et)'u + F,$$
$$q + Dt = Qx - P'u + v,$$
$$r + Et = Px + Ru + y, \tag{18}$$

with $t = x^1$, $x = x^2$, $s = -q^1$, $S = Q_{11}$, $q = q^2$, $D = -Q_{21}$, $Q = Q_{22}$, $P = P_2$, $v = v^2$, $E = -P_1$. (18) is the Kuhn-Tucker equation system of a convex quadratic programming problem with a vector of parameters t. The basic solution of the system for $t = 0$ is an optimal solution since $q > 0$ and $r > 0$. In fact, it is an optimal solution for all values of t satisfying

$$q + Dt \geq 0,$$
$$r + Et \geq 0,$$
$$t \geq 0. \tag{19}$$

These constraints determine a non-empty convex region in the t-space. For each point in this region, the optimal solution is given by the basic solution of (18). To determine the optimal solution also with respect to t and situated in this region, we have to solve the following subproblem: Maximize

$$F = q_0 + 2s't + t'St$$

subject to

$$-Dt \leq q,$$
$$-Et \leq r,$$
$$t \geq 0.$$

This is another general quadratic programming problem in fewer variables than the original one. Moreover, it is less difficult, since if S is positive semi-definite it may be solved by a concave programming method; and if it is not, then a partition or a diagonalization is possible, which results in a multiparametric convex quadratic programming problem with fewer parameters than the number of elements in t.

The subproblem is solved in the same manner as the original problem, thus generating a hierarchy of subproblems, each time involving fewer or no parameters. Each time a subproblem with a positive semi-definite matrix of the quadratic form is met, it is solved by a method for concave quadratic programming, which may be Tui's, Ritter's, or simply the enumeration of extreme points; a subproblem with one parameter is always solved by the last method.

The feasible region of (19) gives only a part of the entire t-space. Regions in the t-space for which other complementary basic solutions of (18) are optimal are generated by means of multiparametric convex quadratic programming. Such a method was treated in chapter 13.

For each constraint in (19) we find out whether it is redundant or not by minimizing the right-hand side subject to the other constraints. If the minimum is nonnegative it is redundant. For each nonredundant constraint, we inspect the corresponding row of (18). If the diagonal element of that row is negative, this element is used as a pivot. If it is zero, we inspect the elements of $-P'$ in that row. If there are no negative elements, the entire half-space for which the constraint is not satisfied contains no feasible solution. If there is one negative element in that row, then a double pivot transformation with this and its corresponding element in P as pivots results in another basic solution of the form (18).

If there are more negative elements in $-P'$ in the row, the situation is

more complicated. In this case there may be more than one adjacent region for the same constraints; these are generated as follows. Let the basic
variable of the constraint be v_k. Consider then the following system

$$q_k + d_{k1}t_1 + \ldots + d_{kl}t_l = 0,$$

$$r_i + e_{i1}t_1 + \ldots + e_{il}t_l - p_{ik}x_k \geq 0, \quad (i|p_{ik} > 0)$$

$$x_k \geq 0. \tag{20}$$

In this system x_k should be made basic. Then the nonredundancy of
each of the inequalities should be tested for x_k remaining basic. Each
nonredundant inequality gives rise to a double pivot operation with p_{ik}
and $-p_{ik}$ as pivots, which results in a new basic solution with corresponding region.

In all cases in which a new solution is generated, it has the same form
as (18), except that q and r are not necessarily nonnegative. The new
basic solution $q^* + D^*t$, $r^* + E^*t$ gives optimal solution of the problem
for a region in the t-space

$$q^* + D^*t \geq 0,$$

$$r^* + E^*t \geq 0,$$

which is adjacent to the previous region. For this region a subproblem
may be formulated and so on.

In this manner the original region (19) is surrounded by adjacent regions or infeasible half-spaces. Each of the resulting adjacent regions is
used to generate new regions adjacent to it which have not yet been
generated. All regions related to the multiparametric convex quadratic
programming problem are generated in this fashion. For each of these
regions, subproblems are formulated and solved in the same manner as
the original problem.

Solution of concave subproblems or of convex subproblems, that is,
solution of subproblems which do not give rise to further subproblems,
are called *final solutions*. After all final subproblems have been worked
out, the final solutions are compared and the best one is chosen.

15.6. Diagonalization

If no suitable partitioning of C or Q can be found, such a partitioning can
be constructed by adding constraints and double pivot transformations.
This will result in a diagonal matrix Q, which can easily be partitioned.

As point of departure a basic complementary solution of the Kuhn-

Tucker equations, augmented by the linearized objective function is taken:

$$q_0 = q'x - r'u + F,$$
$$q = Qx - P'u + v,$$
$$r = Px + Ru + y. \tag{21}$$

R is negative semi-definite, Q is in general indefinite. It is not possible that Q is negative semi-definite if C is indefinite; if the basic solution of (21), $v = q$, $y = r$, is nonnegative in addition to Q negative semi-definite, then this basic solution is a global optimum. The same is true if $-Q$ is copositive[1] instead of positive semi-definite.

Let us assume that Q has at least one positive diagonal element and that this is the first diagonal element; the latter can always be achieved by a suitable rearrangement. The vectors x, v, and q and the matrices P and Q can then be partitioned accordingly. The tableau consisting of the equations of (21) is then partitioned as follows:

B.v.	V.b.v.	x_1	x^2	u
F	q_0	q_1	$q^{2\prime}$	$-r'$
v_1	q_1	q_{11}	Q_{12}	$-P'_1$
v^2	q^2	Q_{21}	Q_{22}	$-P'_2$
y	r	P_1	P_2	R

Let us now add the following constraint to the problem:

$$t_1 = q_{11}x_1 + Q_{12}x^2, \tag{22}$$

where t_1 is considered to be a variable parameter. This constraint is added as a new row and a new column. Since t_1 is a new parameter, an additional t_1 column is introduced in the values of basic variables, while an additional row is introduced into the f-equation. The artificial variable of eq. (22) is named s_1 and its dual variable w_1.

The first tableau of table 1 gives the set-up tableau for this problem. After a double pivot operation on the elements q_{11} and $-q_{11}$ in the added row and column, the second tableau results. Since the constraint (22) was an equation, the s_1-column and the w_1-row may be deleted. The final result is a new tableau with an objective function and values of basic variables, which are dependent on the parameter t_1.

---

[1] For a definition, see p. 462.

Table 1

Transformation after adding constraint

B.v.	c-term	t_1-term	x_1	x^2	u	w_1
f_c	q_0	0	q_1	$q^{2\prime}$	$-r'$	0
f_{t1}	0	0	0	0	0	-1
v_1	q_1	0	q_{11}	Q_{12}	$-P'_1$	$-q_{11}$
v^2	q^2	0	Q_{21}	Q_{22}	$-P'_2$	$-Q'_{12}$
y	r	0	P_1	P_2	R	0
s_1	0	1	$\underline{q_{11}}$	Q_{12}	0	0

	c-term	t_1-term	s_1	x^2	u	v_1
f_c	q_0	$-q_1q_{11}^{-1}$	$-q_1q_{11}^{-1}$	$q^{2\prime}-q_1q_{11}^{-1}Q_{12}$	$-r'$	0
f_{t1}	$-q_{11}^{-1}q_1$	q_{11}^{-1}	q_{11}^{-1}	0	$q_{11}^{-1}P'_1$	$-q_{11}^{-1}$
w_1	$-q_{11}^{-1}q_1$	q_{11}^{-1}	q_{11}^{-1}	0	$q_{11}^{-1}P'_1$	$-q_{11}^{-1}$
v^2	$q^2-Q'_{12}q_{11}^{-1}q_1$	0	0	$Q_{22}-Q_{21}q_{11}^{-1}Q_{12}$	$-P'_2+Q'_{12}q_{11}^{-1}P'_1$	$-Q'_{12}q_{11}^{-1}$
y	r	$-P_1q_{11}^{-1}$	$-P_1q_{11}^{-1}$	$P_2-P_1q_{11}^{-1}Q_{12}$	R	0
x_1	0	q_{11}^{-1}	q_{11}^{-1}	$q_{11}^{-1}Q_{12}$	0	0

If the matrix Q had only one positive root, the matrix $Q_{22}-Q_{21}q_{11}^{-1}Q_{12}$ must be negative semi-definite, so that we can stop here. If it is indefinite and has a positive diagonal element, which after possible rearrangement may be the first one, we may add the constraint

$$t_2 = q_{22}^* x_2 + Q_{23}^* x^3,$$

where $[q_{22}^* \quad Q_{23}^*]$ is the first row of the $Q_{22}-Q_{21}q_{11}^{-1}Q_{12}$ and $x^{2\prime} = [x_2 \quad x^{3\prime}]$. After a double pivot transformation, with q_{22}^* and $-q_{22}^*$ as pivots, another tableau is obtained with t_1 and t_2 as parameters, and so on.

In this manner a multiparametric quadratic programming problem is constructed with a diagonal matrix of the quadratic form S

$$f = q_0 + 2s't + t'St.$$

It is obvious that the procedure is similar to that of a diagonalisation of a quadratic form.

In case the matrix Q or $Q_{22}-Q_{21}q_{11}^{-1}Q_{12}$ or any lower-order matrix in the remaining number of x-variables has no positive diagonal elements but is still indefinite, we may take a negative diagonal element, add a constraint the type (22), but consider the t_i of this constraint an unconstrained variable, which later is made basic again.

If all diagonal elements of the remaining matrix are zero, we may take any nonzero term $q_{ij}x_ix_j$ and add the constraints

$$x_i^* = \tfrac{1}{2}(x_i + x_j),$$
$$x_j^* = \tfrac{1}{2}(x_i - x_j),$$

with x_i^* and x_j^* unconstrained. This results in a matrix with one positive and one negative diagonal element; the positive element can be used as described before.

Since the number of positive latent roots of the matrix $\begin{bmatrix} Q & 0 \\ 0 & R \end{bmatrix}$ remains the same, the diagonalisation procedure is stopped as soon as the same number of parameters is introduced as the number of positive latent roots of C. The remainder of the Q-matrix should then be negative semi-definite.

15.7. Numerical example

Consider the following general quadratic programming problem:
Maximize:

$$f = -22x_1 + 6x_2 + 14x_3 - 2x_1x_2 + 2x_2^2 - 6x_2x_3 + 4x_3^2$$

subject to

$$x_1 + x_2 + x_3 \le 1,$$
$$x_1 \ge 0,\ x_2 \ge 0,\ x_3 \ge 0.$$

The matrix C is in this case:

$$C = \begin{bmatrix} 0 & -1 & 0 \\ -1 & 2 & -3 \\ 0 & -3 & 4 \end{bmatrix}.$$

In order to determine the number of positive roots and minimum number of t-parameters, the principal pivot transformations indicated in table 2 are performed.

From this we conclude there are two positive and one negative roots, so that we need two t-parameters. By inspection we see that C can be rearranged and partitioned as

$$\begin{bmatrix} 2 & -3 & -1 \\ -3 & 4 & 0 \\ -1 & 0 & 0 \end{bmatrix} = \begin{bmatrix} \begin{bmatrix} 2 & -3 \\ -3 & 4 \end{bmatrix} & \begin{bmatrix} -1 \\ 0 \end{bmatrix} \\ [-1 \quad 0] & [0] \end{bmatrix} = \begin{bmatrix} C_{11} & C_{12} \\ C_{21} & C_{22} \end{bmatrix}$$

with C_{22} negative semi-definite. This leads to the selection $t_1 = x_2$, $t_2 = x_3$. The equations indicated in (21) may now be put into table 3.

Table 2

Principal pivots on C

0	0	-1	0
	-1	2	-3
	0	-3	4
1	$-\tfrac{1}{2}$	$\tfrac{1}{2}$	$-1\tfrac{1}{2}$
	$-\tfrac{1}{2}$	$\tfrac{1}{2}$	$-1\tfrac{1}{2}$
	$-1\tfrac{1}{2}$	$1\tfrac{1}{2}$	$-\tfrac{1}{2}$
2	-2	-1	3
	-1	0	0
	-3	0	4

Table 3

Tableau for example

	$c-t$	t_1-t	t_2-t	x_3	u
F	0	3	7	11	-1
F_{t_1}	3	2	-3	1	1
F_{t_2}	7	-3	4	0	1
v_3	11	1	0	0	-1
y	1	-1	-1	1	0

The region in the t-space for which this solution is optimal is described by

$$-t_1 \le 11,$$
$$t_1 + t_2 \le 1,$$
$$t_1 \ge 0, t_2 \ge 0.$$

It is obvious that the first constraint is redundant. The feasible region is indicated in figure 1.

Now we have to solve the sub-problem: Maximize

$$f = 6x_2 + 14x_3 + 2x_2^2 - 6x_2 x_3 + 4x_3^2$$

subject to

$$x_2 + x_3 \le 1$$
$$x_2 \ge 0, \; x_3 \ge 0.$$

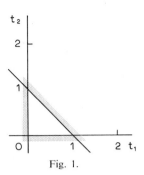

Fig. 1.

The set-up tableau is given in tableau 0 of table 4, together with the diagonalizing constraint

$$t = 2x_2 - 3x_3.$$

Tableau 1 results after diagonalization. After application of the simplex

Table 4

Solution of subproblem

Tableau		$c-t$	x_2	x_3	u	w_1
	f	0	-3	-7	-1	0
0	v_2	-3	2	-3	-1	-2
	v_3	-7	-3	4	-1	3
	y	1	1	1	0	0
	s_1	0	2	-3	0	0
		$c-t$	$t-t$	x_3	u	v_2
	f_c	0	$1\frac{1}{2}$	$-11\frac{1}{2}$	-1	0
	f_t	$1\frac{1}{2}$	$\frac{1}{2}$	0	$\frac{1}{2}$	$-\frac{1}{2}$
1	v_3	$-11\frac{1}{2}$	0	$-\frac{1}{2}$	$-2\frac{1}{2}$	$1\frac{1}{2}$
	y	1	$-\frac{1}{2}$	$2\frac{1}{2}$	0	0
	x_2	0	$\frac{1}{2}$	$-1\frac{1}{2}$	0	0
		$c-t$	$t-t$	y	v_3	v_2
	f_c	$9\frac{3}{25}$	$-\frac{19}{25}$	$4\frac{13}{25}$	$-\frac{2}{5}$	$-\frac{3}{5}$
	f_t	$-\frac{19}{25}$	$\frac{12}{25}$	$\frac{1}{25}$	$\frac{1}{5}$	$-\frac{1}{5}$
2	u	$4\frac{13}{25}$	$\frac{1}{25}$	$-\frac{2}{25}$	$-\frac{2}{5}$	$-\frac{3}{5}$
	x_3	$\frac{2}{5}$	$-\frac{1}{5}$	$\frac{2}{5}$	0	0
	x_2	$\frac{3}{5}$	$\frac{1}{5}$	$\frac{3}{5}$	0	0

method for quadratic programming, tableau 2 is obtained which is optimal for $t = 0$.

The basic solution of tableau 2 is optimal for

$$-\tfrac{1}{25}t \leq 4\tfrac{13}{25},$$
$$\tfrac{1}{5}t \leq \tfrac{2}{5},$$
$$-\tfrac{1}{5}t \leq \tfrac{3}{5}.$$

The first constraint is redundant; the other two amount to

$$-3 \leq t \leq 2.$$

The objective function for these values of t is

$$9\tfrac{3}{25} - 1\tfrac{13}{25}t + \tfrac{12}{25}t^2,$$

and is therefore concave, so that the optimum occurs at an extreme point in the feasible region $-3 \leq t \leq 2$. We find $f = 18$ and $f = 8$, so that $f = 18$, $x_3 = 1$ is the best solution in this region of the subproblem.

Since the rows associated with the boundaries -3 and 2 in tableau 2 contain no negative elements, all other values of t give infeasible solutions. Hence the best solution of the subproblem is $f = 18$, $x_3 = 1$.

We now return to the main tableau, see table 3, and try to generate solutions for adjacent regions in the t_1-, t_2-space. The nonredundant constraints of the region are

$$t_1 + t_2 \leq 1,$$
$$t_1 \geq 0, \; t_2 \geq 0.$$

Since in the row corresponding to the first constraint there is no negative element, the region

$$t_1 + t_2 > 1$$

is infeasible. The same is obviously true for the regions $t_1 < 0$, and $t_2 < 0$, so that there are no more regions to consider. Hence, the problem is solved and the optimal solution is $x_3 = 1$, $f = 18$.

The partitioning of C turned out to lead to just one region in a two-dimensional space. If a suitable partitioning cannot be found, then diagonalization must be employed. Let us solve the same problem in this manner.

Tableau 0 of table 5 gives the set-up tableau. After one double pivot iteration of the simplex method for quadratic programming a local maximum is found, see tableau 1. Then the constraint

$$t_1 = 4x_1 + 6x_2 + 4x_3$$

Table 5

Main tableaux for example

	B.v.	$c-t$	x_1	x_2	x_3	u		
0	f	0	11	-3	-7	-1		
	v_1	11	0	-1	0	-1		
	v_2	-3	-1	2	-3	-1		
	v_3	-7	0	-3	4	$\underline{-1}$		
	y	1	1	1	$\underline{1}$	0		

	B.v.	$c-t$	x_1	x_2	y	v_3		
1	f	18	22	11	11	-1		
	v_1	22	4	6	4	-1		
	v_2	11	6	12	7	-1		
	u	11	4	7	4	-1		
	x_3	1	1	1	1	0		

	B.v.	$c-t$	t_1-t	x_2	y	v_3	v_1	
2	f_c	18	$-5\frac{1}{2}$	-22	-11	-1	0	
	f_{t1}	$-5\frac{1}{2}$	$\frac{1}{4}$	0	0	$\frac{1}{4}$	$-\frac{1}{4}$	
	v_2	-22	0	3	1	$\frac{1}{2}$	$-1\frac{1}{2}$	
	u	-11	0	1	0	0	-1	
	x_3	1	$-\frac{1}{4}$	$-\frac{1}{2}$	0	0	0	
	x_1	0	$\frac{1}{4}$	$1\frac{1}{2}$	1	0	0	

	B.v.	$c-t$	t_1-t	t_2-t	y	v_3	v_1	v_2
3	f_c	18	$-5\frac{1}{2}$	$7\frac{1}{3}$	$-3\frac{2}{3}$	-1	0	0
	f_{t1}	$-5\frac{1}{2}$	$\frac{1}{4}$	0	0	$\frac{1}{4}$	$-\frac{1}{4}$	0
	f_{t2}	$7\frac{1}{3}$	0	$\frac{1}{3}$	0	$-\frac{1}{6}$	$\frac{1}{2}$	$-\frac{1}{3}$
	u	$-3\frac{2}{3}$	0	0	$-\frac{1}{6}$	$-\frac{1}{6}$	$-\frac{1}{2}$	$-\frac{1}{3}$
	x_3	1	$-\frac{1}{4}$	$\frac{1}{6}$	$\frac{1}{6}$	0	0	0
	x_1	0	$\frac{1}{4}$	$-\frac{1}{2}$	$\frac{1}{2}$	0	0	0
	x_2	0	0	$\frac{1}{3}$	$\frac{1}{3}$	0	0	0

	B.v.	$c-t$	t_1-t	t_2-t	x_2	v_3	v_1	u
4	f_c	18	$-5\frac{1}{2}$	11	11	-1	0	0
	f_{t1}	$-5\frac{1}{2}$	$\frac{1}{4}$	0	0	$\frac{1}{4}$	$-\frac{1}{4}$	0
	f_{t2}	11	0	0	-1	0	1	-1
	v_2	11	0	-1	-3	$\frac{1}{2}$	$1\frac{1}{2}$	-3
	x_3	1	$-\frac{1}{4}$	0	$-\frac{1}{2}$	0	0	0
	x_1	0	$\frac{1}{4}$	-1	$-1\frac{1}{2}$	0	0	0
	y	0	0	1	3	0	0	0

Table 5　(*continued*)

		$c-t$	t_1-t	t_2-t	x_3	v_2	v_1	u
	f_c	50	-8	9	10	2	3	-6
	f_{t1}	-8	$-\frac{1}{2}$	$\frac{1}{2}$	3	$-\frac{1}{2}$	-1	$1\frac{1}{2}$
	f_{t2}	9	$\frac{1}{2}$	0	-2	0	1	-1
5	v_3	10	3	-2	-12	2	3	-6
	x_2	-2	$\frac{1}{2}$	0	-2	0	0	0
	x_1	-3	1	-1	-3	0	0	0
	y	6	$-1\frac{1}{2}$	1	6	0	0	0

		$c-t$	t_1-t	t_2-t	x_1	v_3	v_2	u
	f_c	18	$-3\frac{2}{3}$	$3\frac{2}{3}$	$7\frac{1}{3}$	-1	0	0
	f_{t1}	$-3\frac{2}{3}$	$\frac{1}{6}$	$\frac{1}{6}$	$-\frac{1}{3}$	$\frac{1}{3}$	$\frac{1}{6}$	$-\frac{1}{2}$
	f_{t2}	$3\frac{2}{3}$	$\frac{1}{6}$	0	$\frac{2}{3}$	$-\frac{1}{3}$	$-\frac{2}{3}$	1
6	v_1	$7\frac{1}{3}$	$-\frac{1}{3}$	$\frac{2}{3}$	$-1\frac{1}{3}$	$\frac{1}{3}$	$\frac{2}{3}$	-2
	x_3	1	$-\frac{1}{3}$	$\frac{1}{3}$	$-\frac{1}{3}$	0	0	0
	x_2	0	$-\frac{1}{6}$	$\frac{2}{3}$	$-\frac{2}{3}$	0	0	0
	y	0	$\frac{1}{2}$	-1	2	0	0	0

is added both as a row and a column. After a double pivot transformation with 4 and -4 as pivots, tableau 2 is found. Now we add the constraint

$$t_2 = 3x_2 + y$$

and perform a double pivot transformation with 3 and -3 as pivots, which results in tableau 3. The solution of this tableau is not optimal for any value of t_1 and t_2 since

$$u = -3\tfrac{2}{3} + 0t_1 + 0t_2.$$

Application of the simplex method for quadratic programming leads to a double pivot transformation with $\frac{1}{3}$ and $-\frac{1}{3}$ as pivots. The result is tableau 4, which is optimal for $t_1 = t_2 = 0$.

The region in the t_1-, t_2-space, for which this solution is optimal, is described by

$$t_2 \le 11,$$
$$\tfrac{1}{4}t_1 \le 1,$$
$$-\tfrac{1}{4}t_1 + t_2 \le 0,$$
$$t_2 \ge 0.$$

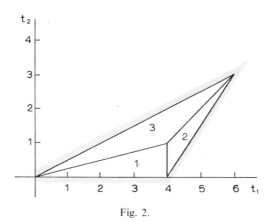

Fig. 2.

Obviously the first constraint is redundant; the others indicate region 1 in figure 2.

Now the following subproblem should be considered: Maximize

$$f = 18 - 11t_1 + 22t_2 + \tfrac{1}{4}t_1^2$$

subject to the constraints indicated above. Since the objective function is concave, we may simply examine all extreme points of region 1, which give

$$f(0,0) = 18, f(4,0) = -22, f(1,4) = 8.$$

Next we find out the adjacent regions of region 1. The constraint $t_2 \le$ 11 is redundant, so it should not be considered. The constraint $\tfrac{1}{4}t_1 \le 1$, is not redundant. The diagonal element in the row of x_3 is zero, but there is one negative off-diagonal element, so that this element is used in a double pivot operation which leads to tableau 5. The feasible region of the basic solution is determined by

$$-3t_1 + 2t_2 \le 10,$$
$$-\tfrac{1}{2}t_1 \le -2,$$
$$-t_1 + t_2 \le -3,$$
$$1\tfrac{1}{2}t_1 - t_2 \le 6.$$

This is region 2 of figure 2. The first constraint turns out to be redundant. The subproblem for this region is: Maximize

$$f = 50 - 16t_1 + 18t_2 - \tfrac{1}{2}t_1^2 + t_1 t_2$$

subject to

$$-\tfrac{1}{2}t_1 \leq -2,$$
$$-t_1 + t_2 \leq -3,$$
$$1\tfrac{1}{2}t_1 - t_2 \leq 6.$$

The matrix of the quadratic form of this problem can be easily partitioned with $t = t_2$. Tableau 0 of table 6 gives the set-up tableau in a standard notation. The first double pivot transformation is made to find an optimal solution for $t = 0$. t is varied upwards to 1, another optimal solution is found, see tableau 2. After this, t cannot be increased any further.

Inspecting the optimal solutions of tableaux 1 and 2, we find that the

Table 6

Solution of subproblem 2

Tableau		$c-t$	$t-t$	x_1	u_1	u_2	u_3
	f_c	50	9	8	2	3	-6
	f_t	9	0	$-\tfrac{1}{2}$	0	1	-1
0	v_1	8	$-\tfrac{1}{2}$	$-\tfrac{1}{2}$	$\tfrac{1}{2}$	1	$-1\tfrac{1}{2}$
	y_1	-2	0	$-\tfrac{1}{2}$	0	0	0
	y_2	-3	-1	-1	0	0	0
	y_3	6	1	$1\tfrac{1}{2}$	0	0	0
		$c-t$	$t-t$	y_1	v_1	u_2	u_3
	f_c	-22	11	20	-4	-1	0
	f_t	11	0	-1	0	1	-1
1	u_1	20	-1	-2	2	2	-3
	x_1	4	0	-2	0	0	0
	y_2	1	-1	-2	0	0	0
	y_3	0	1	3	0	0	0
		$c-t$	$t-t$	y_2	v_1	u_1	u_3
	f_c	$-2\tfrac{1}{2}$	1	$9\tfrac{1}{2}$	-3	$\tfrac{1}{2}$	$-1\tfrac{1}{2}$
	f_t	1	$\tfrac{1}{2}$	0	-1	$-\tfrac{1}{2}$	$\tfrac{1}{2}$
2	u_2	$9\tfrac{1}{2}$	0	$-\tfrac{1}{2}$	1	$\tfrac{1}{2}$	$-1\tfrac{1}{2}$
	x_1	3	1	-1	0	0	0
	y_1	$-\tfrac{1}{2}$	$\tfrac{1}{2}$	$-\tfrac{1}{2}$	0	0	0
	y_3	$1\tfrac{1}{2}$	$-\tfrac{1}{2}$	$1\tfrac{1}{2}$	0	0	0

objective function of both solutions is concave in terms of t_1, so that the optimal solutions in terms of t are extreme points in t. For tableau 1 we find the solutions

$$t = 0, f = -22.$$
$$t = 1, f = 0,$$

and for tableau 2:

$$t = 1, f = 0,$$
$$t = 3, f = 8.$$

Obviously, $t = 3$ gives the best solution. This is point $(6, 3)$ in figure 2.

We now turn to region 3, which gives optimal solution to the basic solution of tableau 6 of table 5. The subproblem is now: Maximize

$$f = 18 - 7\tfrac{1}{3}t_1 + 7\tfrac{1}{3}t_2 + \tfrac{1}{6}t_1^2 + \tfrac{1}{3}t_1 t_2$$

subject to

$$\tfrac{1}{3}t_1 - \tfrac{2}{3}t_2 \le 7\tfrac{1}{3},$$
$$\tfrac{1}{3}t_1 - \tfrac{1}{3}t_2 \le 1,$$
$$\tfrac{1}{6}t_1 - \tfrac{2}{3}t_2 \le 0,$$
$$-\tfrac{1}{2}t_1 + t_2 \le 0.$$

The first constraint is redundant and can therefore be deleted. The matrix C of the problem can be partitioned in such a way that t_1 is parametrically varied. Tableau 0 of table 7 is then the set-up tableau of the problem in a standard notation. After one iteration of the simplex method for quadratic programming an optimal solution is obtained, see tableau 1.

This solution is optimal for $0 \le t \le 6$. Evidently no solutions exist outside this range. Since the objective function of this problem is concave, the maximum solution in terms of t of the solution of tableau 1 occurs at an extreme point. We evaluate

$$t = 0, f = 18,$$
$$t = 6, f = 8.$$

Hence the best solution of region 3 is the point $(0, 0)$ with $f = 18$, which was found before in region 1.

We have now explored the entire feasible region in the t_1-, t_2-space and found that the best solution is $t_1 = 0, t_2 = 0_2$, which implies $x_1 = 0_1, x_2 = 0,$ $x_3 = 1, f = 18$.

Table 7

Solution of subproblem 3

Tableau		$c-t$	$t-t$	x_1	u_1	u_2	u_3
	f_c	18	$-3\frac{2}{3}$	$-3\frac{2}{3}$	-1	0	0
	f_t	$-3\frac{2}{3}$	$\frac{1}{6}$	$-\frac{1}{6}$	$\frac{1}{3}$	$\frac{1}{6}$	$-\frac{1}{2}$
0	v_1	$-3\frac{2}{3}$	$-\frac{1}{6}$	0	$\frac{1}{3}$	$\frac{2}{3}$	-1
	y_1	1	$-\frac{1}{3}$	$-\frac{1}{3}$	0	0	0
	y_2	0	$-\frac{1}{6}$	$-\frac{2}{3}$	0	0	0
	y_3	0	$\frac{1}{2}$	1	0	0	0
		$c-t$	$t-t$	y_3	u_1	u_2	v_1
	f_c	18	$-1\frac{5}{6}$	$3\frac{2}{3}$	-1	0	0
	f_t	$-1\frac{5}{6}$	$\frac{1}{3}$	$\frac{1}{6}$	$\frac{1}{6}$	$-\frac{1}{6}$	$-\frac{1}{2}$
1	v_3	$3\frac{2}{3}$	$\frac{1}{6}$	0	$-\frac{1}{3}$	$-\frac{2}{3}$	-1
	y_1	1	$-\frac{1}{6}$	$\frac{1}{3}$	0	0	0
	y_2	0	$\frac{1}{6}$	$\frac{2}{3}$	0	0	0
	x_1	0	$\frac{1}{2}$	1	0	0	0

15.8. Concluding remarks

The method can be summarized as follows. For the original problem and all resulting subproblems, perform the following steps:

(1) Determine the number of positive and negative roots of C. If there are no negative roots, use a method of concave programming.

(2) Determine a local optimum by means of the simplex method for quadratic programming.

(3) Find a suitable partitioning of Q or apply the diagonalisation procedure. Formulate the resulting multiparametric problem.

(4) Find an optimal solution of the multiparametric problem for $t = 0$ by means of the simplex method for quadratic programming.

(5) Formulate the subproblem resulting from (4).

(6) Find all nonredundant constraints of the subproblem.

(7) For each nonredundant constraint found in (6), find out whether there is an adjacent basic complementary solution; if so, formulate the resulting subproblem.

(8) Repeat (6) and (7) for each adjacent basic complementary solution until no new adjacent basic complementary solutions are found.

After these steps have been performed in all subproblems, the value of

f for all final solutions is compared and the one with the highest value is selected. The optimization steps guarantee that the subproblems generated have a feasible solution.

In order to avoid confusion between primal and dual infeasibility, it is convenient, if the feasible region of the objective function allows infinite solutions, to add a constraint.

$$x_1 + x_2 + \ldots + x_n \leq b_0,$$

where b_0 is a suitably large number.

In step 1, the use of some method for concave programming was prescribed if C is positive semi-definite. However, if the number of positive roots of C is less than its order, the parametric method may be used as in the other cases. Alternative methods for concave programming are then used only if a problem or subproblem is encountered with a positive definite matrix.

The parametric method for general quadratic programming is obviously finite, which does not mean too much, as the number of subproblems to be solved may be very large. Clearly, the method works best if the matrix C has few positive latent roots.

Possible improvements of the method could be thought of. Since this method is basically a branching method, it could be made into a branch-and-bound method if upper bounds of f could be found for each subproblem.

Another possibility is combination with Ritter's cutting plane method. The construction of the cutting plane involves the solution of a general quadratic programming problem with one constraint; this problem could be solved by the parametric method.

CHAPTER 16

THE LINEAR COMPLEMENTARITY PROBLEM, I

16.1. Introduction

In previous chapters we considered the convex quadratic programming problem and concluded that the optimal solution to the problem is obtained if vectors x, y, u and v are obtained which satisfy

$$-p = -Cx - A'u + v,$$
$$b = \quad Ax \qquad + y,$$
$$\begin{bmatrix} x \\ u \end{bmatrix}' \begin{bmatrix} v \\ y \end{bmatrix} = 0,$$
$$x, y, u, v \geq 0. \tag{1}$$

These conditions may be formulated as follows:

$$b^* = A^* x^* + y^*,$$
$$x^{*\prime} y^* = 0,$$
$$x^* \geq 0, \; y^* \geq 0 \tag{2}$$

where

$$b^* = \begin{bmatrix} -p \\ b \end{bmatrix}, \quad A^* = \begin{bmatrix} -C & -A' \\ A & 0 \end{bmatrix}, \quad x^* = \begin{bmatrix} x \\ y \end{bmatrix}, \quad y^* = \begin{bmatrix} v \\ u \end{bmatrix}.$$

The problem (2) is a more general type of problem, which is called *the linear complementarity problem*; it may be written without the asterisks as[1]

$$b = Ax + y,$$
$$x'y = 0,$$
$$x \geq 0, \; y \geq 0. \tag{3}$$

Solution methods for this problem, in which the matrix A does not necessarily have any specific structure, are considered in this and in the following chapter. Apart from linear and quadratic programming prob-

---

[1] In other work, the notation $q = -Mz + w$ is used instead of $b = Ax + y$.

426

lems, bimatrix game problems (see Lemke and Howson[110]) and some engineering problems (see Ingleton[106]) may be formulated as special cases of linear complementarity problems.

It should be noted that there are quite significant differences between the convex quadratic programming problem (1) and the general linear complementarity problem (3). The latter problem has no objective function (though one may be formulated), it has no distinction between primal and dual variables, and, most importantly, no special structure of the matrix A, which in quadratic programming has a pattern of symmetry and skew-symmetry and in convex problems a positive semi-definite matrix C.

If the matrix A of the linear complementarity problem has no specific structure, the methods explained in the following will not necessarily process the problem, that is, find a solution or show that no solution exists. However, if A belongs to a certain class of matrices the methods will do this.

In linear and convex quadratic programming, a problem has either
   (i) no feasible solution to the constraints,
  (ii) one or more optimal solutions, or
 (iii) feasible solutions but no finite optimal solution.

If these problems are interpreted as linear complementarity problems, then for the cases (i) and (iii) there will be no solution to the corresponding linear complementarity problem. If, for a linear or convex quadratic programming problem, the optimal solution is not unique, then one or more dual variables will be zero, which means that the corresponding solution to the linear complementarity problem will be degenerate.

A general linear complementarity problem may have different solutions, which are not degenerate. The methods to be explained in this chapter are aimed at finding one solution to the problem. Methods to find all solutions to the problem involve combinatorial search procedures which are outside the scope of this book. Murty[113] has proposed a method of this kind.

The methods discussed in this chapter are generalizations of Dantzig's self-dual parametric method for linear programming, which obviously can be extended easily to cover convex quadratic programming. For linear and convex quadratic programming problems these methods offer nothing new, a fact which has not been recognized by some authors. For linear programming, usage of the format of the linear complementarity problem doubles the size of the problem without any advantage.

In the following sections the terms complementary tableaux and complementary solutions will be used. They have the same meanings as the terms standard tableaux and standard solutions in the previous chapters; the word complementary seems more appropriate in the context of the linear complementarity problem.

This chapter will deal with generalizations of Dantzig's self-dual parametric method, in the following to be called the parametric method, to the linear complementarity problem. In this chapter, the impossibility of cycling is, as in parametric methods treated in earlier chapters, based on a monotonic variation of λ and certain other variables. Such a variation is possible only if certain tableau-elements are nonpositive, which is assured if A belongs to certain classes of matrices.

Chapter 17 deals with an extension of the parametric method, in which the monotonicity of the variation of λ and other variables can no longer be used to prove that no cycling occurs. However, Lemke has shown that a graph property can be used for the same purpose.

The exposition will be in terms of increasing complexity of the method, so that in this chapter we first deal with cases with negative principal elements, then with negative semi-definite matrices and then with matrices A such that $-A$ has nonnegative principal minors.

In quadratic programming, equivalent symmetric and asymmetric variants of the same parametric methods were distinguished. In the same way we have for the linear complementarity problem two equivalent variants, namely the complementary variant and the nearly complementary variant, which we shall also call the Lemke method. The following treatment concentrates on the complementary variant, but the Lemke method, which is treated by almost all other authors[2] writing on the linear complementarity problem, is also described and the equivalence between both methods is proved.

16.2. The case of negative principal pivots

If the vector b in (3) is nonnegative, a solution is obviously $y = b$, $x = 0$. If some elements of b are negative, the linear equations can be modified by the addition of $a\lambda$ to b:

$$b + a\lambda = Ax + y. \tag{4}$$

---

[2] GRAVES[54] also uses a complementary formulation.

a is assumed to have positive elements, usually taken as 1, in places where b has negative elements; the other elements of a should be nonnegative and are usually taken as zero. In this way the so-called *extended problem* is created. For some positive value of λ, the extended problem has a solution. The parametric method consists in varying λ to 0, while retaining in the successive solutions the complementarity and nonnegativity of the basic solutions.

In this section the simplest case will be attacked, namely the case in which all principal elements in complementary tableaux are negative.

Consider the linear complementarity problem with the following linear equations

$$7 = -3x_1 + 2x_2 + 2x_3 + y_1,$$
$$-2 = 2x_1 - 2x_2 \qquad + y_2,$$
$$-4 = -3x_1 - x_2 - x_3 + y_3.$$

After addition of λ-terms to negative constant terms, a set-up tableau as given in tableau 0 of table 1 is formulated.

Obviously, for $\lambda \geq 4$, the basic solution of this tableau is a solution of the extended problem. For $\lambda = 4$, y_3 is zero and the row of y_3 is the critical row. The principal element in this row, -1, which is negative according to our assumption, is used as a pivot which results in tableau 1. The basic solution of this tableau must be nonnegative for $\lambda = 4$, because it is the same as the basic solution of tableau 0 for $\lambda = 4$ since $y_3 = 0$ for $\lambda = 4$. But in tableau 1, $\lambda = 4$ is now an upper bound.

A lower bound of λ in tableau 1 is found to be $\lambda = 2$ with y_2 as the critical row. The principal element in this row, -2, is used as a pivot and tableau 2 is generated. Now $\lambda = \frac{1}{2}$ is the lower bound and y_1 the critical row. After pivoting on -9, tableau 3 is found, which has a nonnegative solution for $\lambda = 0$, so that this solution is a solution of the original problem.

The method can be described in a very simple manner. Let the solution of any tableau be the same as in the set-up tableau; one rule then suffices to describe the method:

Determine

$$\text{Max}_i (-b_i/a_i | a_i > 0) = -b_r/a_r. \tag{5}$$

Transform with a_{rr} as pivot. Stop if $b_i \geq 0$ for all i.

Termination of the method in a solution is assured if all principal elements are negative. In fact, all that is required is that principal elements

Table 1

The parametric method with negative principal pivots

Tableau	B.v.	Values basic variables			x_1	x_2	x_3
		c-term	λ-term	$\lambda = 4$			
0	y_1	7	0	7	-3	2	2
	y_2	-2	1	2	2	-2	0
	y_3	-4	1	0	-3	-1	-1
			$\lambda = 4$	$\lambda = 2$	x_1	x_2	y_3
1	y_1	-1	2	7	3	-9	0	2
	y_2	-2	1	2	0	2	-2	0
	x_3	4	-1	0	2	3	1	-1
			$\lambda = 2$	$\lambda = \frac{1}{2}$	x_1	y_2	y_3
2	y_1	-1	2	3	0	-9	0	2
	x_2	1	$-\frac{1}{2}$	0	$\frac{3}{4}$	-1	$-\frac{1}{2}$	0
	x_3	3	$-\frac{1}{2}$	2	$2\frac{3}{4}$	4	$\frac{1}{2}$	-1
			$\lambda = \frac{1}{2}$	$\lambda = 0$	y_1	y_2	y_3
3	x_1	$\frac{1}{9}$	$-\frac{2}{9}$	0	$\frac{1}{9}$	$-\frac{1}{9}$	0	$-\frac{2}{9}$
	x_2	$1\frac{1}{9}$	$-\frac{13}{18}$	$\frac{3}{4}$	$1\frac{1}{9}$	$-\frac{1}{9}$	$-\frac{1}{2}$	$-\frac{2}{9}$
	x_3	$2\frac{5}{9}$	$\frac{7}{18}$	$2\frac{3}{4}$	$2\frac{5}{9}$	$\frac{4}{9}$	$\frac{1}{2}$	$-\frac{1}{9}$

in critical rows are negative, but since in almost all cases the critical rows are not known beforehand, this is not of much use.

In chapter 1 it was shown that any element in any tableau may be represented as a ratio of two determinants composed of elements of the matrix of the set-up tableau. Principal elements of a complementary tableau may be represented as ratios of two successive minors of the matrix A:

$$a_{ii}^* = |A_i|/|A_{i-1}|. \tag{6}$$

If the matrix $-A$ has positive principal minors, we have

$$-a_{ii}^* = |(-A)_i|/|(-A)_{i-1}| > 0. \tag{7}$$

This implies that the principal element a_{ii}^* in (6) must be negative. We may therefore conclude that if $-A$ has positive principal minors, principal elements in complementary tableaux are negative and the method must terminate after a finite number of iterations in a solution. In section 16.6 it will be proved that the solution is unique if it is nondegenerate.

Instead of the complementary variant, the nearly complementary variant or Lemke method may be used. In this method, the λ-term is considered a nonbasic variable and is then made basic. After that the corresponding variable of the variable which just left the basis is chosen as the new basic variable while the leaving basic variable is selected as in the simplex method.

Tableau 0 of table 2 gives the set-up tableau; the underlined element -1 is used as a pivot and y_3 becomes nonbasic. In tableau 2 the nonbasic variable corresponding to the variable which just left the basis, y_3, is introduced and the leaving basic variable is determined as in the simplex

Table 2

Example of the nearly complementary variant

Tableau	B.v.	V.b.v.	λ	x_1	x_2	x_3
0	y_1	7	0	-3	2	2
	y_2	-2	-1	2	-2	0
	y_3	-4	$\underline{-1}$	-3	-1	-1
			y_3	x_1	x_2	x_3
1	y_1	7	0	-3	2	2
	y_2	2	-1	5	-1	1
	λ	4	-1	3	1	$\overline{1}$
			y_3	x_1	x_2	y_2
2	y_1	3	2	$\overline{-13}$	4	-2
	x_3	2	-1	5	$-\overline{1}$	1
	λ	2	0	-2	2	-1
			y_3	x_1	y_1	y_2
3	x_2	$\frac{3}{4}$	$\frac{1}{2}$	$-3\frac{1}{4}$	$\frac{1}{4}$	$-\frac{1}{2}$
	x_3	$2\frac{3}{4}$	$-\frac{1}{2}$	$1\frac{3}{4}$	$\frac{1}{4}$	$\frac{1}{2}$
	λ	$\frac{1}{2}$	-1	$4\frac{1}{2}$	$-\frac{1}{2}$	0
			y_3	λ	y_1	y_2
4	x_2	$1\frac{1}{9}$	$-\frac{2}{9}$	$-\frac{13}{18}$	$-\frac{1}{9}$	$-\frac{1}{2}$
	x_3	$2\frac{5}{9}$	$-\frac{1}{9}$	$-\frac{7}{18}$	$\frac{4}{9}$	$\frac{1}{2}$
	x_1	$\frac{1}{9}$	$-\frac{2}{9}$	$\frac{2}{9}$	$-\frac{1}{9}$	0

method for linear programming. After three more iterations, a solution is obtained which is the same as the one obtained before.

The complementary variant and the nearly complementary variant of the method are equivalent to one another in the sense that they generate the same sequence of solutions. For reasons of space, proof of this is postponed until more complicated cases have been treated.

16.3. The negative semi-definite case

We shall now deal with a somewhat more general case, for which the pivoting rules are the same as in parametric linear and convex quadratic programming. It will be shown that pivot selection of this kind is always possible if the matrix A of the linear complementarity problem is negative semi-definite, but not necessarily symmetric. The treatment is the same as in Graves[54], apart from some less essential points.

In parametric quadratic programming, first the critical value of λ is determined by means of

$$\underline{\lambda} = \text{Max}_i (-b_i/a_i | a_i > 0) = -b_k/a_k, \tag{8}$$

where b_i is the constant term of the value of the basic variable in the ith row in the current tableau and a_i is the similar λ-term element. If the principal element in the kth row, a_{kk}, is negative, it is used as a pivot, after which a further lower bound of λ is determined. If $a_{kk} = 0$, we determine

$$\text{Min}_i \left(\frac{b_i + a_i \underline{\lambda}}{a_{ik}} \,\middle|\, a_{ik} > 0 \right) = \frac{b_r + a_r \underline{\lambda}}{a_{rk}}. \tag{9}$$

If $a_{ik} \leq 0$ for all i, there is no solution. If the minimum occurs in the rth row, there is a double pivot operation with a_{rk} and $a_{kr} = -a_{rk} < 0$ as pivots; $a_{kr} = -a_{rk}$ follows from the symmetry relations in a standard tableau.

The same procedure can be followed in all cases in which the principal element in the critical row happens to be negative, or if it is zero and (9) is used we have $a_{kr} = -a_{rk}$. The method can also proceed without difficulty if instead of $a_{kr} = -a_{rk}$ we have $a_{kr} < 0$, because a pivot with a negative element a_{kr} will unblock the decrease of λ.

As a small example consider table 3. Let the row of y_1 be the critical row, so that $\lambda = -b_1/a_1$. If $a_{11} < 0$, we pivot on a_{11}, after which $-b_1/a_1$ is an upper bound of λ. If $a_{11} = 0$, (9) is used. Let the minimum be

Table 3

Basic variables	Values basic variables		x_1	x_2	x_3
	c-term	λ-term			
y_1	b_1	a_1	$a_{11} = 0$	$a_{12} < 0$	a_{13}
y_2	b_2	a_2	$a_{21} > 0$	a_{22}	a_{23}
y_3	b_3	a_3	a_{31}	a_{32}	a_{33}

found in row 2, so that a_{21} must be positive. Let us then increase x_1 to $(b_2 + a_2\underline{\lambda})/a_{21}$. The values y_1, y_2, and y_3 are then

$$b_1 + a_1\underline{\lambda} = b_1 + a_1(-b_1/a_1) = 0$$
$$b_2 + a_2\underline{\lambda} - a_{21}(b_2 + a_2\underline{\lambda})/a_{21} = 0$$
$$b_3 + a_3\underline{\lambda} - a_{31}(b_2 + a_2\underline{\lambda})/a_{21} > 0.$$

If $a_{12} < 0$, both a_{21} and a_{12} can be used in a double pivot operation which do not change the values of basic variable for $\lambda = -b_1/a_1$ and $x_1 = (b_2 + a_2\underline{\lambda})/a_{21}$. Since $a_{11} = 0$, the transformation with a_{21} as pivot has no impact on the first row. The transformation with a_{12} as pivot means a division of the first row by a negative number, which turns the lower bound of λ into an upper bound. If $a_{12} \geq 0$, the method as explained so far is no longer applicable, but generalizations, to be discussed later on, are.

In linear and quadratic programming, if no $a_{ik} > 0$ is found in (9) no solution exists. This can be generalized here in the following manner: if $a_{kk} = 0$ and $a_{ik} \leq 0$ for all i, then $a_{ki} \geq 0$ for all i. If this is satisfied then since $b_k < 0$ and $a_{ki} \geq 0$ for all i, no solution exists.

If A is negative semi-definite we shall always have $a_{ii} \leq 0$ for all i and if $a_{ii} = 0$ for any i, then $a_{ij} = -a_{ji}$ for all j. In order to prove this, let us first express A in terms of a symmetric part C and a skew-symmetric part D:

$$C = \tfrac{1}{2}(A + A'), \qquad D = \tfrac{1}{2}(A - A'), \tag{10}$$

which implies

$$A = C + D. \tag{11}$$

Note that the elements of the main diagonal of D must be zero. Furthermore, we have

$$x'Ax = x'Cx + x'Dx; \tag{12}$$

since $x'Dx = \tfrac{1}{2}x'(A - A')x = 0$, we have $x'Ax = x'Cx$. Hence, if A is negative semi-definite, so must C be. If, therefore, a diagonal element of C is zero, the elements in the same row or column must be zero. Hence, if $a_{ii} = c_{ii} + d_{ii} = c_{ii} = 0$, then $a_{ij} = c_{ij} + d_{ij} = d_{ij}$ for all j and $a_{ji} = c_{ji} + d_{ji} = d_{ji}$ for all j; since $d_{ji} = -d_{ij}$, the negative semi-definite matrix A has the required properties for an application of the method as far as the set-up tableau is concerned.

If we prove that any principal transform of a negative semi-definite matrix yields another negative semi-definite matrix, the proof is complete. This may be proved by performing in a general tableau a transformation with a negative principal pivot and a transformation with a principal 2×2 block pivot in which the principal elements are zero and the two other elements are skew-symmetric; the resulting matrices can then be shown to be negative semi-definite if the original matrix has this property.

However, Wolfe[3] has indicated a simpler proof which runs as follows. Consider the equation system $y = Ax$. Then $x'y = x'Ax \leq 0$ if A is negative semi-definite. Let $y^* = A^*x^*$ be a principal transform of $y = Ax$. Note that some y-variables have been replaced by their corresponding x-variables and vice versa; this means that $x^{*'}y^* = x'y$, since if x_i and y_i have changed places the product $x_i y_i$ remains the same. Hence $x^{*'}y^* \leq 0$. Since $x^{*'}y^* = x^{*'}A^*x^*$, we conclude $x^{*'}A^*x^* \leq 0$ so that A^* must be negative semi-definite.

16.4. The general case with nonincreasing λ

After the rather special case of negative principal pivots, a more general case will be treated. The key concept in all parametric methods treated so far was that of a monotonic decrease or increase of the parameter, on which the convergence of the method is based. This section deals with the most generalized version of the method in which this concept is applied. Such a monotonic variation of a parameter requires that certain tableau-elements have a negative sign. In the next section it is proved that these elements will always have the appropriate sign if $-A$ has nonnegative principal minors.

To explain the basic ideas of the method, the following example will be used.

---
[3] Cottle and Dantzig[102], p. 132.

$$-3 = \qquad\qquad -3x_3 + y_1,$$
$$-2 = \quad x_1 - 2x_2 + 2x_3 + y_2,$$
$$-1 = -3x_1 + 7x_2 - 8x_3 + y_3.$$

After addition of λ-terms to all equations, tableau 0 of table 4 is found. The basic solution of this tableau is feasible for $\lambda \geq 3$. The first solution is $\lambda = 3$, $y_2 = 1$, $y_3 = 2$.

The variable y_1 blocks the decrease of λ. Since the principal element in the row of y_1 is zero, x_1 is increased. It is then found that, while keeping λ at the value 3, that y_2 becomes zero for $x_1 = 1$. A new solution is found by

Table 4

Example for the complementary variant with decreasing λ

Tableau	B.v.	V.b.v. c-term	V.b.v. λ-term	$\lambda = 3$		x_1	x_2	x_3
				$\lambda = 3$	$x_1 = 1$			
0	y_1	-3	1	0	0	0	0	-3
	y_2	-2	1	1	0	1	-2	2
	y_3	-1	1	2	5	-3	7	-8
				$\lambda = 3$ $x_1 = 1$	$\lambda = 3$ $x_1 = 11$	x_1	y_2	x_3
1	y_1	-3	1	0	0	0	0	-3
	x_2	1	$-\frac{1}{2}$	0	5	$-\frac{1}{2}$	$-\frac{1}{2}$	-1
	y_3	-8	$4\frac{1}{2}$	5	0	$\frac{1}{2}$	$3\frac{1}{2}$	-1
						y_3	y_2	x_3
2	y_1	-3	1			0	0	-3
	x_2	-7	4			1	3	-2
	x_1	-16	9			2	7	-2
				$\lambda = 3$	$\lambda = 1\frac{17}{25}$	y_3	y_2	y_1
3	x_3	1	$-\frac{1}{3}$	0	$\frac{11}{25}$	0	0	$\frac{1}{3}$
	x_2	-5	$3\frac{1}{3}$	5	$\frac{3}{5}$	1	3	$-\frac{2}{3}$
	x_1	-14	$8\frac{1}{3}$	11	0	2	7	$-\frac{2}{3}$
				$\lambda = 1\frac{17}{25}$	$\lambda = 0$	y_3	y_2	x_1
4	x_3	8	$-4\frac{1}{2}$	$\frac{11}{25}$	8	-1	$-3\frac{1}{2}$	$-\frac{1}{2}$
	x_2	9	-5	$\frac{3}{5}$	9	-1	-4	-1
	y_1	21	$-12\frac{1}{2}$	0	21	-3	$-10\frac{1}{2}$	$-1\frac{1}{2}$

increasing x_1 parametrically to 1, so that the solution is now $\lambda = 3$, $x_1 = 1$, $y_3 = 5$. No transformation takes place so far.

Since y_2 blocks the increase of x_1, this must be undone. In this case, the row of y_2 has a negative principal pivot, so that it can be used for a transformation which changes the sign of the elements in this row. Tableau 1 results, in which the same solution is still feasible, but now x_1 can be increased further. Now y_3 becomes zero for $x_1 = 1 + 5/\frac{1}{2} = 11$, so that the new solution is $\lambda = 3$, $x_2 = 5$, $x_1 = 11$.

y_3 now blocks the increase of x_1, so that an increase of x_3 is considered. But since x_3 has an element -3 in the row of y_1, which blocked λ, this element can be used to unblock λ. This element is used together with the element $\frac{1}{2}$ in the row of y_3 and the column of x_1 to form the block pivot

$$\begin{bmatrix} 0 & -3 \\ \frac{1}{2} & -1 \end{bmatrix};$$

in practice, successive transformations are made with $\frac{1}{2}$ and -3 as pivots. The result is tableau 3.

If in tableau 2 the element in the y_1-row and the x_3-column had been zero, we would have unblocked x_1 by pivoting on the element in the row of y_3 and the column of x_3. If this element had been zero, we would have increased x_3 until a basic variable would become zero.

The pivot element which was used in tableau 1 to unblock λ in y_1 was negative. If this element had been positive, pivoting on such an element would not lead to a new lower bound of λ; the lower bound would remain exactly the same. Cases of this nature will be treated in the next chapter. In order for λ to be decreased monotonically, it is necessary that the elements in the column of the variable which is increased and in blocked rows are nonpositive. It will be seen that a sufficient condition for this is that $-A$ has nonnegative principal minors.

In tableau 3, λ is unblocked and can be decreased further. The next lower bound found for λ is $\lambda = 1\frac{17}{25}$, with the row of x_1 as the critical row. This row has a negative principal element, which can be used as a pivot. A transformation with this element as a pivot results in tableau 4, which has a nonnegative solution for $\lambda = 0$; it is therefore a solution to the problem.

The method can be explained in general terms by means of table 5. This could be the set-up tableau or any other tableau in which the variables are renamed.

Let the starting-point be the determination of a new lower bound for λ,

Table 5

B.v.	c-term	λ-term	x_1	x_2	x_3	x_4
y_1	b_1	a_1	a_{11}	a_{12}	a_{13}	a_{14}
y_2	b_2	a_2	a_{21}	a_{22}	a_{23}	a_{24}
y_3	b_3	a_3	a_{31}	a_{32}	a_{33}	a_{34}
y_4	b_4	a_4	a_{41}	a_{42}	a_{43}	a_{44}

which is done by

$$\underline{\lambda} = \operatorname*{Max}_i \left(- b_i/a_i \,|\, a_i > 0\right). \tag{13}$$

Let the maximum be found in the row of y_1, so that $\underline{\lambda} = - b_1/a_1$. The new solution is then

$$\underline{\lambda} = - b_1/a_1,$$
$$y_i = b_i + a_i \underline{\lambda}, \qquad i = 1, ..., p.$$

Note that $y_1 = 0$. We say that y_1 blocks λ.

If $a_{11} < 0$, this blocking can be undone by pivoting on a_{11}. Since $a_{11} < 0$, so that the first row is divided by a negative constant, $\lambda = - b_1/a_1$ is in the resulting tableau an upper bound; then a new lower bound should be found as indicated in (13).

Consider now the case $a_{11} = 0$. (In case $a_{11} > 0$, λ will increase, which is treated in the next chapter.) x_1, the corresponding variable of y_1 which blocked λ, is increased parametrically and an upper bound of λ is determined by

$$\bar{x}_1 = \operatorname*{Min}_i \left(\frac{b_i + a_i \underline{\lambda}}{a_{i1}} \,\middle|\, a_{i1} > 0\right). \tag{14}$$

If $a_{i1} \leq 0$ for all i, the method terminates unsuccessfully.

Assume that the minimum in (14) is found in the row of y_2. Then y_2 blocks the increase in x_1. The new solution is now

$$\lambda = \underline{\lambda} = -b_1/a_1,$$
$$x_1 = \bar{x}_1 = (b_2 + a_2 \underline{\lambda})/a_{21},$$
$$y_i = b_i + a_i \underline{\lambda} - a_{i1}\bar{x}_1, \qquad i = 3, 4, ..., p.$$

Note that no tableau transformations have taken place, but that the values of basic variables have been changed by the parametric varations of λ and x_1, which have made y_1 and y_2 equal to zero.

Since x_1 was blocked in the row of y_2, x_2 should be increased. First we see whether any of the blocked variables λ and x_1 can be unblocked. If

$a_{12} < 0$, λ can be unblocked by pivoting on this element and afterwards on a_{21}, so that x_2, x_1, y_3, and y_4 are basic variables. This amounts to using the block pivot

$$\begin{bmatrix} a_{11} = 0 & a_{12} < 0 \\ a_{21} > 0 & a_{22} \end{bmatrix}.$$

Another complementary tableau is found in which the former lower bound of λ is now an upper bound; then a further lower bound is found by means of (13) if the solution does not happen to be feasible for $\lambda = 0$.

In linear and quadratic programming, it follows from the symmetry and skew-symmetry properties of the tableau that, if $a_{11} = 0$, then $a_{12} = -a_{21}$, so that, if $a_{21} > 0$, then $a_{12} < 0$. This means that in these cases, there are scalar pivots in case the principal element in a blocked row is negative (which cannot happen in linear programming) or 2×2 block pivots, as indicated above.

The case $a_{12} > 0$ is considered in the next chapter. If $a_{12} = 0$, so that λ cannot be unblocked, we check whether $a_{22} < 0$, in which case x_1 can be unblocked. In this case a_{22} is used as a pivot and a further upper bound for x_1 should be found in the next tableau. The case $a_{22} > 0$ is again not considered here.

If $a_{12} = 0$ and $a_{22} = 0$, x_2 is increased parametrically to determine an upper bound for x_2:

$$\bar{x}_2 = \operatorname*{Min}_i \left(\frac{b_i + a_i \underline{\lambda} - a_{i1}\bar{x}_1}{a_{i2}} \middle| a_{i2} > 0 \right). \tag{15}$$

If $a_{i2} \leq 0$ for all i, the method terminates unsuccessfully. Assume that the minimum was found in the row of y_3. The next solution is then

$$\lambda = \underline{\lambda} = -b_1/a_1.$$
$$\bar{x}_1 = (b_2 + a_2\underline{\lambda})/a_{21},$$
$$\bar{x}_2 = (b_3 + a_3\underline{\lambda} - a_{31}\bar{x}_1)/a_{32},$$
$$y_i = b_i + a_i \underline{\lambda} - a_{i1}\bar{x}_1 - a_{i2}\bar{x}_2.$$

In this solution, λ, x_1, and x_2 are blocked by y_1, y_2, and y_3. Since x_2 was blocked by y_3, x_3 should now be considered a new basic variable. If the element in the column of x_3 and the row in which λ is blocked, a_{13}, is negative it can be used to unblock λ. This results in the following block pivot

$$\begin{bmatrix} a_{11} = 0 & a_{12} = 0 & a_{13} < 0 \\ a_{21} > 0 & a_{22} = 0 & a_{23} \\ a_{31} & a_{32} > 0 & a_{33} \end{bmatrix}$$

A transformation with such a block pivot can be performed by pivoting successively on a_{13}, a_{21}, and a_{32}. In the resulting solution, x_1, x_2, and x_3 have become basic. Since $a_{13} < 0$, λ is unblocked in the row of y_1 and a new lower bound for λ should be found if a solution to the problem is not yet obtained.

The case $a_{13} > 0$ is considered in the next chapter. If $a_{13} = 0$, λ cannot be unblocked at this moment. We then check whether the next blocked variable, x_1, can be unblocked, which is the case if $a_{23} < 0$. Then the block pivot

$$\begin{bmatrix} a_{22} = 0 & a_{23} < 0 \\ a_{32} > 0 & a_{33} \end{bmatrix}$$

is used and a further upper bound for x_1 should be found. The case $a_{23} > 0$ is considered in the next chapter. If $a_{23} = 0$, x_1 cannot be unblocked now. Then we try whether x_2 can be unblocked, which it can if $a_{33} < 0$. In that case a_{33} is used as a pivot and a further upper bound for x_2 should be found.

If $a_{13} = a_{23} = a_{33} = 0$, we determine

$$\bar{x}_3 = \underset{i}{\text{Min}} \left(\frac{b_i + a_i \underline{\lambda} - a_{i1}\bar{x}_1 - a_{i2}\bar{x}_2}{a_{i3}} \middle| a_{i3} > 0 \right), \qquad (16)$$

and so on.

Let us call the column of the new basic variable, in the previous discussion x_1, x_2 and x_3, the *incoming column*. If the elements of the incoming column in blocking rows are nonpositive and if, when these elements are zero, there is always at least one positive element in this column in other rows, then in each successive solution, either λ decreases monotonically or λ is unchanged, but then either the next blocked variable increases monotonically or remains the same, and so on. This means that after each change of solution, there is a monotonic change, so that no cycling can occur. This means that the method must terminate in a finite number of iterations.

If some of the elements in the incoming column and the blocking rows are positive, the changes in λ and other blocked variables are no longer monotonic; these cases are treated in the next chapter. If the elements in the incoming column and blocked rows are zero and those in the other rows are nonpositive, the algorithm terminates unsuccessfully. In the cases of linear and quadratic programming, it can be shown, by means of the symmetry and skew-symmetry properties of tableaux, that in this case there is no solution to the equation system, because then a row is found with $b_i < 0$ and $a_{ij} \geq 0$ for all j.

In the general case, infeasibility may be found out by testing after each transformation whether an infeasibility sign pattern $b_i < 0$ and $a_{ij} \geq 0$ for all j and for some i exists. Any existing infeasibility is not necessarily found in this way.

16.5. P- and P₀-Matrices

Square matrices with positive principal minors are called P-matrices and square matrices with nonnegative principal minors are called P_0-matrices.[4] In this section we shall investigate the properties of the parametric method if $-A$ is a P- or a P_0-matrix. If $-A$ is a P-matrix, all minors of A of order i have the sign of $(-1)^i$, if $-A$ is a P_0-matrix, they have this sign or are zero.

Since every principal element in a complementary tableau can be expressed as a ratio of two successive principal determinants, these elements are negative if $-A$ is a P-matrix or nonpositive if $-A$ is a P_0-matrix. This means that if $-A$ is a P-matrix, any blocking of λ can always immediately be undone by a negative principal pivot; the parametric method must therefore terminate in a finite number of iterations.

If $-A$ is a P_0-matrix, principal elements in a complementary tableau are nonpositive. If the principal element in a blocking row is negative, it can be used as a pivot; if it is zero, a new incoming column is chosen. This situation leads to a principal minor of the current tableau of the following type

$$\begin{bmatrix} a_{11} = 0 & a_{12} \\ a_{21} > 0 & a_{22} \end{bmatrix}. \tag{17}$$

If $-A$ is a P_0-matrix, the determinant of this matrix should have the sign $(-1)^2 = 1$ or zero. Hence $-a_{21}a_{12} \geq 0$ or $a_{12} \leq 0$. This means that $a_{12} > 0$ cannot occur if A is a P_0-matrix.

A similar situation occurs if there are three blocked variables and three blocking rows. The relevant submatrix is then

$$\begin{bmatrix} a_{11} = 0 & a_{12} = 0 & a_{13} \\ a_{21} > 0 & a_{22} = 0 & a_{23} \\ a_{31} & a_{32} > 0 & a_{33} \end{bmatrix}. \tag{18}$$

---

[4] See FIEDLER and PTAK [105].

The determinant of this matrix should have the sign of $(-1)^3 = -1$ or be zero, so that $a_{13}a_{21}a_{32} \leq 0$ if $-A$ is a P_0-matrix. Hence $a_{13} \leq 0$. By the same reasoning as used in (17), $a_{23} \leq 0$, and since a_{33} is a principal element, $a_{33} \leq 0$.

In general, it can be concluded that the elements in the incoming column and the blocking rows are nonpositive if $-A$ is a P_0-matrix. This ensures a monotonic variation of λ and other nonbasic variables for successive solutions.

However, this does not guarantee that the method will terminate with a solution if $-A$ is a P_0-matrix. The method terminates unsuccessfully if the elements in the incoming column and in the blocking rows are zero and the elements in other rows are nonpositive.

In linear and convex quadratic programming, such a termination indicates that there is no solution to the equations of the linear complementarity problem, a fact which is derived from the symmetry and skew-symmetry properties and negative semi-definiteness of A and its transforms. This is not necessarily so if $-A$ is a P_0-matrix. Consider a minor of order 2 of the current tableau

$$\begin{bmatrix} a_{kk} = 0 & a_{ik} \\ a_{ki} \leq 0 & a_{ii} \end{bmatrix}, \tag{19}$$

where k refers to the incoming column and its corresponding row, and i to another column and its corresponding row. If $-A$ is a P_0-matrix, we have $-a_{ki}a_{ik} \geq 0$. If $a_{ki} < 0$, we conclude $a_{ik} \geq 0$; if $a_{ki} = 0$, no conclusion can be made as to the sign of a_{ik}. Hence, only if $a_{ki} < 0$ for all $i \neq k$, then we have necessarily $a_{ik} \geq 0$ for all i. If $a_{ki} < 0$ for all i, only the kth row can be the blocking row in this case. Then only λ must be blocked, which means $a_i > 0$ and $b_i < 0$. This establishes the infeasibility in this case.

Infeasibility can therefore only be inferred from unsuccessful termination if $-A$ is a P_0-matrix if all elements in the incoming column are negative, except for the principal element, which must be zero. Even if some $a_{ki} = 0$ for some i, there may still be infeasibility if the corresponding elements a_{ik} happen to be nonnegative. Of course, infeasibility tests—as suggested earlier—will find such an infeasibility anyway.

It is interesting to observe that the assumption that $-A$ is a P_0-matrix is a more general one than $-A$ is positive semi-definite. The parametric method is more complicated in the P_0-case than in the positive semi-definite case in two aspects; firstly, apart from λ, other nonbasic variables may be blocked, secondly, unsuccessful termination does not necessarily imply infeasibility.

16.6. Uniqueness of solution for P_0-matrices

If $-A$ has nonnegative principal minors, a solution to the problem, if one is found and it is nondegenerate, must be unique. The proof of this is based on Fiedler and Ptak [105]. Let, after renaming the variables, a solution of the problem be written as

$$b^* = A^*x + y; \tag{20}$$

b^* and A^* are the elements of the current tableau and $y = b^* > 0$ is the solution. Since the matrix $-A$ of the original problem has nonnegative principal minors, $-A^*$ has nonnegative principal minors as well.

Assume that there is another solution in which a number of x-variables are basic instead of the corresponding y-variables. The system (20) may then be partitioned as follows where the vector x^1 contains these basic x-variables:

$$\begin{bmatrix} b^{1*} \\ b^{2*} \end{bmatrix} = \begin{bmatrix} A_{11}^* & A_{12}^* \\ A_{21}^* & A_{22}^* \end{bmatrix} \begin{bmatrix} x^1 \\ x^2 \end{bmatrix} + \begin{bmatrix} y^1 \\ y^2 \end{bmatrix}. \tag{21}$$

Consider now the first set of equations, deleting x^2, since it is zero in both solutions:

$$b^{1*} = A_{11}^* x^1 + y^1. \tag{22}$$

Let the value of x^1 in the assumed second solution be $\bar{x}^1 > 0$. Then there exists a diagonal matrix D with positive principal elements such that $b^{1*} = D\bar{x}^1$. Then both solutions satisfy the equation

$$b^{1*} = Dx^1 + y^1. \tag{23}$$

Subtracting (22) from (23) we find

$$(-A_{11}^* + D)x^1 = 0. \tag{24}$$

The matrix $-A_{11}^* + D$ must therefore be singular which implies

$$|-A_{11}^* + D| = 0. \tag{25}$$

Since $-A_{11}^*$ has nonnegative principal minors, $-A_{11}^* + D$ must have positive principal minors, which contradicts (25). Hence, a nondegenerate solution to the problem, if $-A$ is a P_0-matrix, is unique.

CHAPTER 17

THE LINEAR COMPLEMENTARITY PROBLEM, II

17.1. The Lemke approach to the parametric method

In the previous chapter, the elements in the incoming column and in blocking rows were assumed to be nonpositive. As a consequence, in each successive solution there was either a decrease in λ or an increase in one of the blocked nonbasic variables, given nondegeneracy, or failing that, the same could be proved for lexicographically positive vectors. From this it may be concluded that cycling of solutions is not possible.

If an element in the incoming column and a blocking row is positive and is used as a pivot, then, if λ was at its lower bound in this row, the same row would give the same lower bound after transformation with such a pivot, which means that the complementary variant as explained so far would come to a stop.

In the nearly complementary variant a stop does not occur in these cases, but λ may then vary upwards and downwards. This means that in this case the monotonic decrease in λ cannot be used any more to prove that the method does not cycle. Lemke [108] has come forward with a proof that cycling cannot occur which is not based on the monotonicity of λ, but which is based on the graph structure of the successive solutions.

Consider a graph, in which a number of nodes are joined by a number of arcs, which has the property that each node is joined to other nodes by either one or two arcs. Below is an example of such a set of nodes:

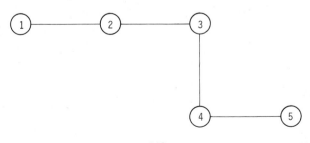

Now assume that we have a graph which has a node with one arc. Such a graph cannot contain a loop. For assume that it would contain a loop, as in the following graph:

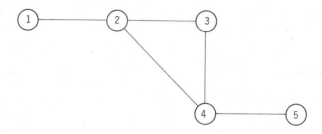

It is then obvious that some nodes must have more than two arcs. Hence, in any connected graph with at most two arcs at each node and with at least one node with one arc, there can be no loop. If in such a graph we start in a node with one arc, go through the arc to the next node and then leave this node through the other arc of this node (if any), and so on, it is not possible to cycle. If, furthermore, the number of nodes is finite, the number of steps must be finite, so that we must arrive, in a finite number of steps, in a node with only one arc.

Let us now consider the set-up tableau and the following tableau of the nearly complementary variant, as given in table 1.

Table 1

Initial tableaux for nearly complementary variant

Tableau	B.v.	V.b.v.	λ	x_1	x_2	x_3
0	y_1	b_1	-1	a_{11}	a_{12}	a_{13}
	y_2	b_2	$\underline{-1}$	a_{21}	a_{22}	a_{23}
	y_3	b_3	-1	a_{31}	a_{32}	a_{33}
			y_1	x_1	x_2	x_3
1	λ	$-b_1$	-1	$-a_{11}$	$-a_{12}$	$-a_{13}$
	y_2	$b_2 - b_1$	-1	$a_{21} - a_{11}$	$a_{22} - a_{12}$	$a_{23} - a_{13}$
	y_3	$b_3 - b_1$	-1	$a_{31} - a_{11}$	$a_{32} - a_{12}$	$a_{33} - a_{13}$

Assume that $b_1 = \text{Min } b_i < 0$. After pivoting on the underlined element -1, tableau 1 is obtained. The solution of this tableau is feasible and has no pair of corresponding basic variables. It has a nonbasic pair, namely y_1

and x_1. In the following iteration, x_1 is introduced into the basis, replacing some basic variable, and so on, until the method terminates with a solution with λ nonbasic (successful termination) or a solution in which no leaving basic variable can be found (unsuccessful termination).

Consider one of the tableaux generated in this manner. It will have a pair of corresponding nonbasic variables. The introduction of one of these variables will give the next tableau and the introduction of the other one will give the previous tableau, if there is a next or previous tableau. Any solution may therefore be considered a node with two arcs, which lead to the previous or the next node. The solution of tableau 1 has x_1 and y_1 as the nonbasic pair, but if y_1 is introduced, no leaving basic variable is found. It is said that the increase of y_1 in the solution of tableau 1 leads to a ray, which is called the *primary ray* of that solution. If the method terminates when after the selection of the new basic variable no leaving basic variable is found because all elements in the column of the new basic variable are nonpositive, it is said that a *secondary ray* is found.

It will now be obvious that the method generates a number of successive solutions which are related to each other as the nodes of a graph with nodes having at most two arcs and with one node with only one arc. In such a graph, no loop is possible, so that the method cannot cycle. Since the number of basic solutions to an equation system with a finite number of equations is finite, the number of iterations must be finite, so that after a finite number of iterations the solution of the problem must be found or a solution with a secondary ray must be generated.

17.2. The complementary variant with nonzero principal elements

Lemke's approach can also be used for the complementary variant. This will first be demonstrated for the easiest case, namely the case in which the principal elements in the critical row are always nonzero.

The following problem will be used as an example:

$$-6 = x_1 - x_2 - x_3 + y_1,$$
$$-5 = 2x_1 - x_2 \qquad + y_2,$$
$$-3 = \qquad\qquad - x_3 + y_3.$$

Tableau 0 of table 2 gives the set-up tableau after addition of the λ-terms. It is obvious that the basic solution of tableau 0 is feasible for $\lambda \geq 6$. The row of y_1 blocks further decrease of λ. But now the principal element in

Table 2

Example with nonzero principal elements

Tableau	B.v.	c-term	λ-term	Values basic variables		x_1	x_2	x_3
				$\lambda = 6$				
0	y_1	-6	1	0		1	-1	-1
	y_2	-5	1	1		2	-1	0
	y_3	-3	1	3		0	0	-1
				$\lambda = 6$	$\lambda = 7$	y_1	x_2	x_3
1	x_1	-6	1	0	1	1	-1	-1
	y_2	7	-1	1	0	-2	1	2
	y_3	-3	1	3	4	0	0	1
				$\lambda = 7$	$\lambda = 3$	y_1	y_2	x_3
2	x_1	1	0	1	1	-1	1	-3
	x_2	7	-1	0	4	-2	1	2
	y_3	-3	1	4	0	0	0	-1
				$\lambda = 3$	$\lambda = 0$	y_1	y_2	y_3
3	x_1	10	-3	1	10	-1	1	-3
	x_2	1	1	4	1	-2	1	2
	x_3	3	-1	0	3	0	0	-1

the critical row is positive, so that it cannot be used to unblock λ. However, this element can still be used as a pivot for a transformation; the result is tableau 1. Here $\lambda = 6$ is, of course, still the lower bound of λ, but we now start looking for an *upper* bound of λ and observe that the solution of tableau 1 is feasible for $6 \leq \lambda \leq 7$. It can be said that, by using a positive pivot in the critical row, λ has been *deflected*, this in contrast with a negative pivot which results in the unblocking of λ.

The upper bound of λ in tableau 3 is found in the row of y_2, which blocks the increase of λ. This increase of λ is unblocked if a negative principal element in this row is used or deflected if a positive principal element is used. The principal element is positive, so that λ is again deflected. After transformation with the underlined element 1 as a pivot, tableau 2 is found.

In this tableau $\lambda = 7$ is still the upper bound, but now we start looking for a lower bound, which is found to be $\lambda = 3$ in the row of y_3. Since the principal element in this row is negative, the decrease of λ is unblocked by a transformation with this element as a pivot. Tableau 3 happens to contain a solution for $\lambda = 0$, which is a solution to the original problem.

Figure 1 gives the successive solutions in terms of the range of values λ, for which a solution is feasible; if the principal element in the critical row is positive, there is a deflection, and if it is negative, there is an unblocking. It should be noted that, if the method is to be successful, there should be an even number of positive principal pivots.

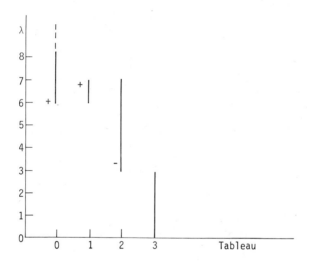

Cycling of solutions is impossible for similar reasons as stated in the previous section: each solution is entered via its lower bound and left in its upper bound or vice versa, so that each node in the corresponding graph has at most two arcs.

The method terminates unsuccessfully if an upper bound of λ is considered and none is found, because all elements in the λ-term column are nonnegative. In contrast with the case of negative principal pivots, such a termination is not impossible with nonzero principal pivots.

17.3. The complementary variant in the general case: examples

In the previous section it was shown that the variation of λ in the general linear complementarity problem is not necessarily monotonic, but that, in addition to pivot transformations which unblock λ for a further decrease, there can be pivot transformations which reverse or deflect the variation of λ.

If all principal elements are nonzero, only λ is varied, but if some principal elements are zero, then, as we have seen in the previous chapter, some nonbasic variables are also parametrically varied. In the last chapter, the variation of these nonbasic variables was upwards only, since only negative, unblocking pivots were considered, but if also positive, deflecting pivots occur, the variation of these variables can be both upwards and downwards.

To explain the complementary variant for the general case, the following example will be used:

$$-5 = \qquad\quad -2x_3 + y_1,$$
$$-3 = 9x_1 + 3x_2 + 2x_3 + y_2,$$
$$-2 = \quad x_1 + 6x_2 + 4x_3 + y_3.$$

Table 3 gives the set-up tableau after the λ-terms have been added.

First a lower bound of λ is considered, which is found to be $\lambda = 5$ in the row of y_1; the corresponding solution is $\lambda = 5$, $y_2 = 2$, $y_3 = 3$. Since the row of y_1 is the critical row, the column of x_1 is the incoming column. The principal element in this column is 0, so that x_1 is increased until one of the basic variables becomes zero. Comparing the ratios $\frac{2}{9}$ and $\frac{3}{1}$, we find that y_2 blocks the increase of x_1. We put $x_1 = \frac{2}{9}$, so that the new solution is $\lambda = 5$, $x_1 = \frac{2}{9}$, $y_3 = 2\frac{7}{9}$.

Since the row of y_2 is blocking the increase of x_1, the column of x_2 is chosen as the incoming column. Then we check whether this column has any nonzero elements in blocking rows, which are the rows of y_1 and y_2 in the order in which they were found blocking. The element in the first row is zero, but that in the second row is positive. This means that this element can be used as a pivot, thereby deflecting the increase of x_1. After transformation with 3 as a pivot, tableau 1 is found.

In this tableau, x_1 should be varied downward from its current value $\frac{2}{9}$, until it becomes zero or until some basic variable becomes zero. The values of basic variables as a function of x_1 can be described as follows:

$$y_1 = 0 - \quad 0(x_1 - \tfrac{2}{9}),$$
$$x_2 = 0 - \quad 3(x_1 - \tfrac{2}{9}),$$
$$y_3 = 2\tfrac{7}{9} + 17(x_1 - \tfrac{2}{9}).$$

The lower bound of x_1 found from this is $x_1 = \frac{1}{17}$, so that the new solution is $\lambda = 5$, $x_2 = \frac{25}{51}$, $x_1 = \frac{1}{17}$.

Table 3

Example of application of the complementary variant

Tableau	B.v.	Values basic variables						
		c-term	λ-term	(1)	(2)	x_1	x_2	x_3
				$\lambda = 5$	$x_1 = \frac{2}{9}$			
0	y_1	-5	1	0	0	0	0	-2
	y_2	-3	1	2	0	9	3	2
	y_3	-2	1	3	$2\frac{7}{9}$	1	6	4
				(3) $x_1 = \frac{2}{9}$	$x_1 = \frac{1}{17}$	x_1	y_2	x_3
1	y_1	-5	1	0	0	0	0	-2
	x_2	-1	$\frac{1}{3}$	0	$\frac{25}{51}$	3	$\frac{1}{3}$	$\frac{2}{3}$
	y_3	4	-1	$2\frac{7}{9}$	0	-17	-2	0
						y_3	y_2	x_3
2	y_1	-5	1			0	0	-2
	x_2	$-\frac{5}{17}$	$\frac{8}{51}$			$\frac{3}{17}$	$-\frac{1}{51}$	$\frac{2}{3}$
	x_1	$-\frac{4}{17}$	$\frac{1}{17}$			$-\frac{1}{17}$	$\frac{2}{17}$	0
				$\lambda = 5$ (4) $x_1 = \frac{1}{17}$ $\lambda = 4$		y_3	y_2	y_1
3	x_3	$2\frac{1}{2}$	$-\frac{1}{2}$	0	$\frac{1}{2}$	0	0	$-\frac{1}{2}$
	x_2	$-1\frac{49}{51}$	$\frac{25}{51}$	$\frac{25}{51}$	0	$\frac{3}{17}$	$-\frac{1}{51}$	$\frac{1}{3}$
	x_1	$-\frac{4}{17}$	$\frac{1}{17}$	0	0	$-\frac{1}{17}$	$\frac{2}{17}$	0
				(5) $\lambda = 4$		y_3	x_2	y_1
4	x_3	$2\frac{1}{2}$	$-\frac{1}{2}$	$\frac{1}{2}$		0	0	$-\frac{1}{2}$
	y_2	100	-25	0		-9	-51	-17
	x_1	-12	3	0		1	6	2
				(6) $\lambda = 4$		y_3	x_2	x_1
5	x_3	$-\frac{1}{2}$	$\frac{1}{4}$	$\frac{1}{2}$		$\frac{1}{4}$	$1\frac{1}{2}$	$\frac{1}{4}$
	y_2	-2	$\frac{1}{2}$	0		$-\frac{1}{2}$	0	$8\frac{1}{2}$
	y_1	-6	$1\frac{1}{2}$	0		$\frac{1}{2}$	3	$\frac{1}{2}$

The decrease of x_1 was blocked in the row of y_3, so that x_3 is chosen as the incoming column. The blocking rows are those of y_1 and y_3 and the element in the column of x_3 and the row of y_1 is negative, so that λ may be unblocked by using this element as a pivot. In this case we have the block pivot

$$\begin{bmatrix} 0 & -2 \\ -17 & 0 \end{bmatrix},$$

which can be executed in two steps with a single pivot. The result is tableau 3.

In this tableau, x_1 is a basic variable. We proceed by further decreasing λ and find a lower bound of 4; the next solution is therefore $x_3 = \frac{1}{2}$, $\lambda = 4$, $x_1 = 0$. Note that here there is no unique blocking row, since both the row of x_2 and of x_1 block λ at the value 4. To find a unique blocking row, lexicographic vectors could be used for the values of basic variables, but it is sufficient to select arbitrarily one row as the blocking row connected with this lower bound. Let us select the row of x_2.

Then the column of y_2 must be selected as the incoming column. The element of this column in the blocking row is negative, so that it can be used as a pivot of a transformation. This results in tableau 4. Since the pivot was negative, the decrease of λ was unblocked and a further lower bound should be found. This lower bound coincides with the previous lower bound due to the nonuniqueness of the blocking row in tableau 3.

In tableau 4, the blocking row is that of x_1 so that the column of y_1 is the incoming column. The element of this column in the blocking row is positive, which means that a transformation with this element as a pivot turns the downward variation of λ into an upward variation. In tableau 5, no upper bound of λ is found, so that the method terminates unsuccessfully.

Table 4 gives the tableaux for an application of the nearly complementary variant or Lemke method. It should be noted that the successive solutions are the same as in the complementary method. The termination is here in a secondary ray.

One particular case deserves separate treatment. Suppose a nonbasic variable is decreased and the lower bound of this variable is zero. How does the complementary variant proceed? In this case the variable should be put at a value zero and the previous blocked variable should be varied as if it is deflected.

The above example may be slightly modified so that the last equation is

$$-\tfrac{1}{2} = 15x_1 + 6x_2 + 4x_3 + y_3.$$

The set-up tableau is then as given in tableau 0 of table 5. λ is varied downwards and is blocked in the row of y_1. x_1 is then increased and is

Table 4

Example of application of the Lemke method

Tableau	B.v.	V.b.v.	x_1	x_2	x_3	λ
0	y_1	-5	0	0	-2	-1
	y_2	-3	9	3	2	-1
	y_3	-2	1	6	4	-1
			x_1	x_2	x_3	y_1
1	λ	5	0	0	2	-1
	y_2	2	9	3	4	-1
	y_3	3	1	6	6	-1
			y_2	x_2	x_3	y_1
2	λ	5	0	0	2	-1
	x_1	$\frac{2}{9}$	$\frac{1}{9}$	$\frac{1}{9}$	$\frac{4}{9}$	$-\frac{1}{9}$
	y_3	$2\frac{7}{9}$	$-\frac{1}{9}$	$5\frac{2}{3}$	$5\frac{5}{9}$	$-\frac{8}{9}$
			y_2	y_3	x_3	y_1
3	λ	5	0	0	2	-1
	x_1	$\frac{1}{17}$	$\frac{2}{17}$	$-\frac{1}{17}$	$\frac{2}{17}$	$-\frac{1}{17}$
	x_2	$\frac{25}{51}$	$-\frac{1}{51}$	$\frac{3}{17}$	$\frac{50}{51}$	$-\frac{8}{51}$
			y_2	y_3	x_2	y_1
4	λ	4	$\frac{1}{25}$	$-\frac{9}{25}$	$-2\frac{1}{25}$	$-\frac{17}{25}$
	x_1	0	$\frac{3}{25}$	$-\frac{2}{25}$	$-\frac{3}{25}$	$-\frac{1}{25}$
	x_3	$\frac{1}{2}$	$-\frac{1}{50}$	$\frac{9}{50}$	$1\frac{1}{50}$	$-\frac{4}{25}$
			x_1	y_3	x_2	y_1
5	λ	4	$-\frac{1}{3}$	$-\frac{1}{3}$	-2	$-\frac{2}{3}$
	y_2	0	$8\frac{1}{3}$	$-\frac{2}{3}$	-1	$-\frac{1}{3}$
	x_3	$\frac{1}{2}$	$\frac{1}{6}$	$\frac{1}{6}$	1	$-\frac{1}{6}$

blocked in the row of y_2. x_1 is then deflected by a pivot on the element 3, so that it should be decreased in the first solution of tableau 1. The values of basic variables are then, expressed in terms of x_1:

$$y_1 = 0 + 0(x_1 - \tfrac{2}{9}),$$
$$x_2 = 0 - 3(x_1 - \tfrac{2}{9}),$$
$$y_3 = 1\tfrac{1}{6} + 3(x_1 - \tfrac{2}{9}).$$

Table 5

Second example of the complementary variant

Tableau	B.v.	c-term	λ-term	Values basic variables				x₁	x₂	x₃
				$\lambda = 5$ $x_1 = \frac{2}{9}$						
0	y_1	-5	1	0	0			0	0	-2
	y_2	-3	1	2	0			9	$\overline{3}$	2
	y_3	$-\frac{1}{2}$	1	$4\frac{1}{2}$	$1\frac{1}{6}$			15	6	4

Tableau	B.v.	c-term	λ-term	$\lambda = 5$ $x_1 = \frac{2}{9}$	$\lambda = 5$	$\lambda = 5\frac{1}{2}$	$\lambda = 5\frac{1}{2}$ $x_3 = 1\frac{1}{4}$	x_1	y_2	x_3
1	y_1	-5	1	0	0	$\frac{1}{2}$	3	0	0	-2
	x_2	-1	$\frac{1}{3}$	0	$\frac{2}{3}$	$\frac{5}{6}$	0	3	$\frac{1}{3}$	$\frac{2}{3}$
	y_3	$5\frac{1}{2}$	-1	$1\frac{1}{6}$	$\frac{1}{2}$	0	0	-3	$\underline{-2}$	0

Tableau	B.v.	c-term	λ-term	$\lambda = 5\frac{1}{2}$ $x_3 = 1\frac{1}{4}$	x_1	y_3	x_3
2	y_1	-5	1	3	0	0	-2
	x_2	$-\frac{1}{12}$	$\frac{1}{6}$	0	$2\frac{1}{2}$	$\frac{1}{6}$	$\frac{2}{3}$
	y_2	$-2\frac{3}{4}$	$\frac{1}{2}$	0	$1\frac{1}{2}$	$-\frac{1}{2}$	0

Tableau	B.v.	c-term	λ-term	$\lambda = 5\frac{1}{2}$ $x_3 = 1\frac{1}{4}$	x_1	y_3	x_2
3	y_1	$-5\frac{1}{4}$	$1\frac{1}{2}$	3	$7\frac{1}{2}$	$\frac{1}{2}$	3
	x_3	$-\frac{1}{8}$	$\frac{1}{4}$	0	$3\frac{3}{4}$	$\frac{1}{4}$	$1\frac{1}{2}$
	y_2	$-2\frac{3}{4}$	$\frac{1}{2}$	0	$1\frac{1}{2}$	$-\frac{1}{2}$	0

The lower bound of x_1 is now 0. Hence we put $x_1 = 0$, thus obtaining the solution $\lambda = 5$, $x_2 = \frac{2}{3}$, $y_3 = \frac{1}{2}$ and consider λ as being deflected, so that now an upper bound of λ should be found, which turns out to be $5\frac{1}{2}$, so that the next solution is $y_1 = \frac{1}{2}$, $x_2 = \frac{5}{6}$, $\lambda = 5\frac{1}{2}$. After a few more iterations the method terminates unsuccessfully.

17.4. General discussion of the complementary variant

In order to formulate the complementary variant in a somewhat more general manner, let us use a general tableau with four rows, as given in table 6.

Table 6

Example of a general complementary tableau

B.v.	c-term	λ-term	x_1	x_2	x_3	x_4
y_1	b_1	a_1	a_{11}	a_{12}	a_{13}	a_{14}
y_2	b_2	a_2	a_{21}	a_{22}	a_{23}	a_{24}
y_3	b_3	a_3	a_{31}	a_{32}	a_{33}	a_{34}
y_4	b_4	a_4	a_{41}	a_{42}	a_{43}	a_{44}

Suppose that in this tableau a lower bound for λ should be found, as would be the case in the set-up tableau. This is done by determining

$$\underline{\lambda} = (\text{Max}_i \, (- b_i/a_i)|a_i > 0). \tag{1}$$

Assume that the maximum is found in the first row, so that $\underline{\lambda} = - b_1/a_1$. The values of basic variables are then

$$y_i = b_i + a_i \underline{\lambda} \qquad i = 1, ..., 4. \tag{2}$$

Since the row of y_1 blocks the decrease of λ, the column of x_1 is to be the incoming column. There is now only one element in this column and in blocked rows, namely a_{11}. If $a_{11} < 0$, it is used as a pivot, the current lower bound of λ has become an upper bound and a new lower bound should be found.

If $a_{11} > 0$, it is used as a pivot, but now λ is deflected, so that in the new tableau an upper bound of λ should be found. If $a_{11} = 0$, we determine

$$\bar{x}_1 = \text{Min}_i \left(\frac{b_i + a_i \underline{\lambda}}{a_{i1}} \middle| a_{i1} > 0 \right). \tag{3}$$

Let the minimum be found in the second row, so that $\bar{x}_1 = (b_2 + a_2\underline{\lambda})/a_{21}$. The values of basic variables are now determined for $x_1 = \bar{x}_1$:

$$y_i = b_i + a_i \underline{\lambda} - a_{i1}\bar{x}_1, \qquad i = 1, 2, ..., 4. \tag{4}$$

Since the row of y_2 blocked the increase in x_1, the column of x_2 is taken as incoming column. First the elements in this column and blocked rows are checked. If $a_{12} < 0$, λ is unblocked in the first row by using the block pivot

$$\begin{bmatrix} a_{11} = 0 & a_{12} < 0 \\ a_{21} > 0 & a_{22} \end{bmatrix}, \tag{5}$$

which can be performed in two transformations with a_{21} and a_{12} as pivots. Then a new lower bound of λ should be found. If $a_{12} > 0$, the same block pivot is used, but afterwards an upper bound of λ should be found. If $a_{12} = 0$, we check a_{22}. If $a_{22} < 0$, it is used as a pivot, x_1 is unblocked and further increased. If $a_{22} > 0$, it is used as a pivot, x_1 is deflected and is decreased. Finally, if $a_{22} = 0$, we determine

$$\bar{x}_2 = \underset{i}{\text{Min}} \left(\frac{b_i + a_i \underline{\lambda} - a_{i1}\bar{x}_1}{a_{i2}} \middle| a_{i2} > 0 \right). \tag{6}$$

Let us suppose that the minimum is found in the third row. The new values of basic variables are now

$$y_i = b_i + a_i \underline{\lambda} - a_{i1}\bar{x}_1 - \bar{a}_{i2}\bar{x}_2. \tag{7}$$

Then the column of x_3 is the incoming column. If $a_{13} \neq 0$, we use the block pivot

$$\begin{bmatrix} a_{11} = 0 & a_{12} = 0 & a_{13} \neq 0 \\ a_{21} > 0 & a_{22} = 0 & a_{23} \\ a_{31} & a_{32} > 0 & a_{33} \end{bmatrix}; \tag{8}$$

in the new tableau λ is unblocked if $a_{13} < 0$ and is deflected if $a_{13} > 0$.

If $a_{13} = 0$, we check a_{23}; if $a_{23} \neq 0$, the following block pivot is used

$$\begin{bmatrix} a_{22} = 0 & a_{23} \neq 0 \\ a_{32} > 0 & a_{33} \end{bmatrix}. \tag{9}$$

If $a_{23} < 0$, x_1 is unblocked and if $a_{23} > 0$, it is deflected. If $a_{23} = 0$, we check a_{33}; if $a_{33} \neq 0$, it is used as a pivot; if $a_{33} < 0$, x_2 is unblocked and if $a_{33} > 0$, it is deflected. If $a_{33} = 0$, we determine a minimum as in (6), and so on. If there is no positive denominator in ratios as in (3) and (6), the method terminates unsuccessfully.

The Lemke approach to prove that cycling cannot occur is also valid in these cases. Each solution has its specific upper and lower bound of λ or some nonbasic variable and each solution is entered or left via the lower and upper bounds.

17.5. Equivalence of the complementary and the nearly complementary variant

In this section it will be proved that the complementary variant and the original Lemke method are equivalent in the sense that the values of basic variables in successive solutions are identical.

The complementary variant is characterized by the fact that λ and possibly a number of nonbasic variables are fixed, that is that they have nonzero values in such a manner that the values of basic variables in blocking or tied rows are zero.

Consider a tableau for the complementary method which has its rows and columns rearranged in such a manner that the columns of blocked columns come first; then columns and their corresponding blocking rows are ordered in such a manner that a submatrix is obtained in which the elements on the diagonal below the main diagonal are nonzero, and those on the main diagonal and above it are all zero. Furthermore, it is convenient to move the λ-term column of the values of basic variables over to the other side of the equality sign. After renaming the variables, the current tableau has the following structure:

B.v.	V.b.v.		λ	x_1	x_2	x_3	x_4	x_i
y_1	b_1	b_1^*	$-a_1$	0	0	0	0	a_{1i}
y_2	b_2	b_2^*	$-a_2$	a_{21}	0	0	a_{24}	a_{2i}
y_3	b_3	b_3^*	$-a_3$	a_{31}	a_{32}	0	a_{34}	a_{3i}
y_4	b_4	b_4^*	$-a_4$	a_{41}	a_{42}	a_{43}	a_{44}	a_{4i}
y_i	b_i	b_i^*	$-a_i$	a_{i1}	a_{i2}	a_{i3}	a_{i4}	a_{ii}

(10)

The row of y_i and the column of x_i stand for all remaining rows and columns which are not directly involved at this moment. b_1, b_2, ..., b_i stand for the values of basic variables with λ and fixed nonbasic variables equal to zero, while b_1^*, b_2^*, ..., b_i^* gives the values after λ and the fixed nonbasic variables are put equal to their fixed values.

Let us assume that λ, x_1 and x_2 are fixed and that x_2 has just been changed in such a manner that $b_3^* = 0$. Hence the row of y_3 is the critical row. In the complementary variant x_3 is now going to be increased. Then two cases may be distinguished:

(i) all elements of the column of x_3 in blocking rows are zero,

(ii) at least one of these elements is nonzero.

Case (i) is considered first. The complementary variant would now determine

$$\text{Min} (b_i^* / a_{i3} | a_{i3} > 0)$$
$$\scriptstyle i$$

(11)

for all free rows i. Let us assume that this minimum occurs in row 4. The values of basic variables in the next solution are then

$$b_i^* - a_{i3}(b_4^*/a_{43})$$

and in the next iteration x_4 is increased.

Consider now the Lemke method. First the current tableau for this method is generated from (10) by pivoting on the elements $- a_1$, a_{21}, and a_{32}. The elements b_1, b_2, and b_3 are then transformed into the fixed values of λ, x_1 and x_2, while the elements in free rows, b_4 and b_i, are transformed into b_4^* and b_i^*. The new basic variable in the Lemke method is then x_3. Since the elements of the column of x_3 in the first three rows are still zero, the leaving basic variable is determined as in (11). Hence both methods are equivalent in case (i).

To show the equivalence in case (ii), assume that we are one step further in (10) and that x_3 has also become fixed and that the row of y_4 has become tied and is the critical row. x_4 is now increased. Let us assume that $a_{24} \neq 0$. The complementary variant would now pivot on a_{24}, a_{32} and a_{43}. The resulting tableau would have λ and x_1 as fixed nonbasic variables and could be represented as follows:

B.v.	V.b.v.		λ	x_1	y_3	y_4	y_2	x_5
y_1	$\bar{b}_1$	$\bar{b}_1^*$	$- \bar{a}_1$	0	0	0	0	$\bar{a}_{15}$
x_4	$\bar{b}_2$	$\bar{b}_2^*$	$- \bar{a}_2$	$\bar{a}_{21}$	0	0	$\bar{a}_{24}$	$\bar{a}_{25}$
x_2	$\bar{b}_3$	$\bar{b}_3^*$	$- \bar{a}_3$	$\bar{a}_{31}$	$\bar{a}_{32}$	0	$\bar{a}_{34}$	$\bar{a}_{35}$
x_3	$\bar{b}_4$	$\bar{b}_4^*$	$- \bar{a}_4$	$\bar{a}_{41}$	$\bar{a}_{42}$	$\bar{a}_{43}$	$\bar{a}_{44}$	$\bar{a}_{45}$
y_i	$\bar{b}_i$	$\bar{b}_i^*$	$- \bar{a}_i$	$\bar{a}_{i1}$	$\bar{a}_{i2}$	$\bar{a}_{i3}$	$\bar{a}_{i4}$	$\bar{a}_{i5}$

$$(12)$$

Since (12) is obtained from (10) by pivoting on a_{24}, a_{32} and a_{43}, we have $\bar{a}_{21} = a_{21}/a_{24}$. If $a_{21} > 0$, x_1 was at its upper bound in (10), if $a_{21} < 0$, it was at its lower bound. If $a_{24} < 0$, the direction in which x_1 should be varied in (12) is that same as in (10), if $a_{24} > 0$, this direction is reversed. Hence for finding the critical row we determine

$$\underset{i \neq 2}{\text{Min}} \left(\frac{\bar{b}_i^*}{\text{sgn} \, (- \bar{a}_{21}) \bar{a}_{i1}} \Big| \, \text{sgn} \, (- \bar{a}_{21}) \bar{a}_{i1} > 0 \right). \quad (13)$$

If $\bar{a}_{21} < 0$ and $\bar{a}_{i1} \leq 0$ for all i, the method terminates unsuccessfully. If $\bar{a}_{21} > 0$ and

$$\text{Min} \, (\bar{b}_i^*/- \bar{a}_{i1} | \bar{a}_{i1} < 0) = \bar{b}_r^*/- \bar{a}_{r1} > \bar{x}_1$$

or $\bar{a}_{i1} \geq 0$ for all i, where $\bar{x}_1$ is the fixed value of x_1, then x_1 is put equal to zero and λ is varied opposite to its previous direction.

Consider now the Lemke method. The current tableau at the start of the iteration can be found by pivoting in (10) on $-a_1, a_{21}, a_{32}$ and a_{43}, but it can also be found by pivoting in (12) on $-\bar{a}_1$ and $\bar{a}_{21}$. x_4 is then the new basic variable and the leaving basic variable is then determined by

$$\text{Min}_i \left(\frac{\bar{b}_i^*}{\bar{a}_{i1}/-\bar{a}_{21}} \middle| \bar{a}_{i1}/-\bar{a}_{21} > 0 \right) \tag{14}$$

where the b_i^* now are the values of basic variables after pivoting on $-\bar{a}_1$ and $\bar{a}_{21}$. It is easily found that the results are the same as in (13) and the line following it.

In spite of its appearance as an example this proof is general. There may be more or fewer fixed variables and blocking rows involved in any particular case, but the relations involved remain the same.

17.6. Unfeasibility tests

If in the linear complementarity problem the parametric method terminates unsuccessfully, it is of importance to know whether such a termination means that the problem has no solution or merely that the method cannot find a solution. For a certain class of matrices of which the most general formulation is given by Eaves[103], who calls these $\mathscr{L}$-matrices, the first statement holds, as will be shown later; we know already that this statement is true for A negative semi-definite.

In a number of cases, especially in large problems, it is difficult to establish whether A belongs to this class of matrices or not; it may in fact be more difficult to determine whether A belongs to this class or not than to apply the parametric method. In the following a simple infeasibility test will be indicated which in cases of unsuccessful termination will reveal infeasibility for a wider range of cases than those included in the class of matrices $\mathscr{L}$. After that, the application of general infeasibility tests both for the complementary and nearly complementary variants of the parametric method will be considered.

If there is an unsuccessful termination in the complementary variant in the column of a nonbasic variable, the incoming column must have elements which are zero in blocking rows which must include the principal element, and it must have nonpositive elements in other rows. If the incoming column is the kth column, then after rearrangement the kth column and the kth row may be represented as follows:

$$
\left.
\begin{array}{c}
a_{ik} = 0 \\
\cdot \\
\cdot \\
\cdot \\
b_k\ a_k\ \big|\ a_{k1}\ \cdots\ a_{k,k-1}\quad a_{kk} = 0\quad a_{k,k+1}\ \cdots\ a_{kp} \\
a_{k+1,k} \leq 0 \\
\cdot \\
\cdot \\
\cdot \\
\underbrace{\hphantom{aaaaaaaaaa}}\quad a_{pk} \leq 0. \\
\text{fixed columns}
\end{array}
\right\}\ \begin{array}{l}\text{blocking}\\\text{rows}\end{array}
$$

If A is negative semi-definite, if $a_{kk} = 0$, then $a_{kj} = -a_{jk}$ for all j, which means that we can conclude from $a_{kk} = 0$ and $a_{jk} \leq 0$ that $a_{kj} \geq 0$ for all j. In this case it can also be proved that $b_k < 0$. The equation corresponding to the kth row can then be written as

$$b_k + a_k\lambda - a_{k1}x_1 - \ldots - a_{kp}x_p = y_k \tag{15}$$

For $\lambda = 0$ and $x_1, \ldots, x_p \geq 0$, the right-hand side is negative, so that no feasible solution exists.

For $-A$ belonging to the class $\mathscr{L}$ we have $a_{kj} = -d_j a_{jk}$ with $d_j \geq 0$, while further the argument is similar, as will be shown later. Instead of ascertaining whether A is negative semi-definite or $-A$ belongs to $\mathscr{L}$, we may just as well apply the complementary variant and if an unsuccessful termination occurs in the kth column, check whether the infeasibility sign pattern $b_k < 0$ and $a_{kj} \geq 0$ for all j occurs. Note that the infeasibility sign pattern in the kth row may occur even if $-A$ does not belong to $\mathscr{L}$, which means that the complementary variant supplemented by this infeasibility test will process (find a solution or show that none exists) a more general class of problems than the nearly complementary variant which does not use such a test.

The infeasibility test can be applied to any row in any complementary tableau, which means that for each $b_i < 0$ in such a tableau we check whether $a_{ij} \geq 0$ for all j. If such a sign pattern is found, the problem has no solution. In this way an even wider class of problems may be processed. Furthermore, infeasibility found in earlier iterations will save computation time. For example, in table 3, tableau 0, infeasibilities are found in the rows of y_2 and y_3, so that we could have stopped computations there.

Above we considered an infeasibility check on the kth row if there was an unsuccessful termination in the kth column. Such a check cannot be performed if an unsuccessful termination occurs with no upper bound of λ, since there is no row associated with λ. However, if there is no such row, one may be created in the following manner.

To the equations of the extended problem, the following equation may be added:

$$y_0 = a'x. \tag{16}$$

Since $a \geq 0$ and $x \geq 0$, y_0 should be nonnegative for all feasible solutions. (16) may be transformed together with the other equations; its general form in any tableau will be

$$b_0 + a_0\lambda = a_{01}x_1^* + a_{02}x_2^* + \ldots + y_0, \tag{17}$$

where $x_1^*, x_2^*, \ldots$ indicate the nonbasic variables in the tableau. This equation may be checked for infeasibility: if $b_0 < 0$ and $a_{0j} \geq 0$, for all j, then there is no feasible solution. In particular, if there is no upper bound for λ, this equation may be checked first.

Let us now consider infeasibility tests for the nearly complementary variant. Let the equation of the row in which λ is a basic variable be

$$b_l = a_{l1}x_1 + a_{l2}y_1 + a_{l3}x_2^* + \ldots + \lambda, \tag{18}$$

where x_1, y_1 is the nonbasic pair and $x_2^*, \ldots$ are the other nonbasic variables. Since for a solution to the problem we should have $\lambda = 0$ and since $b_l > 0$ for any solution generated by the method, we conclude that there is no solution to the problem if $a_{lj} \leq 0$ for all j.

Assume now that a secondary ray is generated by the introduction of x_1 into the basis; this implies $a_{i1} \leq 0$ for all rows i, the equations of which can be represented as in (18). If $a_{l1} < 0$, it can be used as a pivot, and we obtain

$$a_{l1}^{-1}b_l - a_{l1}^{-1}\lambda = a_{l1}^{-1}a_{l2}y_1 + a_{l1}^{-1}a_{l3}x_2^* + x_1, \tag{19}$$

$$b_i + a_{i1}a_{l1}^{-1}b_l + a_{i1}a_{l1}^{-1}\lambda = \qquad \ldots \qquad + y_i. \tag{20}$$

Since $-a_{l1}^{-1} \geq 0$ and $a_{i1}a_{l1}^{-1} \geq 0$ for all i, there must be no upper bound of λ. The infeasibility test in (18) is then equivalent to an infeasibility test in row (19) of the complementary variant, since eqs. (19) and (20) are equations of the rows of a complementary tableau. Eqs. (20) for any i may also be tested for infeasibility.

If $a_{l1} = 0$, the coefficient a_{l2} of the other member of the nonbasic pair may be used as a pivot if it is negative. The equations of the other rows may then be used for infeasibility tests.

If $a_{i1} = 0$, a complementary tableau may be generated for the nearly complementary tableau by a series of nonzero pivots, of which one is in the column of x_1 and the others are in rows for which $a_{i1} = 0$. The resulting tableau is not uniquely determined, but there is always one such complementary tableau since the complementary variant generates such a tableau. Infeasibility tests may be performed on each row.

The infeasibility tests proposed so far related to tableaux generated by the parametric method in which the complementarity conditions $x'y = 0$ were satisfied. It is possible to test the feasibility of the problem without these conditions; if there is no solution to the problem without the conditions, then there must be no solution to the problem in which also the complementarity conditions are satisfied.

The most obvious way of doing this is to use the extended problem and vary λ downwards, but pivot on any negative element in the critical row. If there is no negative element in a critical row, there must be no solution to the problem.

17.7. Rules for the complementary variant

For a convenient statement of the rules, table 7 is used. The λ-term elements have been moved to the other side of the equality sign. In the tableau, the basic variables are represented as y-variables and the non-basic ones as x-variables as in the set-up tableau, but this can be different in other tableaux.

In any tableau, the rows and columns are arranged in such a way that the tied rows and the fixed variables come first, ordered according to the hierarchy of fixed variables. The values of the fixed variables are indicated as $\bar{x}_0, \bar{x}_1, \bar{x}_2, ...,$ which are initially zero.

Table 7

B.v.	Values basic variables		$\lambda = x_0$	x_1	x_2	x_3	x_4
			$\bar{x}_0$	$\bar{x}_1$	$\bar{x}_2$	$\bar{x}_3$	$\bar{x}_4$
y_1	b_1	b_1^*	a_{10}	a_{11}	a_{12}	a_{13}	a_{14}
y_2	b_2	b_2^*	a_{20}	a_{21}	a_{22}	a_{23}	a_{24}
y_3	b_3	b_3^*	a_{30}	a_{31}	a_{32}	a_{33}	a_{34}
y_4	b_4	b_4^*	a_{40}	a_{41}	a_{42}	a_{43}	a_{44}

The b_i^* indicate the values of basic variables after the fixed nonbasic variables are put equal to their fixed values. $b_i^* = 0$ for tied rows and $b_i^* \geq 0$ for other rows. The b_i^* are initially 0; after step 0 all but one have become positive (in nondegenerate cases). If, after a tableau transformation, a row which has been tied has become free, the corresponding b_i^* takes the fixed value of the formerly fixed variable, which has become basic in that row.

The b_i are the values of basic variables for fixed nonbasic variables equal to zero, and are used for infeasibility testing only.

The direction in which the nonbasic variable of the active column is varied is positive if this column was not fixed before and is otherwise opposite to the sign of $a_{k+1,k}$; this direction is indicated by $S = 1$ or -1.

Step 0: Determine Min $b_i = b_r$; put $b_i^* = b_i - b_r$, $x_0 = b_r$, $k = r$; rearrange rows and columns and go to step 1.

Step 1: Determine the first $a_{ik} \neq 0$, $i = 1, ..., k$; let this be a_{rk}. If $a_{ik} = 0$, $i = 1, ..., k$, go to step 3. Transform with pivots $a_{rk}, a_{r+1,r}, a_{r+2,r+1}, ..., a_{k,k+1}$; put $S = \text{sgn}(-a_{r,r-1})$ and $k = r - 1$; rearrange rows and columns. Go to step 2.

Step 2: (Testing for infeasibility.) For all i with $b_i < 0$, determine

$$\text{Max } a_{ij} = \bar{a}_i.$$

If any $\bar{a}_i \leq 0$, stop; there is no solution to the problem. If $S = 1$, go to step 3, if $S = -1$, go to step 4.

Step 3: Determine

$$\text{Min } (b_i^*/a_{ik} | a_{ik} > 0) = b_r^*/a_{rk}.$$

If $a_{ik} \leq 0$ for all i, stop; the method has terminated unsuccessfully. Put $x_k = b_r^*/a_{rk}$, $b_i^* = b_i^* - a_{ik}x_k$, $k = r$ and rearrange rows and columns. Go to step 1.

Step 4: Determine

$$\text{Min } (b_i^*/-a_{ik} | a_{ik} < 0, \bar{x}_k) = b_r^*/-a_{rk} \text{ or } \bar{x}_k.$$

If the minimum is $\bar{x}_k$, put $b_i^* = b_i^* + a_{ik}\bar{x}_k$; if $k = 0$, stop; a solution of the problem has been obtained; otherwise, put $k = r - 1$. If $a_{k+1,k} > 0$, repeat step 4, otherwise go to step 3. If the minimum occurs in row r, put $\bar{x}_k = \bar{x}_k + b_r^*/a_{rk}$, $b_i^* = b_i^* + a_{ik}(b_r^*/a_{rk})$, $k = r$, rearrange rows and columns, and go to step 1.

CHAPTER 18

TERMINATION FOR COPOSITIVE PLUS AND $\mathcal{L}$-MATRICES

18.1. Introduction

The complementary variant can terminate in three ways:

(1) The solution is feasible for $\lambda = 0$; this means that a solution to the problem has been found,

(2) No upper bound of λ for the current solution can be found,

(3) No upper bound for some nonbasic variable can be found.

In Lemke's method, a successful termination takes place after λ leaves the basis; this corresponds to termination 1 of the complementary variant. Lemke's method terminates unsuccessfully if there is no positive element in the column of the new basic variable; this is called termination in a secondary ray and it corresponds to terminations 2 and 3 of the complementary variant.

Lemke[107] has proved that, if $-A$ belongs to a class of matrices called copositive plus, termination in a secondary ray implies that no solution to the problem exists. A matrix M is called copositive if

$$y'My \geq 0 \quad \text{for} \quad y \geq 0$$

and copositive plus if, in addition to the above, we have that $y'My = 0$ for $y \geq 0$ implies $M'y = -My$. Positive matrices, that is matrices M with $m_{ij} > 0$ for all i, j, are evidently a special case of copositive matrices.

Eaves[103] has extended this for a class of matrices $\mathcal{L}$ which is defined as $\mathcal{L} = \mathcal{L}_1 \cap \mathcal{L}_2$, where M belongs to $\mathcal{L}_1$ if, for each vector $y \geq 0$ and $\neq 0$, for some i we have $y_i > 0$ and $(My)_i \geq 0$; M belongs to $\mathcal{L}_2$ if for each $y \geq 0$ and $\neq 0$ such that $My \geq 0$ and $y'My = 0$, there are diagonal matrices $\Lambda \geq 0$ and $\Omega \geq 0$ such that $\Omega y \neq 0$ and $M'\Omega y = -\Lambda My$.

To prove that copositive plus matrices are a subclass of $\mathcal{L}$-matrices, consider the definition of a copositive matrix

$$y'My = \sum_i y_i(My_i)_i \geq 0 \qquad \text{for any } y \geq 0. \tag{1}$$

Take any $y \neq 0$, so that it must have at least one positive element. If it has one positive element let it be the kth element, so that (1) reduces to

$y_k (My)_k \geq 0$, which means that M must be in $\mathscr{L}_1$. If there are more positive elements of y, but the corresponding elements of My are nonnegative, M must also belong to $\mathscr{L}_1$. If some elements of My corresponding to positive elements of y are negative, there must be other elements of My corresponding to positive elements of y which are positive, because otherwise the sum of the products cannot be nonnegative. Hence also in this case M belongs to $\mathscr{L}_1$.

It is obvious that the "plus" condition: if $y'My = 0$ for $y \geq 0$ then $M'y = -My$, is the particular case of $\mathscr{L}_2$ in which $\Lambda = \Omega = I$.

In the following an alternative proof will be given based on complementary tableaux. It will also become clear that the infeasibility implied by an unsuccessful termination for these classes of matrices will show up as a sign configuration in the critical row indicating infeasibility, so that the infeasibility test as suggested before would find this infeasibility.

This means that the complementary variant, supplemented by the infeasibility test, can process (which means finding a solution or showing that no solution exists) problems with a matrix belonging to a more general class of matrices than $\mathscr{L}$. A formal description of this class is probably impossible and does not seem to be useful.

The next section deals with a termination of the complementary variant in which no upper bound for λ can be found, while section 18.3 deals with the case of no upper bound for a nonbasic variable. Both copositive plus and $\mathscr{L}$-matrices are considered.

18.2. Termination in the λ-column

First a general complementary tableau will be generated. It is convenient to add to the equation system the equation

$$y_0 = a'x. \tag{2}$$

Tableau 0 of table 1 gives the set-up tableau of the resulting system. A general tableau in which the x-variables which are basic in the current solution are put first is then generated by block-pivoting on the matrix A_{11}. The result is tableau 1 of table 1.

Let us assume that in this tableau no upper bound for λ can be found. This means that

$$A_{11}^{-1}a^1 \geq 0,$$
$$a^2 - A_{21}A_{11}^{-1}a^1 \geq 0. \tag{3}$$

Table 1

Generation of a general complementary tableau

Tableau	Bas. var.	Values basic variables		x^1	x^2
		c-term	λ-term		
0	y_0	0	0	$-a''$	$-a^{2'}$
	y^1	b^1	a^1	A_{11}	A_{12}
	y^2	b^2	a^2	A_{21}	A_{22}
				y^1	x^2
1	y_0	$a''A_{11}^{-1}b^1$	$a''A_{11}^{-1}a^1$	$a''A_{11}^{-1}$	$-a^{2'}+a''A_{11}^{-1}A_{12}$
	x^1	$A_{11}^{-1}b^1$	$A_{11}^{-1}a^1$	A_{11}^{-1}	$A_{11}^{-1}A_{12}$
	y^2	$b^2-A_{21}A_{11}^{-1}b^1$	$a^2-A_{21}A_{11}^{-1}a^1$	$-A_{21}A_{11}^{-1}$	$A_{22}-A_{21}A_{11}^{-1}A_{12}$

Furthermore, since the solution is feasible for $\lambda \geq \underline{\lambda}$, we have that if any element of $A_{11}^{-1}a^1$ is zero, then the corresponding element of $A_{11}^{-1}b^1$ must be nonnegative.

Consider the λ-term element of the value y_0. Since $a^1 \geq 0$ and $A_{11}^{-1}a^1 \geq 0$, we have

$$a''A_{11}^{-1}a^1 \geq 0. \tag{4}$$

On the other hand, we have

$$a''A_{11}^{-1}a^1 = (A_{11}^{-1}a^1)'A_{11}'(A_{11}^{-1}a^1) =$$

$$\begin{bmatrix} A_{11}^{-1}a^1 \\ 0 \end{bmatrix}' \begin{bmatrix} A_{11} & A_{12} \\ A_{21} & A_{22} \end{bmatrix} \begin{bmatrix} A_{11}^{-1}a_1 \\ 0 \end{bmatrix} = -\begin{bmatrix} a^{*1} \\ 0 \end{bmatrix}'(-A)\begin{bmatrix} a^{*1} \\ 0 \end{bmatrix}, \tag{5}$$

where $a^{*1} = A_{11}^{-1}a^1$. If $-A$ is copositive plus, this must be nonpositive. Hence $a''A_{11}^{-1}a^1 = 0$. Then according to the "plus" condition we have

$$\begin{bmatrix} -A'_{11} & -A'_{21} \\ -A'_{12} & -A'_{22} \end{bmatrix}\begin{bmatrix} A_{11}^{-1}a^1 \\ 0 \end{bmatrix} = \begin{bmatrix} A_{11} & A_{12} \\ A_{21} & A_{22} \end{bmatrix}\begin{bmatrix} A_{11}^{-1}a^1 \\ 0 \end{bmatrix}, \tag{6}$$

from which we deduce

$$A_{11}'^{-1}a^1 = -A_{11}^{-1}a^1, \tag{7}$$

$$A'_{12}A_{11}'^{-1}a^1 = A_{21}A_{11}^{-1}a^1. \tag{8}$$

Let us use the notation $a^{1*} = A_{11}'^{-1}a^1$. Let us further rearrange the equations and the variables of the system in such a way that in the vector a^1 the positive elements appear first. The vectors a^1, b^1, a^{*1} and a^{1*} can then be partitioned as follows

$$a^1 = \begin{bmatrix} a_1^1 \\ a_2^1 \end{bmatrix}, \qquad b^1 = \begin{bmatrix} b_1^1 \\ b_2^1 \end{bmatrix}, \qquad a^{*1} = \begin{bmatrix} a_1^{*1} \\ a_2^{*1} \end{bmatrix}, \qquad a^{1*} = \begin{bmatrix} a_1^{1*} \\ a_2^{1*} \end{bmatrix},$$

with $a_1^1 > 0$ and $a_2^1 = 0$; also $a_1^{*1} \geq 0$ and $a_2^{*1} \geq 0$.

We have

$$a^{1\prime} A_{11}^{-1} a^1 = a^{1\prime} A_{11}^{\prime -1} a^1$$

or

$$a^{1\prime} (A_{11}^{-1} a^1 - A_{11}^{\prime -1} a^1) = 0. \tag{9}$$

Substitution of (7) leads to

$$2a^{1\prime} a^{*1} = 0. \tag{10}$$

or

$$a_1^{1\prime} a_1^{*1} + a_2^{1\prime} a_2^{*1} = 0.$$

Since $a_2^1 = 0$, $a_1^{1\prime} a^{*1} = 0$. Then, since $a_1^{*1} \geq 0$ and $a_1^1 > 0$, we find $a_1^{*1} = 0$, which implies $a_1^{1*} = 0$. Furthermore, $a_2^{*1} \neq 0$ and $a_2^{1*} \neq 0$, because otherwise $A_{11}^{-1} a^1 = 0$ and $A_{11}^{\prime -1} a^1 = 0$. Hence a_2^{*1} must contain some positive elements and a_2^{1*} some negative elements.

Consider now the constant-term element of y_0,

$$a^{1\prime} A_{11}^{-1} b^1 = a_1^{1*\prime} b_1^1 + a_2^{1*\prime} b_2^1. \tag{11}$$

Since $a_1^{1*} = 0$, $a_2^{1*} \leq 0$ and $\neq 0$, and $b_2^1 > 0$ (because of nondegeneracy) this must be negative. On the other hand, (11) may be written as

$$a^{1\prime} A_{11}^{-1} b^1 = a^{1\prime} b^{*1} = a_1^{1\prime} b_1^{*1} + a_2^{1\prime} b_2^{*1}, \tag{12}$$

where we define

$$A_{11}^{-1} b^1 = b^{*1} = \begin{bmatrix} b_1^{*1} \\ b_2^{*1} \end{bmatrix}.$$

Now $a_2^1 = 0$ and $b_1^{*1} > 0$ because $a_1^{*1} = 0$, so that we find that $a^{1\prime} A_{11}^{-1} b^1 > 0$, which contradicts what we found before. This means that, if $-A$ is copositive plus, termination with no upper bound of λ cannot happen.

The same result can be proved for $-A$ belonging to the class of matrices $\mathcal{L} = \mathcal{L}_1 \cap \mathcal{L}_2$ if $a > 0$ is taken. Consider

$$y = \begin{bmatrix} A_{11}^{-1} a^1 \\ 0 \end{bmatrix}. \tag{13}$$

$A_{11}^{-1} a^1 \geq 0$ and $\neq 0$. Then, if $-A$ belongs to $\mathcal{L}_1$, there should exist a positive element of y, which has a nonnegative corresponding element of

My. But we have

$$My = \begin{bmatrix} -A_{11} & -A_{12} \\ -A_{21} & -A_{22} \end{bmatrix} \begin{bmatrix} A_{11}^{-1} a^1 \\ 0 \end{bmatrix} = \begin{bmatrix} -a^1 \\ -A_{21} A_{11}^{-1} a^1 \end{bmatrix}. \tag{14}$$

Since $-a^1 < 0$, $-A$ does not belong to $\mathscr{L}_1$ and hence not to $\mathscr{L}$.

This means that if $-A$ belongs to $\mathscr{L}$, a termination with no upper bound for λ cannot happen. The proof for $\mathscr{L}$ is slightly less general than the one given by Eaves[103], since $a > 0$ rather than $a \geq 0$ has to be assumed.

18.3. Termination in a column of a nonbasic variable

The representation of a general complementary tableau as given in tableau 1 of table 1 is used again. First assume that a nonbasic y-variable, say y_i in the ith column, should be increased. If the method terminates, we have

  (i) all elements of the column in tied rows are zero,

  (ii) all other elements in that column are nonpositive.

The principal element in the column belongs to a tied row and must therefore be zero. This implies

$$0 = e_i' A_{11}^{-1} e_i = e_i' A_{11}^{-1'} A_{11}' A_{11}^{-1} e_i =$$
$$\begin{bmatrix} -A_{11}^{-1} e_i \\ 0 \end{bmatrix}' \begin{bmatrix} -A_{11} & -A_{12} \\ -A_{21} & -A_{22} \end{bmatrix} \begin{bmatrix} -A_{11}^{-1} e_i \\ 0 \end{bmatrix}. \tag{15}$$

We shall now verify whether the conditions of class $\mathscr{L}_2$ apply for

$$y = \begin{bmatrix} -A_{11}^{-1} e_i \\ 0 \end{bmatrix}.$$

According to (15) we have $y'My = 0$. Also $y \geq 0$ since $-A_{11}^{-1} e_i \geq 0$, because no positive element was found in $A_{11}^{-1} e_i$. Furthermore, $y \neq 0$ because otherwise $A_{11}^{-1} e_i = 0$. Finally,

$$My = \begin{bmatrix} -A_{11} & -A_{12} \\ -A_{21} & -A_{22} \end{bmatrix} \begin{bmatrix} -A_{11}^{-1} e_i \\ 0 \end{bmatrix} = \begin{bmatrix} e_i \\ A_{21} A_{11}^{-1} e_i \end{bmatrix} \geq 0; \tag{16}$$

the latter inequality follows because no positive element could be found in the ith column and therefore $-A_{21} A_{11}^{-1} e_i \leq 0$. Hence the conditions of $\mathscr{L}_2$ are satisfied.

We have therefore

$$M' \Omega y = -\Lambda My$$

or

$$\begin{bmatrix} -A'_{11} & -A'_{21} \\ -A'_{12} & -A'_{22} \end{bmatrix} \begin{bmatrix} \Omega_1 & 0 \\ 0 & \Omega_2 \end{bmatrix} \begin{bmatrix} -A_{11}^{-1}e_i \\ 0 \end{bmatrix} =$$

$$\begin{bmatrix} \Lambda_1 & 0 \\ 0 & \Lambda_2 \end{bmatrix} \begin{bmatrix} A_{11} & A_{12} \\ A_{21} & A_{22} \end{bmatrix} \begin{bmatrix} A_{11}^{-1}e_i \\ 0 \end{bmatrix}. \quad (17)$$

The first set of relations reduces to

$$A'^{-1}_{11}\Lambda_1 e_i = -\Omega_1 A_{11}^{-1}e_i. \quad (18)$$

Since $\Omega y \neq 0$, we have $-\Omega_1 A_{11}^{-1}e_i \neq 0$. Hence $A'^{-1}_{11}\Lambda_1 e_i \neq 0$, so that $\lambda_i > 0$. Eq. (18) may therefore be written as

$$A'^{-1}_{11}e_i = -\lambda_i^{-1}\Omega_1 A_{11}^{-1}e_i. \quad (19)$$

From this, we may conclude that, if $A_{11}^{-1}e_i$, which is the ith column of A_{11}^{-1}, contains no positive elements, then $A'^{-1}_{11}e_i$, which is the ith row of A_{11}^{-1}, contains no negative elements.

Consider now the second set of relations of (17); these may be written as

$$A'_{12}\Omega_1 A_{11}^{-1}e_i = -\Lambda_2 A_{21}A_{11}^{-1}e_i. \quad (20)$$

Substitution of eq. (19) into the left-hand side leads to

$$A'_{12}A'^{-1}_{11}e_i = \lambda_i^{-1}\Lambda_2 A_{21}A_{11}^{-1}e_i. \quad (21)$$

This means that if no positive element can be found in the ith column of $-A_{21}A_{11}^{-1}$, no negative element can be found in the ith row of $A_{11}^{-1}A_{12}$.

Assume now that termination occurs in the column of a nonbasic x-variable, say in the ith column. We have then

$$e'_i(A_{22} - A_{21}A_{11}^{-1}A_{12})e_i = 0, \quad (22)$$

$$(A_{22} - A_{21}A_{11}^{-1}A_{12})e_i \leq 0, \quad (23)$$

$$A_{11}^{-1}A_{12}e_i \leq 0. \quad (24)$$

Eq. (22) may be written as

$$0 = e'_i \begin{bmatrix} -A_{11}^{-1}A_{12} \\ I \end{bmatrix}' \begin{bmatrix} -A_{11} & -A_{12} \\ -A_{21} & -A_{22} \end{bmatrix} \begin{bmatrix} -A_{11}^{-1}A_{12} \\ I \end{bmatrix} e_i. \quad (25)$$

Take

$$y = \begin{bmatrix} -A_{11}^{-1}A_{12} \\ I \end{bmatrix} e_i.$$

Then we have $y \geq 0$ (from (24)), $y \neq 0$, $y'My = 0$ (from (25)), and

$$My = \begin{bmatrix} -A_{11} & -A_{12} \\ -A_{21} & -A_{22} \end{bmatrix} \begin{bmatrix} -A_{11}^{-1}A_{12} \\ I \end{bmatrix} e_i =$$

$$\begin{bmatrix} 0 \\ -A_{22} + A_{21}A_{11}^{-1}A_{12} \end{bmatrix} e_i \geq 0, \qquad (26)$$

(from (23)). Hence the conditions of $\mathscr{L}_2$ apply, so that

$$\begin{bmatrix} -A'_{11} & -A'_{21} \\ -A'_{12} & -A'_{22} \end{bmatrix} \begin{bmatrix} \Omega_1 & 0 \\ 0 & \Omega_2 \end{bmatrix} \begin{bmatrix} -A_{11}^{-1}A_{12} \\ I \end{bmatrix} e_i$$

$$= \begin{bmatrix} \Lambda_1 & 0 \\ 0 & \Lambda_2 \end{bmatrix} \begin{bmatrix} A_{11} & A_{12} \\ A_{21} & A_{22} \end{bmatrix} \begin{bmatrix} -A_{11}^{-1}A_{12} \\ I \end{bmatrix} e_i. \qquad (27)$$

From the first set of equalities we deduce

$$A_{11}'^{-1}A'_{21}\Omega_2 e_i = \Omega_1 A_{11}^{-1}A_{12} e_i. \qquad (28)$$

Now if $\Omega_2 e_i = \omega_i = 0$, then $-\Omega_1 A_{11}^{-1}A_{12}e_i = 0$, so that $\Omega y = 0$, which is impossible if $-A$ is in $\mathscr{L}_2$. We conclude that $\omega_i > 0$. Hence (28) may be written as

$$A_{11}'^{-1}A'_{21}e_i = \omega_i^{-1}\Omega_1 A_{11}^{-1}A_{12}e_i. \qquad (29)$$

The second set of relations may be written as

$$A'_{12}\Omega_1 A_{11}^{-1}A_{12}e_i - A'_{22}\Omega_2 e_i = \Lambda_2(A_{22} - A_{21}A_{11}^{-1}A_{12})e_i. \qquad (30)$$

Substitution of (29) and $\Omega_2 e_i = \omega_i$ leads to

$$(A'_{22} - A'_{12}A_{11}'^{-1}A'_{21})e_i = -\omega_i^{-1}\Lambda_2(A_{22} - A_{21}A_{11}^{-1}A_{12})e_i. \qquad (31)$$

According to (19), (21), (29) and (31) we have that if a certain column contains nonpositive elements, then the corresponding row has nonnegative elements at places corresponding with negative elements in the column and zero elements at places corresponding with zero elements in the column. Since elements in blocking rows in the column are zero, rows and columns in a tableau with a column with nonpositive elements may be rearranged to give the following configuration of the signs of elements:

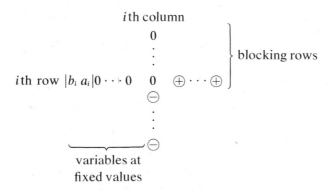

$\oplus$ and $\ominus$ indicate nonnegative and nonpositive elements.

Since the elements in the ith column in blocking rows are zero, the elements in the ith row for nonbasic variables at fixed values should be zero. But the latter should have at least one nonzero element, because the variable in the ith column was increased because a bound for a nonbasic variable was found in the ith row. From this it follows that no nonbasic variables are at fixed values. Hence the ith row must have been the critical row for a variation in λ.

Then there are two possibilities: $a_i > 0$ and $a_i < 0$; $a_i = 0$ is excluded because $\lambda^* = b_i / - a_i$.

If the ith row has a y-variable as basic variable, we may write $a_i = e'_i A_{11}^{-1} a^1 = a^{1'} A_{11}^{-1} e_i$. Since $A_{11}^{-1} e_i \geq 0$ (see (19)) we conclude $a_i > 0$. If the ith row has an x-variable as basic variable, we have

$$e'_i(a^2 - A_{21} A_{11}^{-1} a^1) = e'_i a^2 - a^{1'} A_{11}^{-1} A'_{21} e_i. \tag{32}$$

Since $a^1 \geq 0$, $a^2 \leq 0$, and $A_{11}^{-1} A'_{21} e_i \leq 0$ (see (29)), we conclude $a_i > 0$. Since $\lambda^* = b_i / - a_i > 0$ and $a_i > 0$, we conclude $b_i < 0$. Then the sign-configuration of the ith row indicates that there is no solution to the problem for $\lambda = 0$.

This means that for $- A$ belonging to $\mathcal{L}$, a termination in a column can only occur if λ is at a lower bound and no nonbasic variables are at fixed values.

It should be noted that for the complementary variant it is not necessary to establish beforehand whether $- A$ belongs to $\mathcal{L}$ or not in order to find out whether termination implies infeasibility; the method itself will unearth such a situation as a sign configuration in the critical row. Furthermore, the method allows infeasibility tests to be performed at each row, so that there is a much greater chance to detect any existing infeasibility.

BIBLIOGRAPHY

Linear programming (chs. 1-7):

[1] M.L. BALINSKI, An algorithm for finding all vertices of convex polyhedral sets. *Journal of S.I.A.M.*, Vol. 9 (1961), pp. 72–88.

[2] C.-A. BURDET, Generating all the faces of a polyhedron. Management Sciences Research Report No. 271, G.S.I.A., Carnegie-Mellon University, June 1971.

[3] A. CHARNES, Optimality and degeneracy in linear programming. *Econometrica*, Vol. 20 (1952), pp. 160–170.

[4] ___ and W.W. COOPER, *Management Models and Industrial Applications of Linear Programming*, Vols. I and II. New York, Wiley, 1961.

[5] G.B. DANTZIG, Maximization of a linear function of variables subject to linear inequalities. In: *Activity Analysis of Production and Allocation*, edited by T.C. Koopmans. New York, Wiley, 1951.

[6] ___, *Linear Programming and Extensions*. Princeton, N.J., Princeton University Press, 1963.

[7] ___, L.R. FORD and D.R. FULKERSON, A primal-dual algorithm for linear programs. In: *Linear Equalities and Related Systems*, edited by Kuhn and Tucker. Princeton, N.J., Princeton University Press, 1956, pp. 171–181.

[8] ___, A. ORDEN and P. WOLFE, Notes on linear programming, Part I. *Pacific Journal of Mathematics*, Vol. 5 (1955), pp. 183–195.

[9] G. DEBREU, Nonnegative solutions of linear inequalities. *International Economic Review*, Vol. 5 (1964), pp. 178–184.

[10] W. DINKELBACH, *Sensitivitätsanalysen und parametrische Programmierung*. Berlin, Springer-Verlag, 1969.

[11] T. GAL, A method for systematic simultaneous parametrization of vectors b and c in LP-problems. *Ekonomicko-Matematický Obzor*, Vol. 6 (1970), pp. 161–175.

[12] ___, *Betriebliche Entschiedungsprobleme, Sensitivitätsanalyse und parametrische Programmierung*. Berlin, W. de Gruyter, 1973.

[13] ___, Homogene mehrparametrische lineaire Programmierung. *Zeitschrift für Operations Research*, Vol. 16 (1972), pp. 115–136.

[14] ___ and J. NEDOMA, Multiparametric linear programming. *Management Science*, Vol. 18, No. 7 (1972), pp. 406–422.

[15] S.I. GASS, *Linear Programming, Methods and Applications*. New York, McGraw-Hill Book Company, 1958.

[16] ___ and T.L. SAATY, The computational algorithm for the parametric objective function. *Naval Research Logistics Quarterly*, Vol. 2 (1955), pp. 39–45.

[17] ___, The parametric objective function, I. *Operations Research*, Vol. 2 (1954), pp. 316–319; The parametric objective function, II. *Operations Research*, Vol. 3 (1955), pp. 395–401.

[18] A.S. GONCALVES, A unified theory of primal-dual techniques in linear programming. *Revue Belge de Statistique et de Recherche Opérationnelle*, 1972/3.

[19] T. C. KOOPMANS (ed.), *Activity Analysis of Production and Allocation*. New York, J. Wiley and Sons, 1951.

[20] H.W. KUHN and R.E. QUANDT, An experimental study of the simplex method. *Proceedings of Symposium in Applied Mathematics*. XV. Providence, R.I., Am. Math. Soc., (1963), pp. 107–124.

[21] C.E. LEMKE, The dual method of solving the linear programming problem. *Naval Research Logistics Quarterly*, Vol. 1 (1954), No. 1, pp. 36–47.

[22] M. MAŇAS and J. NEDOMA, Finding all vertices of a convex polyhedron. *Numerische Mathematik*, Vol. 12 (1968), pp. 226–229.

[23] M.B. MORONEY and R. K. DIMOND, A 'reverse simplex' tree search technique for producing multiple solutions to a blending problem. Paper presented at the Annual Conference of the Canadian Operations Research Society, May 1973.

[24] C. VAN DE PANNE, *Linear Programming and Related Techniques*. Amsterdam, North-Holland Publishing Company, 1971.

[25] ——, An index method for linear programming. Discussion Paper Series No. 25, The University of Calgary, Department of Economics, July 1972.

[26] ——, A node method for multiparametric linear programming. Discussion Papers Series No. 29, The University of Calgary, Department of Economics, December 1973.

[27] —— and A. WHINSTON, An alternative interpretation of the primal-dual method and some related parametric methods. *International Economic Review*, Vol. 9 (1968), No. 1, pp. 87–99.

[28] T.L. SAATY, Coefficient perturbation of a constrained extremum. *Operations Research*, Vol. 7 (1959), pp. 284–303.

[29] S. VAJDA, *Mathematical Programming*. Reading, Mass., Addison-Wesley Publishing Company, 1961.

[30] P. WOLFE and L. CUTLER, Experiments in linear programming. In: *Recent Advances in Mathematical Programming*, edited by R.L. Graves and P. Wolfe. New York, McGraw-Hill, 1963.

Convex quadratic programming (chs. 8–13):

[31] E. BARANKIN and R. DORFMAN, On quadratic programming. *University of California Publications in Statistics*, Vol. 2 (1958), pp. 285–318.

[32] E.M.L. BEALE, On minimizing a convex function subject to linear equalities. *Journal of the Royal Statistical Society*, Vol. 17 (1955), Series B, pp. 173–184.

[33] ——, On quadratic programming. *Naval Research Logistics Quarterly*, Vol. 6 (1959), pp. 227–243.

[34] ——, Note on a comparison of two methods of quadratic programming. *Operations Research*, Vol. 14 (1966), pp. 442–443.

[35] ——, Numerical methods. In: *Nonlinear Programming*, edited by J. Abadie. Amsterdam, North-Holland Publishing Company, 1967.

[36] ——, *Mathematical Programming in Practice*. London, Pitman, 1968.

[37] ——, Computational methods for least squares. In: *Integer and Nonlinear Programming*, edited by J. Abadie. Amsterdam, North-Holland Publishing Company, 1970.

[38] C. BERGTHALER, Minimum risk problems and quadratic programming. Discussion Paper No. 7115, Louvain, Center for Operations Research and Econometrics, June 1971.

[39] J.C.G. BOOT, Binding constraint procedures of quadratic programming. *Econometrica*, Vol. 31 (1963), pp. 464–498.

[40] ——, *Quadratic Programming, Algorithms-Anomalies-Applications*. Amsterdam, North-Holland Publishing Company, 1964.

[41] R.E. BOVE, The one-stage deformation method: an algorithm for quadratic programming. *Econometrica*, Vol. 38 (1970) pp. 225–230.

[42] R.J. BRAITSCH, JR., A computer comparison of four quadratic programming algorithms. *Management Science*, Vol. 18 (1972), pp. 632–643.

[43] R.W. COTTLE, Symmetric dual quadratic programs. *Quarterly of Applied Mathematics*, Vol. 21 (1963), pp. 237–243.

[44] ——, The principal pivoting method of quadratic programming. In: *Mathematics of the Decision Sciences* (Part 1), edited by Dantzig and Veinott. Providence, R.I., Am. Math. Soc., 1968.

[45] ——, G.J. HABETLER, and C.E. LEMKE, Quadratic forms semi-definite over convex cones. In: *Proceedings of the Princeton Symposium on Mathematical Programming*, edited by H.W. Kuhn. Princeton, N.J., Princeton University Press, 1970.

[46] G.B. DANZIG and R.W. COTTLE, Positive (semi-) definite programming. In: *Nonlinear Programming*, edited by J. Abadie. Amsterdam, North-Holland Publishing Company, 1967, pp. 57–73.

[47] W.S. DORN, Duality in quadratic programming. *Quarterly of Applied Mathematics*, Vol. 18 (1960), pp. 155–162.

[48] ——, Self-dual quadratic programs. *Journal of S.I.A.M.*, Vol. 9 (1961), pp. 51–54.

[49] M. FRANK and P. WOLFE, An algorithm for quadratic programming. *Naval Research Logistics Quarterly*, Vol. 3 (1956), pp. 95–110.

[50] G.H. GOLUB and M.A. SAUNDERS, Linear least squares and quadratic programming. In: *Integer and Nonlinear Programming*, edited by J. Abadie. Amsterdam, North-Holland Publishing Company, 1970.

[51] A.S. GONCALVES, A version of Beale's method avoiding the 'free variables'. *Proceedings of the A.C.M. National Conference*, Chicago, 1971, pp. 433–441.

[52] ——, A primal-dual method for quadratic programming. *Revista da Fac. Ciencias Univ. Coimbra*, Vol. 47.

[53] ——, A primal-dual method for quadratic programming with bounded variables. In: *Numerical Methods for Nonlinear Optimization*, edited by A. Lootsma. New York, Academic Press, 1972.

[54] R.L. GRAVES, A principal pivoting simplex algorithm for linear and quadratic programming. *Operations Research*, Vol. 15 (1967), pp. 482–494.

[55] G. HADLEY, *Nonlinear and Dynamic Programming*. Reading, Mass., Addison-Wesley Publishing Company, 1964.

[56] C. HILDRETH, A quadratic programming procedure. *Naval Research Logistics Quarterly*, Vol. 4 (1957), pp. 79–85.

[57] H.S. HOUTHAKKER, The capacity method of quadratic programming. *Econometrica*, Vol. 28 (1960), pp. 62–87.

[58] R. JAGANNATHAN, A simplex-type algorithm for linear and quadratic programming–a parametric procedure. *Econometrica*, Vol. 34 (1966), pp. 460–471.

[59] H.W. KUHN and A.W. TUCKER, Nonlinear programming. In: *Second Berkeley Symposium on Mathematical Statistics and Probability*, edited by J. Neyman. Berkeley, Calif., University of California Press, 1951.

[60] H.P. KÜNZI AND W. KRELLE, *Nichtlineare Programmierung*. Berlin, Springer-Verlag, 1962. An English translation under the title *Nonlinear Programming* was published by Blaisdell in 1966.

[61] A.H. LAND and G. MORTON, An inverse-basis method for Beale's quadratic programming algorithm. *Management Science*, Vol. 19 (1973), pp. 510–516.

[62] C.E. LEMKE, A method of solution for quadratic programs. *Management Science*, Vol. 8 (1962), pp. 442–453.

[63] H.M. MARKOWITZ, The optimization of a quadratic function subject to linear con-

[64] ——, *Portfolio Selection: Efficient Diversification of Investments*. New York, Wiley, 1959. Reprinted in 1970 by Yale University Press, New Haven, Conn.

[65] J.H. MOORE and A.B. WHINSTON, Experimental methods in quadratic programming. *Management Science*, Vol. 13 (1966), pp. 58–76.

[66] W.C. MYLANDER, Finite algorithms for solving quasi-convex quadratic problems. *Operations Research,* Vol. 20 (1972), pp. 167–173.

[67] C. VAN DE PANNE, A non-artificial simplex method for quadratic programming. Rotterdam, Report 22 of the International Center for Management Science, 1962.

[68] ——, Programming with a quadratic constraint. *Management Science*, Vol. 12 (1966). No. 11, pp. 798–815.

[69] —— and W. POPP, Minimum-cost cattlefeed under probabilistic protein constraints. *Management Science*, Vol. 9 (1963), No. 3, pp. 405–430.

[70] —— and A. WHINSTON, The simplex and the dual method for quadratic programming. *Operational Research Quarterly*, Vol. 15 (1964), pp. 355–388.

[71] —— and ——, Simplicial methods for quadratic programming. *Naval Research Logistics Quarterly*, Vol. 11 (1964), pp. 273–302.

[72] —— and ——, A parametric simplicial formulation of Houthakker's capacity method. *Econometrica*, Vol. 33 (1965), pp. 354–380.

[73] —— and ——, A comparison of two methods for quadratic programming. *Operations Research*, Vol. 14 (1966), pp. 422–441.

[74] —— and ——, The symmetric formulation of the simplex method for quadratic programming. *Econometrica*, Vol. 37 (1969), No. 3, pp. 507–527.

[75] T.D. PARSONS, *A Combinatorial Approach to Convex Quadratic Programming*. Doctoral Dissertation, Princeton University, Department of Mathematics, 1966.

[76] M.H. RUSIN, A revised simplex method for quadratic programming. *SIAM Journal of Applied Math.*, Vol. 20 (1971), pp. 143–160.

[77] H. THEIL, *Optimal Decision Rules for Government and Industry*. Amsterdam, North-Holland Publishing Company, 1964.

[78] —— and C. VAN DE PANNE, Quadratic programming as an extension of conventional quadratic maximization. *Management Science*, Vol. 7 (1960), No. 1, pp. 1–20.

[79] A. WHINSTON, The bounded variable problem—an application of the dual method for quadratic programming. *Naval Research Logistics Quarterly*, Vol. 12 (1965), pp. 173–180.

[80] P. WOLFE, The simplex method for quadratic programming. *Econometrica*, Vol. 27 (1959), pp. 382–398.

[81] S. ZAHL, A deformation method for quadratic programming. *Journal of the Royal Statistical Society*, Series B, Vol. 26 (1964), pp. 141–160.

[82] ——, Supplement to a deformation method for quadratic programming. *Journal of the Royal Statistical Society*, Series B, Vol. 27 (1965), pp. 166–168.

[83] G. ZOUTENDIJK, *Methods of Feasible Directions*. Amsterdam, Elsevier Publishing Company, 1960.

General quadratic programming (chs. 14 and 15):

[84] E. BALAS, Nonconvex quadratic programming via generalized polars. Management Sciences Research Report No. 278(R), G.S.I.A., Carnegie-Mellon University, October 1973.

[85] E. BALAS and C.-A. BURDET, Maximizing a convex quadratic function subject to linear constraints. Management Science Research Report No. 299, G.S.I.A., Carnegie-Mellon University, September 1972–July 1973.

[86] C.-A. BURDET, General quadratic programming. Management Sciences Research Report No. 272, G.S.I.A., Carnegie-Mellon University, November 1971.

[87] A.V. CABOT and R.L. FRANCIS, Solving certain nonconvex quadratic minimization problems by ranking extreme points. *Operations Research*, Vol. 18 (1970), pp. 82–86.

[88] W. CHANDLER and R.J. TOWNSLEY, The maximization of a quadratic function of variables subject to linear inequalities. *Management Science*, Vol. 10 (1964), pp. 515–523.

[89] J. COFFMAN, A. MAJTHAY and A. WHINSTON, Local optimization for nonconvex quadratic programming. Unpublished paper.

[90] R.W. COTTLE and W.C. MYLANDER, Ritter's cutting plane method for nonconvex quadratic programming. In: *Integer and Nonlinear Programming*, edited by J. Abadie. Amsterdam, North-Holland Publishing Company, 1970.

[91] R. FLETCHER, A general quadratic programming algorithm. *Journal of the Inst. for Maths Applic.*, Vol. 7 (1971), pp. 76–91.

[92] A.S. GONCALVES, A nonconvex quadratic programming algorithm. In: *Mathematical Programming in Theory and Practice*, edited by A. S. Goncalves. Dordrecht (Holland), Reidel Publishing Company, 1973.

[93] H. KONNO, Bilinear programming: Part I: Algorithm for solving bilinear programs; Part II: Application of bilinear programming. Technical Report No. 71–9/10, Operations Research House, Stanford University, August 1971.

[94] A. MAJTHAY, Optimality conditions for quadratic programming. *Mathematical Programming*, Vol. 1 (1971), pp. 359–365.

[95] R.K. MUELLER, A method for solving the indefinite quadratic programming problem. *Management Science*, Vol. 16 (1970), pp. 333–339.

[96] C. VAN DE PANNE, Local optima of quadratic programming problems by means of the simplex method for quadratic programming. Discussion Paper No. 7335, Heverlee (Belgium), Center for Operations Research and Econometrics.

[97] ——, A parametric method for general quadratic programming. Discussion Papers Series No. 28, The University of Calgary, Department of Economics, November 1973.

[98] K. RITTER, Stationery points of quadratic maximum problems. *Zeitschrift für Wahrscheinlichkeitstheorie und verwandte Gebiete*, Vol. 4 (1965), pp. 149–158.

[99] ——, A method for solving maximum problems with a nonconcave quadratic objective function. *Zeitschrift für Wahrscheinlichkeitstheorie und verwandte Gebiete*, Vol. 4 (1966), pp. 340–351.

[100] H. TUI, Concave programming under linear constraints. *Soviet Mathematics*, 1965, pp. 1437–1440.

[101] P.B. ZWART, Nonlinear programming: counterexamples to two global optimization algorithms. *Operations Research*, Vol. 2 (1973), pp. 1260–1266.

The linear complementarity problem (chs. 16–18):

[102] R.W. COTTLE and G.B. DANTZIG, Complementary pivot theory. In: *Mathematics of the Decision Sciences* (Part 1), edited by Dantzig and Veinott. Providence, R.I., Am. Math. Soc., 1968.

[103] B.C. EAVES, The linear complementarity problem. *Management Science*, Vol. 17 (1971), pp. 612–634.

[104] ——, On quadratic programming. *Management Science*, Vol. 17 (1971), pp. 698–711.

[105] W. FIEDLER and V. PTAK, Some generalizations of positive definiteness and monotonicity. *Numerische Mathematik*, Vol. 9 (1966), pp. 163–172.

[106] A.W. INGLETON, A problem in linear inequalities. *Proceedings of the London Mathematical Society*, Vol. 16 (1966), pp. 519–536.

[107] C.E. LEMKE, Bimatrix equilibrium points and mathematical programming. *Management Science*, Vol. 11 (1965), pp. 681–689.

[108] ——, On complementary pivot theory. In: *Mathematics of the Decision Sciences* (Part 1), edited by Dantzig and Veinott. Providence, R.I., Am. Math. Soc., 1968.

[109] ——, Recent results on complementarity problems. In: *Nonlinear Programming*, edited by Rosen, Mangasarian and Ritter. New York, Academic Press, 1970.

[110] —— and J.T. HOWSON, Jr., Equilibrium points of bimatrix games. *Journal of the S.I.A.M.*, Vol. 12 (1964), pp. 413–423.

[111] B. MARTOS, Subdefinite matrices and quadratic forms. *S.I.A.M. Journal of Applied Mathematics*, Vol. 7 (1969), pp. 1215–1223.

[112] S.R. McCAMMON, *Complementary Pivoting*. Ph.D. Thesis, Rensselaer Polytechnic Institute, 1970.

[113] K.G. MURTY, Algorithm for finding all the feasible complementary bases for a linear complementarity problem. Technical Report No. 72-2, Ann Arbor, The University of Michigan, Department of Industrial Engineering.

[114] ——, On a characterization of P-matrices. *S.I.A.M. Journal of Applied Mathematics*, Vol. 20 (1971), pp. 378–384.

[115] C. VAN DE PANNE, A complementary variant of Lemke's method for the linear complementarity problem. *Mathematical Programming* (1974), to appear.

[116] T.D. PARSONS, Applications of principal pivoting. In: *Proceedings of the Princeton Symposium in Mathematical Programming*, edited by H.W. Kuhn. Princeton, N.J., Princeton University Press, 1970.

[117] A. RAVINDRAN, Computational aspects of Lemke's complementarity algorithms applied to linear programs. *Opsearch*, Vol. 7 (1970), pp. 241–262.

[118] R. SAIGAL, A characterization of the constant parity property of the number of solutions to the linear complementarity problem. *S.I.A.M. Journal of Applied Mathematics*, Vol. 23, (1972), pp. 40–45.

[119] R. SAIGAL, On the class of complementary cones and Lemke's algorithm. *S.I.A.M. Journal of Applied Mathematics*, Vol. 23 (1972), pp. 46–60.

INDEX